Regulation of Water and Wastewater Services

Regulation of Water and Wastewater Services

An International Comparison

Rui Cunha Marques

Contributors
Pedro Simões, João Simão Pires,
João Almeida and Tiago Neves

Publishing
London • New York

Published by **IWA Publishing**
Alliance House
12 Caxton Street
London SW1H 0QS, UK
Telephone: +44 (0)20 7654 5500
Fax: +44 (0)20 7654 5555
Email: publications@iwap.co.uk
Web: www.iwapublishing.com

First published 2010

Typeset in India by OKS Prepress Services.

British Library Cataloguing in Publication Data
A CIP catalogue record for this book is available from the British Library

Library of Congress Cataloging-in-Publication Data
A catalog record for this book is available from the Library of Congress

ISBN 10: 1843393417
ISBN 13: 9781843393412

Contents

ISBN: 9781843393412. Published by IWA Publishing, London, UK.

Notes about the author and contributors

Author

Dr Rui Cunha Marques started his academic career in 1996. He has an academic background in engineering, law and economics. His Ph.D. thesis focused on performance measurement and regulation of water utilities. Currently he is a Professor at Instituto Superior Técnico of Technical University of Lisbon in Portugal but he has been an invited professor in other universities abroad in post-graduation programs. Dr Rui Cunha Marques has carried out research for about 15 years in the scope of performance evaluation, regulation and public procurement both in Portugal and abroad. He is a research fellow of the Centre of Urban and Regional Systems (CESUR) at the Technical University of Lisbon and of the Public Utility Research Center (PURC) at the University of Florida. His specialization areas are public service regulation, performance evaluation and public procurement, particularly public-private partnerships related to public utilities, transportation and other public sectors. He is the author of more than 200 technical and scientific publications in about 20 countries in different languages, including four books, two academic theses and more than 40 articles in international journals. He has been a consultant for several entities, both public and private, in Portugal and abroad in countries such as the USA, France, Morocco, Mozambique, Cape Verde and Brazil, in the areas of project management, public-private partnerships, regulation and performance evaluation of urban infrastructures.

Dr Rui Cunha Marques
Center of Urban and Regional Systems (CESUR)
Instituto Superior Técnico, Technical University of Lisbon
Avenida Rovisco Pais, 1049-001 Lisbon, Portugal
Tel: +351-218-418-305
Fax: +351-218-409-884
Email: rcmar@civil.ist.utl.pt

ERSAR (The Water and Waste Services Regulation Authority) regulates the provision of public water supply, urban wastewater and waste management services and is the public authority for the supervision of drinking water quality in Portugal. Through its intervention, ERSAR seeks to protect the interests of service users, while safeguarding the financial sustainability of service provision, encouraging continuous improvement and efficiency and promoting service and environmental quality.

ERSAR
Centro Empresarial Torres de Lisboa
Rua Tomás da Fonseca, Torre G
1600-209 Lisbon, Portugal
Tel: +351-210-052-2 00
Fax: +351-210-052-259
Email: geral@ersar.pt
www.ersar.pt

ISBN: 9781843393412. Published by IWA Publishing, London, UK.

Contributors

Pedro Tiago Francisco Simões holds a Civil Engineering and a Master's degree from the Instituto Superior Técnico (IST) of the Technical University of Lisbon. Currently he is a doctorate student at the Technical University of Lisbon. He is a researcher at the Centre of Urban and Regional Systems (CESUR) of IST. Pedro Simões has authored several technical articles and papers in the field of regulation and performance evaluation of public utilities and transportation. He has experience in consultancy in the area of performance evaluation, project management, public procurement and regulation and in management and operation of urban infrastructure services.

João Paulo Simão Pires was born in Lisbon in 1969. He completed his first degree in Economics at the Universidade Católica Portuguesa in 1992. He followed this with an MBA specializing in Finance and Strategy at the William E. Simon Graduate School of Business Administration (University of Rochester, USA), which he completed in 1997. He began his lecturing career at the Universidade Católica Portuguesa in 1992 and has been an Assistant Chair at this university since 2004, responsible for the undergraduate Management Course. He began his professional career at the Bank of Portugal in 1992 as a member of the Economic Research Unit of the Economic Studies Department. As a consultant based in the London office of the strategy consulting firm The Monitor Company between 1993 and 1999, he contributed to and led projects in the United Kingdom, the USA, Germany, Spain and Greece. The client sectors with which he has worked include brewing, aluminium, pay-TV, retail banking, wines and spirits, forestry management, wood products and telecommunications. He was a manager at the Lisbon office of the management consultancy company Arthur D. Little, principally concerned with the utilities and telecommunications sectors. For three years, he was Head of Strategic Planning and Control at Empresa Portuguesa das Águas Livres, S.A. (EPAL). He was a member of the Executive Committee of Sonae Novobord, the South African subsidiary of Sonae Indústria, based in Johannesburg. He was a Partner of Bright Partners – Gestão, Tecnologia e Capital, S.A. He has been a member of the Board of Directors of the Portuguese Water and Waste Services Regulation Authority (ERSAR), formerly Institute for the Regulation of Water and Solid Waste (IRAR), since November 2006.

João Miguel Tiago de Almeida is currently an adviser to the Board at DHV, S.A. in Portugal, managing the Studies, Projects and Consultancy Unit. Before joining DHV he was Head of Projects and Studies Department at the Portuguese Institute for the Regulation of Water and Solid Waste (IRAR) since June 2006, having joined IRAR in November 2004. From 2002 to 2004 he was an adviser to the Secretary of State for the Environment, working on several areas including Environmental Impact Assessment and water. Until 2002 he mainly worked as a consultant at Azorean Regional Directorate for Land Planning and Water Resources, participating in the water resources planning and management in the region. He graduated in Environmental Engineering and has a specialization in Project Management. His main fields of interest are environmental technologies and policies, information technologies and change management.

Tiago Julião Lourenço Pinto da Cruz Neves was born in Lisbon in 1976. He holds a Master of Science degree in Environmental Technology from Imperial College London, a Post-Graduation in Public Regulation from Universidade de Coimbra and a B.Sc. in Economics from Universidade Nova de Lisboa. He has seven years of working experience as an economist and a project manager in the Portuguese Institute for the Regulation of Water and Solid Waste (IRAR), and previously worked as a business consultant for Cap Gemini Ernst & Young. Currently he is an adviser to the Minister of Economy, Innovation and Development and a Ph.D. candidate at the London School of Economics and Political Science.

Acknowledgements

In the period between 2006 and 2008, when this research was carried out and this book was written, many people gave me everyday support and advice. Special thanks, however, are due to my colleagues at the Centre of Urban and Regional Systems (CESUR), the research body to which I am affiliated in my university (Instituto Superior Técnico of Technical University of Lisbon), particularly to Pedro Simões, Pedro Carvalho, Manuel Carvalho, Nuno Cruz, Tânia Correia, Thelma Ohari, Ana Catana, Rui Oliveira, Antunes Ferreira and Patrícia Nunes. Also in CESUR, my thanks to its President, João Quinhones Levy, and to Professor Emeritus Luís Valadares Tavares who is always a source of inspiration.

I would especially like to thank the Institute for the Regulation of Water and Solid Waste (IRAR), nowadays renamed ERSAR, especially to its board of directors, including the President Jaime Melo Baptista and the directors João Simão Pires and Dulce Pássaro but also to other people of ERSAR such as David Alves, Tiago Neves, João Almeida and Edgar Carvalho. I am grateful for their trust in me to carry out this stimulating challenge and for all their support during this period.

Pedro Simões from CESUR and João Simão Pires, Tiago Neves and João Almeida from ERSAR deserve a redoubled 'thank you' for their valuable contributions for the book, revision and comments and for the help in the definition of its structure.

This research would not have been possible without all the regulators and organizations in the five continents which participated in this research. I am indebted to them. From Papua New Guinea to Uruguay, including Indonesia, Armenia and Ghana, more than 60 entities encompassing 39 countries cooperated in this study. Without naming anybody individually so that no one is forgotten, we draw attention to the dedication and interest of all these entities in the regulation of water and wastewater services

I am also grateful to the International Water Association (IWA) for their interest in publishing this document, especially to its editors Maggie Smith and Michael Dunn and, of course, to the IWA Executive Director, Paul Reiter.

Finally, I would like to thank Ângela Marques for the revision of the document and all the support in this period.

ISBN: 9781843393412. Published by IWA Publishing, London, UK.

Acronyms

AAC	Average Annual Costs
AATO	*Autorità di Ambito Territoriale Ottimale*
ABAR	Brazilian Association of Regulatory Agencies
ABCON	Brazilian Association of Concessionaires
ABES	Brazilian Association of Water Supply, Wastewater and Waste
ABRH	Brazilian Association of Water Resources
AC	Competition Authority
ACCC	Australian Competition and Consumer Commission
ACEA	*Azienda Governatoriale Elettricità e Acque*
ACTEW	Australian Capital Territory Electricity and Water
ADASA	*Agência Reguladora de Águas e Saneamento do Distrito Federal*
ADERASA	*Asociación de Entes Reguladores de Aguas de las Américas*
AdP	*Águas de Portugal*
ADWG	Australian Drinking Water Guidelines
AED	*Administration de l'Equipement et des Déplacements*
AESBE	Brazilian Association of State-Owned Companies
AFNOR	*Association Française de Normalisation*
AGENERSA	*Agência Reguladora de Energia e Saneamento Básico do Estado do Rio de Janeiro*
AGERSA	*Agência Municipal de Regulação dos Serviços Públicos Delegados de Cachoeiro de Itapemirim*
AGL	Australian Gas Light Company
AGR	*Agência Goiana de Regulação*
AMAE	*Agência Municipal de Água e Esgoto*
AMINAL	*Administratie Milieu, Natuurbeheer en Landinrichting*
AMWA	Association of Metropolitan Water Agencies
ANA	*Agência Nacional de Águas*
ANEA	*Associazione Nazionale Autorità e Enti di Ambito*
ANEAS	*Asociación Nacional de Empresas de Agua y Saneamiento* do México
ANPA	*Agenzia Nazionale por la Protezione dell'Ambiente*
ANRSC	*Autoritatea Nationala de Reglementare pentru Serviciile Publice de Gospodarie Comunala*
APA	*Agência Portuguesa do Ambiente*
APDA	*Associação Portuguesa de Distribuição e Drenagem de Águas*
APR	*Agua Potable Rural*
APSC	Arkansas Public Service Commission
AQUABRU	*Association Bruxelloise de l'Eau*
AQUAWAL	*Union professionnelle des Opérateurs publics du cycle de l'eau en Wallonie*
ARA	*Administrações Regionais de Águas*
ARA	*Asociatia Romana a Apei*
ARCE	*Agência Reguladora de Serviços Públicos Delegados do Estado do Ceará*

ISBN: 9781843393412. Published by IWA Publishing, London, UK.

ARE	*Agência de Regulação Económica*
ARESEP	*Autoridad Reguladora de Servicios Públicos*
ARH	*Administrações de Região Hidrográfica*
ARM	*Autorité de Régulation Multisectorielle*
ARPA	*Agenzia Regionale per la Protezione dell'Ambiente*
ARPB	*Agência de Regulação do Estado da Paraíba*
ARPE	*Agência de Regulação de Pernambuco*
ARSAE	*Agência Reguladora dos Serviços de Água e Esgoto de Mauá*
ARSAL	*Agência Reguladora de Serviços Públicos do Estado de Alagoas*
ARSAM	*Agência Reguladora de Serviços Públicos Concedidos*
ARSBAN	*Agência Reguladora de Serviços de Saneamento Básico do Município do Natal*
ARSESP	*Agência Reguladora de Saneamento e Energia do Estado de São Paulo*
ASADAS	*Asociaciones Administradoras de Acueductos y Alcantarillados Rurales*
ASEP	*Autoridad Nacional de los Servicios Públicos*
ASSEMAE	*Associação Nacional dos Servidores Municipais de Saneamento*
ATB	*Adhya Tirta Batam*
ATO	*Ambito Territoriale Ottimale*
AWA	Australian Water Association
AWWA	American Water Works Association
AWWARF	American Water Works Association Research Foundation
AyA	*Instituto Costarricense de Acueductos y Alcantarillados*
BAPPENAS	*Badan Perencanaan dan Pembangunan Nasional*
BELGAQUA	*Fédération Belge du Secteur de l'Eau*
BNH	*Banco Nacional de Habitação*
BOOT	Build-Own-Operate-Transfer
BOT	Build-Own-Transfer
BPD	Building Partnerships for Development
CA	Cooperation Agreement
CAAR	*Comités Administradores de Acueductos Rurales*
CABEI	Central American Bank for Economic Integration
CACE	*Coordination Nationale des Associations de Consommateurs d'Eau*
CAPEX	Capital Expenses
CAPM	Capital Asset Pricing Model
CAR	*Corporación Autónoma Regional*
CAS	*Communal Social Assistance Committees*
CC	Competition Commission
CCDR	*Comissões de Coordenação e Desenvolvimento Regional*
CCE	*Comité de Contrôle de l'Eau*
CCSPL	*Commission Consultative des Services Publics Locaux*
CCW	Consumer Council for Water
CEDAE	*Companhia de Águas e Esgotos do Rio de Janeiro*
CESB	*Companhia Estadual de Saneamento Básico*
CIBE	*Compagnie Intercommunale Bruxelloise des Eaux*
CIEAU	*Centre d'Information sur l'Eau*
CIP	*Comitato Interministeriale Prezzi*
CLCV	*Association de la Consommation éco-citoyenne*
CMP	*Costo Medio de Mediano Plazo*
CNP	*Comissão Nacional de Preços*
CoAG	Council of Australian Governments

COIL	Community Opinion Investigation Committee
CONAMA	*Comissão Nacional do Meio Ambiente*
COREMAS	*Departamentos Regionais*
CORFO	*Corporação de Fomento da Produção*
COVIRI	*Comitato di Vigilanza sull'uso delle Risorse Idriche*
CPI	Consumer Price Index
CPP	*Comitato Provinciale Prezzi*
CPSSW	Code of Practice on Sewerage and Sanitary Works
CPSWD	Code of Practice on Surface Water Drainage
CPUC	California Public Utilities Commission
CPWS	Code of Practice for Water Services
CRA	*Comisión de Regulación de Agua Potable y Saneamiento Básico*
CRA	*Conselho de Regulação do Abastecimento de Água*
CSC	Customer Service Code
CSS	Customer Service Standard
CWA	Clean Water Act
DAPSBA	*Direcção para o Serviço de Água Potável e Saneamento Básico*
DDAF	*Direction Départementale de l'Agriculture et de la Forêt*
DDASS	*Direction Départementale des Affaires Sanitaires et Sociales*
DDCCRF	*Direction Départementale de la Consommation, de la Concurrence et de la Répression des Fraudes*
DDE	*Direction Départementale de l'Equipement*
DEA	Data Envelopment Analysis
DEFRA	Department for Environment, Food and Rural Affairs
DEWHA	Department of the Environment, Water, Heritage and the Arts
DG	Director-General of Water Services
DGA	*Direcção-Geral das Águas*
DGC	*Direcção-Geral do Consumidor*
DGRNE	*Direction Général dês Ressources Naturelles et de l'Environnement*
DHS	Department of Human Services
DISS	Department of Infrastructure and Support Services
DNA	*Direcção Nacional de Águas*
DNP	*Departamento Nacional do Planeamento*
DNS	*Direcção Nacional de Saúde*
DOH	*Direcção Nacional de Obras Hidráulicas*
DRIRE	*Directions Régionales de l'Industrie, de la Recherche et de l'Environnement*
DRIVA	Benchmarking System in Sweden
DSE	Department of Sustainability and Environment
DTe	*Energiekamer*
DTF	Devolution Trust Fund
DTSS	Deep Tunnel Sewerage System
DWA	Department of Water Affairs
DWI	Drinking Water Inspectorate
E&W	England and Wales
EA	Environment Agency
EAAB	*Empresa de Acueducto y Alcantarillado de Bogotá*
ECZ	Environment Council of Zambia
EHS	Environment and Heritage Service
EICE	*Empresas Industriales y Comerciales del Estado*
EPA	Environment Protection Agency

EPA	Environment Protection Authority
EPAL	*Empresa Portuguesa das Águas Livres*
EPM	*Empresas Públicas de Medellín*
EPS	*Empresa Prestadora de Servicios de Saneamiento*
EPU	Environmental Pollution Unit
ERA	Economic Regulation Authority
ERSAPS	*Ente Regulador de los Servicios de Agua Potable y Saneamiento*
ERSAR	*Entidade Reguladora dos Serviços de Águas e Resíduos*
ESC	Essential Services Commission
ESPH	Empresa de Servicios Públicos de Heredia
ESSBIO	*Empresa de Servicios Sanitarios del Bio-Bío*
ESVAL	*Empresa Sanitária de Valparaíso*
EU	European Union
EWI	Environment and Water Industry Development Council
EWOV	Energy and Water Ombudsman
EWURA	Energy and Water Utilities Regulatory Authority
EYDAP	*Εταιρεία 'Υδρευσης και Αποχέτευσης Πρωτεύουσας*
FAE	*Fundo de Financiamento para Águas e Esgotos*
Federutility	*Federazione delle Imprese Energetiche e Idriche*
FGTS	*Fundo de Garantia por Tempo de Serviço*
FINDETER	*Financiera de Desarrollo Territorial S.A.*
FIPAG	*Fundo de Investimento e Património do Abastecimento de Água*
FKPM	Public and Consumer Communication Forum
FNCCR	*Fédération Nationale des Collectivités Concédantes et Régies*
FNDAE	*Fonds National pour le Developpement des Adductions d'Eau*
FP2E	*Fédération professionnelle des entreprises de l'eau*
FPSC	Florida Public Service Commission
FTC	Fair Trading Commission
GDP	Gross Domestic Product
GSS	Guaranteed Service Standards
GWCL	Ghana Water Company Ltd
IADB	Inter-American Development Bank
IBAMA	*Instituto Brasileiro do Meio Ambiente e dos Recursos Naturais Renováveis*
IBDE	*Intercommunale Bruxelloise de Distribution d'Eau*
IBGE	*Institut Bruxellois pour la Gestion de l'Environnement*
IBrA	*Intercommunale Bruxelloise pour l'Assainissement*
IDAAN	*Instituto de Acueductos y Alcantarillados Nacionales*
IDEA	*Intercommunale de Développement Economique et d'Aménagement de la Région Mons-Borinage-Centre*
IFEN	*Institut Français de l'Environnement*
IGA	*Instituto de Gestão da Água*
INAG	*Instituto da Água*
INASEP	*Intercommunale Namuroise de Services Publics*
INHGA	*Institutul National De Hidrologie Si Gospodarire A Apelor*
INSFOPAL	*Instituto de Fomento Municipal*
IPART	Independent Price Authority Regulatory Tribunal
IPAs	*Inspectores de Protecção Ambiental Locais*
IRAC	PEI Regulatory and Appeals Commission
IRAR	Institute for the Regulation of Water and Solid Waste

IUIDP	*Programa Integrado de Desenvolvimento de Infra-estruturas*
IURC	Indiana Utility Regulatory Commission
IWA	International Water Association
JWSRB	Jakarta Water Supply Regulatory Body
KPPM	Public Opinion Committee
LGWSC	Lukanga Water and Sewerage Company
LNEC	*Laboratório Nacional de Engenharia Civil*
LRIC	Long-run Incremental Cost
LWSC	Lusaka Water and Sewerage Company
MAOTDR	*Ministério do Ambiente, do Ordenamento do Território e do Desenvolvimento Regional*
MAVDT	*Ministerio de Ambiente, Vivienda y Desarrollo Territorial*
MEWD	Ministry of Energy and Water Development
MEWR	Ministry of the Environment and Water Resources
MINAE	*Ministerio del Ambiente y Energía*
MINSA	*Ministerio de Salud*
MLGH	Ministry of Local Government and Housing
MMC	Monopolies and Mergers Commission
MMGA	*Ministerul Mediului şi Gospodăririi Apelor*
MoH	Ministry of Home Affairs
MOP	*Ministério das Obras Públicas*
MOPH	*Ministério das Obras Públicas e Habitação*
MPS	*Ministerio de la Protección Social*
MPSC	Montana Public Service Commission
MSP	*Ministerul Sănătății Publice*
MWD	Metropolitan Water District of Southern California
MWSSRO	Metropolitan Waterworks and Sewerage System – Regulatory Office
NARUC	National Association of Regulatory Utility Commissioners
NAW	National Assembly for Wales
NAWC	National Association of Water Companies
NCC	National Competition Council
NCCPE	National Control Commission for Prices and Energy
NEA	National Environment Agency
NGO	Non-governmental Organization
NIAUR	Northern Ireland Authority for Utility Regulation
NPC	National Price Commission
NRA	National Rivers Authority
NRWA	National Rural Water Association
NSW	New South Wales
NUT	*Nomenclatura de Unidades Territoriais para Fins Estatísticos*
NWASCO	National Water Supply and Sanitation Council
NWC	National Water Commission
NWI	Intergovernmental Agreement on a National Water Initiative
NWRB	National Water Resources Board
O/M	Operation and Maintenance
OECD	Organization for Economic Co-operation and Development
OFWAT	Water Services Regulation Authority
OLS	Ordinary Least Squares
ONEMA	*Office National de l'Eau et des Milieux Aquatiques*
OPA	Overall Performance Assessment

OPEX	Operational Expenses
ORG	Office of Regulator General
OUR	Office of Utilities Regulation
PAC	*Programa de Aceleração de Crescimento*
PDAM	*Perusahaan Daerah Air Minum*
PDO	*Plan Maestro*
PEAASAR	*Plano Estratégico de Abastecimento de Água e Saneamento de Águas Residuais*
PEI	Prince Edward Island
PERPAMSI	*Persatuan Perusahaan Air Minum di Seluruh Indonesia*
PJT	*Perum Jasa Tirta*
PLANASA	*Plano Nacional de Saneamento*
PLC	Public Limited Company
PMSS	*Programa de Modernização do Sector de Saneamento*
PNDA	*Programa Nacional de Desenvolvimento de Águas*
PNG	Papua New Guinea
PPP	Public-Private Partnership
PrGWS	'Private' Group Water Schemes
PSC	Public Service Commission
PSRC	Public Service Regulatory Commission
PUB	Public Utilities Board
PUC	Public Utility Commission
PuGWS	Public Group Water Schemes
PURA	*Programa de Uso Racional da Água*
PURC	Public Utilities Regulatory Commission
PWS	Public Water Supplies
QGD	*Quadro da Gestão Delegada*
RCA	Restated Cooperation Agreement
RIC	Regulated Industries Commission
RIZA	*Rijksinstituut voor Integraal Zoetwaterbeheer en Afvalwaterbehandeling*
RPI	Retail Price Index
RSP	*Ministerul Sănătății Publice*
RV	Rateable Value
RWA	Regional Water Authorities
SABESP	*Saneamento Básico do Estado de São Paulo*
SBGE	*Société Bruxelloise de Gestion de l'Eau*
SDWA	Safe Drinking Water Act
SEDAPAL	*Servicio de Agua Potable y Alcantarillado*
SEPA	Scottish Environment Protection Agency
SFA	Stochastic Frontier Analysis
SGEI	Services of General Economic Interest
SII	*Servizio idrico intergrato*
SISS	*Superintendencia de Servicios Sanitarios*
SNDE	*Société Nationale de Distribuition d'Eau*
SNIS	*Sistema Nacional de Informação sobre Saneamento*
SNSA	*Secretaria Nacional de Saneamento Ambiental*
SOVAK	*Sdružení oboru vodovodů a kanalizací*
SPA	*Programa de Saneamento para Assentamientos*
SPDE	*Syndicat Professionnel des Entreprises de Service d'Eau et d'Assainissement*
SPGE	*Société Publique de Géstion de l'Eau*

SPS	Small Private Supplies
SSE	Secretaries of State for the Environment of E&W
SSEFRA	Secretary of State for Environment, Food and Rural Affairs
SSPD	*Superintendencia de Servicios Públicos Domiciliarios*
SUNASS	*Superintendencia Nacional de Servicios de Saneamiento*
SVW	*Samenwerking Vlaams Water*
SWA	Singapore Water Association
SWD	Swedish Water Development
SWDE	*Société Wallonne de Distribution d'Eau*
TFP	Total Factor Productivity
TRA	Tennessee Regulatory Authority
TRM	*Tarifa Real Média*
UK	United Kingdom
UNICEF	United Nations Children's Fund
URSEA	*Unidad Reguladora de Servicios de Energía y Agua*
URSO	Regulatory Office for Network Industries
USA	United States of América
USL	*Unità Sanitaria Locale*
V&W	*Ministerie van Verkeer en Waterstaat*
VAT	Value Added Tax
VEWIN	*Vereniging van waterbedrijven in Nederland*
VMM	*Vlaamse Milieumaatschappij*
VMW	*Vlaamse Maatschappij voor Watervoorziening*
VNG	*Vereniging van Nederlandse Gemeenten*
VROM	*Ministerie van Volkshuisvesting, Ruimtelijke Ordening en Milieubeheer*
WACC	Weighted Average Cost of Capital
Wadata	*Association de Défense des Droits des Consommateurs*
WASAZA	Water and Sanitation Association of Zambia
WaSCs	Water-and-Sewerage Companies
WCC	Water Supply Customer Advisory Committees
WEF	Water Environment Federation
WHO	World Health Organization
WICS	The Water Industry Commissioner for Scotland
WIRO	Water Industry Regulatory Order
WoCs	Water-only Companies
WPSC	Wisconsin Public Service Commission
WSAA	Water Services Association of Australia
WSB	Water Service Boards
WSP	Water Service Providers
WTP	Water Treatment Plant
WVPSC	Western Virginia Public Service Commission
WWG	Water Watch Groups
WWRO	Water and Waste Regulatory Office
WWS	Water and Wastewater Services
WWTP	Wastewater Treatment Plant

Foreword

Drinking water supply to domestic consumers and to economic activities, as well as urban wastewater collection and treatment, are both major structural challenges in the development of modern societies.

These services share some common features – they are essential to consumer well-being, to public health, to economic activities and to environmental protection. For these reasons, a set of principles should be applied to their provision, namely universality, continuity and quality of service, efficiency and equity in pricing. Ensuring transparent compliance with these principles is the primary duty of ERSAR, as the regulation authority of these public services. Regulatory mechanisms aim at acting as a substitute for typical competitive market incentives, reducing monopoly power and protecting the interests of service users. This is achieved through quality of service and price oversight procedures and by benchmarking schemes ('yardstick competition' or 'sunshine regulation'), thus encouraging performance improvements in the provision of water and wastewater services, regardless of public or private utility ownership.

In the last decade, the water supply and urban wastewater sectors in Portugal have exhibited a remarkable evolution accompanied by a permanent search for a more adequate regulation model supportive of the national goals for these services. This success has been recognized nationally and internationally, and it is therefore important to build on the strength of this experience by identifying and understanding other good practices at the international level.

This publication seeks to gather experiences from the five continents, bringing forward the diversity observed on the provision of water supply and sanitation services, often reflected in the regulatory models adopted in different countries. Relying on the valuable collaboration of various entities, regulators and others, a set of information on some 60 countries is now available to all stakeholders, allowing comparisons and identification of interesting case studies and potential best practices. This study has the advantage of aggregating, in a single publication, systematic and comparable information of each country's institutional and legal frameworks, the structure of their markets, the identification of the key players and figures for the sector, including tariff structures, regulation models adopted, and the various mechanisms used in assessing the quality of service and the fulfilment of public service obligations.

The partnership established by ERSAR with Instituto Superior Técnico (Technical University of Lisbon), through CESUR – Centro de Sistemas Urbanos e Regionais (Centre for Urban and Regional Systems) – an interdisciplinary unit devoted to advanced research projects, was decisive in the design of this study and in the collection and analysis of the data. In particular, we would like to express our appreciation of the dedication that the author, Prof. Rui Cunha Marques, has devoted to this ambitious project. Even though the opinions and conclusions expressed in this work are the author's responsibility, from our standpoint it will be a fundamental contribution for our future reflection on the potential avenues of improvement in the Portuguese regulatory model.

Given the importance and innovative nature of this study, particularly due to its worldwide scope, an English edition of this study is now being released by IWA – the International Water Association – to whom we would also like to express our thanks for their enthusiastic support and cooperation. We hope this publication becomes a valuable contribution to multiple stakeholders in different countries in promoting better water services for all.

The Board of Directors
Jaime Melo Baptista (President)
João Simão Pires (Director)
Fernada Maçãs (Director)

ISBN: 9781843393412. Published by IWA Publishing, London, UK.

Preface

Regulation of public utilities and in particular of water and wastewater services is a matter of huge economic and social importance as they are essential to the development and cohesion of society. In fact we are now living in the age of regulation. All over the world countries are creating new agencies to provide for the regulation of their water utilities. This is a current affair nowadays.

This book provides a snapshot of the water regulation situation in the world today. It is based on a questionnaire distributed worldwide encompassing more than 60 water regulators and 39 countries. It is made up of three parts and three annexes. The first part includes some theoretical perspectives about the regulatory issues. It has six sections comprising the characteristics of the water and wastewater services, the need and aim of regulation, the governance models of water, the public services obligations, the regulation of quality of service and the economic regulation. The second part presents 18 case studies about regulation of water and wastewater services. It includes eight European countries (Belgium, France, Holland, Italy, Portugal, the United Kingdom and Sweden), five American countries (Brazil, Chile, Colombia, Costa Rica and the United States of America), two African countries (Mozambique and Zambia), two Asian countries (Indonesia and Singapore) and Australia. Each case study consists of different sections considering the historical evolution where the major facts and laws are presented, the legal and institutional framework, including the main responsibilities in the water sector and the regulation framework, the market structure which comprises the ownership and management of the water services, the players of the sector, the verticalization and scope of the activities provided as well as the major figures of the sector. Also outlined in each case study are the rules of the game, covering the setting of prices and tariffs, the quality of service, public service obligations and the performance of the water sector. The third part provides an overview of the water regulation worldwide. In the first section the summary of the research is presented, analysing the results of the questionnaire and the investigation made. Section 2 provides the outcomes related to the public service obligations and section 3 presents the aspects regarding the quality of service regulation. Section 4 deals with the results concerning the economic regulation and the last section highlights the best practices and the final recommendations. The book is complemented by three annexes where the contacts of the regulators worldwide are to be found, along with the questionnaire sent to the competent authorities and the results derived from each answer provided.

The target audience of the book is potentially broad, encompassing all the countries in the world. Any community interested in governance and regulation of water and wastewater services might find this a useful publication. It may also be of interest to non-regulated countries since it presents worldwide experiences and the best practices regarding the three vertices of the water and wastewater services, respectively, the tariffs, the quality of service and the public services obligations. The book is also aimed at water professionals, including economists, engineers, lawyers and academics, in both developed and developing countries. Other professionals from other sectors or academics interested in regulation and governance of network industries are potential readers as well.

All the countries included in this book had the opportunity to inspect and comment on the data and observations sent back to them in order to provide more accuracy to the elements presented. The research refers to the year 2008. The findings and views expressed in this book are those of the author and should not be attributed to the ERSAR, his university or other affiliated institutions.

Finally, hoping that the readers find this book useful in their work, studies and research the author is looking forward to receiving comments and feedback about it.

Rui Cunha Marques
Center of Urban and Regional Studies (CESUR)
Instituto Superior Técnico
Technical University of Lisbon

ISBN: 9781843393412. Published by IWA Publishing, London, UK.

Part 1
The regulation of drinking water and wastewater services

1.1 CHARACTERISTICS OF WATER UTILITIES

Industries that are characterized by market failures must, in the public interest, be subject to a visible external hand that allows them to be mitigated. This intervention becomes even more important when infrastructure services that provide collective and essential services to citizens are at stake, or when they have a network structure and require the use of property rights for the purpose of public interest (public utilities) and therefore cannot perform under a self-regulating regime or without any regulation at all. As water utilities (drinking water and wastewater services – WWS) are characterized by significant market failures and as their provision is associated with the fulfilment of the public interest, they are a classic example of services of general interest that require public regulation.

When supplied to customers through a fixed network structure, the WWS represent typical examples of monopolist industries. These services can even be considered to be ***natural monopolies***, since the costs of production are lesser in the case of a single producer for a given spectrum of demand. Thus, it becomes necessary to regulate the monopoly power to protect customer interests, especially to control prices and the quality of service in situations. Even without considering the aspects of quality, in the absence of regulation customers are doubly vulnerable in situations of a natural monopoly by consuming less (or using the service less) and paying more for the service. However, it must be noted that in developing countries, where door-to-door water supply infrastructures (networks) do not exist, the prevailing supply regime is not a monopoly and one can even find a great deal of competition between water vendors.

These services also provide evidence of ***economies of scale*** (increasing returns to scale), i.e. the unit cost of production decreases in proportion to increased demand, albeit only up to a certain extent, after which diseconomies occur. In effect, the entrance of new players into the WWS market is strongly restricted. In practice, the economies of scale associated with a natural monopoly result in the fact that each market only bears a single operator, thus diminishing the possibility of competition and the absence of self-regulation which characterizes competitive markets. However, it must be noted that there is an optimum level for the WWS in each country or region, which, when surpassed, creates diseconomies of scale (decreasing returns to scale). This fact, which is more relevant in water distribution and wastewater networks (retail component), is associated with the growing complexity and difficulties of managing infrastructures as they grow in size and also with the large investments that are necessary for interconnections and infrastructure expansion.

WWS also have ***economies of scope***, which further hinder the presence of competition in such markets. ***Scope economies*** occur when an operator performs two distinct activities which could have been carried out by two independent operators. Such situations result in a reduction in the costs of providing both activities. Generally, the synergies that emerge from sharing human resources, equipment and facilities lead to lower costs with the presence of a single operator. For example, a water service presents scope economies if, in addition to providing drinking water supply, it also makes bulk water available for irrigation purposes or if it is simultaneously responsible for the wastewater service. In fact, this is the main reason why in many countries the WWS are provided by the same entity. This is also the basis for the vertical integration of these services, i.e. the same operator provides all the phases of the production chain including the 'wholesale' and 'retail' components (sometimes called verticalization or vertical integration economies). There are also doctrinal divergences with regard to scope economies, since specialization and, consequently, a greater efficiency and transparency can surpass the benefits of shared costs and thus give rise to scope diseconomies. However, this is not the view of many governmental entities that favour the multi-utility model (e.g. Germany).

ISBN: 9781843393412. Published by IWA Publishing, London, UK.

Owing to the existence of high fixed costs, WWS also have increasing ***economies of density*** that are worthy of note. The concept of economies of density is defined as the proportional increase in the total cost resulting from a likewise proportional increase in output while the structure of the system remains constant, as in the case of network expansion, and is closely linked to the systems' peak factor. The WWS with high peak factors are penalized when compared to those that have low peak factors, since they have to invest in infrastructures with excess capacity that only correspond to a suitable productivity during the peak period.

WWS infrastructures entail ***very capital intensive and long-term assets***. These assets are, or were, often also subsidized. For example, for the water service, the assets include abstraction facilities, water treatment plants (WTP), reservoirs, lift stations and pipes, for both transportation and water distribution networks. These facilities and equipments have a project horizon of several decades. A substantial component of the assets is also located underground, and its state and quality cannot be assessed. After having been installed or built, these components do not have alternative uses and cannot be sold or transferred to another site. They are therefore considered to be irrecoverable or ***sunk costs***. The irrecoverable nature of these infrastructures corresponds to an additional risk while operating these services, which have to be suitably allocated and carefully preserved, constraining the presence of new entrants into the WWS market.

WWS have inherent ***problems of asymmetric information*** related to moral hazard and adverse selection. The former refers to the variables which are endogenous to the regulated operator, to which the stakeholders do not have access, i.e. hidden information, such as the distortion of costs and the respective allocation (e.g. the allocation of hours to a given task or the acquisition of goods at excessive prices). The latter occurs when the operator has more information than the regulator (and the other stakeholders) about exogenous variables, i.e. when the operator is better acquainted with the market than the regulator. Agents that act outside the operators are limited by restrictions of information about the quality of their services and products, their performance and results and, above all, about the value for money involved. This incomplete and inaccurate information may lead to market abuse of the monopolist operators and cause mistrust amongst customers with regard to the quality of goods and services provided. In developed countries, the drinking water quality is a typical example of the presence of asymmetric information because customers mainly consume bottled water even though the water supplied is generally of good quality. Such asymmetric information, which affects all the stakeholders, is a regulatory problem as well.

WWS are also associated with negative or positive ***externalities***, which have important consequences for third parties. ***Negative*** externalities might have a strong impact on public health and the environment. An example of this is pollution, i.e. the absence of upstream wastewater treatment for the water lines, which, apart from the consequences for the environment and public health, results in a higher cost for treating drinking water in downstream abstractions. ***Positive*** externalities are no less significant. For example, the social benefits for public health resulting from the expansion of water supply or sewer facilities are extremely important and must be internalized.

Water was once considered to be a free good (commodity). Today, although it is widely accepted to be an economic commodity, water can be considered to be an ***almost public good*** (see Marques, 2005 and Moran & Dann, 2008). In fact, the WWS market is characterized by a non-rivalry in terms of consumption and by an almost non-excludable good. On the one hand, the cost of extending the service is insignificant, i.e. the marginal cost is close to zero and, on the other hand, even though it is possible to disconnect services in most countries, their social value is far superior to their economic value and therefore the exclusion of customers from the services should only occur in very specific situations. In fact, the cost of exclusion (for example in terms of public health or social segregation) is far higher than the cost of subsidizing more low-income customers.

The presence of external factors that are exogenous to WWS management means that no two operators are identical. For example, the availability of water resources and their quality largely limits the water service cost. There are many operators that do not have water within their area of jurisdiction or if they do the water is available only at a very high cost (e.g. desalinized water). There are various other constraining factors, such as topographical conditions, the geology of the area, the population density and its variation and the social characteristics of the region served, which reinforce the specific nature of the ***operational environment*** of each operator. The structure of the WWS costs is strongly influenced by the contextual factors that characterize the operational and institutional environment where they operate. Consequently, given the importance and specific nature of the operational environment of each operator, it becomes difficult to establish technical, political, social and economic criteria that avoid some external control (i.e. regulation).

WWS are services of general interest or, to be more precise, ***services of general economic interest*** (SGEI). Services of general interest are those services that satisfy the basic needs of most citizens; they are economic, social or cultural, and their existence is essential for the lives, health or social participation of citizens (Gouveia, 2001). Thus, SGEI are a subgroup of services of general interest, such as those services that satisfy basic economic needs. Although the classification of the WWS as SGEI is not peaceful, especially at the heart of the European Union (EU), even causing a fair degree of controversy, WWS are in fact SGEI as long as they are subject to the condition of self-sustainability, which is the case (or efforts are being made to ensure it) in most countries. Countries where this does not happen, at least within Europe, have very special situations, such as water supply virtually free of charge owing to abundant water resources (e.g. in Ireland and in Austria), but even in such situations they are breaking EU legislation, namely the Water Framework Directive (WFD). The SGEI are also fundamental elements to promote economic and social cohesion, for both the welfare of the population and the economy on the whole. Consequently, they should conform to a set of public service obligations, such as universality, equality of access, continuity, affordability and adaptability. Likewise, they should adopt rules of good conduct, including transparency in terms of management, the tariff system and funding, which are essential requisites for WWS in any society. However, it must also be noted that, as subsequent sections and this document generally show, public service obligations are often divergent. It is thus necessary to have a visible hand that can ensure compatibility between the principles of economic efficiency and those of social efficiency. Figure 1.1 provides a summary of the main characteristics of WWS.

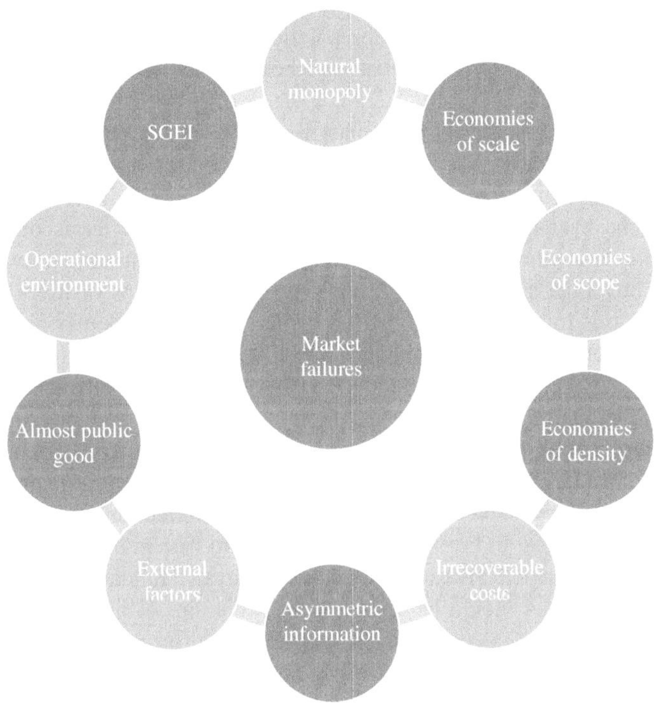

Figure 1.1 Main WWS market failures

1.2 THE NEED FOR REGULATION AND ITS OBJECTIVES

The WWS market failures outlined above justify state intervention in terms of regulating these services to protect public interest. The invisible hand of the market in the context of network industry services is not effective and, consequently, the state has to intervene by defining rules, controlling and supervising the way they work. As in the case of other monopolist network infrastructures, WWS are prone to a quiet life and inefficiency and thus require incentives to become more efficient and innovative. Although regulation is instituted in the public interest, it also inevitably generates winners and losers and can thus be sought by some interest groups, including that of the bureaucrats themselves. The objectives for regulating WWS vary considerably from country to country (or from region to region) and are sometimes outlined in different degrees depending, for example, on organizational arrangements or the level of development that exists. Many countries also impose obligations on private companies that are different from the obligations imposed on public enterprises. On the other hand, the imposition of very ambitious targets with regard to the quality of service does not make much sense if the operator is limited in its coverage level. In some countries, the aims of WWS regulation are very close to the purposes of other network infrastructures, such as gas and electricity, in which creating a market and promoting competition are the primary objectives (e.g. in the United Kingdom – UK), although in the vast majority of countries the exclusive monopolist rights of WWS continue, for better or worse, to be guaranteed and regulated.

The main objectives of economic regulation are related to WWS providing value for money and, above all, are as follows:

- To protect the interests of customers with regard to public service obligations;
- To promote efficiency and innovation;
- To ensure the WWS stability, sustainability and robustness.

a) Protecting the interests of customers with regard to public service obligations

Considering that WWS are essential services, the role they play concerning the ***socio-economic cohesion of the population and customer protection*** and the fact that they are strictly necessary, they require the fulfilment of diverse principles which are commonly known as the general interest or public service obligations. Among these requirements, frequently applied to all infrastructure services, the principles of universality, continuity, quality of service, equality of access, affordability, health protection, the security of goods and services, transparency, choice of suppliers and services, options for different forms of payment, representativeness and customer participation in decisions and the existence of mechanisms for conciliation and conflict resolution are particularly relevant. Public service obligations will be analysed in further detail in Section 1.3. Satisfying these obligations in a suitable manner is thus one of the main objectives of WWS regulation.

Although public service obligations have not been classified uniformly in international terms in the context of the WWS, they are based on a common matrix, varying with regard to the intensity and the ways in which they are applied in each country. For example, the obligation to provide a universal service in the context of the water utilities implies that all potential customers have access to WWS in suitable conditions in terms of price and quality of service irrespective of their geographical location or social characteristics. The obligation to provide a universal service entails a comprehensive territorial coverage and special care vis-à-vis certain groups of users (e.g. the disabled and senior citizens), guaranteeing conditions for accessibility that are suited to their particular needs. Regardless of the models that are adopted for the implementation of this principle, which will be presented shortly (see Section 1.3) and which vary from country to country or even from operator to operator, and irrespective of the final objective, which is naturally ensuring universality, countries have framed it in distinct legal and institutional ways. Thus, there are countries where the public service obligation related to the issue of universal service is unequivocally enshrined in the main legislation, others that consider it in secondary legislation and others where it is not even compulsory. It must be emphasized that, unlike other utilities, such as electricity or landline telephone services, where universal service is always considered in legislation, at least in Europe, this does not hold true in the specific case of the WWS due to the significant impact of the cost of such a measure. Regulation plays a fundamental role in this context since the fulfilment of public service obligations should be regulated with the aim of not only guaranteeing them, but also of determining the extent to which they have been achieved, namely, the effort that is necessary and that has an impact on the economic performance of the WWS and their stability and sustainability. Universal service is by itself a good example, since possible subsidies or cross-subsidization could send the wrong messages to the market, causing wastage and inefficiency that can be avoided. In addition, operators can also be reluctant to serve certain customers, since the additional cost of making WWS available to such customers is very high. These are clearly problems that come under the scope of regulatory issues. The quality of service, especially in terms of its multidimensional nature, is another interesting aspect. Providing WWS with a very high quality (gold plate practices) does not necessarily safeguard the interests of customers, since they (or someone) will always have to pay for the cost of the service. Moreover, due to their very nature, WWS have to be provided with a high level of quality, which means, for example, providing water at least in a sufficient quantity, quality and pressure. It is not easy to institute this balance, so it requires the intervention of regulation. Very often, the regulatory authorities impose this responsibility upon customers, who are the main interested parties, asking them how they would like the quality of their rivers or beaches to be or, to formulate the question in another way, how much they are willing to pay to have a certain environmental quality or quality of service. Naturally, this option can only be effective when the customers are extremely well informed about the matters discussed, which is not the case in most situations.

b) Promoting efficiency and innovation

Promoting efficiency and innovation should be one of the primary objectives of regulation. This is how the customer interests will, in fact, be safeguarded. Quoting the first Scottish regulator, '*Ultimately the best way of promoting customer interests in a public sector model is by improving the economic efficiency of the industry, and thereby the value for money generated*' (WICS, 2002). However, it must be emphasized that the objectives of maximizing (economic) efficiency can clash with some of the objectives regarding public service obligations, as well as with some associated interests. This is the situation, in fact, and it could not be any other way. For the regulator and with the aim of promoting the public interest, the market should function and social aspects should be naturally guaranteed, but by those legally responsible to do so. For example, free of charge supply or political tariffs lead to high usage and wastage, which will be paid for by all the customers or, even more critically, by all taxpayers. Especially in the context of the WWS, the customer-pays and polluter-pays principles are fundamental for their functioning and sustainability.

Efficiency corresponds to a comparison between the inputs consumed and the outputs produced and their optimum values. Efficiency can be decomposed into allocative efficiency and productive or economic efficiency. The former corresponds to the circumstance where the market performs under perfect competition and, consequently, with tariffs at marginal prices, which can not be applied to the WWS market, and the latter is associated with producing at the lowest possible production costs. Productive efficiency can be further decomposed into input or output technical efficiency, which measures the intensity of the wastage observed during the production process (ratio between inputs consumed and outputs produced) and into allocative price efficiency, which involves combining inputs (i.e. work and capital) that produce a given quantity of outputs at the minimum cost. Technical efficiency can also be decomposed into various components, the most relevant of which refers to efficiency of scale, which measure the impact of the size of the operator concerning efficiency. It must be stressed that the concept of efficiency essentially has a static nature. Although there are diverging opinions with regard to this issue, some experts include in efficiency decomposition a part relative to the dynamic efficiency associated with innovation and production technology change (function) over time. In this case, we are in fact talking about productivity, which is dynamic and corresponds to the variation, over time, of efficiency, the operational and institutional environment and the innovation that has occurred. The aim of the regulator is to motivate a productivity increase. Figure 1.2 provides a summary of the decomposition of efficiency.

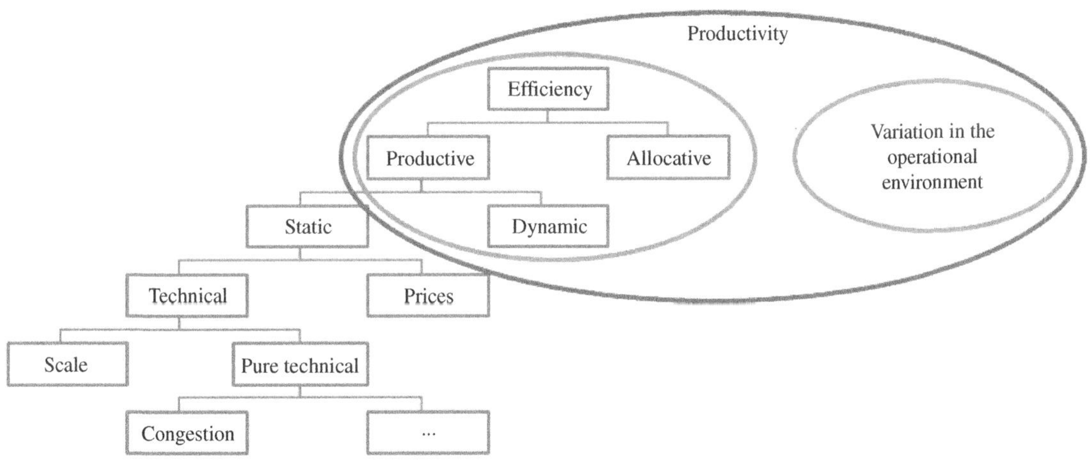

Figure 1.2 Decomposition of efficiency

While promoting efficiency, it is necessary to keep in mind the incentives that are being provided. In fact, the promotion of a certain kind of efficiency can prove to be detrimental to other kinds of efficiency. These conflicts between different types of efficiency are quite evident in the scope of the WWS. In this case, the maximization of static efficiency requires the presence of a single enterprise, while a monopolist company will adopt a second-best policy (and not one of marginal prices) for price choices, thus causing allocative inefficiency. In contrast, to maximize allocative efficiency it would be necessary to have several operators, so that the competition generated would result in the establishment of marginal prices (first choice), which would lead, actually, to static inefficiency (e.g. inefficiency of scale). From another perspective, at least in the short term, the regulated operators need incentives to greater productivity gains in order to become more efficient and innovative. Such stimuli (retaining a larger share of profits) are contrary to allocative efficiency, which requires prices to be closer to the costs involved. Another relevant question corresponds to the dichotomy between the short and long term. The accomplishment of investments that mitigate long-term costs (dynamic efficiency) can collude with short-term objectives of reducing costs, and populist and comfortable measures, which makes the entire regulatory process an extremely complex and controversial one. In fact, it must be stressed that while the problem of a monopoly is, above all, a short-term problem, regulatory difficulties are mainly associated with the long term.

c) Ensuring the WWS stability, sustainability and robustness

Regulation should promote ***stability*** in the water sector, namely stability of the policies that are applied, and should seek to develop the sector gradually and in a balanced manner, while simultaneously striving to ensure that the sector is sufficiently versatile and robust to meet the challenges of any potential adversities.

One of the fundamental contributions for achieving this objective is the WWS self-financing. If the principles of customer-pays and polluter-pays were instituted, in which customers pay the real cost of the service (participation constraint), encompassing both operational costs and capital costs (including a return on the capital invested), a large part of the problems would be solved. In such circumstances, operators have stability and can carry out the necessary investments, especially non-urgent investments, and develop correct management policies, without political concerns or without considering immediate results. In this scenario, regulation is fundamental and curbs short-term populism with regard to the non-fulfilment of the aforesaid principles and the postponement of the investments. It further ensures rational and technically appropriate decisions, although the population and/or vested interests might not be pleased with it. It must be noted that satisfying public service obligations does not hinder these principles, but merely allows certain groups of low-income users to benefit from some kind of assistance which does not exclude them from enjoying these essential services.

In the long term, regulation also plays an important role in guaranteeing the ***sustainability and robustness*** of the WWS. Thus, it is essential that the objectives concerning economic efficiency and public service obligations are pursued with an integrated and long-term vision and not just as short-term objectives. Sometimes, these short-term objectives, sought by the population and offered by politicians, make the long-term objectives unfeasible and incompatible owing to the popularity of these measures. As emphasized above, WWS should not be managed or planned according to the immediate and the short term but rather to the whole life cycles of the infrastructures, which are much greater.

In the past, in other infrastructure services such as electricity and landline telecommunications, regulation sought to achieve the same objectives mentioned above in the context of the WWS. Currently, owing to innumerable factors, the central purpose of the regulation of these utilities is to create a market and promote competition, where technological development and the size and value of the business itself have relevance, believing that the interests of customers are thus better safeguarded if the free market works. The ultimate objective, i.e. protecting customers in the provision of essential services, continues to be the same, although the way of guaranteeing it is different. Actually, even concerning the WWS, the functions of the regulator are increasingly also those of replacing and mimicking the market, seeking to ensure the existence of incentives even when there is no competition as in these monopolist infrastructures. For this reason, yardstick competition and the potential of benchmarking are more and more used, thus trying to replace competition in the market by competition by comparison. In the case of WWS, the great difference is the supposition that competition by comparison will always be a less onerous instrument and therefore less efficient to achieve these objectives. Consequently, regulation is currently a 'friend' of the market and not hostile to the market, unlike in the past.

In the energy (gas and electricity) and telecommunications infrastructure services, apart from a great substitutability (i.e. electricity by gas) and a large number of competition alternatives fostered in the market (i.e. mobile phones), third-party access to networks has also become possible. In fact, in the aftermath of EU legislation, and despite the reduced practical outcome, which fell short of the desired and expected results with the incumbent, keeping a dominant position, this liberalization of the markets generated very beneficial competitive incentives, both in terms of prices practised as well as in terms of the quality of services provided.

In the WWS, this same liberalization is currently underway in some countries, namely in the UK (i.e. common carriage) and in some Australian and American states, although the results, while being positive (particularly in terms of the industrial segment) have lagged far behind the results achieved with regard to other infrastructure services. This has been essentially due to technical reasons, such as the cost of transport associated with the weight of water, the high degree of fragmentation of these services and the lack of inter-connections; problems pertaining to the quality of water when there are diverse sources and the responsibility for their supervision have hindered the success of this initiative. Apart from this, the business value of the WWS is still quite inferior to that of other public utilities and in most countries the local administrative authorities are responsible for these services, increasing the difficulties of coordination.

Figure 1.3 presents a hypothetical scheme of a competitive WWS market. In any case, all the great reforms that have been implemented in other infrastructure services have also been implemented to a certain extent in WWS. For example, if in energy or telecommunications steps were taken to attempt to integrate the markets at the level of a group of countries or even globally, in the case of WWS, the regulation, supervision and constitution of partnerships have become increasingly important, even though these services are often supplied by the local administration. This has been achieved through regional associations or organizations in countries, of a more or less voluntary nature (e.g. in the EU, with the publishing of the Water Framework Directive or, in a voluntary manner, by the ADERASA in South America). In Europe, it has even been advocated that a European regulator or observatory should be created for the EU (Marques & De Witte, 2010).

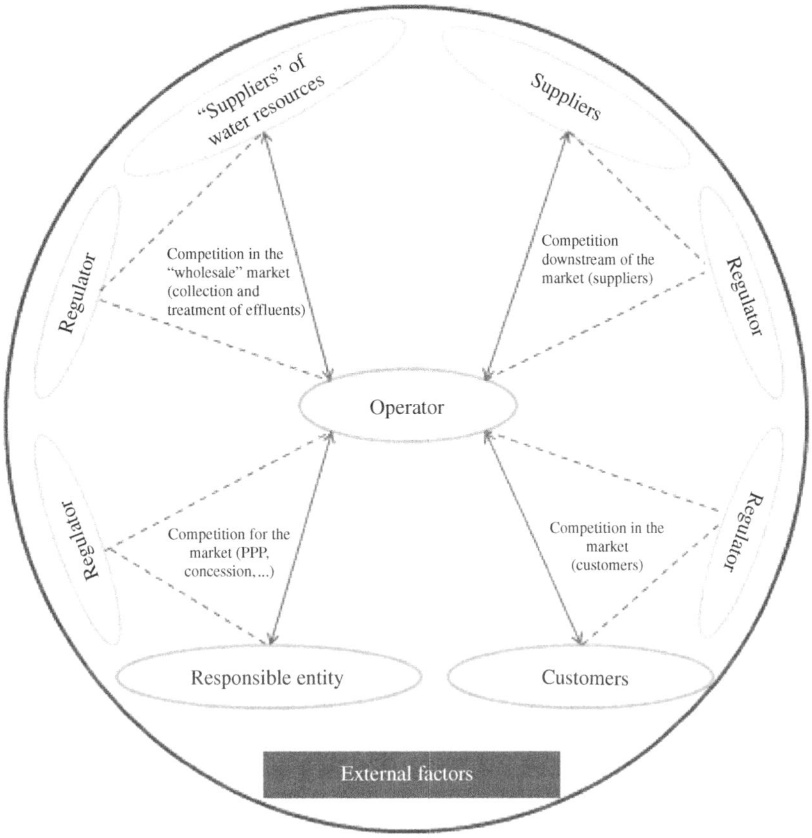

Figure 1.3 Types of competition within the water industry

The dismantling of monopolies, which occurred in the sectors of energy and telecommunication services, has not occurred with the same intensity in the WWS. Even so, its impact was, in a certain way, attenuated as a result of its commercialization and corporatization, in the sense that there is a limit to profits and that the management has to be transparent and subject to the public judgement of customers and other stakeholders.

In the WWS, regulation is also increasingly being contracted and agreed upon with different stakeholders and is not instituted in a unilateral manner. With a view to ensuring greater accountability of all stakeholders, the concept of so-called 'procedural democracy' has been applied along with measures to fight asymmetrical information within the WWS, to promote an effective participation and coordinate all the stakeholders in the regulatory process, namely through self regulation mechanisms or the existence of advisory councils. Finally, it is important to mention the increase in 'social' regulation, which is a transversal element common to the entire economy and all infrastructure services, including the WWS. In this context, one can highlight the aspects pertaining to the environment, health and working conditions, amongst others, as well as the transversal regulation of the competition.

1.3 MODELS OF ORGANIZATION IN THE WATER SECTOR

WWS models of organization are not very different from the models of other infrastructures. The main difference lies in the fact that these models are known not on the basis of the respective underlying institutional framework, but based on the countries that were the forerunners of the implementation of such models. According to Moreira (1997), public utilities can be provided through entities with private management and ownership, which are especially regulated, through concession arrangements (or other form of delegation) to private companies under public control and through public companies.

In the context of WWS, the first institutional configuration is known as the ***English model***, the second as the ***French model*** and the third as the ***public operator model***, which, while not exactly corresponding to a single country is implicitly associated with the Dutch model (Marques, 2005). Amongst the most relevant aspects that can be highlighted, it must be noted that there are no perfect models and that in the cases mentioned here, with their virtues and, naturally, some problems, these models have good results and have been in effect for at least about two decades, which is a significant fact in itself. The stability of these models has allowed one to ascertain their main merits and less successful aspects with a fair degree of certainty. The different models are associated with a certain degree of privatization. This ranges from the model of public enterprise, which only includes the juridical-formal privatization of WWS, to the English model, which encompasses the full divesture (ownership, management and juridical-formal aspects) of WWS, likewise encompassing the French model, which corresponds to the privatization of the management (management and juridical-formal aspects) of the WWS (Marques & Monteiro, 2004). It must also be mentioned that other models exist amidst these broad systems. These usually include the characteristics of more than one model or present specific trait, which almost always relate to the degree of the vertical and horizontal integration of the production chain and the scope of the services being provided. Apart from this, in global terms, the predominant institutional framework is associated with the absence of a specific operator to provide WWS. In this case, the WWS management is carried out along with other activities and is directly undertaken by the public entities. Owing to the absence of autonomy with regard to its functioning and the fact that it is almost always economically unsustainable, the author does not believe that this 'model' of providing WWS is viable and that it is an option for the future.

a) The English model

The ***English model*** of organization is intrinsically linked to the experiment that was developed in England and Wales (E&W) from 1989 onwards, when the WWS were fully privatized (ownership). This model is based on WWS provision through private operators who, apart from being responsible for managing the systems, also own the assets (including the fixed infrastructures). Nevertheless, the public interest is guaranteed and supervised by the state, by means of an independent regulatory agency. In this model, there is an unequivocal separation between the state, the regulator, the operators and the customers. The managerial scale of the WWS also assumes a regional dimension, preferably at the level of the river basin, which can encompass the entire water cycle. The success of the English model is closely associated with the leading role played by regulation and its effectiveness, as well as the capacity that the regulatory authorities do, in fact, have to remain independent with regard to the government, the operators and other stakeholders. The mission of regulation is to defend and protect users (quality of service and tariffs setting) and to simultaneously guarantee the sustainability of operators, seeking to create and substitute the market and adopting mechanisms of yardstick competition, which are very important for this purpose. Apart from E&W, other countries have also sought to follow this model of organization (e.g. Chile, Sweden and the Czech Republic), albeit without the same level of visibility. Figure 1.4 presents an overview of the English model of WWS organization.

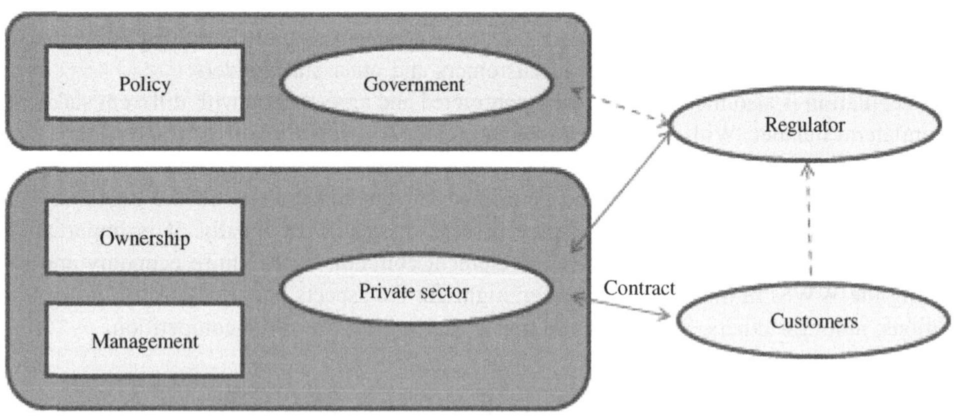

Figure 1.4 English model of WWS organization

b) The French model

In the French model, the WWS management is awarded to private companies through public tenders that, by means of a contract arrangement (generally a concession or lease), establish a set of rights and obligations. The WWS asset ownership remains always within the public sector. The existence of true competition for the field and the completeness of the contract design are the key elements for the success of this model. It was originated in France during the eighteenth century and rapidly expanded to other Mediterranean countries (e.g. Spain and Portugal). Nowadays, it is the most common form for the private sector participation in the WWS management, especially when their responsibility is attributed to the municipalities. Numerous bidders promote the elimination of excessive rents and profits. A contract that has been correctly designed bears a rational sharing of risks, enables effective monitoring and suitably protects the partners concerning the adversities. It must be highlighted that in this model there is a regulation by contract, even though the countries that are following this model are increasingly creating regulatory agencies focused on the WWS to supervise the quality of service and to intervene in the case of conflicts and unforeseen circumstances. In fact, it is impossible to design comprehensive contracts that foresee all possible future scenarios. This problem is accentuated with an increase in the duration of the contract. Here, the functions of the diverse intervening parties are not as clear as in the case of the English model and there is a tenuous line between the awarding authority and the concessionaire with regard to the public service obligations that have been established and the corresponding responsibility for them, thus penalizing customers. The governmental authorities (mayors) generally assume an excessively visible role, which is sometimes harmful in this model. Nevertheless, the problems associated with the French model concerning public tenders and contracts, and especially public-private partnerships (PPP), which in practice result in the fact that the contracts are not suitably upheld and in economic and financial rebalancing, have not hindered this model from proliferating throughout the world (Guasch, 2004). Figure 1.5 provides an overview of the French model.

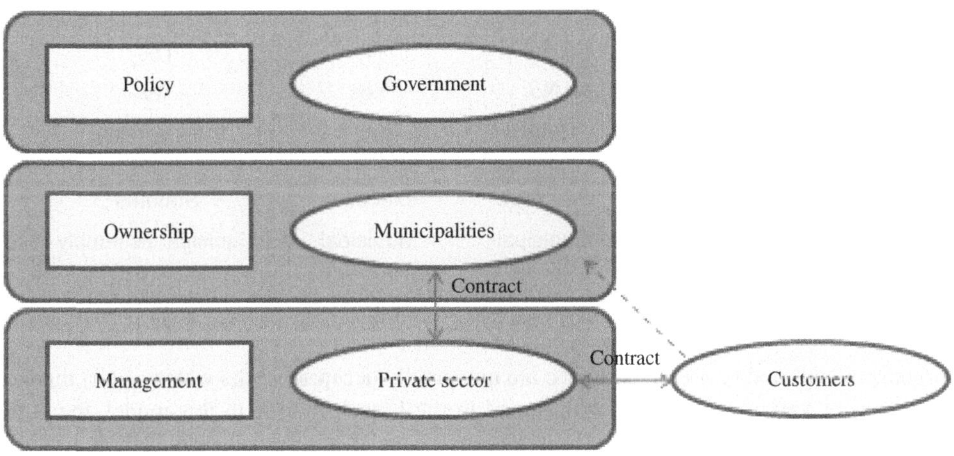

Figure 1.5 French model of WWS organization

c) Public operator model

In the public operator model, the public sector is responsible for the management of the service and owns the assets. The state, at its different levels (central, regional or municipal), regulates itself by directly intervening in the market, thus maximizing social well-being. It must be noted that the success of this model in some countries is associated with the juridical-formal privatization of the WWS, with the streamlining of procedures by means of a clear tendency towards private law. Apart from this, this model has been successful only when there are no state failures, which generally only takes place in the most developed countries. In such nations, the social regulation that is transversal to the entire economy provides a very significant contribution towards the success of this model, as it is proved by the self-regulation that has been implemented by associations in this sector. In this kind of situation, the importance of regulating the monopolist infrastructure services has been widely recognized. Even in cases of exclusively public intervention such situations tend to be regulated by independent regulatory agencies (e.g. Scotland and Australia) as a result of the long life cycles and the great temptation to adopt short-term policies. In the public operator model, the WWS management is frequently a little transparent, in an attempt to avoid responsibility for any possible inefficiency and reduced productivity that might exist (Stern & Holder, 1999). Be that as it may, the public operator model is predominant in the global context. Figure 1.6 illustrates the public operator model. Note, however, that this model is not the same as the public management model, as emphasized above. Thus, WWS are still frequently supplied by operators that do not have an entrepreneurial arrangement, namely by means of direct *régie* and indirect *régie*.

However, it is believed that these latter institutional arrangements will tend to disappear in the future. Table 1.1 presents the main organizational models of WWS.

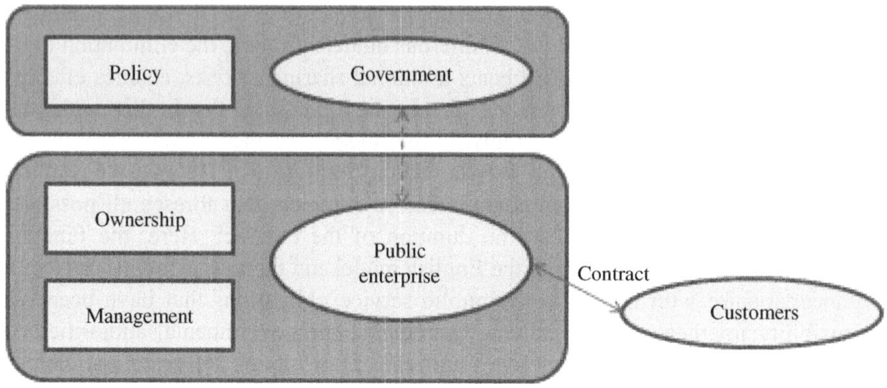

Figure 1.6 Model of WWS organization based on public enterprises

Table 1.1 Organizational models of WWS

	Direct régie	Indirect régie	Public company/ municipal authority	Private company
Juridical entity	No	No	Yes	Yes
Administrative and financial autonomy	No	Yes	Yes	Yes
Rules of law	Public	Public	Public/private	Private
Attribution of the definition of tasks	Municipal executive	Municipal executive	General assembly/ Statutes	Contract
Supervision of the quality of service	Municipal executive	Municipal executive	Municipal assembly	Contract/ regulator

d) Other models

Although the organizational models presented above are universally accepted in the water sector, there are many other intermediary situations or variants, especially with regard to the French model. In this model, the purely contractual PPP type is favoured. This can correspond to the traditional type of purely contractual PPP, i.e. of concession or leasing (*affermage*) arrangements, according to whether or not investments are made by the private partner or of an institutional type of PPP, where a joint venture is constituted to manage a given WWS (e.g. in Berlin, in Germany). Table 1.2 presents the main forms by which the private sector can participate in the water sector.

Table 1.2 Participation of the private sector in WWS provision

	Management contract	Lease	Concession	Joint-venture	Full divesture
Service ownership	Public	Public	Public	Shared	Private
Fixed assets ownership	Public	Public	Public	Shared	Private
Investment	Public	Public	Private	Shared	Private
Assumption of risks	Public	Public	Private	Shared	Private
Remuneration	Public	Public	Private	Shared	Private
Duration	Short or medium	Medium	Long	Long or without duration	Without duration

The specific characteristics of the different models in various countries/regions also frequently occur at the level of the vertical or horizontal integration of the services provided by the WWS. There are countries/regions where WWS are vertically integrated, i.e. the ‘wholesale’ elements (wholesale market), which corresponds to the abstraction, production, treatment and transport elements of the drinking water service and the transportation and treatment of wastewater in the wastewater service and the 'retail' elements (retail market), which corresponds to the distribution of water in the drinking water service and the collection of wastewater in the wastewater services, are provided by the same entity (e.g. Italy and Sweden) and there are countries/regions where this is not the case (e.g. Portugal and Holland). Apart from this, while in certain countries the WWS have a limited degree of horizontal integration, serving only a few hundred inhabitants on average (e.g. in Denmark or Finland), in other countries they are characterized by a high horizontal level, serving hundreds of thousands (millions) of inhabitants (e.g. in the UK), and can provide water services in conjunction with wastewater services (e.g. in Greece) or separately (e.g. in Belgium), or can even include other activities, such as gas, electricity or other public utilities or transportation services (e.g. in Germany).

Finally, there are some atypical situations, both in terms of self-regulation, for example in community systems of WWS in low-income countries, and in terms of deregulation in which the market performs in an open manner and without established rules (e.g. water tankers). In the first case, i.e. self-regulation, the regulation is developed by the entities that are being regulated by means of suitable collective mechanisms, especially decisions or agreements established by and between the professional associations or their representatives. In the second case, deregulation consists in the abolition of the main legislation, statutes and other rules that constrain the behaviour of operators, allowing the market to function in an autonomous manner. However, while these two models can provide great advantages in certain circumstances, they are not expected to have a significant presence in the future of worldwide WWS.

1.4 PUBLIC SERVICE OBLIGATIONS

As they are SGEI, WWS are crucial to promote social and economic cohesion for the well-being of the population due to the important role they play in all sectors of the economy, such as, for example, in industry (see COM2003/270). Thus, given their essential nature, such services must satisfy a set of requirements that, as mentioned above, are known as public service or general (economic) interest obligations. Universality, equality, accessibility, continuity and adaptability, as well as the adoption of rules of good conduct, which include managerial transparency, the tariff and funding system, participation, conflict resolution mechanisms and access to information, are the main public service obligations. The quality of service provided is also an integral element of these obligations, which will be analysed in a specific section dedicated to this aspect (see Section 1.5). The main objectives underlying customer protection and the assurance of public service obligations seek to guarantee that the market progressively ensures better results and that any shortcomings liable to be detrimental to customers are minimized. In this scope, it is critical that there is transparency and probity in terms of commercial practices. It must be underlined that although public service obligations constitute an essential element of the model of social development (especially in Europe), market forces alone cannot guarantee these obligations in the WWS. It is for this precise reason that public authorities, and especially regulation, are profoundly involved at all levels in order to provide, regulate, organize or, to different degrees, to finance or support these services. These public service obligations, which can sometimes even be discordant in regard to each other, have significant consequences in terms of economic regulation and, therefore, cannot be dissociated from it. The following paragraphs provide some insights about each public service obligation that should be guaranteed by the WWS.

a) Universality

When the principle of universal service is fulfilled, all citizens have access to WWS in equal conditions in terms of price and quality of service, irrespective of their geographical location or social characteristics. Hence, a universal service encompasses the entire area of an operator's jurisdiction and pays particular attention to certain groups of customers, especially the low-income and disadvantaged segments of society, which can include the poor, sick or other customers with special needs. In order to guarantee this public service obligation, governments or operators (or customers) in each country can adopt the policies they deem to be most suitable, although there are at least three distinct ways of achieving this. These range from the existence of a compensation or tariff equilibrium fund, by which the richer operators finance the poorer operators, to cross-subsidization, where different goods, services and customers finance the services of the less profitable customers (e.g. water services subsidize wastewater and industrial users, or those with a greater consumption subsidize domestic users or users who belong to the lowest consumption scales) and the attribution of subsidies directly to the operator (both by directly transferring pecuniary amounts and by financing investments). The objective behind instituting subsidies is to establish the price of a good or a service close to its marginal cost, thus generating additional revenues. Its attribution is justified when the surplus revenues are superior to the losses that occur by transferring the costs of the operator to the users. Subsidies can be attributed in different forms, from supplying transfers or compensation indemnities, or can be given directly to customers in kind (e.g. a certain volume of free-of-charge water) or take a pecuniary form. In many countries, the principle of universal service is enshrined in the law. The concept of universal service is characterized by its dynamic nature, owing to which it evolves over the course of time, not just with technological developments but also with the changes in the different kinds of needs on the part of customers and society.

b) Equality

Equality is one of the fundamental requisites for providing WWS. Dealing with customers on an equal basis is an essential principle. However, while satisfying the condition of universality, the principle of equality can sometimes be overlooked. Thus, WWS must always be characterized by an objective, transparent and non-discriminatory intervention, and must respect the principle of proportionality, especially with regard to tariffs and the quality of service.

The main objective of the presence of equality in the regulatory process is to guarantee equal treatment for all customers, namely in terms of right to benefit from WWS under conditions that do not discriminate between similar categories of users, which can refrain from obtaining the universal service mentioned above. Thus, while seeking to eliminate situations of discrimination, the principle of equality also seeks, from an economic point of view, to mitigate situations of cross-subsidization, since the benefits resulting from the respective availability are the same for all those who are covered by the service, irrespective of the level of use.

c) Accessibility

Controlling the accessibility of prices and tariffs determines that a SGEI must be offered at a reasonable price, so as to be accessible to everybody, contributing towards socio-economic cohesion. In some cases, accessibility in terms of prices can imply providing services free for some kinds of users, with tariff blocks for other customers, so as to guarantee that low-income populations or communities with special social needs are not deprived of these services, by means of cross-subsidization. The promotion of accessibility should, however, also promote an efficient use of water, avoiding granting subsidies to non-essential consumption.

It is an obligation of governments to ensure affordable prices, creating price control mechanisms such as the regulation of maximum prices and revenues (profits), the geographical levelling of prices and/or subsidizing users with special needs. Accessibility is clearly linked to the universality of services, although it is normally against equality.

d) Continuity

Continuity is related to the permanent supply of WWS, whereby it is interrupted only in clearly defined situations or after due and timely justification. For the drinking water, continuity implies the supply not just in suitable quantities but also with the desired quality and adequate pressure. Customers must be compensated when the causes of a lack of continuity can be imputed to the operators or to third parties, as long as the latter can be identified and can be held responsible. These dissuasive measures result in a reduction of the service interruption and a consequent increase in the reliability and quality of the systems. Certain customers can have priority over others, such as hospitals, schools and other public facilities, and the conditions of their supply should be defined in the WWS regulations. In specific situations, it is not necessary to impose the requirement of continuity, since an uninterrupted supply is of commercial interest.

e) Adaptability

Adaptability is one of the many attributes that WWS, as a SGEI, should present. This is demonstrated by the capacity of WWS to progress according to the economic, technological and even social evolution of the context in which these services are provided. WWS should also adapt and respond to the individual and particular needs of each customer, avoiding their exclusion.

f) Transparency

The activities of a WWS should be characterized by transparency. Decision-making processes should be public, publicized and easily accessed. The justification of the tariffs and the prices adopted should be clear, comprehensible, justifiable and easily verifiable (by a third entity if necessary). WWS should make available to the public information regarding the management and operation of the systems and, especially, all possible information concerning the quality of water supplied and the quality of the environment. The quality of service provided should likewise be clear and transparent for the different stakeholders. Access to (free and public) information is also a fundamental element in this context.

Transparency is a rather important principle of good governance and it is widely accepted that WWS activities should be managed with maximum clarity and openness, as should the actions of the different stakeholders that intervene in these services (e.g. government and regulators). Transparency is also extremely relevant throughout the entire regulatory process, all the more to safeguard against vested interests and reduce the risk of capturing and gaining an undue influence in the markets. Apart from this, transparency promotes greater confidence amongst customers, guarantees the confidence and satisfaction of investors, holding the entire industry accountable for the management and evolution of services. Concerning the WWS quality, the principle of transparency is an underlying element of the need for clear, simple and accessible regulation, as far as it is possible in terms of the preparation, promulgation, codification and dissemination of such regulation.

g) Participation

The decision-making process of the operators (and regulators) must, as far as possible, include the participation of the different stakeholders and particularly that of customers. In addition to enabling the entity to receive suggestions, criticisms and comments, which can be beneficial and assist in the decision-making process, this contributes towards a better acceptance of and fewer complaints about the decisions implemented. If customers (and the operators being regulated) could participate in the process of defining the tariffs and not receive them as a *fait accompli*, it would be better accepted. Participation also increases the accountability of the various stakeholders, making them equally responsible for decisions. Participation can assume different forms, ranging from the most formal procedures, with defined time frames, to informal participation on the initiative of both parties. Advisory councils may also exist (e.g. in Portugal), where all the stakeholders can participate and make known their opinions about the decision-making process.

h) Resolution of conflicts

Customers and other stakeholders must be able to access conflict resolution mechanisms, to be used whenever appropriate. Generally, conflicts have a minor economic dimension and do not justify making use of judicial measures. Customers often end up giving up the complaint, thus harming their own interests as well as the overall quality of service. Options to solve conflicts should include mediation, conciliation and voluntary arbitration. These measures can be promoted by regulators, by associations in the sector, by quasi-judicial entities (e.g. justices of the peace) or by other entities, which preferably should be independent from the interests that are involved in the conflict.

1.5 REGULATING THE QUALITY OF SERVICE

1.5.1 General considerations

As mentioned above, many of the WWS regulatory agencies worldwide do not perform functions of economic regulation. Instead, their role is simply to supervise and foster the quality of service. There are various reasons for this. The first is the fact that the attribution of these SGEI falls under the purview of the local administration. Thus, on the one hand, the large number of operators makes economic regulation in the traditional sense of the word unfeasible, since it is neither possible nor desirable to institute a regulatory agency in each municipality. On the other hand, these local bodies enjoy a specific autonomy, which is normally protected by the fundamental law of each nation (the Constitution), owing to which a supra-municipal administration (e.g. at a regional, state or federal level) cannot intervene in their sphere, at least not very easily. Customer defence and protection, which includes the quality of service being provided, are almost always attributed to the aforesaid supra-municipal administration. The second reason has to do with the participation of the private sector, with the frequent existence of regulation by contract, where the principles of tariff structures and their establishment are defined in the contract and cannot be changed. Although some norms relative to the quality of service could be defined in such a contract, there are no watertight safeguards in this regard, nor are all the terms in this regard subject to a tender, which justifies the existence of a regulator to monitor the quality of service provided. Finally, the third reason is associated with the nature and essence of WWS.

The creation of an economic regulator can have two distinct significances or consequences, namely, paving the way for privatization and increasing prices and tariffs to ensure the total recovery of the costs involved, which constrains its implementation, especially by the political authorities (i.e. avoiding 'political' prices, ensuring that the principle of customer-pays and polluter-pays are fulfilled). Very often, the regulators that monitor the quality of service are instituted as an intermediary step towards creating a more demanding and broad service, with powers related to the economic regulation of WWS (e.g. in some Australian states).

1.5.2 Approaches to regulation

Regulation of quality of service will never cease to be a controversial issue. The multidimensionality of quality and the fact that customers cannot choose other suppliers makes it more difficult to check and regulate the quality of service and reinforce the need for such regulation. There is no single, perfect and consensual methodology; instead there are various approaches, each of which has its pros and cons. One of the main ambitions of the operators, especially when methodologies based on performance incentives are adopted, is to reduce costs during the regulatory period so as to increase their profits. This is only possible either by increasing productivity or reducing the quality of service provided. Since there is no effective competition (actual or potential), the interests of customers could be jeopardized if the latter were the case, if provisions had not been made in this regard in the regulatory process. When the tariff setting is based on rate-of-return regulation the reverse occurs, which is no less worrying, since operators could intend to make a service available with very high standards of quality (gold plate practices), making significant investments (over-investment as a result of the A–J effect, a phenomenon which will be examined in Section 1.6), with a view to obtaining a greater level of remuneration, which is, presumably, against the interests of customers (and against the public interest).

There are usually two approaches while considering the issue of regulating the quality of service, a direct approach and an indirect approach [see Rovizzi & Thompson (1995) and Williamson (2001)]. The direct approach is applicable only when economic regulation exists simultaneously and consists of the inclusion of a representative parameter regarding the quality of service, inserting it into the equations of regulatory formulae based on performance incentives and rate-of-return regulation, penalizing and rewarding operators whose standards of quality deteriorate or improve during the regulatory process. In such circumstances, the quality index that is defined includes the various elements that represent the quality of service and a relationship is established between this index and the prices to be practised in the corresponding formula, agreeing beforehand upon the monitoring process. The advantages of this methodology include the fact that its application is an automatic process with lower transaction costs, while it gives the operator the freedom to manage the quality of service and its interaction with the price limits or rate-of-return that have been defined.

The indirect approach penalizes (or rewards) occasionally and periodically the operators that evidence shortcomings or excellence in terms of their performance (e.g. a failure to address complaints), according to the minimum standards of quality that had been defined. This process can include the application of overall sanctions (which, in extreme cases, can result in the forfeiture or revoking of the licence or permit) and individual indemnities to the customers who were affected. In the latter situation, despite the significant role of transaction costs (if the system is not automated) and the difficulties involved in establishing some of the parameters that define the quality of service (e.g. interference during telephone calls), this approach has the advantage of indicating precise information about shortcomings in quality and its monitoring. When the penalties are not automatic and do not directly involve the customers, defining the amounts to be established and their respective distribution becomes a complex and even arbitrary issue. Besides, it has significant transaction costs and, very often, the opportunity of applying the corresponding penalty is lost (e.g. for interruptions in the water supply).

Complementing this indirect method, the regulatory agency can and should publish and divulge information about the quality of service provided by the operators, and hope that the pressure of customers, the media and other stakeholders constitutes a suitable method of coercion (sunshine regulation). It must also be noted that 'carrot-and-stick' methodologies (e.g. in E&W) are an established feature, rewarding the operators who show out-performance and penalizing those who perform worse than the average. One can highlight the fact that this meritocracy can be instituted and reinforced by allowing positive or negative variations of the tariff system (i.e. factor X in terms of price cap regulation or the cost of capital in terms of rate-of-return regulation).

1.5.3 Sunshine regulation

The regulatory process known as sunshine regulation is based on measuring and comparing the results of the performance of WWS. This awareness of its performance is obtained by pressure through the customers, the media, the political classes (government and political parties) and NGOs. This strategy of 'naming and shaming' ensures that operators who evidence a mediocre performance get 'embarrassed' and consequently tend to correct the shortcomings that have been highlighted. This method does not establish tariffs and its coercive power is almost always limited. However, the exposure and the public discussion of the activities of the regulated operator have fairly positive effects, introducing competitiveness amongst operators and results in the progressive improvement of performance throughout the sector.

This regulatory tool can be implemented separately or as a complement to other regulation techniques. The power of sunshine regulation is less effective when applied separately than when it is applied as a complement to the prevailing regulatory process. In any case, as mentioned above, it is often the easiest and even the most suitable mechanism to implement, especially because it has added benefits in the case of WWS. In this sector, the standards for the quality of service are of the utmost importance and traditional explicit regulation (economic regulation) is, in some circumstances, more difficult to implement (given the fact that WWS fall under the purview of municipalities and that there are consequently a significant number of players).

Sunshine regulation always encompasses the use of benchmarking mechanisms, namely through performance indicators (or other methods), to a greater or lesser degree. While presenting the results of economic performance, sunshine regulation is not very effective (since improvements imply a reduction in profits in the same relationship). However, concerning the quality of service the consequences are almost always very positive.

For sunshine regulation to be effective it is necessary to respect the following requirements: a) publicizing the operator's performance must involve not just the operator in question but also other WWS or players from other similar industries (utilities); b) the performance results must be made available along with optimum or reference values, as well as with any eventual factors regarding the context that might be relevant; c) the results must be publicized in a manner that is accessible and easily available to customers; and d) a public debate about the results must be promoted with the presence of the different stakeholders. It must also be emphasized that if improvements in the quality of service translate into an increase in prices and tariffs, the effects of sunshine regulation will offer few or even no benefits. Thus, if possible, steps must be taken to ensure that the improvement in the quality of service is not achieved based on a costly service.

It must also be mentioned that the implementation of sunshine regulation raises some difficulties and needs some requirements, such as the institution of a body at a national level to implement this task, preferably one that is independent from the state, since the operators are very often state operators themselves. Insufficient information owing to a lack of awareness on the part of operators, the fear that providing information about performance may put their image at risk and uneasiness with regard to the possible consequences of revealing their performance are other difficulties brought about by this regulatory method.

1.5.4 The potential of benchmarking

The use of benchmarking in the context of regulation has an enormous potential, especially in the field of WWS, where it is more difficult to implement competition in the market. Although various operators exist, they generally behave according to a regime of natural regional (and local) monopolies. The following factors can be highlighted amongst the main advantages offered by the use of benchmarking (Marques, 2006a): a) strongly motivates operators to be efficient and innovative, mitigating their operating costs and capital expenses; b) results in continuous pressure on WWS to improve the quality of service; c) ensures a 'fairer' recovery of costs and capital investments and d) increases the sharing and transparency of information, minimizing the asymmetry of information that exists between the different stakeholders (particularly between the regulator and the operators).

The use of benchmarking comprises various approaches, namely, usage in the sphere of yardstick competition (see Section 1.6.3), employment in the referred sunshine regulation and a proactive role in the industry, which is the regulator's task to ensure it. Benchmarking can essentially be summarized as the application of comparative and quantitative methods used to assess and measure the performance of operators over the course of time (metric benchmarking), thus allowing the regulator to have a conclusive overview of the scores, which will be considered while making decisions during the regulatory process.

The advantages of using benchmarking in the WWS context are so significant that in certain countries the merger and acquisition of operators is forbidden (e.g. in the UK by the Competition Authority) to maintain a sufficient number of players so as to enable benchmarking exercises to be carried out, even though this implies losing out on the efficiency gains that could result from economies of scale and scope economies. In other countries, the restructuring and reform of the water sector resulted in operators being split into smaller entities so as to foster gains in efficiency that outweighed the effects of losing out on economies of scale and economies of scope, as a result of the incentives promoted by benchmarking. Among other examples, this situation occurred in Melbourne (Australia), in Jakarta (Indonesia), in Manila (Philippines) and in Mexico City (Mexico).

Despite the advantages of using this tool, governments (and regulators) must seek to mitigate the problems of benchmarking, namely those associated with the comparability of WWS. To this end, governments can divide regions or municipalities into more than one zone, each of which is served by a different WWS (loss of horizontality), without, however, putting at risk the minimum reasonable dimensions (which reflect economies of scale), and impose a separation of activities (loss of verticality) and functions in the WWS (unbundling), even though such services might remain in the hands of a single operator. Apart from this, they must define and implement information systems that make it possible to share elements between different WWS (e.g. Brazil) and to associate the comparison of the performance of WWS with incentives and penalties, including, *inter alia*, tax benefits, the renewal of contracts, rewards and fines. Governments should also constrain the accomplishment of mergers and acquisitions between the different players in the context of an evident public interest (significant reduction of tariff systems and/or maintenance of a sufficient number of players).

It must be noted that in any regulatory process, the implementation of benchmarking schemes always has added advantages. One of the main problems, as referred to, which occurs in regulatory processes in monopolistic and oligopoly markets is related to the asymmetry of information that exists between the regulator and operators. The application of benchmarking promotes the sharing of information and transparency in the processes. The use of standard costs or the calculation of the costs of capital, through the Capital Asset Pricing Model (CAPM), among diverse possible examples, also constitute circumstances in which the carrying out of benchmarking studies increases the credibility of results.

Finally, one can highlight the advantages of the use of benchmarking in and for the industry. It is widely known that the use of benchmarking by operators offers innumerable advantages, irrespective of the kind of benchmarking implemented (metric or process based). Consequently, it is the responsibility of governments and regulators to stimulate and promote benchmarking initiatives, both metric and process based, so as to increase efficiency and innovation on the part of operators. Although it has frequently been affirmed that the regulator's role in the benchmarking process must be limited, since it is developed at the heart of companies and at a level of detail and information that are not compatible with the functions of the regulator, regulators can play a proactive role in the sense of fostering best practices in the respective sector. Moreover, regulators can always participate as neutral observers in this exercise and contribute significantly, by means of mediation with regulators and companies of other sectors, to implement benchmarking in activities that are common to diverse industries. Regulators should also publish documents that show successful case studies in the sector, which can even serve to establish suitable norms for procedures.

1.6 ECONOMIC REGULATION

1.6.1 General considerations

For the reasons stated above, economic regulation plays a fundamental role throughout the functioning of WWS. Although, in fact, explicit economic regulation does not yet exist in the context of many WWS, this absence can only be viewed as a failure of the state itself, which must be rectified in the future. Irrespective of the WWS ownership (public or private) and the institutional framework in which the services are supplied, the existence of economic regulation is necessary and essential for a sustainable and balanced progress of the WWS. It must also be stressed that the term economic regulation is used here in a broad sense and does not have to be imposed by a regulatory agency created specifically for the purpose. It can, for example, be instituted via a contract or a normative document (regulation by contract).

Thus, economic regulation can be defined as the establishment and implementation of a set of specific rules for the institution of tariffs and prices or, in more technical terms, economic regulation seeks to maximize social well-being, inducing operators to produce what is desirable so as to achieve optimum results in terms of the prices adopted, the quantities produced and the standards of quality offered.

Although different methods of economic regulation are used throughout the world, there is no pre-established form of classification. However, these methods can be divided into two broad groups according to the incentives they offer operators to reduce their costs (Marques, 2005). The situation of supplying a reduced stimulus to mitigate their costs corresponds to rate-of-return regulation, whereas the opposite situation results in regulation based on performance incentives. The latter group includes various methods, namely, price cap regulation, revenue cap regulation or even yardstick competition in a limited sense, since this is often an integral part of the first two methods. These two broad groups can also be called cost-plus regulation, in the case of the first group, and price-based regulation, in the second case. The main underlying idea in the first case is that, as part of the cost structure that is presented, a predefined rate-of-return is guaranteed for providing the service, according to the costs involved. In the second case, the prices and tariffs are established per head, the operator assuming the risk for operating the service, which could result in higher profits or losses.

In reality, this separation is not so evident in practical terms since the two methods can be fairly close, not just because the methods of regulation based on performance incentives presuppose the attribution of a fair and reasonable remuneration (to ensure the operator's sustainability), but also because they allow some costs (revenues) to be passed on to customers. These two groups will be analysed in greater detail in the following points.

1.6.2 Rate-of-return regulation

Rate-of-return regulation allows the operator to establish a level of remuneration on the investments carried out, which is approved by the regulatory authority. In most situations, the establishment of prices and tariffs is instituted through this method, even when there is no explicit economic regulation of the WWS. As this method was developed in the USA as early as the nineteenth century and currently is very popular and applied, it is known as the American method, even though there are two distinct ways of implementing it. In the first scenario, a regulatory period is established, after which prices are re-evaluated on the basis of cost and demand projections. In the second approach, the prices are reviewed only when a process of revision is set in motion (rate case), which can occur by the initiative in situations of the company regulated or by any other stakeholder. In this scenario, no regulatory period is defined. Sometimes, in situations of rate-of-return regulation it is said to be 'pure' or 'impure' according to whether the rate-of-return is revised periodically or not, so that its value corresponds precisely to what has been defined beforehand. In such situations, additional profits or the recovery of unforeseen costs are shared.

The rate-of-return regulation method includes the following steps. First, the investments that are required (based on a business case) and the assets to be remunerated (rate base) are defined. Then the costs of the service to be regulated are approved (or revised), stipulating a rate-of-return that is deemed to be adequate (fair and reasonable return). Finally, the prices and their structure are determined, so as to generate revenues that cover costs and also allow the remuneration that has been defined. The formula that translates this method of regulation is as below:

$$\sum_{i=1}^{n} P_i \times q_i = \sum_{j=1}^{m} C_j + r \times V$$

In this formula:

P_i corresponds to the unit sale price of a given service i;
q_i represents the expected quantity of sales of the service units i;
C_j represents the share of the costs j to be borne by the company;
r corresponds to the remuneration of the investments and/or existing assets;
V represents the value of the investments and assets.

The key aspects of rate-of-return regulation are establishing the cost of capital and the amount of the investments and assets that are to be remunerated.

Rate-of-return regulation is characterized by a short duration, after which tariffs are revised (commonly, a regulation interval or period of one year) and by the fact that the costs of the subsequent period are based on the preceding period. Besides, the regulator determines the level of prices and their structure, and this regulatory method presents a relatively low risk for the operator.

The great advantage of rate-of-return regulation consists of the establishment of a direct relationship between costs, prices and the subsequent profits of the operator during the regulatory period, in a very ***simple manner***. In fact, the regulator effectively defines the prices and not a rate-of-return and thus ***facilitates the definition of tariff systems*** that reflect the elasticity of the market and, hence, enable and motivate the ***accomplishment of non-economic objectives*** and, especially, cross-subsidization between users or between the goods (services) supplied.

The operators that are under-regulated by this regulatory method ***do not have incentives to reduce costs***. The remuneration is, *a priori*, independent of its performance, although there can be stipulations for penalties (or rewards) in cases of lesser (or better) performance. The elimination of wastage and the consequent gains in efficiency, as well as innovation, are hindered. The sole incentive for any mitigation consists of the gains that could occur between the periods, in which tariffs are revised and the tariff system is fixed (residual claimant), which is not really relevant since the periods are generally fairly short (normally one year). It makes little difference to a WWS, for example, to present 15% or 50% of water losses given that the remuneration does not depend on its performance.

Another problem associated with this method of regulation occurs when the rate-of-return is higher than the cost of capital, whereby ***there can be over-investment***. In such a scenario, making investments by making use of external capital (debt) compensates and, consequently, the operator seeks to augment investments, obtaining an additional profit for each supplemental unit of capital spent. This phenomenon, which has been called the Averch–Johnson effect (or A–J effect), is particularly serious in an industry that is very close to the construction industry and where gold plate practices are well accepted by the government, due to their political importance and visibility. In the reverse sense, if the rate-of-return is inferior to the cost of capital, ***this could result in sub-investment***.

Likewise, if the regulated operator provides more than one service and not all the services are regulated in the same way, an ***imputation of costs*** (share of common costs) can occur from the ***competitive market to the regulated market*** or vice versa, i.e. revenues can be diverted (cost padding).

In the context of rate-of-return regulation and, above all, as the result of the asymmetric information that exists, the regulatory agency requires a great deal of information with a view to ascertaining the full distributed costs, existing assets, necessary future investments and a credible forecast of demand. This situation means that regulation is expensive, both for the regulator and for the operator being regulated.

With regard to the quality of service, although the method of rate-of-return regulation is not associated, at the outset, with performance incentives, it is in the interest of the operators being regulated to carry out investments to improve the quality of service and the environment, keeping in mind the A–J effect. Consequently, due care should be taken to safeguard the services offered against a disproportionate quality. In any case, if the rate-of-return is low or if the managerial quality is not the most suitable, there can also be a decline in the quality of service provided.

Finally, it must be mentioned that, according to the value of the rate-of-return that is adopted and the respective relationship with the cost of capital, there could be situations of excess or shortfalls in terms of the debt (leverage) indicator and hence ***inefficiencies from the financial point of view*** in terms of the capital structure of the operator being regulated.

1.6.3 Regulation based on performance incentives

Regulation based on performance incentives introduces the component of productivity along with the reduction of costs as the main element of the regulatory method. This regulatory model encompasses different approaches such as price cap regulation, revenue cap regulation and yardstick competition. Although they were first developed a long time ago, these methodologies only began to be applied in the empirical world during the 1980s, namely in the telecommunications sector regulation (British Telecom) in the UK (Littlechild, 1983). The first time it was used in the water sector was in 1989, also in the UK. Models of economic regulation can also include mechanisms to share profits (revenues) or costs between customers and the regulated operator, keeping in mind external factors, closer to rate-of-return regulation. In such cases, they are also called 'impure' methods. Regulation based on performance incentives, and especially price cap regulation, has become extremely popular in recent years and has been applied to the WWS in a number of countries. Initially and above all in the UK, very long periods for tariff revisions were a standard practice; however, improved productivity, the unpredictable functioning of the market, the inefficiency of liberalization to date (predominance of incumbents) and the growing role of social regulation have proved that this is neither possible nor desirable.

a) Price cap regulation

Price cap regulation consists of establishing an average maximum limit (cap) for the prices of the services or goods being supplied during the regulatory period. Thus, the regulated operators retain the profits corresponding to the over-reduction of costs that occurs during this regulatory period, in addition to the expected gains in terms of productivity. In this way, the regulated operators are motivated to improve their (static) efficiency and increase innovation, with a view to reducing costs and consequently augmenting their profits. At the end of each regulatory period, the benefits of the cost reductions are transferred to customers by means of the reduction of prices during the following period.

This regulatory method comprises two parts (CPI-X), in which one corresponds to the Consumer Price Index (CPI) and the other to the expected change of productivity, which the operator is expected to obtain during the regulatory period (X factor). This factor, while generally positive, can also be negative, depending on the level of future investment that is necessary. The formula for price adjustments is as follows:

$$P_{i,t} = P_{i,t-1} \times \left(1 + \frac{I_{i,t,t-1} - X_{i,t,t-1}}{100}\right)$$

In this equation:

$P_{i,t}$ corresponds to the maximum unit price of service i during the period t;

$P_{i,t-1}$ represents the maximum unit price of service i during the period t–1;

$I_{i,t,t-1}$ refers to the CPI associated with supplying service i, expressed in percentages, between the periods t and t–1;

$X_{i,t,t-1}$ represents a factor, in a percentage, of the expected productivity gains between the periods t and t–1 while providing service i.

This formula can be complemented by a positive or negative adjustment coefficient, which enables the incorporation of extraordinary or unpredictable situations or circumstances that can have a substantial impact on the performance of the operator being regulated such as, for example, changes in the macroeconomic scenario, energy crises, unusual climatic factors or, alternatively, the accomplishment of specific objectives, such as improvements in the drinking water quality or the service reliability (e.g. as happens in the UK).

Price cap regulation is characterized by a long interval between the revision of tariffs (between three and ten years), by forecasts for projected future costs and estimated gains in efficiency and by a relatively high risk. In this type of regulation, despite the fact that the maximum prices are defined at the outset, the operator is completely free to manage these prices. The unpredictable oscillations of the exogenous variables which affect costs are not encompassed in the formula referred to. This results in increased risks, which are further aggravated by any increase in the regulatory period.

Figure 1.7 displays this regulatory model for an interval of four years. In each regulatory period the operator's profits grow over time. At the end of each period, the profits are reduced to a defined value and the benefits of the gains in productivity are passed on to customers, but only for the following period. As the operator becomes increasingly efficient, the maximum value of the profits will be lower, i.e. the gains in efficiency that can be expected will be lower, as will transfers to customers, since the operator will draw increasingly closer to the frontier of efficiency (best practices).

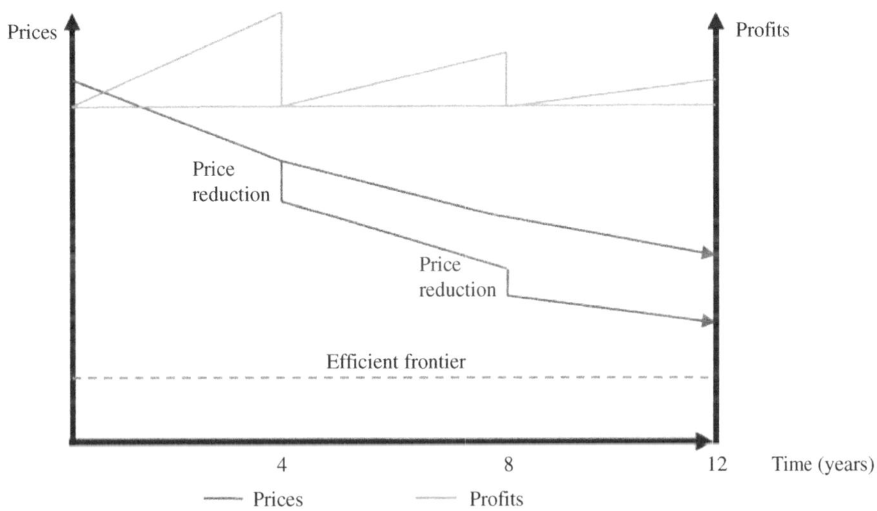

Figure 1.7 Incentives and transfers to customers in price cap regulation

When compared to rate-of-return regulation, price cap regulation presents a ***far higher risk***, which is borne by the operator. This situation leads to some instability for the operators, increasing their cost of capital. Since the costs were defined for a lengthy regulatory interval, the presuppositions that were assumed and any eventual deviations have more serious consequences. For customers, the scenario is precisely the opposite, as there is a greater stability in terms of prices, at least during the regulatory period.

One of the greatest difficulties, and one that is simultaneously a controversial issue in the context of this regulatory method application, consists of defining the ***X factor***, especially its initial value. Theoretically, the X factor comprises two different elements: one specifically for the sector to be regulated (efficiency gains obtained by means of technological changes in the sector) and another element related to the operator in question (static and allocative efficiency of inputs and outputs, i.e. the correction of existing inefficiencies, drawing closer to more efficient practices). Consequently, the tool of benchmarking is increasingly used to determine the X factor, especially to determine the element of efficiency concerning the individual operator regulated. Apart from the difficulties involved in estimating the amount that each inefficient operator can catch-up during the regulatory period, inefficient operators disagree with the results of the assessments, arguing on the basis of their particular characteristics (operational environment), in which they operate and do not control. It must be emphasized that factors that cannot be controlled do, in fact, exist and they must be taken into account, starting with change in the scale (of consumption) of the operators that are being regulated, which can lead to (dis)economies. Whether a single X factor or various X factors should be defined, with the associated pros and cons, is another relevant question.

The ***CPI element*** that is adopted can (and should) also be discussed in depth. The adoption of a national index, which reflects all the prices, may prove to be inadequate for the WWS, whose cost structure does not coincide with the structure used to calculate the CPI. In this particular case, there are elements whose price variations have a significant impact, such as variations in the costs of 'wholesale' services or of energy.

Another noteworthy aspect is the ***revision period of the regulatory process***. Generally, the regulatory interval is quite long, ranging from three to ten years, although there are no legal impediments to the term being interrupted, which can occur for diverse reasons (unforeseen costs or higher profits). While this possibility offers a positive aspect, since it reduces the risk involved, it serves as a disincentive for efforts to increase efficiency and innovation, reducing the gap between price cap regulation and rate-of-return regulation. It must also be highlighted that the conduct of the regulated operators is influenced by the phase of the regulatory period (regulatory gambling), showing lower profits in the respective final phase than the profits of the initial phase. This conduct, which is associated with the so-called *ratchet effect*, reflects the fact that the regulated operators are influenced by the expectation that the prices in the next regulatory period will be based on the last known elements and can therefore result in lower margins for gains in terms of efficiency and innovation in the next period.

Indeed, the great advantage of price cap regulation is the fact that it stimulates operators to reduce costs, which can be achieved by ***increasing productivity*** (static efficiency and innovation). However, in this method the prices could be out of alignment with the costs, which means into ***allocative inefficiency***. These distortions of prices can be detrimental to some activities and customers.

The ***quality of service*** is another aspect that deserves a great deal of attention in the context of this regulatory method. As the maximum prices are defined in the outset, the operator's profits increase in direct proportion to the reduction of costs achieved. If the standards of the quality of service are not safeguarded, the operator can try to mitigate costs by reducing the quality of service. As mentioned in the last section, there are various ways of integrating the issue of quality of service, although all of them entail certain difficulties.

In price cap regulation, ***public service obligations should be well defined***, since operators are only interested in serving more profitable clients, which could jeopardize the universality and equality of WWS or even subsidize, through more monopolistic segments, the more competitive activities, by means of, for example, cross-subsidization (e.g. phenomena of cream-skimming or by-passes). In some countries, WWS for industrial users have lower tariffs than domestic customers. In this sense, the definition of minimum prices (floors), and not just maximum prices (ceilings), can be a suitable option. In such scenarios, the regulator plays a decisive role in controlling abusive situations.

To ensure the operator's economic sustainability, price cap regulation must ***guarantee an adequate remuneration on the capital invested***. Thus, while defining the costs for the new regulatory period, all expenditure must be taken into account, in terms of capital, operating expenses and the respective interaction, including long-term interaction, along with an adequate remuneration for the operators. Note that price cap regulation entails a high level of risk, which could deter potential investors.

The complexity and the requirements of the entire regulatory process, which guarantee the independence and reliability of the regulatory act, ***give rise to significant costs***, both for the regulator and for the operators, which, inevitably, sooner or later, will be paid for by customers.

The ***investment*** to be carried out during the regulatory period is generally ***insufficient***. Its level is defined and proposed by the operator and approved by the regulator. Even so, during the phase of negotiations and the revisions of the regulatory process, the regulated operators can seek to display and propose very high levels of investment. This intention is not entirely genuine and rarely produces real effects. In a final analysis, the investments to be made in the infrastructure services are normally below the necessary levels. Investments must be paid for and, owing to their characteristics, the risk increases, which might have a substantial effect on tariffs. On the other hand, regulators are often consciously complacent, since the added costs derived from substantial investments would be borne by users, which, in turn, would result in distress and regulatory instability and would postpone the problem. This scenario, which is systematically repeated, is one of the greatest problems and challenges experienced by price cap regulation, especially in the medium and long term.

b) Revenue cap regulation

In revenue cap regulation, operators are limited to a maximum average value for their revenues. The revenue ceiling is defined according to a CPI and a factor that translates variations in terms of productivity (X). Thus, like price cap regulation, the regulated operator is motivated to increase its efficiency and innovation more than expected, reducing costs in order to fulfil the objectives that have been established and to increase its profit margins. At the end of the regulatory period, the gains achieved are transferred to customers.

Despite being similar to price cap regulation, in terms of the main objective of fostering productivity, revenue cap regulation does evidence some differences. In this method, the regulated operator generally has a greater freedom to define new tariffs, since there is no control over neither unit prices nor their structure. Instead, the limitations are imposed over total revenues and some services can subsidize others (allowing for the cross-subsidization). Apart from this, as the revenues depend directly on demand in the market, being subject to the prices of the services and the volume supplied, they do not stimulate the consumption of the services or goods supplied and are especially suitable for sectors where management policies and the rationalization of demand (demand side management) are relevant, as in the case of the water and energy sectors. Finally, it must be noted that revenue cap regulation is also beneficial when there is a high component of fixed costs, when revenues do not depend greatly on the fluctuation of consumption.

c) Yardstick competition

Yardstick competition, or regulation by comparison, is based on an assessment of the performance results of the regulated operator in comparison with those of other operators from the same sector (i.e. the use of benchmarking) and the respective financial consequences. The key element of this regulatory method consists of redirecting incentives to improve efficiency for a given operator by means of information extracted from other operators. This simultaneously constitutes an artificial form of competition between the regulated operators.

Although the use of yardstick competition can be implemented in a pure form, i.e. the regulation of a certain tariff is calculated by the average (or according to the best practices) of all the tariffs (costs), for example, it becomes extremely relevant in the context of determining the X factor of price cap regulation or revenue cap regulation.

Yardstick competition is aimed, above all, at sectors where competition mechanisms are limited and asymmetric information prevails, as in the case of the WWS. This regulatory method has a great deal of potential and provides strong incentives to the operators to innovate and be efficient and mitigate their operating costs and costs of capital. Apart from this, it improves the sharing and transparency of information, minimizing asymmetric information between the different stakeholders.

However, this regulatory method has been criticized owing to the inherent difficulties in fulfilling the assumptions for its use and to the techniques adopted (e.g. the quality and quantity of information, discretion and heterogeneity), the possibility of collusion and behavioural manipulation between operators, and the capacity for commitment that the regulator is capable of assuming in this regard.

In order to overcome some of the complexity of the different available methodologies and to strengthen decision-making, regulators sometimes use more than one benchmarking technique to consolidate the results. Figure 1.8 presents the main benchmarking techniques used in economic regulation. These have been classified into ***parametric* and *non-parametric***, according to whether or not they allow a functional form, which is defined beforehand, for the production (or cost) technology, which represents the productive process. Parametric techniques enable errors to be measured but introduce the additional difficulty of having to know the specification associated with the allowed behaviour. Of the most used methods, the data envelopment analysis (DEA) and Index Numbers are considered to be non-parametric methods, while stochastic frontiers (SFA) and regression models (e.g. OLS) are parametric. Index numbers serve to determine the total factor productivity (TFP), which measures the variation of productivity over the course of time. The TFP is very important in determining the X factor when performance incentive regulation methods are adopted. Parametric or non-parametric methods can also be classified as ***non-frontier*** or ***frontier***, according to whether or not they presuppose that the organizations are technically efficient or, in other words, that the benchmarks are based on average approximations or on the best practices of the samples. The DEA and the SFA are frontier methods while index numbers (with the exception of the Malmquist productivity index) and regression models (e.g. OLS) are non-frontier.

Even using the various techniques, the regulated operators criticize the results and the decisions made, arguing that the aforesaid available techniques are subjective and discretionary and could result in inconsistencies and opportunism by the regulator, which weakens incentives and increases the cost of capital (CAPEX); in heterogeneity ('idiosyncratic shocks'), which is impossible to completely eliminate (irrespective of the model adopted), thus contributing for discretionarity in the regulatory process (comparing 'apples with oranges'); that not all types of costs are known or suitably contemplated and integrated into the analysis (e.g. the quality parameter is not easy to insert); that the differences between the costs of the different pairs can be imputed to inefficiency as well as to intrinsic errors in applying the techniques themselves and that the determination of parameter X, based only on the differences in efficiency found between the regulated operators, can be inadequate.

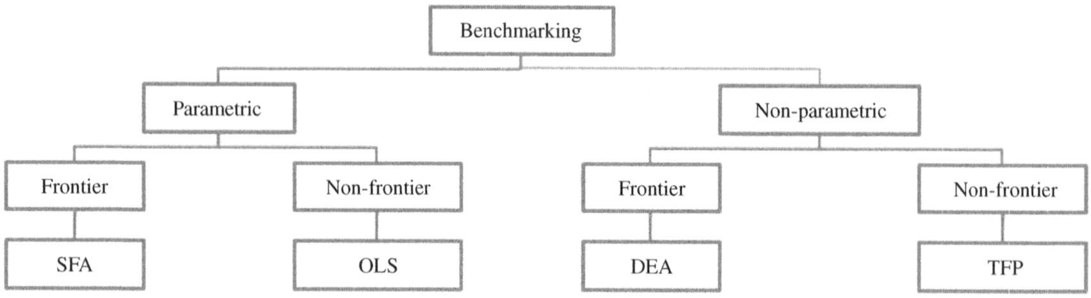

Figure 1.8 Benchmarking methodologies

d) Other regulatory methods

Economic regulation is frequently based on more than one of the aforesaid methods. These so-called hybrid methods refer to situations in which the method used has the characteristics of various methods. One of the most frequently used hybrid methods consists of the sliding scale, which has a profit (cost) sharing scheme, allowing the regulated operator to increase the rate-of-return above the pre-established value, in case prices are simultaneously reduced. Figure 1.9 presents the context of the application of these methods of economic regulation.

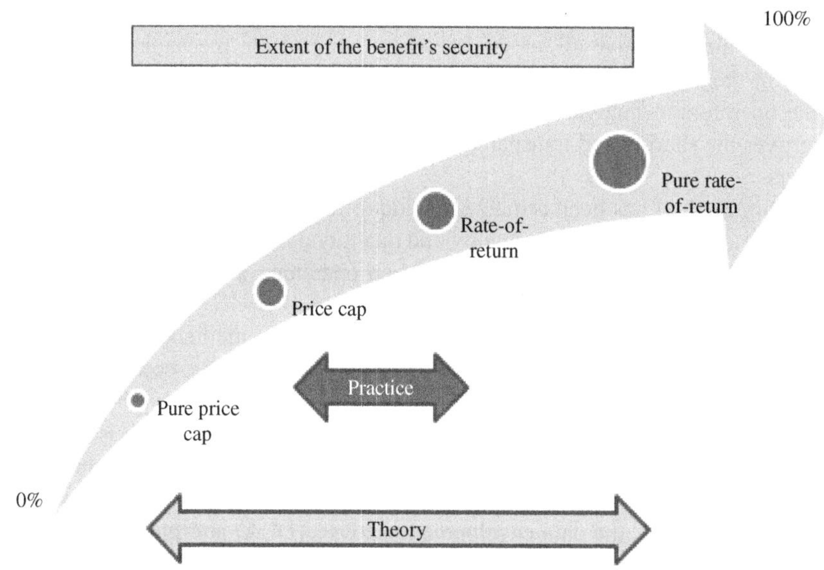

Figure 1.9 Comparison of rate-of-return and price cap regulation

Part 2
Regulation case studies

2.1 AUSTRALIA

2.1.1 Historical background

Australia, officially known as the Community of Australia, is a constitutional democracy founded on the federal division of powers. The form of government that characterizes the country is a constitutional monarchy with a parliamentary system, whose monarch is represented at a federal level by the governor-general and by governors at a state level. Legislative power is held by the Parliament of the Community, constituted by the monarch (governor-general), the Senate and the House of Representatives. In its turn, executive power is exercised by the Federal Executive Council (the governor-general advised by the executive counsellors who, in practice, are the prime minister and the ministers).

The country has a population of about 21 million inhabitants over 7.7 million km^2 of land and is organized administratively into six states, two main territories and other secondary territories. Considering their independence and autonomy, each state and territory has its own legislation.

Australia is the world's driest inhabited continent. There is a great deal of variation at a regional level in terms of rainfall, drought being a common pattern in the country. Environmental pressures impact on the economy led, in 2004, the Australian government and all the state and territorial governments to sign the Intergovernmental Agreement on a National Water Initiative (NWI) aiming at guaranteeing better management of Australia's scarce and valuable water resources.

Since the early 1990s, WWS in Australia have undergone significant changes. Unbundling, independent regulation, the cost recovery principle and increased transparency were some of the transformations and principles introduced in the course of this restructuring. The reforms that have been implemented have provided a great impetus to increase efficiency and effectiveness and the quality of the services provided, while reinforcing economic and social cohesion. The following steps aimed, above all, to improve economic efficiency, to promote an even greater environmental sustainability and to extend the principles applied in the metropolitan and the most populated areas to the rural ones.

The 1994 reforms, implemented by the Council of Australian Governments (CoAG), established a strategic guideline for the water sector, in which the key elements were based on an institutional reform, the management of water resources at the level of river basins, an optimized allocation of water resources, the ecological sustainability of new enterprises, the protection of underground water resources and the full cost recovery.

One of the most remarkable aspects of the reforms that took place in the infrastructure sectors in Australia, which usually exhibit monopolistic features and where the WWS are included, consisted of separating the State operator from the State regulator, stipulating that all the states and territories should be endowed with independent regulatory entities, irrespective of assets ownership and service management (private or public sector).

Notwithstanding the legal autonomy of Australian states and territories, diverse federal regulations exist in Australia, such as the Australian Drinking Water Guidelines (ADWG), established in 2004 by the Health and Medical Research Council, in collaboration with the Ministerial Council for Managing Water Resources. The ADWG include minimum principles and recommendations for managing water services and provide the Australian community and, especially, the operators, with guidelines for a high-quality water service.

Recently, the federal government approved the Water Act 2007, which focuses on water resources management of the Murray-Darling river basin and refers to diverse matters of national interest concerning the water sector. In this scope, the act created an independent body to manage the Murray-Darling river basin, with the necessary roles and powers to promote an integrated and sustainable management of water resources.

ISBN: 9781843393412. Published by IWA Publishing, London, UK.

The Australian Constitution stipulates the management of WWS within the competence of the state and territorial authorities. In this context, the sector has proved to be quite complex and many laws and agencies (approximately 800) co-exist, with shared responsibilities regarding the management of water resources, from a federal to a municipal level.

The scenario in Victoria is not very different from other states and territories in Australia concerning the implementation of policies and strategies defined by the federal government during the last decade. It must be mentioned, however, that the changes introduced in Victoria were more significant and, to a certain extent, even pioneering, while achieving the same or an even a greater degree of success when compared to other Australian regions.

The water industry in the state of Victoria is highly regulated and diverse entities intervene in the process, with different powers and functions. There is a sector-specific – although multisector – regulatory agency and various government bodies responsible for implementing policies and preparing documents which stipulate the obligations and conditions the operators are required to comply with. Each body performs important functions, influencing or regulating different aspects of the water sector, thus playing an essential role in assisting the regulatory agency to carry out its task of supervising the quality of service, guaranteeing the public service obligations and regulating the sector in economic terms.

The water market in the state of Victoria includes the WWS operators and is structured into metropolitan, rural and regional entities. Its operators vary considerably in terms of size, the services they provide and the environment in which they develop their activities. Water resources are scarce and the drinking water supply is mainly derived from surface water sources.

2.1.2 Legal and institutional framework

2.1.2.1 Responsibilities

Apart from the various entities at a state level, which include the regulatory entities, in Australia, responsibility for the water sector is associated, at a federal level, with the Department of the Environment, Water, Heritage and the Arts (DEWHA), the National Competition Council (NCC), the Australian Competition and Consumer Commission (ACCC), the CoAG, the Murray-Darling Basin Authority, the Water Services Association of Australia (WSAA), the Australian Water Association (AWA) and the National Water Commission (NWC).

The role of the DEWHA is primarily related to environmental protection at a national level, which includes a suitable and sustainable management of water resources, advising the Australian government in this domain.

The NCC was established in 1995 and brings together all the Australian governments to play an advisory role, aiming to supervise the implementation of the National Competition Policy. However, the implementation of this policy is the task of the various governments. Although this body is financed by the Commonwealth government, it is accountable to all the states and territories of Australia, through the CoAG. As a statutory body, the NCC is independent from the executive power of the governments.

The ACCC was created in 1995, in the form of an independent regulatory agency. It promotes competition and equity in the market economy, so as to provide benefits for customers, the market and the community. The ACCC also regulates the pan-state service infrastructure and its primary responsibility is to ensure that individuals and the markets uphold the norms and legislation concerning competition and consumer defence and that there is no discrimination between states and territories. The ACCC also has important functions in terms of arbitration and conflict resolution and raising awareness about competition and consumption issues.

The CoAG is an intergovernmental forum where all matters of common interest are discussed, including water management. The objective of CoAG is to safeguard security with regard to right of access, ensuring the best use of water by allocating it between states and developing interstate water markets.

The Murray-Darling Basin Authority has the necessary functions and powers to ensure the management of water resources in an integrated and sustainable way.

The WSAA is the main association of operators in the urban water sector in Australia. Its 30 members are responsible for supplying WWS to about 15 million Australians and a vast number of industries. The WSAA was created in 1995 and aims to foster development and cooperation in the water sector, so as to increase performance and productivity and ensure that the regulatory framework is suitable for the interests of the community.

The AWA, created in 1961, is an independent, non-profit association for professionals of the water sector. The AWA is, in fact, the largest and most broadly based association, representing organizations and individuals throughout the country. The AWA contributes towards the development of the sector by bringing together knowledge and ideas from people involved in research, regulation, consultancy and the management and operation of water and wastewater infrastructure.

The main role of the NWC is to assist the governmental authorities in carrying out national reforms in the water sector and implementing the NWI in Australian territory. The NWC is also responsible for advising the federal government and the state and territorial governments on all matters concerning the water sector.

With a great deal of influence in the Australian water sector, each state and territory has a dedicated (independent and multisector) regulatory agency, with functions of economic regulation and ensuring the quality of service and public service obligations. Table 2.1 provides a summary of the main bodies responsible for this area in the various states and territories in Australia (Crase *et al.*, 2008).

Table 2.1 Operators and the main bodies with regulatory functions in Australia

State	Management of water resources	Water and wastewater services	Environmental regulation	Economic regulation
Queensland	Department of Natural Resources and Mines	Local government Two wholesale companies	Environmental Protection Agency	Queensland Competition Authority
New South Wales	Department for Planning, Infrastructure and Natural Resources	Sydney water company Hunter water company Local government	Department for the Environment and Conservation	Independent Pricing and Regulatory Tribunal
Victoria	Department of Sustainability and Environment	Melbourne water company Three retail operators Regional water operators	Department of Sustainability and Environment	Essential Services Commission
South Australia	Department of Water, Land and Biodiversity Conservation	South Australia Water Company	Department for Environment and Heritage	Council of Ministers
Western Australia	Department of the Environment and Water Resources	Water company and various small operators	Department of the Environment and Water Resources	Economic Regulation Authority
Tasmania	Department of Primary Industries and Water	Local government Three 'wholesale' companies	Department of Environment, Parks, Heritage and the Arts	Government Prices Oversight Commission
Australian Capital Territory	ACT Environment	ACT energy and water company	ACT Environment	Independent Pricing and Regulatory Commission
Northern Territory	Department of Natural Resources, Environment and the Arts	Water and Energy Company	Department of Natural Resources, Environment and the Arts	Utilities Commission of the Northern Territory

In the state of Victoria, apart from the Essential Services Commission (ESC), which is the multisector regulator, there are other entities with relevant regulatory functions. With regard to consumer defence, Australia has an ombudsman for each state and territory. This credible and independent entity is responsible for investigating and resolving disputes between customers and the operators of water, gas and electricity services. In the case of the state of Victoria, this entity is called the Energy and Water Ombudsman (EWOV). In general, the WWS providers have agreed to join the process of mediation, conciliation and conflict resolution promoted by the EWOV.

Through the Department of Sustainability and Environment (DSE), the Victoria Minister for Water is responsible for developing policies for the water sector. The Environment Protection Authority (EPA) is an autonomous public institute which was established by means of the Environment Protection Act and is responsible for preventing pollution and promoting environmental protection in the state of Victoria.

Another important regulatory agency in the water sector in the state of Victoria is the Department of Human Services (DHS), which, amongst other functions, oversees the safety and quality of drinking water supply, ensuring that prevailing legislation is respected, namely the 1958 Health Act, the 2002 Health Quality of Drinking Water

Regulations, the 1973 Fluoridation Health Act, the 1984 Food Act and, more recently, the Safe Drinking Water Act, dating from 2003. Concerning the last regulation, its purpose includes reinforcing the monitoring, publicizing the performance of WWS operators and promoting awareness with regard to the quality of drinking water. The DHS also has essential functions concerning the discharge of wastewater and its re-use and guarantees financing for low-income families by managing a fund that has been specifically constituted for this purpose.

2.1.2.2 Regulation

Over the last two decades, the water sector in Australia has undergone numerous transformations. Since the early 1990s, state and territorial governments have tried to develop the sector, modernizing it and making it more balanced and sustainable, namely by means of applying the principles of consumer-pays and polluter-pays, while nevertheless keeping in mind the essential nature and social importance of this resource. Generally, each state and territory has a multisector regulatory agency, which includes the regulation of WWS operators. Their area and scope of intervention differ between states or territories and can encompass just the metropolitan operators or all other entities. They can also regulate prices and/or the quality of service.

In the context of economic regulation, the different regulators use similar, although not uniform, methods, normally based on price cap regulation. For example, the regulator for New South Wales, the Independent Price Authority Regulatory Tribunal (IPART), uses the formula CPI-X, applying benchmarking methodologies to estimate the X factor to calculate the tariffs for Sydney Water (and other operators). The same principles, related to incentives and performance, have been instituted in other states/territories, such as Western Australia or the Northern Territory.

In Victoria, the Office of the Regulator General (ORG) was responsible for regulating the water sector. In 2002, the ORG was substituted by the ESC. The ESC is a multisector regulatory agency which is responsible for regulating public services in the state of Victoria: electricity, gas, ports, road and rail transport, insurance, renewable energy and WWS. However, the regulation frameworks are quite distinct, depending on the sector being regulated. The regulatory agency was created primarily to protect the interest of customers in the state of Victoria with regard to prices, quality and reliability of public services. In order to achieve its purposes, the regulatory agency has to provide an efficient and economically or financially viable service, competitive and strong enough, while promoting economic and social cohesion within the territory.

Up to 1 January 2004, the ESC (ORG), in the scope of the water sector, was also responsible for regulating the quality of service and guaranteeing the public service obligations, but only for the metropolitan entities. The activity conducted during almost ten years was based on sunshine regulation. From that date onwards, its functions were broadened to include economic regulation and all the water operators of the state of Victoria.

According to the tariff regulations (Water Industry Regulatory Order – WIRO), the ESC establishes the prices of all services, namely water distribution, the collection and drainage of wastewater, reclaimed water, storage, raw water, reclaimed raw water, the drainage of stormwater in metropolitan services, water for irrigation and service connections, among others.

The WIRO also regulates the functions performed by the ESC concerning these services, *inter alia*, the approval or specifications of the prices of each operator for a period of three years (beginning in July 2005) and establishing levels for the quality of service provided by the operators (e.g. contracting the levels of service). Apart from this, the ESC monitors and publicly presents the performance of operators – examining whether they are in conformance with what is expected and has been defined – as well as having regulatory information and public service obligations. Finally, the ESC also has arbitration functions to resolve conflicts between the various operators (e.g. 'wholesale' operator versus 'end-user' operators).

The ESC performs its functions independently, in both functional and organic terms, and is also financially independent. Its funding is determined according to its provisional budget and paid for by the regulated operators in proportion to their revenues. The ESC also has vast regulatory powers and the authority to apply penalties. All these procedures, including the possibility of appeals, are envisaged in its statutes (the 2001 ESC Act). In 2005/2006, the ESC had a budget of approximately 9 million euros, from which 1.8 million euros were allocated to regulating the water sector, which comprised 60 employees.

In accordance with its statutory obligations, the ESC has developed a Memorandum of Understanding with the various stakeholders, including other regulators, with the purpose of coordinating regulatory processes and decisions.

Another noteworthy aspect is the fact that it is compulsory to prepare a Charter of Consultation and Regulatory Practices, which is reviewed every three years. It defines the work programmes for each regulated industry and the procedures for notifications, inquiries, decisions and resolutions and public consultation.

The method of economic regulation adopted by the ESC for the water sector was the price cap regulation. During the regulatory process to establish prices, the creation of a Water Plan by the operators was of fundamental importance. The ESC evaluates the necessary operating and capital expenses, keeping in mind past figures, a comparison between operators (benchmarking), expected consumption, new obligations and legislation to be upheld and the arguments of the different stakeholders. Costs are fully recovered and there is a tendency to define tariffs based on the long-term marginal cost. The value of the X factor for the first regulatory period was equivalent for all the operators and set at 1%. Since it was the first regulatory period and owing to the changes expected in the sector, the X factor was estimated on the basis of the variation in productivity for the less-productive sectors of the economy, considering a moderate scenario. Even though other kinds of incentives were discussed, they were eventually limited to the gains in efficiency that took place during the regulatory period. The quality of service variation was not considered while establishing prices and a commitment was made to integrate unforeseeable events, in case they occurred, at the end of the regulatory period.

All the documentation and its quality, including the water plans, is audited by credible and accredited external entities (auditors). The ESC has developed a set of guidelines (*Approving, Conducting and Reporting Audits*) which intends to ensure the audits' independence and rigour, promoting incentives for the operators to reduce asymmetric and inaccurate information and maximize the relationship between the cost and the effectiveness of the audits.

2.1.3 Market structure

2.1.3.1 Ownership and management of services

In Australia, WWS are mainly promoted by public operators. The ownership of the assets is always public, even though they may belong to the state or territorial governments or to the local administration. With regard to the management of the WWS, there are occasional cases where the private sector provides these services, as it happens, for example, with ACTEW in the Australian Capital Territory. Until recently, in Southern Australia, SA Water (the sole operator) had a (short-term) contract to provide services for all the activities carried out with a private entity (United Water). There are also various PPP concerning the WWTP and the WTP in the different states and territories of Australia. In fact, it must be noted that in Australia there are no legal constraints for privatization and that the private sector is normally very active in WWS provision (outsourcing).

2.1.3.2 Players in the sector

The large companies in the Australian water sector are always publicly owned by the state or territorial government. The small operators are, however, usually owned by the municipalities.

In Victoria, in 1994, Melbourne Water, the main operator in this state, was restructured and was divided into four companies (publicly owned and managed). One company was responsible for the wholesale water service and the main WWTP and the other three WWS provide end-user services, namely City West Water, South East Water and Yarra Valley Water. Currently, apart from these three metropolitan operators, Melbourne Water also provides wholesale service to the operators Western Water, Gippsland Water and Southern Rural Water.

Melbourne Water is a commercial company which operates according to the Melbourne Water Corporation Act and the Melbourne and Metropolitan Board of Works Act. Thus, the Minister for the Environment and Water is responsible for the services being provided. The three metropolitan entities are owned by the state of Victoria and are subject to the Corporation Act. Their powers were established by the Water Industry Act (1994).

Apart from the metropolitan operators, there are also regional and rural operators. As far as regional operators are concerned, there are 13 operators that provide WWS in the state of Victoria, encompassing about 400 cities. Their responsibilities have been defined by the 1989 Water Act. Unlike the metropolitan operators, which were subjected to unbundling, regional operators are generally vertically integrated.

The regional operators act within their service area which is clearly defined according to the region's geography. They are characterized by a lower density of customers and some heterogeneity in terms of size. For example, in 2006, Berwon Water provided water services to 125,044 customers, while Westenport Water supplied water to only 13,098 customers.

The state of Victoria is also divided into regions which are served by five rural operators (Goulburn Murray Water, Southern Rural Water, Sunraysia Rural Water Authority, Wimmera Mallee Water and First Mildura Irrigation Trust), operating according to the statutes established under the 1989 Water Act. These operators are responsible for managing and operating irrigation perimeters, including dams, lagoons and canals. They periodically provide the wholesale service to regional operators and the rural properties in their area of operations.

In the context of private sector operators, the ACT Energy and Water Company (ACTEW) in the Australian Capital Territory is especially worthy of note. It was created in 1988 after a merger between the ACT Electricity Authority and ACT Water and transformed into a corporate entity in 1995. In 2000, the ActewAGL company was created as a joint venture between ACTEW and AGL (Australian Gas Light Company). ActewAGL is the owner of and operates ACTEW, the gas and electricity assets of AGL and the water supply and wastewater service assets of ACTEW.

The largest Australian operator, Sydney Water, is located in New South Wales and serves more than four million inhabitants. It provides the WWS and is also responsible for draining rainwater, encompassing the areas of Sydney, Illawarra and the Blue Mountains. Sydney Water buys untreated water from the Sydney Abstraction Authority. Hunter Water is another significant operator in New South Wales.

In the Northern Territory, the Power and Water Corporation is the most important company. This multi-utility operator provides electricity, water and wastewater services to more than 100,000 inhabitants in urban and rural areas.

In the state of South Australia, SA Water, which serves more than 1.4 million inhabitants, is the main WWS operator.

There are 37 operators in Western Australia but almost all of them are quite small. The operator for Perth (Water Corporation), which is the most representative in this state, provides WWS to 1.4 million inhabitants.

In Queensland, the majority of the 170 operators (local authorities) vertically integrate the WWS. There are two public operators that provide the wholesale water services at a regional scale.

Finally, there are three wholesale water companies in Tasmania that make water available to local operators, which subsequently distribute this water to the end-user.

2.1.3.3 Integration in the sector and services offered

In Australia, there are sharp differences between the different states and territories in terms of WWS integration. There are operators that manage both services integrated vertically, from water abstraction to distribution and from collection to the wastewater treatment. However, there are also operators that exclusively provide the wholesale service (e.g. the Sydney Abstraction Authority, in New South Wales, or Hobart Water, in Tasmania) or that are responsible only for the treatment of wastewater (e.g. Melbourne Water). In some states and territories there is only a single operator (South Australia and the Australian Capital Territory), while there are various operators in others (e.g. Victoria and New South Wales). For example, in New South Wales, apart from the metropolitan companies (which include Sydney Water and Hunter Water), there are 107 local operators which serve 1.7 million inhabitants.

In any case, there is a high level of integration, i.e. the 30 largest companies serve over 70% of the population (~15 million). Another relevant feature is the existence of various multi-utilities operators, i.e. operators which provide other services apart from WWS operations, such as energy and gas. The stormwater service is almost always provided together with the water supply and wastewater service.

2.1.3.4 Numbers for the sector

About 19,000 million m^3 of water are produced every year in Australia. Agriculture is the sector with the highest consumption, accounting for 65% of the total water usage. The use of drinking water supply for human consumption represents approximately 11.2% of the total water production. In terms of the total consumption of water, the main source is surface water, which accounts for about 79% of production. Figure 2.1 provides a summary of the consumption by different water customers (NWC, 2005).

In 2005, the rate of water losses in Australia concerning the water service was approximately 12% of the total value produced. Figure 2.2 provides an overview of the evolution of water losses in the sector between 1997 and 2005.

Australian operators sell about 2.1 billion m^3 of water every year. Considering the Australian population, the average consumption is equivalent to 104 m^3 of water per inhabitant every year, i.e. 280 litres per inhabitant, per day.

The level of WWS coverage (including treatment) in urban areas is close to 100%. However, this is not the case in rural areas, where there is a very significant percentage of scattered population which resorts to individual solutions.

In 2005, the WWS sector in Australia directly employed about 12,000 people.

In Australia, the water service has a network of about 143,000 km, while the wastewater service comprises approximately 91,000 km of sewers. The urban drainage is always provided by separate systems.

The volume of wastewater that is treated corresponds to 83% of the volume of water consumed. From the total amount of the water consumed in Australia, about 16% is re-used. The greatest percentage of re-use occurs in the Australian Capital Territory, with 28% of the total volume.

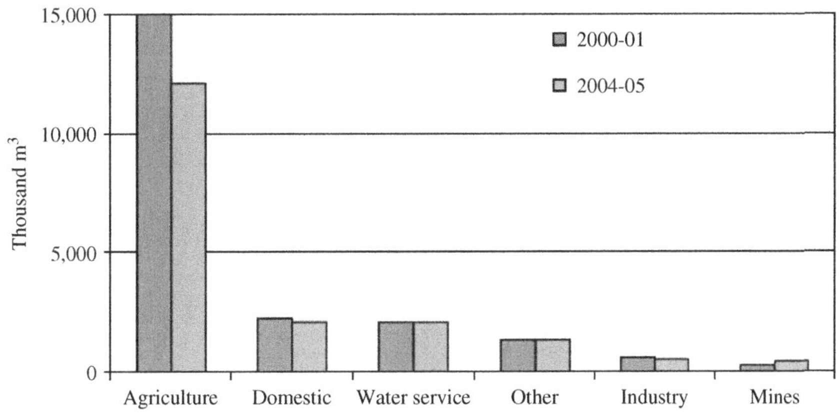

Figure 2.1 Use of water in Australia

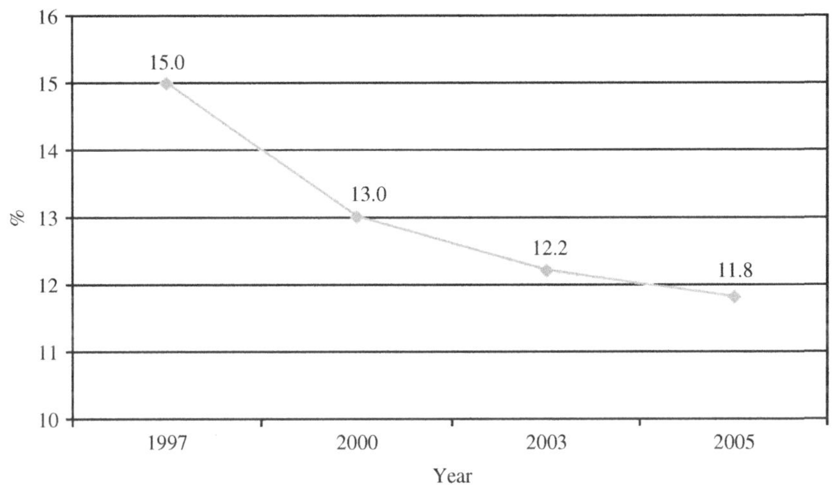

Figure 2.2 Annual evolution of the water losses indicator in Australia

2.1.4 Rules of the game

2.1.4.1 Establishment of tariffs

In Australia, the prices of the main WWS operators are regulated by independent regulatory agencies. However, in some states (e.g. New South Wales), small operators establish their own prices, in the absence of regulation. There are also exceptions in South Australia and in Queensland where, even though tariff regulation exists, this is developed *ex-post* and not *ex-ante*.

The tariff system that is usually implemented by the Australian operators consists of two fixed parts for domestic customers, one for the drinking water supply service and the other for the wastewater service. It also has two proportional parts, one associated with water consumption, divided into increasing blocks, and the other related to wastewater. Both services are billed on the basis of measured values. For the metropolitan operators there is also a fixed component that corresponds to the stormwater service. Table 2.2 provides a summary of the tariff systems for domestic customers of the five largest operators in the state of Victoria, as well as their average billing.

The tariff values do not differ greatly in the case of non-domestic customers and cross-subsidization between groups of customers does not occur. However, in the context of the wastewater service there is generally only one block tariff. In the case of treatment of effluents from industrial customers, this service is billed according to the organic load.

In the tariff system for domestic and non-domestic customers a predefined value is included for the fire services, which varies between a fixed sum charged annually and a sum that depends on the type of water meter.

The ESC also establishes other components of the tariff system, when applicable, such as, *inter alia*, elements pertaining to reclaimed water, storage, untreated water, wholesale water, water for irrigation and the implementation of service connections.

Table 2.2 Tariff system for domestic customers of five operators in Victoria

Operator	City	Water Service Availability Charge (€/year)	Water Usage Charge (€/m³) (quarterly)		Wastewater Service Availability Charge (€/year)	Wastewater Service Usage Charge (€/m³)	Average annual bill (2004–05)	Average annual bill (2005–06)
City West	Melbourne	57.93	0–40 m³	0.47	61.61	0.61	274.20	286.20
			40–80 m³	0.55				
			>80 m³	0.82				
South East	Melbourne	26.3	0–40 m³	0.47	88.18	0.57	277.80	288.60
			40–80 m³	0.55				
			>80 m³	0.86				
Yarra Valley	Melbourne	34.8	0–40 m³	0.47	84.98	0.61	310.20	323.40
			40–80 m³	0.55				
			>80 m³	0.86				
Barwon	Geelong	77.2		0.50	96.73	0.6	325.20	348.60
Coliban	Bendigo	66.8	0–50 m³	0.37	168.32	N/A	301.80	314.40
			50–100 m³	0.44				
			>100 m³	0.73				

On average, the price of water in Australian cities is 0.59 €/m³, representing a very low value when compared to most developed countries on a global scale. In fact, some domestic customers pay far less than the 0.59 €/m³ referred to depending on the area in question. In 2005/2006, the average bill of Australian domestic customers concerning the water services was 188.5 euros and when combined with the wastewater service it amounted to 340.9 euros.

In 2005, the water service represented a cost of 193.3 euros for domestic customers in Sydney and 203.2 euros for the wastewater service, corresponding to a total of 396.5 euros. The average WWS price in Melbourne was 314.5 euros, while in Perth it represented 422.9 euros.

2.1.4.2 Quality of service

There is an association of operators in Australia (WSAA), which covers about 75% of the Australian population. Its functions include the implementation of an annual benchmarking project, which also includes two operators from New Zealand, as a way of improving the sector and it has proved to be an important tool. This activity consists of developing a set of 55 performance indicators from diverse areas, namely water resources, customers, assets, quality of service, prices and finances and health. Currently, this association is developing a voluntary project known as the *2008 IWA-WSAA Asset Management Process Benchmarking Project*, with a view to improving the quality of service even further.

Apart from this, each state/territorial regulatory agency, such as the ESC or the ERA, compares and publicizes the performance of operators (sunshine regulation) and specifies the performance levels and conditions in which the operator provides the services. These requirements are voluntarily included in the Water Plan or established as public service obligations.

In order to monitor and provide incentives for the quality of service, operators compensate customers in situations where the quality of service has declined, as mentioned above. This is especially so in the context of the service levels, which are described in further detail in the next subsection.

With the purpose of achieving the objectives (and results) associated with the quality of service, the regulatory agency calculates a set of performance indicators, which it later publishes and compares in an annual assessment report (sunshine regulation). Table 2.3 presents the indicators that have been adopted by the ESC. Figure 2.3 reflects the positive role of sunshine regulation, providing the evolution of the indicator of the interruption of water supplies over time. Adopting the same indicator, Figure 2.4 compares the performance of all the operators in the state of Victoria for 2005/2006 (ESC, 2008).

Table 2.3 Performance indicators used by the ESC

WWS performance indicators	
Affordability	**Network Reliability and Efficiency**
Average household consumption	<u>Water supply</u>
Average household bills	Water supply interruptions (per 100 km of water mains)
Domestic customers with instalment plans (per 100 customers)	Average customer interruption frequency (interruptions per customer) (number per customer)
Non-domestic customers with instalment plans (per 100 customers)	Planned water supply customer interruptions frequency in peak hours (interruptions per customer)
Domestic restrictions for non-payment of bills (per 100 customers)	Average duration of planned interruptions (minutes)
Non-domestic restrictions for non-payment of bills (per 100 customers)	Average duration of unplanned interruptions (minutes)
Restrictions restored within three days (per cent, domestic only)	Average customer minutes off supply (minutes)
Restrictions over 14 days (per cent, domestic only)	Bursts and leaks (per 100 km of water mains)
Domestic legal actions (per 100 customers)	Average response times to bursts and leaks – priority one
Non-domestic legal actions (per 100 customers)	Average response times to bursts and leaks – priority two
Average debt level – restrictions (€)	Average time to rectify bursts and leaks – priority one (minutes)
Average debt level – legal action (€)	Average time to rectify bursts and leaks – priority two (minutes)
Hardship grants and applications (per cent)	Planned customer interruptions not restored within 5 hours (per cent)
Average value of hardship grants (€)	Unplanned customer interruptions not restored within 5 hours (per cent)
Customer Responsiveness and Service	Water losses
Average time taken to connect to an operator (seconds)	<u>Wastewater service</u>
Calls answered within 30 seconds (per cent)	Sewer blockages (per 100 km of sewer main)
Complaints received by water businesses (per cent)	Customers experiencing a single sewer blockage
Water quality complaints (per cent)	Sewer spills from reticulation and branch sewers (per 100 km)
Information statements processed within five days (per cent)	Containment of sewer spills within 5 hours (per cent)
Drinking Water Quality	Sewer spills to customer property (per 1,000 customers)
Microbiological water quality (per cent)	Sewer spill volume of percentage transported (%)
Turbidity (per cent)	
Disinfection by products (per cent)	
Water quality complaints – all causes (per 100 customers)	
Water quality complaints – by cause	
Melbourne Water compliance with water quality requirements at interface points (per cent)	
Environmental Performance	
Overall sewage treatment plant compliance (per cent)	
Proportion of effluent re-used (per cent)	
Proportion of biosolids re-used (per cent)	
Net historic greenhouse gas emissions	

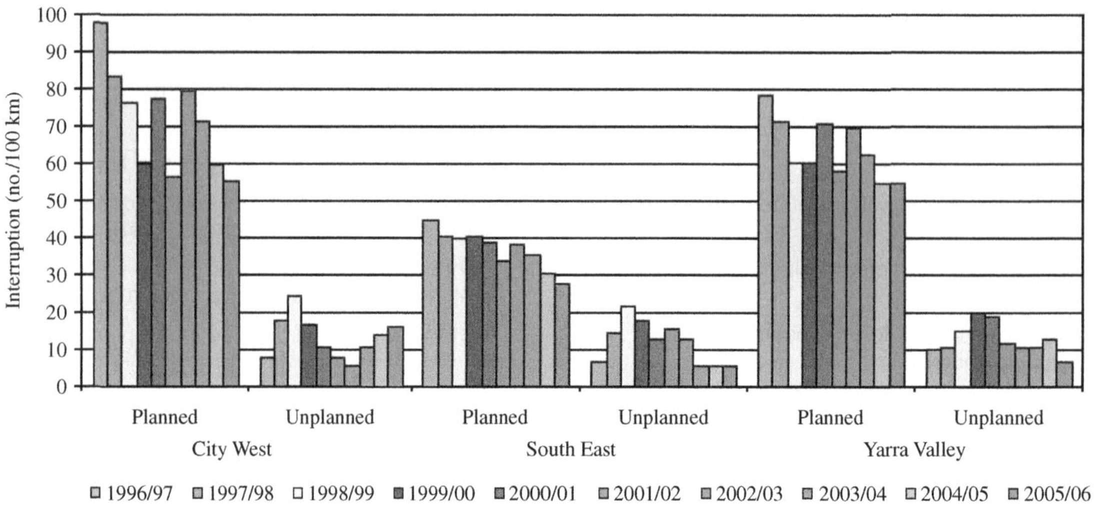

Figure 2.3 Indicator water supply interruption for metropolitan operators

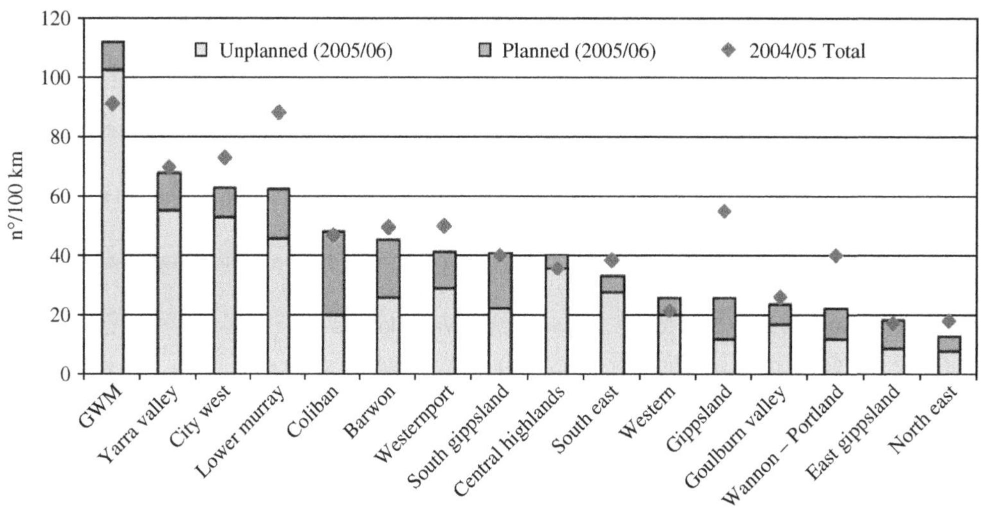

Figure 2.4 Indicator water service interruptions for regulated firms in Victoria

It must be mentioned that even in the case of local WWS operators that are not covered by a regulator, the state governments promote the compulsory implementation of benchmarking, e.g. in New South Wales. Moreover, NSW was the first state to carry out a benchmarking exercise in the water sector in 1992/1993. This was also the first such exercise in the world in this sector.

2.1.4.3 Public service obligations

In order to better defend the interests of customers, the regulatory entities have prepared a Customer Service Code (CSC), which defines the levels and conditions of service for the WWS sector. Table 2.4 provides an overview of the set of public service obligations that have been developed by the ESC, which each operator commits to fulfil.

One of the main measures adopted by the ESC corresponded to the implementation of a Customer Consultative Committee, which advises the ESC about different subjects and consumer needs. One of its initiatives was to create a forum where customers could express their opinions. Likewise, there are non-governmental customer defence associations whose opinions are also usually considered, such as the Consumer Utilities Advocacy Centre.

Another important aspect of regulating essential public services in Victoria is that of the Utility Relief Grant Scheme, administered by the DHS, which provides financial assistance to customers with temporary financial problems.

The subsidy is attributed for the electricity, gas and water bills and covers costs for a period that could extend up to six consecutive months.

Table 2.4 Principles of public service and operators' obligations

Principle	Operators' obligations
Connection and service provision	The operator is compelled to provide connections within a maximum period of 10 business days.
Charges	Publicize tariffs and rates approved by the regulator as well as any changes.
Complaints and disputes	Accept mediation, arbitration and conciliation in the case of conflicts. Defined and approved procedure for complaints.
Billing	Quarterly reading or other duration agreed upon beforehand. The bill must contain all the services that have been charged and the respective tariff. In case the user has been under- or over-charged by the operator, certain conditions have to be respected. In case of under-billing, the billing must be done within 12 months and in case of over-billing, the operator must notify the customer within 10 business days and agree upon a form of repayment.
Payments	Various forms of payment (e.g. agencies, post office and direct debit), including flexible payment plans, keeping in mind the financial capacity of the customer.
Collection	Obligation to notify the customer in case the customer has not paid the respective dues for the service, at least 7 days before taking any action.
Actions for non-payment	The operator will apply legal measures regarding restricting the supply of its services in case of failure to make due payments. There are, however, certain conditions that must be respected.
Quality of services	In addition to acting compulsorily in accordance with health and environmental regulations, the operator must provide the service in conformance with approved service standards. It must periodically demonstrate that it is fulfilling its commitments, compensating customers when this is not the case (levels of service).
Reliability of services	The operator will develop and implement plans to manage its assets with a view to ensuring reliability. It is the operator's responsibility to plan procedures that can respond suitably to unforeseen circumstances such as bursts, illegal discharges or obstructions in its systems, informing customers in cases of planned interruption.
Reconnection	The operator must immediately reconnect a former consumer if requested to do so.
Works and maintenance	The operator is obliged to establish a programme that will maintain its system in accordance with the established service standard.
Information	The operator is obliged to provide access to information free of cost, namely, with regard to excess billing, payment options and the EWOV, amongst others. Apart from this, the operator must inform its customers about regulatory acts if requested to do so. Operators must also allow access through an interpretation service for customers with speech and/or hearing disabilities.
Guaranteed service standards (GSS)	The operator must develop and publicize through its website a charter of consumer rights in order to provide information about its services and the respective rights and responsibilities of the operator and customers.

In the context of regulating the water sector, the ESC also approves a set of service levels proposed by the operators, which they commit to fulfil during the regulatory period. If they fail to fulfil these levels (Guaranteed Service Standards – GSS) they also commit themselves to pay compensation to customers. Unlike the regulation of other sectors (e.g. electricity and gas), which include an index integrating the variation of quality and of the service levels in the price cap formula, the ESC, at least for the first regulatory period, has allowed some freedom in the definition of service levels. It has also allowed that these levels would be imposed only for domestic customers and for the largest operators. Table 2.5 presents the levels of service which were approved for the first regulatory period.

2.1.5 Performance

The states and territories in Australia have a well-conceived, modern and very robust regulatory model, prepared to face any possible adversities. Regulation is essentially focused on customers without disregarding the sustainability of the operators. The different regulators have successfully developed ways of supervising and controlling the quality of service, as can be seen by the improvements year after year.

Table 2.5 GSS for domestic customers

Level of service	Amount payable for failures				
	City West	South East	Yarra Valley	Barwon	Central Highlands
Water					
Unplanned interruptions not restored in specific time	15	15	15	–	15
More than five unplanned interruptions in any 12 months	15	15	15	30	15
Failure to notify a planned interruption	–	–	15	–	–
Planned interruptions during peak hours (5 a.m. to 9 a.m. and 5 p.m. to 11 p.m.)	–	–	15	–	–
Planned interruption longer than advised	–	–	15	–	–
Planned interruption longer than 5 hours	–	–	15	–	–
Repair of leaking service pipes within five days	–	–	–	–	15
Wastewater					
More than three interruptions in 12 months	15	15	15	30	15
Interruptions not restored in specified time	15	15	15	–	15
Spills not contained in specified time of notification	300	300	300	–	–
Spills not contained in a house in one hour of notification	300	300	300	–	–

The regulators have ensured the different public service obligations, which are contracted between the various stakeholders, and have maintained the operators' revenues by suitable economic regulation, i.e. guaranteeing economic and financial viability, while not attaining unreasonable profits. Finally, it must be mentioned that concerning economic, social and environmental aspects, regulation has been developed from a long-term perspective aimed at sustainability. In short, the regulation of the WWS sector throughout Australia – and particularly in the state of Victoria (which has been analysed here in greater detail) – thus constitutes an example worth bearing in mind.

2.2 BELGIUM

2.2.1 Historical background

Belgium, officially known as the Kingdom of Belgium, is a constitutional monarchy and a parliamentary democracy, where the king is the representative of the State and the prime minister is the head of government, in a plural and multiparty system. The government has executive power. Federal legislative power is divided between the government and the two parliamentary chambers, namely the Senate and the House of Representatives.

Belgium has a population of about 10.5 million inhabitants and encompasses an area of about 30,500 km^2. It is divided into three regions, particularly Flanders and Wallonia, with five provinces each, and the Brussels-Capital Region, which includes the Belgian capital. At a local level, the country is organized into 589 municipalities, of which 308 belong to the region of Flanders and 262 to the region of Wallonia. There are 19 municipalities in the Brussels region.

Belgium is one of the European countries with the fewest water resources. Perhaps this is why it has carried out various reforms for the water sector over time. Regionalization, introduced in 1980, was a very significant landmark and greatly altered the institutional homogeneity that had existed until then. The WWS, under the responsibility of the national government until then, were transferred to the regional governments. The regions are characterized by a great heterogeneity at a structural and geographical level, which has an impact in the water sector management. This is even reflected in the availability of water resources. Both the Flanders region (25%) and the Brussels-Capital Region (100%) depend on the region of Wallonia for drinking water supply.

Despite this, the water market in Belgium is not very regulated. Notwithstanding the changes that have taken place in recent years and the fact that rules have been defined regarding the management of water resources and the quality of water and environment in general, WWS have traditionally worked under self-regulation. This situation has been somewhat reversed mainly due to pressure from the EU as well as from customers, who are increasingly more aware and demanding. Apart from this, the water industry is characterized by an enormous lack of economic and financial sustainability, caused by heavy state funding.

The tendency towards a greater government interference in managing the water sector was reflected in the creation and operationalization of the Water Control Committee (*Comité de Contrôle de l'Eau* – CCE) in Wallonia and in the creation of a regulatory agency in Flanders, although it has not yet started operations to date (2007). For example, in Wallonia, among other measures that were implemented, the CCE sought to ensure greater equality and sustainability for the WWS, by introducing uniform tariff systems.

The first stage of the water service development in Belgium took place during the mid-nineteenth century. In this context, the initial investments were derived from the private sector but the municipalities soon took control of the initiative and the monopoly that characterizes the water sector. However, observing the slow evolution of the sector, the Belgian state tried to provide incentives to ensure its progress, funding the construction of infrastructures. On the one hand, the state recognized the role of the municipal authorities in the sector. They started to operate integrated WWS and subsidize the investments in these intermunicipal systems (*intercommunales*), according to prevailing legislation (law dated 18 August 1907, about the association of municipalities, which was later changed by a law dated 1 March 1922). On the other hand, the state created the National Water Distribution Company (*Société Nationale de Distribuition d'Eau* – SNDE) in 1913, entrusted with the task of ensuring the construction and management of the infrastructures whenever the municipal authorities did not have the capacity to do so. This situation continued until 1990, when the water sector was regionalized. Until 1950, wastewater was discharged directly into rivers. This practice, along with the growth and development of the population, gave rise to serious environmental problems of pollution, which the government tried to resolve by implementing the first law concerning environmental protection, namely the law dated 11 March 1950. This normative document was replaced by the law dated 26 March 1971, which referred more specifically to protecting surface water from pollution.

The responsibility of the regions for the WWS was recognized in 1980, as well as in 1988 by means of the norms regarding the drinking water quality. Since then, the evolution of the water sector management has diverged widely in the three regions. The first visible change was the extinction of the SNDE in 1983 and the creation of the *Vlaamse Maatschappij voor Watervoorziening* (VMW) in Flanders and the *Société Wallonne de Distribution d'Eau* (SWDE) in the region of Wallonia.

The region of Flanders, in particular, faced diverse obstacles with these changes, especially due to its great dependence on Wallonia in 1982, since 60% of its drinking water originated in the latter region. In order to counter this situation, in 1984 the government of Flanders developed a policy of greater independence in the area of water supply, publishing a Regional Law of Flanders, dated 24 January 1984, whose objective was to manage underground water resources.

In Wallonia, the management of water evolved in a very different way. In 1977, the wastewater treatment service was entrusted to the *intercommunales*, a competence that was confirmed by a Regional Law of Wallonia, dated 7 October 1985, regarding the protection of surface water against pollution. The organization of the sector was restructured in 1999, by means of a Regional Law of Wallonia, dated 15 April 1999, regarding the management of the water cycles. This regional law also created the Public Water Management Company (*Société Publique de Géstion de l'Eau* – SPGE). The government sought to protect water resources and, by means of a Regional Law of Wallonia, dated 30 March 1990, instituted a tax on the discharge of domestic and industrial wastewater.

The water policy in Brussels region was developed later, based on a comparison with the other two regions. The Brussels region faces a unique situation since it encompasses virtually only an urban area and is totally dependent on water supply from Wallonia and is relatively dependent on the wastewater treatment service of Flanders, given that the effluents produced in Brussels are sent there.

Subsequently, in order to promote greater and better efficiency and interconnection between the different operators and bodies responsible for the water cycle, the government of the Brussels region reformulated the entire sector, by means of a norm dated 17 October 2006. The reforms merged the two bodies that were responsible for the water sector, namely the Belgian Institute for Environmental Management (*Institut Bruxellois pour la Gestion de l'Environnement* – IBGE) and the Administration for Equipment and Decentralisation (*Administration de l'Equipement et des Déplacements* – AED). In addition, it created a regional operator to collect and drain wastewater, the Brussels Water Management Company (*Société Bruxelloise de Gestion de l'Eau* – SBGE), to the detriment of the existing *intercommunale* system, the IBrA (*Intercommunale Bruxelloise pour l'Assainissement*), which had proved to be ineffective.

Up to the early 1980s, the norms related to drinking water quality were implemented on a national scale but after the regionalization they were transferred to a regional level. The 1984 law incorporated the 1980 Water Directive about the drinking water quality. Specific legislation including minimum parameters for the quality of water was enacted only in 1988. Additionally, consistent with new EU directives, the regions implemented new legislation between 2001 and 2002 concerning the drinking water quality.

2.2.2 Legal and institutional framework

2.2.2.1 Responsibilities

The regions have decision-making powers regarding the organizational arrangements of the WWS and their possible privatization, in accordance with the powers of the municipalities to decide about the possibility of contracting a private operator for supplies. The regions ensure the supervision of the sector while the operators monitor their own activities by means of self-regulation, namely with regard to the drinking water quality.

Since 1980, water management and environmental policies have made significant progress. One of the first governmental measures involved the water sector regionalization. The inherent consequence of this policy is the great heterogeneity that has been witnessed in the evolution of the three regions, especially owing to the corresponding regional legislation and the dismantling of the national water company.

The three regions (Flanders, Wallonia and Brussels) operate as the entities responsible for the water sector, even though some municipalities are equally responsible. Belgium has a fragmented structure and is characterized by a heterogeneous development in its regions. In spite of this, it is possible to state that public management is a dominant feature in the three regions.

Although there is no sector-specific regulatory agency for the water sector in Belgium, its professional association, BELGAQUA, plays an important role in defending water as a common resource while promoting the collective interests of its members at a national and international level. Besides, the association encourages the development of scientific studies regarding technical and economic matters, thus increasing the information and recommendations available to operators and customers about the water supplied and its quality.

BELGAQUA brings together the three regional associations of the water sector: AQUABRO (Brussels-Capital), AQUAWAL (Wallonia) and SVW (Flanders), with a total of 48 associated operators.

In the region of Flanders, the water sector comprises eight main organizations, namely BELGAQUA, SVW, TNAV, Aquavisie, ISWa, Aqualink, VLW and VIWC/CIW. Moreover, there are two governmental agencies, the *Administratie Milieu, Natuurbeheer en Landinrichting* (AMINAL), which has limited regulatory and supervisory powers over the sector, and the *Vlaamse Milieumaatschappij* (VMM), a public company whose function is to control the drinking water quality. The latter also has decision-making powers about the location of the WWTP in the region. Figure 2.5 shows the main players in the water sector (adapted from Euromarket, 2003).

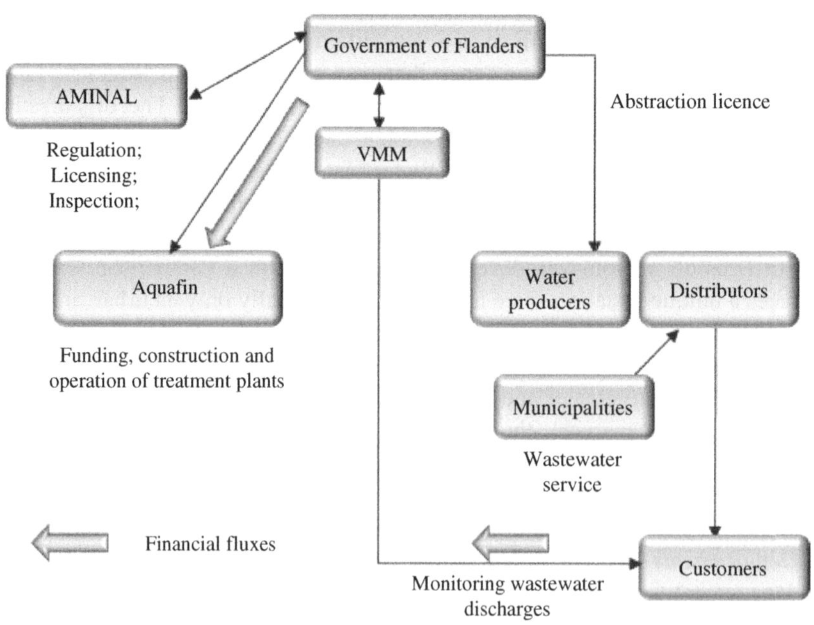

Figure 2.5 Institutional framework of the water sector in Flanders

In the Brussels region, before its autonomy, the water sector was supervised by a single body, the IBGE, which had powers of licensing and monitoring the quality of water in the region. As a result of recent legislation, the IBGE has also acquired the functions of the now abolished AED with regard to wastewater. Figure 2.6 illustrates the main players in the water sector in Brussels (adapted from Euromarket, 2003).

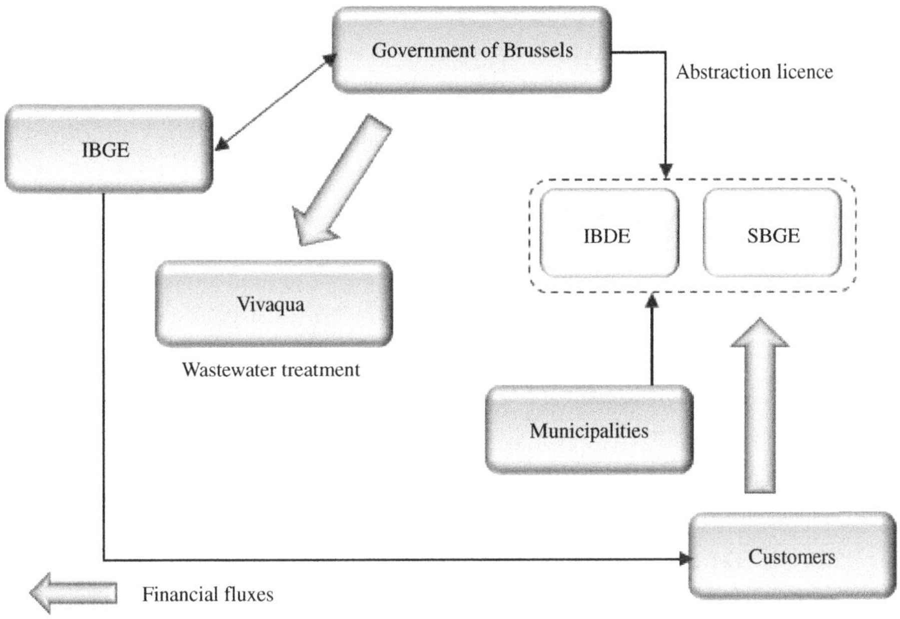

Figure 2.6 Institutional framework of the water sector in Brussels

In Wallonia, much of the authority over the water sector has been entrusted to a single department, the Directorate General for Natural Resources and the Environment (*Direction Général dês Ressources Naturelles et de l'Environnement* – DGRNE), which belongs to the Ministry of the Wallonia region. The DGRNE is responsible for compiling information about the quality of surface and underground water, licensing abstractions, authorizing the discharge of wastewater and increasing the respective rates and, above all, monitoring the drinking water quality.

The SPGE has been a public limited company since 1999. Its functions include controlling the coordination and financing of the water sector in the region of Wallonia. Despite being a public entity, the SPGE frequently relies on the (private) market for providing services.

The CCE is responsible for establishing and ensuring that the water cycle and the tariff system are upheld by the operators, in accordance with the general interest and water policies for the region of Wallonia. All changes in water prices must compulsorily be subjected to the opinion of the CCE before any other procedures as well. Thus the CCE plays a dominant role in the evolution of the tariff systems applied from 1 January 2005 onwards, contributing towards the implementation of the 'truth-cost' policy for the water sector, harmonizing tariffs. Figure 2.7 shows the main players in the water sector in the region of Wallonia (adapted from Euromarket, 2003).

2.2.2.2 Regulation

Although there is no sector-specific regulatory agency in Belgium state for the WWS, each region has entities with important regulatory functions. The water sector is explicitly regulated in Wallonia, albeit only recently, through a body that is integrated into the traditional public administration, the CCE. In Flanders, legislation has also drafted plans for the institution of a regulatory agency, but this body has not yet become operational (2007). The fact that the local authorities frequently provide water services has proved to be detrimental for the creation of regulatory agencies.

The CCE has important regulatory functions in the Wallonia region, including controlling and supervising tariff systems and implementing performance indicators, which seek to increase the efficiency and effectiveness of operators and uphold the legal missions that have been entrusted to them.

In addition, the Wallonia region has an Ombudsman (*Médiateur de la Région Wallonne*), created in 1994, whose main function is to defend the interests of public services customers. It is fully empowered to resolve the disputes that might arise between customers and operators.

In the Flanders region, a law dated 24 May 2002 created a sector-specific regulatory agency which was meant to be instituted up to the end of July 2003. Its functions were to list, provide advice for and report upon all matters concerning the water sector, with the specific objective of ensuring transparency, the fulfilment of public service obligations, the approval of investments, monitoring accounts and performing yardstick comparison. However, until the end of 2007, this entity had not yet been established.

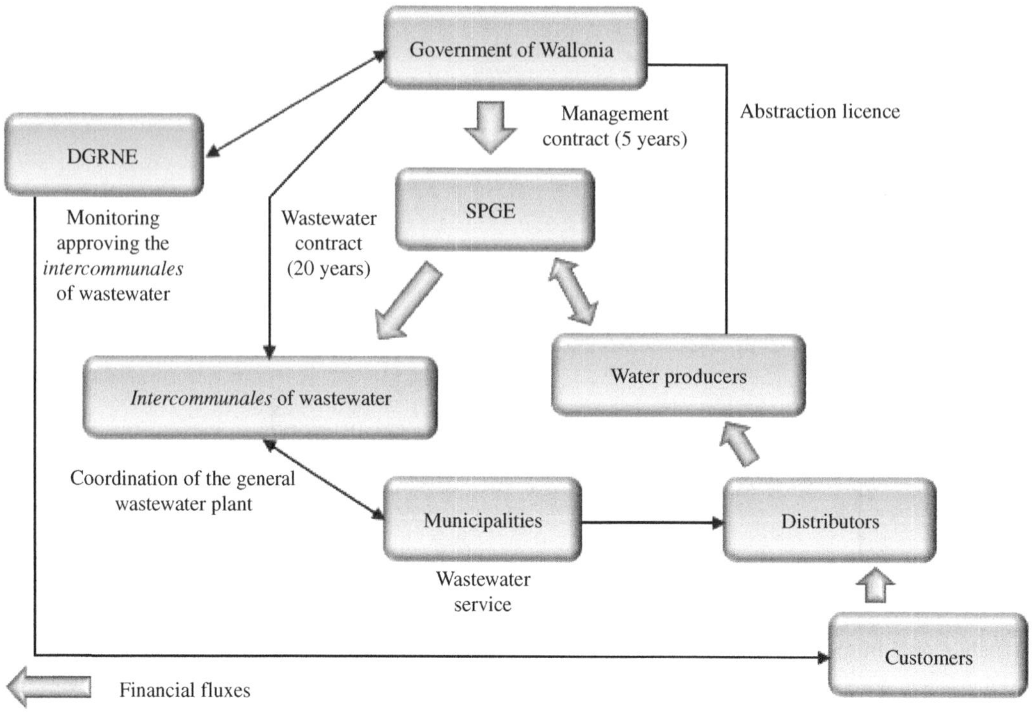

Figure 2.7 Institutional framework of the water sector in Wallonia

In contrast to other countries, and even the region of Wallonia, few initiatives have been developed with the aim of promoting efficiency and effectiveness in the WWS. Despite the fact that most operators make tariff systems available on their websites, there is no overall view about the different WWS concerning the quality of service and tariff systems implemented. Besides, no entity has promoted any study to determine the factors that contribute to the total cost of the service.

In 2002, the association for the water sector in the Flanders region, the SVW, organized a benchmarking study between operators, following the Dutch example. Each participant in the exercise received a document with their performance as compared to the benchmark reference group. However, these results were never publicized. In contrast to the absence of formal regulatory initiatives, the WWS in the Flanders region have created diverse bodies in this sector in order to collaborate on various matters (e.g. self-regulation).

2.2.3 Market structure

2.2.3.1 Ownership and management of the services

In overall terms, the water sector in Belgium is controlled by the public sector. Regional and municipal companies, as well as inter-municipal associations (*intercommunales*), provide services in the area of producing and distributing water. The services related to wastewater are organized differently within Belgium, where the respective services are provided by a regional company (Flanders), specific *intercommunales* companies (Wallonia) or are even under the direct control of the operators (Brussels). The private sector plays an important role in treating wastewater by means of PPPs. For example, the WWTP North for Brussels was constructed using this public procurement arrangement leading to the creation of Aquiris, in which the multinational firm Veolia has a majority stake. It must also be mentioned that during the 1990s the private sector was more active, both through Suez, which operated various systems, and through the Sever Trent company, which had a 49% stake in the Aquafin PPP.

2.2.3.2 Players in the sector

The operators of water services in this region vary considerably in terms of size. There are many companies operating in small territories. The largest company, VMW, provides water services to about 40% of the population of Flanders, i.e. 2.4 million inhabitants, and the three largest operators (AWW, PIDPA and VMW) provide water services to 80% of the population. The wastewater service is supplied by a single operator, Aquafin, a regional company, which was a PPP in the past. Aquafin has a monopoly of the wastewater in the Flanders region, except in districts with less than 2,000 inhabitants. Figure 2.8 illustrates the geographical distribution of the operators.

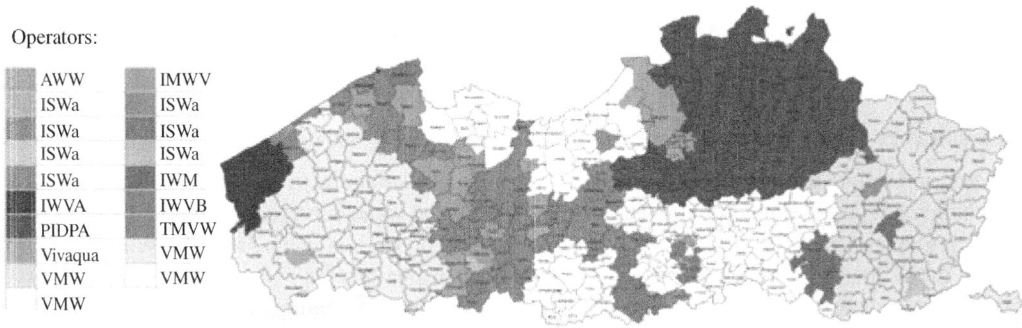

Figure 2.8 Geographical distribution of the water service operators in Flanders

In the Brussels region, the wastewater collection and treatment service is provided by the SBGE, a regional company responsible for large investments (e.g. treatment plants and drainage basins), which was created in October 2006. Just like in the case of wastewater services, the water service is provided by a single operator, i.e. by the *Intercommunale Bruxelloise de Distribution d'Eau* (IBDE).

One of the two existing WWTPs (WWTP South) is managed by Vivaqua, after it merged with the *Compagnie Intercommunale Bruxelloise des Eaux* (CIBE), a company that provides all the water cycle in the three regions of Belgium (water and wastewater delegated by the municipal authorities). The WWTP North is managed by Aquiris, as a result of a PPP in which the multinational Veolia has a majority stake.

In the Wallonia region, the drinking water supply is fragmented into 18 operators. The four largest companies cover more than 75% of water customers in the region. The largest water company in Wallonia, SWDE, provides water services to about 70% of the population of Wallonia, i.e. 2.3 million inhabitants. Apart from the regional company, the segment of the water supply encompasses 8 intermunicipal systems and 2 small municipal operators, while the segment of wastewater service is limited to 7 intermunicipal systems, two of which also supply drinking water. Figure 2.9 shows the geographical distribution of the operators.

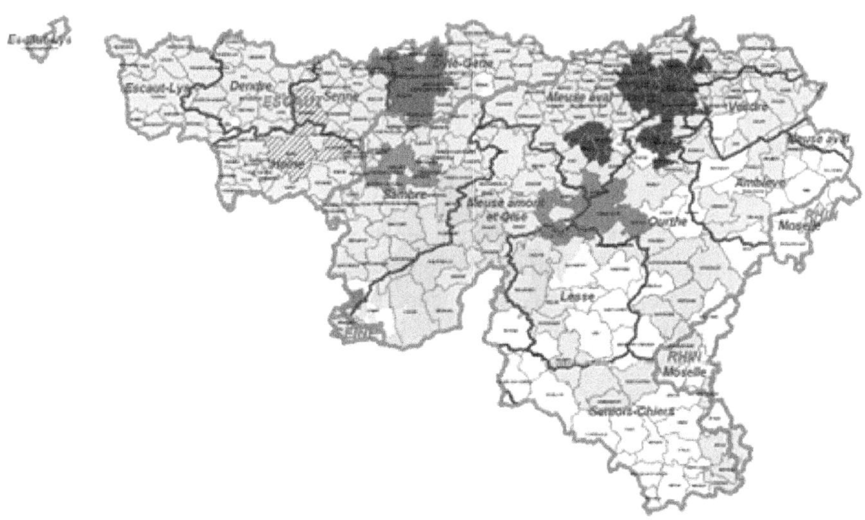

Figure 2.9 Geographical distribution of the water service operators in Wallonia

Direct public management, through municipal services, is mainly integrated into the policy for public service management in the region of Wallonia. However, delegated public management, through independent municipal departments, intermunicipal systems and regional companies, is the key arrangement in Belgium.

Independent municipal departments exist both in the region of Flanders (12) and in the Wallonia region (2). In their turn, the intermunicipal systems cover a large part of the water service in Belgium, with 7 systems in Flanders, 15 in Wallonia (7 intermunicipal systems encompass the entire wastewater service in Wallonia) and 2 in Brussels (IBDE and Vivaqua). Finally, there are two regional companies in Flanders (VWW provides water services to 40% of the population of Flanders and Aquafin provides wastewater service to 100% of the Flanders population) and one in Wallonia (SWDE supplies water services to 70% of the population of Wallonia).

2.2.3.3 Integration of the sector and services provided

The water sector in Belgium is characterized by a partial division between the operators performing integrated services and the ones that do not do it, both vertically and horizontally. In the water service, despite the fact that some operators in the Flanders region and the sole operator in the Brussels region depend on water from the Wallonia region, there are other operators in the Flanders region that wholly integrate the water service.

The wastewater services are characterized by a complete vertical disintegration, since there are operators throughout Belgium exclusively dedicated to the wastewater collection services and other operators specialized in WWTP.

In addition, it is usual to have specialized operators for the different WWS and there is a horizontal disintegration of these services. However, Vivaqua is one of the exceptions, since it supplies all the services related to the water cycle, namely providing drinking water to some *intercommunales*, managing the WWS of other *intercommunales* as well as the WWTP South in the Brussels-Capital Region. This is also the case of INASEP (*Intercommunale Namuroise de Services Publics*) and IDEA (*Intercommunale de Développement Economique et d'Aménagement de la Région Mons-Borinage-Centre*), which provides integrated WWS in the Wallonia region.

2.2.3.4 Numbers for the sector

About 730 million m^3 of water are abstracted in Belgium every year for drinking water supply, of which 400 million m^3 are used by domestic customers. The main source is groundwater, representing about 470 million m^3 of water, i.e. 65% of all the water that is tapped.

In 2005, the Belgians had a per capita consumption of drinking water of about 106 litres per inhabitant per day.

Water losses in the water sector in Belgium are close to 20% (BELGAQUA, 2005).

Water services are supplied to about 99% of the Belgian population, whereas wastewater collection and treatment services are only provided to 81.5% of the population. The level of wastewater treatment is about 50% (2004 data). The wastewater systems are still predominantly combined systems.

In 2005, the WWS sector in Belgium employed about 7,200 people.

The water sector in Belgium encompasses about 4 million customers. The patrimony includes approximately 101,000 km of water mains (BELGAQUA, 2005).

2.2.4 Rules of the game

2.2.4.1 Tariff setting

The tariff systems of WWS vary significantly from region to region and reflect the legitimate choices of local decision-makers, since water prices are under the supervision of the regional authorities. Diverse factors contribute towards this situation, namely the large number of operators in the Flanders region that buy wholesale water from other operators. This results in higher prices as a consequence of transportation costs. Tapping groundwater leads to lower treatment costs, despite being subject to high taxes. Apart from this, certain issues associated with each operator, such as the network length, the degree of urbanization, the kind of customers and the number of employees among others, give rise to various differences. In the meantime, for several years now there has been an intense debate in Belgian society about introducing uniform tariffs in the various regions and even throughout the country. This discussion, which is based on the argument of essential services provision and their importance for the social and economic cohesion of the population, has already resulted in the definition of a single structure for the tariff system in the region of Wallonia.

There is a uniform tariff system throughout the region of Wallonia implemented by operators since 1 January 2005, irrespective of the nature of the consumption (domestic or non-domestic). It comprises a fixed component and a variable component for customers, which encompasses the WWS. Both components have been established keeping in mind the so-called 'truth-cost' factor, which reflects the real operating and maintenance costs of the different services.

The fixed component is determined by adding $20 \times \text{CVD}$ and $30 \times \text{CVA}$, in which CVD corresponds to the 'truth-cost' estimated by the operator for the service of water distribution (including the abstraction taxes and costs related to production and distribution), which varies between 0.77 euros and 1.86 euros. The CVA represents the 'truth-cost' related to the wastewater treatment (0.7950 euros in 2007), which varies every year.

The variable component, which is linked to consumption, has been established according to four regressive blocks, which are included in Table 2.6. Figure 2.10 illustrates the tariff system of Wallonia (CCE, 2006). This also includes a component whose objective is to constitute a social fund to help low-income families pay for the WWS. The contribution to this fund was 0.0125 €/m^3 at the beginning of 2007.

Table 2.6 Tariff system for household customers in Belgium

Region	Fixed component	Water sales (€/m^3)		Social fund (€/m^3)	Wastewater services tariff (€/m^3)	
Flanders	39–57	0–15 m^3 >15 m^3	0 1.881–2.931	0.01	Aquafin 0.7730	Municipalities Variable
Brussels	11.9–23.80	0–15 m^3 15–30 m^3 30–60 m^3 >60 m^3	0.80 1.39 2.06 3.06	–	IBDE Regional 0.3275	IBDE Municipalities 0.43
Wallonia	20CVD+30CVA	0–30 m^3 30–5,000 m^3 5,000–25,000 m^3 >25,000 m^3	0.5CVD CVD+CVA 0.9 CVD+CVA 0.7CVD+CVA	0.0125	Jointly supplied with water sales	

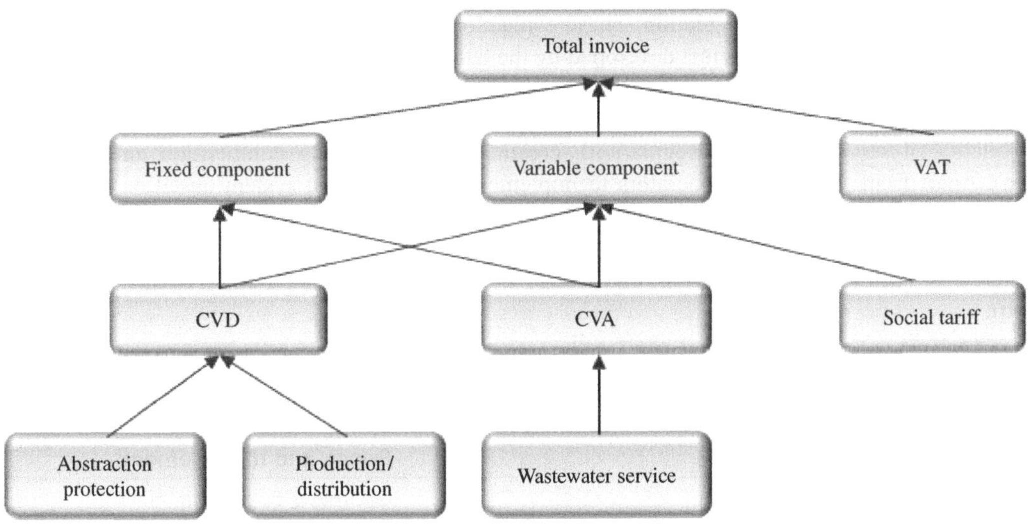

Figure 2.10 Methodology used for defining tariffs in the region of Wallonia

As noted earlier, the prevailing tariff systems in the region of Flanders differ considerably from operator to operator. By law (*Vlaams Decreet betreffende water bestemd voor menselijke aanwending*), each domestic customer is entitled to receive a supply of 15 m^3 of water free of charge. In case this quantity is exceeded, the operator implements a variable component associated with consumption, which varied between 1.881 €/m^3 and 2.931 €/m^3 in 2006. The tariff system also has a fixed component, which represents a fixed subscription, the meter rental charge or a combination of both. In 2006, this component varied between 39 euros and 57 euros, according to the operator. The tariff varies progressively according to consumption in only three situations and is generally fixed or regressive. The wastewater service is also heavily subsidized. However, there is a tariff for treating wastewater for the entire region, 0.7730 €/m^3 (in 2007), imposed by Aquafin, while a sum is already charged for the services provided by the majority of operators that collect wastewater, which, in some cases, is 1.5 times the value of the wastewater treatment. Normally, the tariff is not linked to the nature of the customer and there are some operators that negotiate bilaterally with large customers. Concerning the water supply, many operators do not change their tariff system according to the customer (domestic or non-domestic), while others practise lower prices for large customers (above 1,000 m^3 of consumption). In cases of very high-volume customers, the operators negotiate directly to define a lower price. The definition of business customers varies with the different operators.

In the Brussels region, the water service is characterized by a single tariff system, comprising two main components, which change according to the type of customer. For domestic customers, the tariff system has been divided into four blocks and the charges rise in proportion to consumption. A linear tariff exists for business and non-industrial customers while there are two blocks for industrial customers. The prevailing tariffs seem to indicate that there is no subsidization between different customers. The fixed component for the water service is implemented according to the operator and varies between 11.9 euros and 23.80 euros (2007). The operator also established an additional tariff of 0.01 €/m^3 on the quantity billed, in order to provide a universal service, by creating a fund that subsidizes families in need. There are different tariffs for the wastewater service, depending on local conditions if they are relative to municipalities, or a single tariff for the regional system. All of them include two components, one of which varies with consumption, without predefined blocks, while the other component is fixed, but changes according to the location. No differentiation is made on the basis of the type of customer. Finally, apart from VAT (charged at a rate of 6% for the two services), customers also pay a tax (0.35 €/m^3) to fund the construction of wastewater treatment infrastructures.

2.2.4.2 Quality of service

Until very recently the water sector in the Wallonia region worked on the basis of self-regulation. Prices were not supervised (just like public service obligations) and the quality of service provided by the operators was neither controlled nor even known in some cases. However, the association of operators in Wallonia, AQUAWAL, contributed towards establishing norms and spreading best practices and also promoted the dissemination of some statistical information. In addition to the indicators to monitor the legal obligations mentioned above, the CCE in the region of Wallonia developed a set of 15 indicators about the operators performance, with a view to resolving the problems associated with the existing lack of regulation. These evaluation measures, which are calculated and supplied by the operators to the CCE on a calculation sheet developed by the latter body, have been sorted into six classes, namely, the quality of service, the protection of abstractions, the assets management, prices and management, coverage and solidarity and customer satisfaction and communications. These indicators seek to estimate, compare and publicize the operators' quality of service (sunshine regulation), creating and stimulating competition among them (yardstick competition). Apart from allowing the operators to implement a management system based on objectives, they also indicate the values deemed to be suitable for the indicators as well as the factors that could influence the respective performance. The calculation sheet also includes some predefined contextual factors. Table 2.7 shows the performance indicators that have been defined by the CCE. Figure 2.11 illustrates the CCE method for quality of service regulation (CCE, 2006).

Table 2.7 Performance indicators adopted by the CCE

Performance Indicators	
Drinking water quality	**Prices and management**
Compliance with national standards (%)	Operating cost (€ m^{-3})
Compliance with recommended standards (%)	Total cost (€ m^{-3})
Number of parameter tests (%)	Average bill (€ util.$^{-1}$)
Protection of abstractions	**Coverage and solidarity**
Volume abstracted from protected sources (%)	Average period for receipts (days)
Volume abstracted from licensed sources (%)	Debt customers (%)
Assets management	**Customer satisfaction/communications**
Volume of non-revenue water by mains length (m^3 day^{-1} km^{-1})	Technical delegations (no. 10^{-3} cust.$^{-1}$)
Volume of non-revenue water (m^3 year^{-1} meter^{-1})	
Network productivity (%)	
Network rehabilitation (%)	

In the Flanders region, the drinking water quality is regulated by EU legislation, transposed to the Belgian law and to the regional laws. In 2002, the region of Flanders published a law (*Betreffende de kwaliteit van het voor menselijke consumptie bestemde water*) establishing penalties for operators that do not fulfil the specified norms. The drinking water quality has remained stable over the years. Some water service operators work together so as to promote specific projects regarding the quality of service. However, there is no information about the quality of service of the whole sector and no procedures have been defined for its supervision. Until now, there has not been any comparison of the quality of service, at least not publicly, although the SVW (Association of Operators) has carried out, in the past, a benchmarking study entitled '*Samenwerking Vlaams Water*'. In fact, one of the main reasons behind the political decision to create a regulatory agency for the sector was precisely the need to have a body to regulate the WWS quality of service.

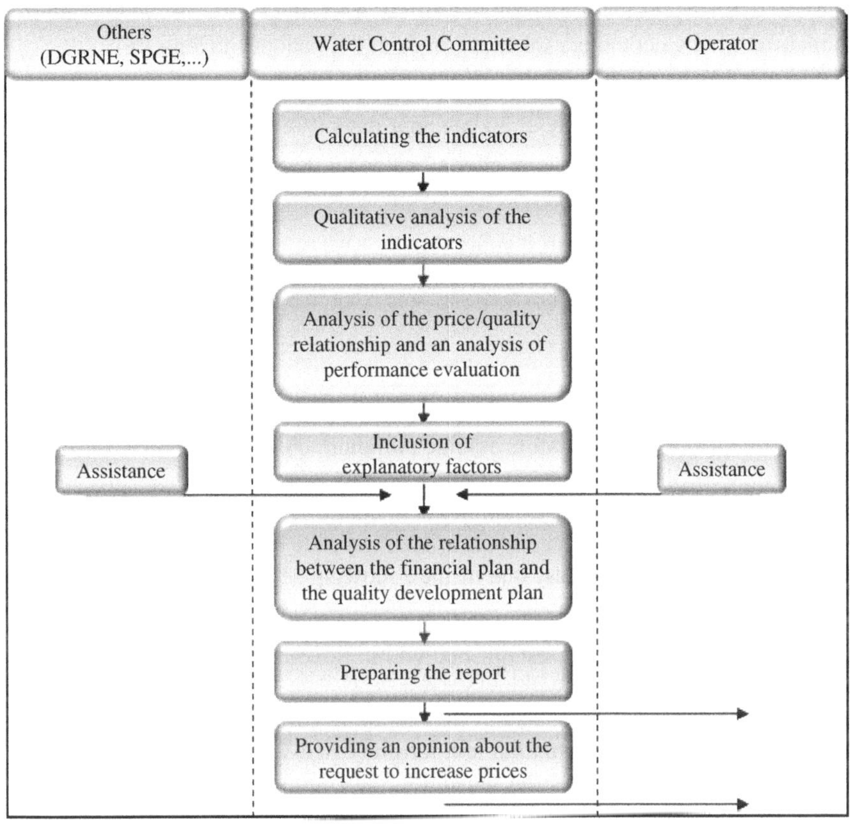

Figure 2.11 Regulatory process for the water sector in Wallonia

No information exists about the quality of service provided by the operator for the Brussels region, except for issues related to the drinking water quality defined by law and for the aspects mentioned in the *Charte de Service*, which is outlined below (Table 2.8). Hence, there are no official bodies responsible for this activity.

2.2.4.3 Public service obligations

Keeping in mind that there is no federal legislation in Belgium regarding public service obligations, the three regions recognize the essential nature of the WWS and have defined a minimum set of conditions for providing these services. Regional regulations exist in certain situations while in other cases the operators themselves – either individually or through their respective associations – voluntarily define these minimum requirements. For example, in the context of universal service, on the one hand funds have been stipulated for the purpose (Wallonia and Brussels) while, on the other hand, a certain quantity of water is provided free of charge (up to 15 m^3 in Flanders). In Belgium, the water sector is subsidized fairly heavily by the respective governments but not between similar or different kinds of customers, thus maintaining a high level of equality for customers.

In the Wallonia region, the majority of the public service obligations have historically been implemented in the WWS by means of norms. Examples of these are the obligations that were imposed by means of European legislation, such as the obligation to protect abstractions or safeguard the drinking water quality, or by the regional parliament of Wallonia, which established the conditions for the public water supply in the *Code de l'Eau* or defined the Social Fund (via a law dated 20 February 2003). The CCE has developed a set of indicators related to the legal obligations of each operator. The 13 indicators that were created (see Table 2.8) enable the CCE and the other stakeholders to analyse the operators' performance and ascertain whether these legal stipulations are being fulfilled.

Table 2.8 Indicators regarding legal obligations in the Wallonia region

Legal obligations
Connections established in conformance (pressure and flow rate)
Renovation of lead connections (–)
Abstractions which were subject to authorization (%)
Abstractions whose delimitation perimeters were accepted by the competent entity (%)
Abstractions whose perimeters were approved in the national gazette (%)
Meters which have not been read in the last 5 years (%)
Connections made after a delay of 30 calendar days (%)
Appointments to install connections after 10 calendar days (%)
Meters older than 16 years (%)
Percentage of meters older than 8 years (%)
Average age of the meters (years)
Water quality tests (%)
Routine water quality tests (%)

In the Flanders region, the majority of operators assume minimum obligations with regard to customers, which they commit to fulfil (GSS). Many of them also compensate customers when the services are not provided in the agreed manner. For example, the VMW has defined a minimum standard of service to be provided to customers and has committed to compensate them with a 25 euro compensation in case the minimum standard is not achieved (GSS). The VMW will compensate its customers if at least one of the following seven conditions is not fulfilled:

- If the quality of drinking water is changed, it will make alternative supplies of drinking water available;
- If a connection to the network is not re-established within three days;
- If a new connection is not established within 15 days;
- If individualized service is not provided within two hours, in cases of malfunctions;
- If all complaints and requests for information are not processed within 10 business days;
- If the operators' employees do not fulfil the schedules and commitments made to customers;
- If customers are not given at least one week's notice in cases of planned interruptions.

In the Brussels region, the IBDE has defined standard measures, listed in its 'service charter', with a view to defending the interests of customers, although no financial compensation is envisaged if they are not upheld. Table 2.9 provides a summary of the IBDE's commitments.

2.2.5 Performance

The water sector in Belgium is characterized by complexity and diversity amongst the three regions. There is little information about the sector, especially for the Flanders region, where there are no elements concerning the performance of this sector. The form of delegation to public companies and their respective functioning is also not very transparent. Belgium has a dearth of water resources and while it has a low water abstraction and a high level of environmental awareness, the tariff system does not transmit a suitable message to the market, encouraging consumption and misuse. The WWS are not very fragmented, but show potential gains of economies of scale and economies of scope.

In any case, the sector has an interesting price/quality ratio and great care is taken with regard to customer protection and fulfilment of public service obligations. In the region of Wallonia, the regulator will prove to be a valuable asset if it becomes effective, as will that of Flanders, if it becomes operational. Finally, it must be mentioned that the wastewater treatment is still poor. There is presently (2007) a very low level of coverage, which is not common in developed countries. In the future the private sector might be able to play a role in this area.

Table 2.9 Commitments assumed by the IBDE

Principle	Operator commitment
Access to information	Access to complete, clear, accurate and updated information.
Respect for private lives	Respect the private lives of customers.
Compulsory to carry identification	Each agent must present identification when interacting with the public.
Respect for commitments	Technical assistance, at households as well as at the relative central office.
Response time	Respond to complaints/requests within 10 days.
Quality of service/complaints	Immediately contact the individual making the complaint when the matter concerns the quality of water. Show the results of the samples to the respective customers within 7 working days.
Billing	Send a clear and legible bill to the customers.
Payment plans	Propose payment plans if necessary. Make available a permanent social fund to the Public Social Action Centre for individuals with specific needs (e.g. disabled individuals) and those who are experiencing financial difficulties, subject to certain conditions.
Unspecified administrative services in bills	No charge for duplicate bills if requested nor for visits by employees to carry out meter readings.
Subscriptions	Cancel the contract if requested to do so by registered letter. Make any change in customers on the same day at no additional cost.
Permanent service	Provide a customer assistance service 24 hours per day. Work over regular business hours if necessary.
Duration of interventions	In case of normal atmospheric conditions, intervene in less than two hours and start repairs in 48 hours. In case of system malfunctions and if the interruption of the water supply affects at least ten customers, provide emergency alternatives. Apart from this, the service should be re-established within 24 hours in case of central malfunctions.
Requests for connections	Establish a connection within 15 days, from the date of the request.
Water supply interruptions	No interruptions in the service allowed for more than 24 hours. Notify customers at least one day before a planned interruption.
Guarantee of disconnection	Guarantee that a connection which has been closed by an agent will be permanently disconnected.
Lead eradication	Eradicate lead connections by 2013.

2.3 BRAZIL

2.3.1 Historical background

Brazil is a federative republic organized into five geographical regions and formed by the union of 26 federated states which are divided into 5,564 municipalities and one Federal District (Brasilia). The country has an area of 8,514,877 km^2 and a population of about 190 million inhabitants. In Brazil executive power is exercised by the President, elected for a period of four years, who simultaneously holds the functions of Head of State and head of the government. Concomitantly to the presidential elections there are elections for the National Congress, the seat of legislative power, which is divided into two parliamentary houses: the Chamber of Deputies, elected for a four-year term, and the Federal Senate, elected in one- and two-thirds, alternating every four years and whose members serve for an eight-year term.

The availability of water resources in Brazil varies from region to region. The Northern Region (Amazonia biome) is rich in water resources but there are areas of relative scarcity such as in the Semi-Arid Northeast Region (Caatinga biome). The availability of water plays a significant role in the economic and social development of each region. The population increase and economic growth that took place during the twentieth century, especially after the Second World War, combined with foreign investment, resulted in an unbridled exploitation of natural resources in general and endangered the quality of water resources, in particular in the south and southeast regions. In this context, the first normative documents related to environmental supervision were approved during the 1980s and private industry was persuaded to treat their effluents. However, control of domestic effluents remained outside the scope of this policy and proved to be difficult to implement since it was a task under the responsibility of the public sector. In their turn, the 1990s were characterized by enormous challenges in the water sector, *inter alia*, defining the institutional aspects of the management of water resources, controlling water resources in large Brazilian metropolises, preserving the environment, the use and control of rural soil and mitigating the impact of diffuse pollution. A rational vision was adopted, which envisaged the use of natural resources coupled with environmental preservation.

Bearing in mind the continental scenario where Brazil is included, the legislation concerning the use, management and regulation of water resources and respective services adopted by the country can be considered adequate and robust. Ever since the institution of the Water Code, by means of Decree No. 24.643 of 1934, which is still in effect, Brazilian legislation has undergone significant changes. In particular, after the promulgation of the Federal Constitution of 1988, with a new separation of powers between the federated entities, it was established that the Brazilian Union would be responsible for defining the general directives and principles that govern the policy of water resources and for establishing norms, by means of federal legislation, for the organization of systems to control the use of water resources. On the other hand, the states are responsible for organizing the water resources management system in their specific territory, such as water courses that start and end within the territory of the same state and the groundwater sources.

The National Water Resources Management System (SNGRH) was created in 1993, by means of Law No. 9433 and was derived from Art. 21 of the Federal Constitution. Article 21 determined that it was the responsibility of the Brazilian Union to implement a national system to manage water resources and define the criteria for attributing the rights for their use. The SNGRH is based on the following fundamental principles: water is a public resource; water is a limited natural resource, with an economic value; human consumption and cattle are to be given priority in terms of using water resources in situations of scarcity; the management of water resources must always allow the multiple use of water; the territorial unit for the implementation of the National Policy for Water Resources and for the SNGRH is the river basin; the management of water resources must be decentralized and include the participation of the public authorities, customers and communities.

The National Policy for Water Resources implemented through the Law No. 9433/97 establishes that the executive authorities of the states and the Federal District should attribute the rights to use water resources and regulate and monitor their use. They should also carry out the technical monitoring of water supply, implement and manage the Water Resources Information System, and promote the integration of water resource and environmental management. Apart from this, while implementing this national policy, the municipalities (and federal district authorities) must promote the integration of local policies for the WWS, the use, occupation and conservation of the soil and the environment, as well as federal and state policies regarding water resources.

Law No. 9984 was approved in 2000. It defined the functions of the National Council for Water Resources, which included finding a balance between different users (consumers) and the national, regional and state plans, prepared by the entities that comprise the SNGRH and formulating the National Policy for Water Resources. The National Water Agency (ANA) was created at the same time and is a corporation linked to the Ministry for the Environment with the aim of implementing the National Policy for Water Resources. Each state (and the Federal District) approved its own legislation concerning water resources or updated existing legislation in this area after the enactment of Law No. 9433/97.

As the first legal document aimed specifically at the water sector, the Water Code determined that any water source or course could be used free of charge. Consequently, there is no private water in Brazil. Even the use of water sources located within the boundaries of a private property, as well as rivers that serve as boundaries between two private properties, must be subordinated to the public interest.

Until recently, in the WWS, the main measure of the federal government had been the establishment of the National Sanitation Plan (PLANASA), created in 1978. The PLANASA was supported by the Financial System for Sanitation managed by the National Housing Bank (BNH), which was an entity that administered the resources generated by the Time of Service Guarantee Fund (FGTS). FGTS was constituted by resources collected by companies corresponding to 8% of monthly salaries paid. In terms of its operational structure, the PLANASA stipulated that WWS should no longer be under municipal responsibility, transferring this competence to the respective states. Each state of the federation should create its own Water and Wastewater State Company (CESB). To this end the CESBs needed to obtain a concession from the municipal authorities of the respective state, so that they could act in the municipalities as a monopoly. The PLANASA further established that each state should create, with specifically allocated budgetary resources, a WWS funding (Fundo de Financiamento para Águas e Esgotos – FAE) and invest at least 50% of the overall amount of its resources. The remaining 50% were made available through the BNH as well as by means of external loans, mainly from the Inter-American Development Bank (IADB) and the World Bank, in addition to the financial resources of the federal government. Significant resources were invested this way, for a total of about 5 billion euros between 1971 and 1991. In 1991 the FAE came to an end, but it allowed increasing the levels of service, mainly in urban areas. It incorporated over 66 million inhabitants into the water service and 43 million inhabitants into the wastewater service. Various plans were subsequently implemented, such as the Pro-Sanitation Programme, the Pro-Hygiene Programme, the Pro-Community Programme, the FCP/PAN Programme, the Rural Sanitation Programme, the 'Sanitation is Life' Programme and the 'Sanitation for All' Programme. These programmes, which tried to achieve specific objectives, were supported by both governmental entities and Brazilian and international funding entities such as the World Bank. The results were generally positive and reflected a significant increase in the access to WWS by the groups covered by the respective programmes.

In 2001, the legislative proposal No. 4147 sought to enable greater access by the private sector to the WWS. This legislative project envisaged the states as concession granting authorities in metropolitan areas. In the meantime, the constitutional reforms of 1988 attributed the right to establish concessions for public services of local interest to the municipal authorities. Owing to this controversy about the interpretation of the Constitution, this legislative proposal never became a law. The question about WWS ownership is currently being judged in the High Court.

In 2005, a legislative proposal (No. 5296) tried to establish a new regulatory landmark for the water sector in Brazil. The Parliamentary Commission approved this project in July 2006, after many changes, and the proposal was transformed into Law No. 11,445, in January 2007. This normative document establishes the national directives for the WWS, altering existing legislation. Since it is very recent and the WWS ownership has not yet been defined, the implications of this law are still to be determined. The objective of Law No. 11,445 is to increase investments aimed at improving universal access to WWS, while considering the specific local circumstances and the use of appropriate technology, combined with customers' affordability. The law establishes the role of the regulators and also intends to increase transparency and accountability in the sector. Moreover, it seeks to harmonize the operational rules and relations between the different stakeholders. However, this law has still left some issues unsolved, namely the responsibility for providing WWS in large metropolitan centres, where some municipal authorities have challenged the legitimacy of state companies operating WWS.

Law No. 11,079, dating from 2004, has played a significant role in the present organization of the sector, instituting general norms to establish PPPs within the scope of the public administration. This has also been replicated by state laws, thus disciplining the participation of private initiative in projects for public services. However, it must be noted that this law only includes the concession contracts where the investment is carried out by the private partner. All the other PPP (e.g. public service concession) are regulated by the Law No. 8987/95 of 13 February.

Likewise in 2007, the President announced the new Growth Acceleration Programme (PAC), which included significant investments in roads, airports, ports, energy and housing as well as in the WWS, thus benefiting the poor sections of society. The programme allocated a total of 159 billion euros for investments up to 2010, of which 139 billion euros will be supplied by state companies and by the private sector while the remaining will be obtained from the federal government.

Apart from these federal laws instituted by Brasília, the state governments can also establish their own laws concerning water resources management. Currently, the majority of the states have adopted their own policy for water resources.

2.3.2 Legal and institutional framework

2.3.2.1 Responsibilities

In Brazil, although there are entities responsible for almost all the activities associated with the management of water resources, the relationship between the different levels of government is still fragile and at an early stage, owing to structural and circumstantial factors largely derived from the political and administrative organization of the country. The absence of coordinated actions to manage water can be attributed to the limitations inherent to running a sector in which three administrative spheres with decision-making powers co-exist (federation, states and municipalities), along with factors such as the historical absence of political attention given to the water sector and the phase of institutional changes that the country is currently undergoing.

The water sector is, above all, regulated by the Ministry for the Environment (*Ministério do Ambiente*), the Ministry for National Integration (*Ministério da Integração Nacional*), the Ministry for the Cities (*Ministério das Cidades*) and the Ministry of Health (Ministério da Saúde). With regard to the Ministry for the Cities, it is possible to highlight the role of the National Water and Wastewater Information System (*Sistema Nacional de Informação sobre Saneamento –*

SNIS), which is part of the Modernization Programme for the Water and Wastewater Sector (*Programa de Modernização do Sector de Saneamento – PMSS*). Despite having lost some of its former prominence to the detriment of the state authorities, the ANA still plays an important role in the water sector. Associations in this sector, such as the Brazilian Association of Water Supply, Wastewater and Waste (ABES), the Brazilian Association of Water Resources (ABRH), the Brazilian Association of State-Owned Companies (AESBE), the Brazilian Association of Concessionaires (ABCON), the National Association of Municipal Water and Wastewater Service Providers (*Associação Nacional dos Servidores Municipais de Saneamento – ASSEMAE*) and the Brazilian Association of Regulatory Agencies (ABAR), also have important functions in defending the interests of their respective members. Finally, perhaps the most important bodies for the defence of stakeholders' interests are the state entities that regulate the WWS at different levels of power, as it will be described next.

At a national level, the Ministry for the Cities coordinates the sector policies for the WWS implemented by various ministries. For example, the Ministry of Health has diverse functions related to the water sector, namely with regard to the quality of water, while the Ministry for National Integration, the Ministry of Agriculture (*Ministério da Agricultura*) and the Ministry for Agrarian Reform (*Ministério da Reforma Agrária*) perform functions in rural areas. At a federal level, the management of water resources has been entrusted to the ANA, which is overseen by the Ministry for the Environment.

Federal Law No. 9984, dated 17 July 2000, established that the ANA is responsible for encouraging and supporting initiatives aimed at creating bodies to manage water resources, i.e. River Basin Committees and, whenever necessary, Basin Agencies. The ANA has sought to fulfil these responsibilities in a comprehensive and transparent way. In short, the ANA's mission is to regulate the use of water from rivers and lakes which fall within federal territory and to implement the SNGRH, ensuring that these resources are used sustainably, preventing pollution and waste and guaranteeing good water quality in sufficient quantities to promote the nation's development, both for the present and future generations.

The institutional mission of the Brazilian Institute for the Environment and Renewable Natural Resources (IBAMA), which is subordinated to the Ministry for the Environment, is to implement national policies concerning the environment. The basic functions of the IBAMA include, among others, monitoring and supervising the use of renewable natural resources, promoting the implementation of conservation units, monitoring the environment and protecting and preserving ecosystems. The state superintendencies are an integral part of the basic structure of this institution, thus ensuring that the IBAMA has a presence in all the units of the federation.

The SNIS is the largest database in the water sector in Brazil. It constitutes an initiative promoted by the Ministry for the Cities through the National Secretary of Environmental Sanitation (SNSA) and is administered within the scope of the PMSS. In its turn, the PMSS, the programme that created and administers the SNIS, has developed numerous actions in the area of WWS structuring, therefore contributing towards the development of the sector's management.

2.3.2.2 Regulation

In Brazil, Law No. 11,445 establishes that the WWS must be regulated. Although it does not specify the level of the government's involvement in these entities and even though the majority of the regulatory agencies date from before the publication of this law, it defines the following objectives for these bodies:

- To establish standards and norms so that suitable services are provided to the satisfaction of customers;
- To ensure that the conditions and targets that have been established are fulfilled;
- To prevent and suppress the abuse of economic power, guaranteeing the capacity of the bodies that make up the national system for safeguarding competition;
- To define tariffs that ensure both an economic and financial equilibrium of contracts as well as affordable tariffs, by means of mechanisms that promote the efficiency and effectiveness of the WWS and enable the social transfer of productivity gains.

The regulatory agency is responsible for publishing norms related to the technical, economic and social aspects of the services provided, which will encompass, at least, the following issues: a) quality standards and indicators for the services provided; b) operational and systems maintenance requirements; c) progressive targets for expansion and quality of service, along with the respective deadlines; d) tariff regime, structure and levels, as well as procedures and deadlines for establishing them; e) adjustments and revisions; f) measurements, bills and service charges; g) cost monitoring; h) evaluation of the efficiency and effectiveness of the services provided; i) accounting plan and information, audit and certification mechanisms; j) tariff and non-tariff subsidies; k) customer service standards and mechanisms for participation and information; and l) contingency and emergency measures, including restrictions of consumption.

In general, the regulatory agencies in Brazil carry out their functions independently and have administrative, financial and budgetary autonomy, even though they have to fulfil strict rules with regard to accountability. They must be governed in accordance with the principles of transparency, technical competence, swiftness and objectivity while making decisions.

In 1999, the regulatory agencies formed an association, ABAR, which has diverse and leading functions (Article 2 of its statutes) related both to the evolution of the regulated sectors and the interests of its members. At the end of 2007 the ABAR had 29 associated regulatory agencies at different government levels – national, state and municipal – encompassing various sectors such as energy, gas, transport, water and wastewater.

Table 2.10 provides an overview of the regulatory agencies in the water sector in Brazil and their respective area of operations. It must be noted that some of these regulators are state and municipal bodies while others are multisector agencies.

Table 2.10 Regulatory agencies in Brazil and their scope of operations

Regulator	State/region	Scope of operations
ARCE	Ceará	Electricity, gas, water, wastewater and transport
ARPE	Pernambuco	Electricity, gas, water, wastewater, transport, lotteries, non-profit-making institutes
AGR	Goiás	Electricity, gas, water, wastewater, transport and water resources
ADASA	Brasília (Federal District)	Water and wastewater
ATR	Tocantins	Electricity, water, wastewater and transport
ARSBAN	Natal (Municipal)	Water and wastewater
AGENERSA	Rio de Janeiro	Water, wastewater, electricity and gas
ARSAM	Amazonas	Water, wastewater and transport
AGERSA	Cachoeiro de Itapemirim	Water and wastewater
AGEPAN	Mato Grosso do Sul	Electricity, water, wastewater, gas and transport
ARSESP	São Paulo	Electricity, water, wastewater and gas
ARPB	Paraíba	Electricity, water, wastewater and gas
AMAE	Joinville (Municipal)	Water and wastewater
ARSAL	Alagoas	Electricity, water, wastewater, gas and transport
AGESC	Santa Catarina	Electricity, water, wastewater, gas and transport
AGEAC	Acre	Electricity, water, wastewater and telecommunications
AGER	Mato Grosso	Electricity, water, wastewater, gas, roadways and transport
AGERGS	Rio Grande do Sul	Electricity, water, wastewater, roadways, irrigation and passenger road transport
ARSAE	Mauá (Municipal)	Water and wastewater

In spite of the vast number of WWS that exist in Brazil, only a very limited number of concessionaires are regulated. This situation, which tends to be criticized, does not diminish the rather positive role of the majority of the regulatory agencies in Brazil such as the ARCE and the ARSBAN.

All the regulators have important functions to defend the interests of customers. Some already apply benchmarking schemes both to the quality of service and tariff regulation. In compliance with Law No. 11,445, regulatory processes have already begun to be adopted to set prices based on performance incentives (e.g. price caps). In terms of quality of service regulation, many of the regulators (e.g. ARPE, AGENERSA, ADASA and ARCE) establish, compare and publicize a set of performance indicators for the WWS. Table 2.11 presents the scope of operations of the regulatory agencies and the methodologies adopted. Several of these bodies have only begun to function recently and hence, to a certain extent, the regulatory processes have not yet been very clearly defined.

The different associations in the water sector mentioned above and some NGOs, such as SOS Mata Atlântica and the Brazilian Consumer Defence Institute, also have important regulatory functions, although by means of self-regulation and advertising. They simultaneously promote the dissemination of best practices among their members and discuss problems within civil society.

Table 2.11 Methods used in Brazil to ensure economic regulation and quality of service

Regulatory agency	Quality of service	Economic regulation
ARCE	Performance indicators	–
ARPE	Performance indicators	Rate-of-return
AGR	Performance indicators	Price cap
ADASA	Performance indicators	Return cap
ATR	–	–
ARSBAN	–	–
AGENERSA	Performance indicators	Rate-of-return
ARSAM	–	–
AGERSA	Performance indicators	Rate-of-return
AGEPAN	–	–
ARSESP	–	–
ARPB	Performance indicators	–
AMAE	Performance indicators	Rate-of-return
ARSAL	–	–
AGESC	–	–
AGEAC	–	–
AGER	–	–
AGERGS	Performance indicators	Rate-of-return
ARSAE	–	Rate-of-return

2.3.3 Market structure

2.3.3.1 Ownership and management of services

In Brazil the WWS are mainly provided by the public sector. In this context the private sector only provides these services to a very limited segment of the Brazilian population, mainly in rural areas. 25% of the population is supplied by private companies, of which 5% own the infrastructure assets. ABCON has argued that it will only be possible to make the heavy investments that the Brazilian WWS need if private sector participation increases.

In addition to the eternal debate that is also common to other countries regarding public versus private ownership and management, Brazil is characterized by another major controversy, i.e. the separation of powers between municipal and state authorities concerning WWS provision. The position of the diverse stakeholders depends on their members. The AESBE, for example, has argued that the WWS should be provided at a state level, owing to the greater scale of operations which would yield economies of scale, and to the possibility of cross-subsidization between larger and smaller municipalities. In contrast, the ASSEMAE favours municipal operations, since the WWS are essential services and operate in a local context.

There are almost 40 private companies in the water sector in Brazil. These companies are spread out all over the country, from the north in Manaus, in the state of Amazonas, and Novo Progresso, in the state of Pará, to the south, in Paraná and Santa Catarina, but also in the centre-west region, in Mato Grosso, Mato Grosso do Sul and Tocantins and the southeast in the states of Rio de Janeiro, São Paulo, Espírito Santo and Minas Gerais.

The private sector participation in the WWS includes not only the provision of water services in rural areas where, by means of consortia with other entities from different sectors, it has the capacity of self-sustaining the service, but also of problem solving in specific and strategic regions (e.g. the region of Búzios in Rio de Janeiro).

2.3.3.2 Players in the sector

In Brazil there are two types of authorities responsible for the WWS, namely the municipalities and the state. Although the municipalities prevail in terms of numerical strength, more than 80% of the Brazilian population is served by public state operators through concession contracts. Given the instability of the regulatory structure when these concession contracts were designed, they were not very complete in terms of objectives, tariff system and transparency concerning cross-subsidization.

It has been estimated that there are about 1,350 WWS in the water sector (Oliveira, 2008) and five different types of operators (organizational arrangements) co-exist in Brazil. The first arrangement consists of public companies (25), known as CESBs, which provide services at the state level and are governed by private law. The second arrangement comprises micro-regional companies, some of which are governed by private law while others are governed by public law, bringing together various municipalities (associations of municipalities), being about 20 altogether. The other three types of operators provide services at a local level (municipal). The first local arrangement consists of companies governed by private law (municipal and mixed companies). The second one is associated with local operators governed by public law (departments, secretaries and local authorities) and is the most numerous of all (encompassing about 1,200 operators). Finally, the third arrangement corresponds to private operators that might or might not have signed concession contracts with the public sector (almost 40 operators).

In the context of Brazilian operators, the Water and Wastewater Company of the State of São Paulo (SABESP) has relevance due to its size, comprising about 23 million inhabitants. The SABESP is a mixed economy company with open capital whose main shareholder is the government of the state of São Paulo. It presently operates in 367 municipalities of the state of São Paulo and is also responsible for planning, constructing and operating urban and industrial WWS.

2.3.3.3 Integration in the sector and services provided

Usually, in the main urban centres the WWS are provided by the same operator, which is generally responsible for the entire chain of production.

In other municipalities, wastewater services are relatively scarce and in areas where these services do exist they are usually provided by the operator responsible for managing the water service.

Although this is the most common scenario, there are occasional cases in which the WWS are provided by different operators (e.g. the operator in Itajai in Santa Catarina) and where wholesale water companies exist (e.g. SABESP).

2.3.3.4 Numbers for the sector

As one of the most populous countries in the world, Brazil annually abstracts a volume of water of about 15,400 million m^3 for drinking water supply. The main water source is groundwater, which accounts for about 68% of the total value.

In Brazil, the agricultural sector is the largest customer of water, accounting for 65% of consumption, while drinking water supply represents only about 16% of the total consumption of water. The industrial sector consumes approximately 14% of the total volume while the remaining 5% corresponds to the consumption of the livestock sector.

In 2005, the average per capita consumption in Brazil was about 145 litres per inhabitant per day.

The values for water losses are quite significant, ranging between 21% and 81%, with an average of about 40%. Values vary considerably in spatial terms as well. The highest figures are located in the northern region, with about 58%, and minimum values are recorded in the south, with about 30%.

In Brazil, in 2005, the water sector had about 200,000 employees.

In 2006, the level of the water service coverage in Brazil was nearly 81% of the population, while the level of the wastewater service coverage (collection) was about 42%. However, coverage was significantly greater in urban areas, where 84% of the Brazilian population lives. Here, the level of coverage was close to 93% for the water service and 48% for the wastewater service. Coverage in rural areas, where 16% of the Brazilian population lives, was far lower, that is, 18% for the water supply and only 13% for the wastewater service. Geographically, the level of coverage was lower in poor regions in northern Brazil and higher in the south. The wastewater systems tend to be separated and stormwater is under the responsibility of the municipal authorities. Table 2.12 shows the levels of WWS coverage for the year 2006.

Table 2.12 Level of WWS coverage in Brazil

Services	Urban population (84%)	Rural population (16%)	Total
Water	93.1%	18%	81%
Wastewater	48.3%	13%	42%

According to the data provided by operators to the SNIS, in 2005 the turnover of the Brazilian WWS was about 1,740 million euros, divided equally among the different activities provided. Figure 2.12 provides a comparison of WWS revenues and the respective costs of the CESBs.

It has been estimated that there are around 1,350 operators. Table 2.13 summarizes the organizational arrangements of Brazilian operators belonging to the SNIS, which covers more than 81% of the total population. Note, however, that less than 50% of the total number of operators are members of SNIS.

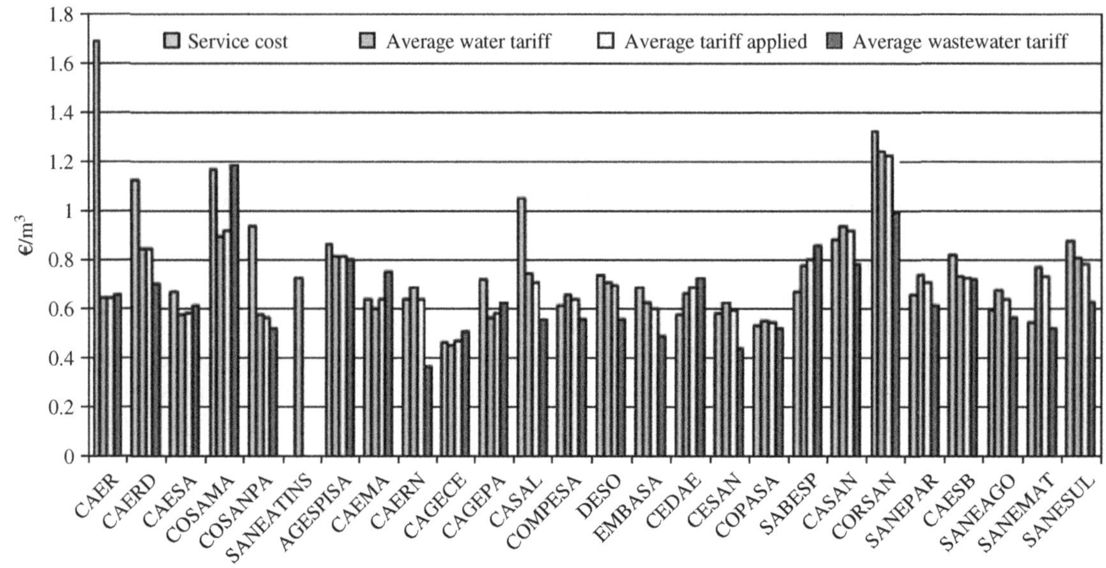

Figure 2.12 Comparison of the costs and tariffs of the CESBs

Table 2.13 Organizational arrangements for the WWS in Brazil

Operator	Regional	Micro-regional	Local	Total
Direct public administration	0	0	189	189
Municipality	1	4	310	315
Public company	0	0	10	10
Mixed commercial company with public administration	23	0	11	32
Mixed commercial company with private administration	1	0	0	1
Private company	1	3	39	43
Total	26	7	559	**592**

2.3.4 Rules of the game

2.3.4.1 Tariff setting

In Brazil, prices vary considerably between WWS and different regions. Like the vast majority of other operators, the CESBs maintain a homogeneous tariff structure. In Brazil tariffs are established by the CESBs and by municipal companies, although they follow the guidelines of the state governments and their majority shareholders, whenever the latter have private capital. However, considering the need to fulfil the principle of universal service, a system of cross-subsidization was instituted in tariff structures based on the assumption that customers with a higher purchasing power would subsidize those with less economic power, while bigger customers subsidize smaller customers. The following criteria were adopted to set tariffs (MPO-SEPURB, 1995):

- Differentiated tariffs should be established according to customer type (domestic, commercial, industrial and public) and the average tariff for commercial and industrial use should be higher than the average tariff of the concessionaire;
- The tariffs for domestic customers should be dependent on consumption blocks and should increase according to the volume consumed. In some states this differentiation is also used for the commercial and industrial sectors;
- The existence of a minimum volume for domestic consumption, comprising water consumption and the production of wastewater, by means of a monthly value of 10 m^3.

Law No. 11,445 identified the principles to be followed by the tariff systems. These principles were defined to ensure the economic and financial sustainability of operators, while keeping in mind the objective of a universal service. Measures have been introduced to protect low-income customers, such as the possibility to apply for social subsidies and cross-subsidization between activities and customers. The law also establishes the key regulatory functions in this sector and clearly indicates that one of the objectives of economic regulation – considering the essential nature of the service and its respective universality – is to provide incentives for economic efficiency.

In many Brazilian cities WWS tariffs are quite high, keeping in mind the context of Latin American countries and the respective burden in terms of costs for households. According to the SNIS, the average WWS tariff is 0.68 €/m^3, as can be seen in Table 2.14.

Table 2.14 Average WWS tariffs by type of operator in Brazil

Dimension	Tariff variation (€/m^3)
Regional	0.48–1.72
Micro-regional	0.26–1.39
Local	0.04–1.75

WWS charges are billed on a monthly basis. Fines are applicable when bills are not paid by the due date. If the delay in payment is greater than one month the operator might interrupt the WWS provided.

For example, the SABESP charges for the WWS in conformance with Decree No. 41,446, dating from 1996, which established the tariff regulations. This norm envisages a standard minimum consumption of 10 m^3 of water, and after that establishes progressive blocks based on variable consumption. Within the area of the SABESP's jurisdiction the tariffs vary according to the different systems. The water that is consumed is always measured, irrespective of the type of customer. The types of customers have been divided into domestic, social and slum and commercial, industrial and public consumptions. Consumption values have been designed for each type of customer, ranging between consumption up to 10 m^3, from 11 to 20 m^3, from 21 to 50 m^3 and above 50 m^3, except for the social and slum tariffs, which have five consumption blocks, i.e. up to 10 m^3, from 11 to 20 m^3, from 21 to 30 m^3, 31 to 50 m^3 and above 50 m^3. A domestic customer pays a tariff about two times higher than a social customer or the residents of slums. The creation of the social domestic tariff reflected the SABESP's concern for this intermediate section between ordinary customers and customers living in the slums. Thus, since 1997, the SABESP has benefited over 600,000 underprivileged customers, with payment options that consider factors such as income, size of the dwelling unit and consumption of electric energy. The public entities that have joined the Programme for the Rational Use of Water (PURA) and social assistance bodies also have differentiated tariffs.

A study developed in 2005 by ADERASA provided evidence that WWS bills in Brazil are about 1.5 times higher than the average of the 21 most important cities in Latin America. Moreover, an analysis of the SNIS data revealed that tariffs increased considerably during the period between 2002 and 2005. The average tariff for water services rose by about 57% and the tariffs for the wastewater services rose by about 54%. Figure 2.13 illustrates the evolution of the average WWS tariffs in Brazil (SNIS, 2006).

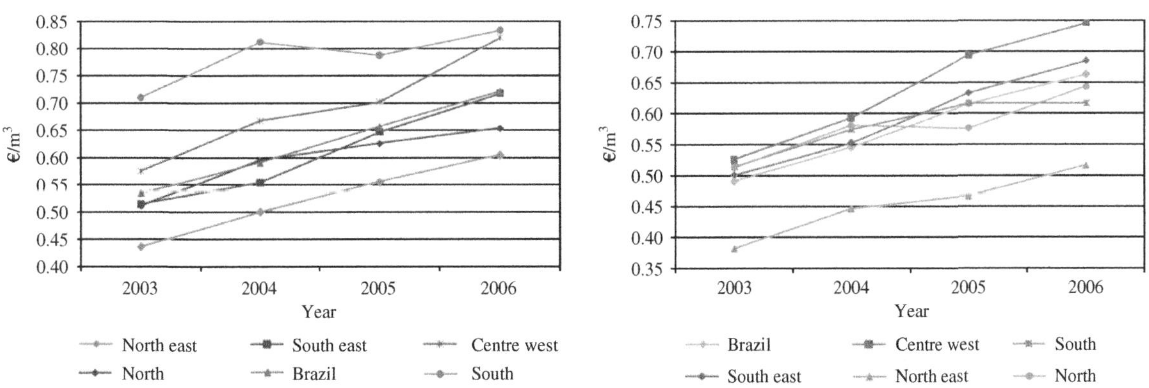

Figure 2.13 Annual evolution of the average WWS tariff in Brazil

2.3.4.2 Quality of service

The water sector in Brazil does not have a compulsory federal system of a predefined quality of service assessment, except with regard to the water quality. The SNIS has played a significant role in this area. The activities of the SNIS are developed by means of gathering information provided on a voluntary basis by operators which are given incentives to join this programme. If operators intend to be funded by public capital, they are compelled to supply the respective data to the SNIS. In accordance with the fundamental principles for the provision of public services, the implementation of Law No. 11,445 seeks to ensure '*the transparency of actions, based on information systems and institutionalised decision-making processes*'.

Despite being merely a body that collects and compiles information about the water sector, the SNIS plays an extremely important role in the water service progress. The regulatory agencies are responsible for preparing an annual report, thus publicizing the information that has been compiled. Such information makes it possible to calculate 85 standardized performance indicators, divided into economic, financial, administrative and operational classes, which seek, above all, to reflect the evolution and quality of services provided. Figure 2.14 presents two of these indicators: water losses and metering (SNIS, 2006).

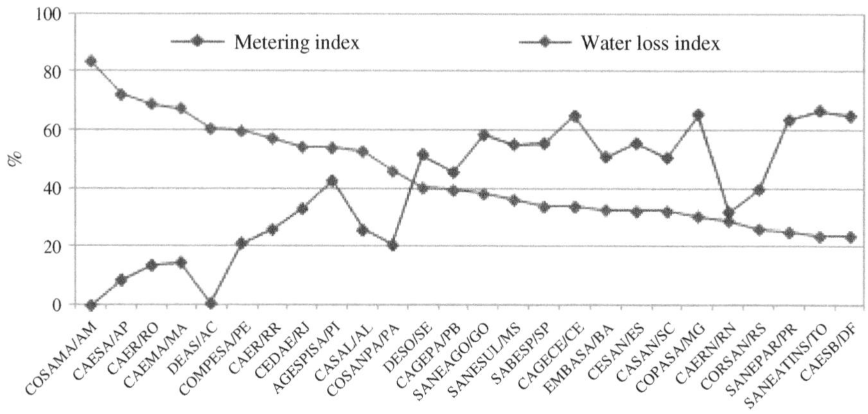

Figure 2.14 Metering and water losses indicators of the CESBs

Concerning the drinking water quality, operators must comply with the parameters established by Regulation No. 514, dated 2004, issued by the Ministry of Health, which establishes norms and standards for supplying water services. In this regard, the studies carried out by the SNIS have revealed that not all operators fulfil these minimum standards, which has caused some concern among the responsible entities.

The process of providing WWS to poor urban neighbourhoods continues to be a great challenge, which frequently requires the use of unconventional methods. Although Brazil has focused on using low-cost technology (individual systems for treating wastewater) as well as community participation in processes of urbanization, much of this process continues to be developed informally, with serious problems at the level of water quality and the discharge of wastewater.

In any case, the quality of service supervision is a question that has been assigned to the regulatory agencies. The majority of these operators are still in the phase of implementing their own systems to analyse the quality of service performance. ABAR recently released a publication that seeks to provide a uniform set of performance indicators to be calculated by the regulators (ABAR, 2006). Some of the regulatory agencies have already begun to use these tools. Table 2.15 illustrates the set of performance indicators proposed and applied by the ARCE. Figure 2.15 shows the evolution of the indicators for complaints and requests for information in the case of the ARCE.

Table 2.15 Performance indicators used by the ARCE

Performance indicators
Annual evolution of inspections
Compliance with the turbidity quality index
Compliance with the residual chlorine quality index
Compliance with the coliforms quality index
Metering
Number of meters installed
Water losses index
Level of WWS coverage
Evolution of requests made to the ARCE (complaints, requests for information, etc.)

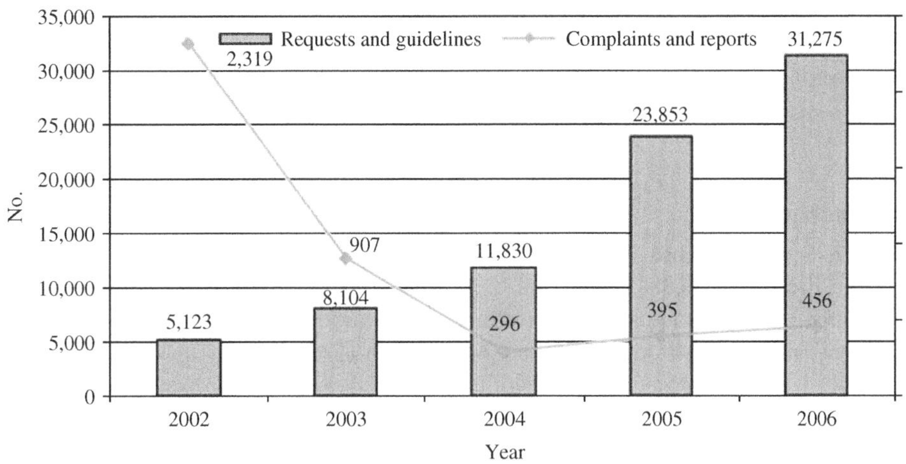

Figure 2.15 Indicators for complaints and requests for information (ARCE)

2.3.4.3 Public service obligations

Public service obligations have been unequivocally defined for the Brazilian water sector, particularly in the Law No. 11,445 mentioned above. The principles of universality and accessibility stand out among the fundamental principles for providing public WWS: '*Providing the population with access to the service according to the respective needs and maximising the effectiveness of actions and results.*' Other essential principles include safety, quality and regularity while providing services. This law also envisages the integration of the infrastructure and services and an efficient management of water resources. It must be emphasized that, according to the normative document, water resources are not considered to be an integral part of the WWS.

Legislation stipulates that the owners of the services must prepare master plans for the WWS, which must be revised periodically (at most every four years, before the preparation of a multiannual plan). These plans must encompass, at least, a diagnosis of the situation and the respective impact on the living conditions of the population (by means of a set of performance indicators); the causes of shortcomings which have been detected; short-, medium- and long-term goals and objectives to ensure a universal service (gradual and progressive solutions are allowed); the programmes, projects and actions necessary to achieve objectives and goals; the possible sources of funding; actions for emergency and contingency situations and the mechanisms and procedures to be adopted while systematically evaluating the effectiveness and efficiency of the scheduled actions. The law further emphasizes that the master plans should be compatible with the plans for the river basin where the owners of the WWS are located.

The regulatory approach for the water sector seems to be following the global trend to define suitable and increasingly effective and transparent mechanisms to evaluate the WWS quality. The next step consists of instituting, in practical terms, an efficient and effective regulatory system, which guarantees the adequate use of resources and an expansion of the levels of coverage and the quality of service, balancing the problem of the operators' sustainability with the issue of promoting universal access to services. It must also be mentioned that the normative document referred to clearly establishes the principles of access and free information in the water sector.

2.3.5 Performance

The water sector in Brazil faces various problems, some of which have been highlighted below. Firstly, Brazil has to deal with a shortage of water in various regions and with floods in other regions. Secondly, the existing conflicts between the different levels of the government regarding the management and attribution of these services seriously hinder the decision-making process and, consequently, the water sector. Thirdly, the extremely high commercial losses in the water sector have an impact on the self-financing capacity of operators, which is reflected in the tariffs charged. Fourthly, enormous levels of investment are necessary to provide the sector with new infrastructures, both for the water and for the wastewater services. Great investments are also required to rehabilitate the systems that are currently operational, which are characterized by a poor overall state translated into high water losses. It might be necessary to rely on private sector participation for the investment required. Finally, it is important to emphasize Brazil's limited credibility to attract private capital for the water sector.

In terms of positive aspects, the quality of the prevailing legislation is especially worthy of note, as well as the existence of qualified labour that characterizes the water sector. The regulatory agencies are also an added value and the efforts they are making will have a considerable impact in the future.

2.4 CHILE

2.4.1 Historical background

Chile is a democratic republic whose President is the head of executive power. Legislative power comprises a bicameral parliament, consisting of the Senate and the Chamber of Deputies. The country has about 16.6 million inhabitants over an area, excluding the Antarctic territory, of 756,950 km^2. In administrative terms, it is organized into 15 regions, 51 provinces and 346 municipalities. Each region is overseen by a superintendent, appointed by the President, and is divided into provinces, each of them headed by a governor who is also appointed by the President. Finally, each province is divided into municipalities individually administered by the respective mayor, which is democratically elected.

The water sector in Chile has evolved considerably, especially in recent years. During the early 1990s, Chile had some problems (e.g. reduced coverage level of wastewater treatment service) but in just ten years became a success story and an example in this sector throughout Latin America. Various international and independent organizations have corroborated this (e.g. the WHO).

During the late 1980s, in response to the growing need for investments to build new infrastructures and to improve the functioning of the existing ones (greater continuity and better quality of service), the Chilean government developed new legislation, by which the old WWS sector was transformed into various state-owned companies. The objective of this legislative measure was to promote a more efficient management of the services, to ensure equality and to draw a clear separation between service management and the state's regulatory activities. With the latter objective in mind, a sector-specific regulatory authority was created, the *Superintendencia de Servicios Sanitarios* (SISS).

In Chile, the water sector is regulated by means of diverse normative documents, including the Water Code, of 1981, which defines the rights to use water, the Law which creates and regulates the SISS (Law No. 18,902 of 1990), the Tariff Law (Decree No. 70 of 1988, by the Ministry of Public Works – MOP) and the respective rules (Decree No. 453 of 1988, by the Ministry of Economy), the General Law for Sanitary Services (Decree No. 382 of 1988, by the MOP) and its rules (Decree No. 1199 of 2004, by the MOP), the Law for WWS Subsidization (Law No. 18,778 of 1989) and the respective rules (Decree No. 195 of 1998).

The period between 1988 and 1990 witnessed some of the most intensive reforms in the history of the water sector in Chile. Various legal documents were approved and new institutions were created, which would have an enormous influence on the sector.

The Tariff Law established the normative basis for the tariff-setting procedures for the WWS, harmonizing the objectives of efficiency, self-financing and equity. Furthermore, this normative document also enabled the creation of the SISS (Law No. 18,902) along with the introduction of the social subsidy (Law No. 18,778) aimed at poor domestic customers.

The legislation that was implemented also made it possible to create public service companies, characterized by a high degree of transparency and autonomy, focused on the sustainability and profitability of their assets. Legislation was simultaneously established for the sector, separating the functions of the state as service provider from the state as regulatory entity. This scenario resulted in significant progress in terms of the commercial and operational management of the public companies, as well as a substantial increase in the quality of WWS.

Owing to concerns about the expansion of WWS coverage, improving the quality of the services provided and ensuring greater sustainability, the government carried out new reforms for the water sector in February 1998. The aspects covered by these reforms were modernizing the regulatory framework for the sector and promoting a greater involvement by the private sector. This included the almost total privatization of the WWS sector and the establishment of tariffs, development and business plans and targets to guarantee the quality of service. Moreover, this law expanded the powers of the SISS, which became responsible for supervising the operators (development and business plans and quality of service levels), apart from implementing directives concerning the quality of service and substantially increasing the fines and penalties for operators that do not comply with prevailing norms.

This reform also introduced significant improvements in the tariff-setting processes, reducing the fixed component of the tariff system and establishing principles to prevent the concentration of ownership within a restricted group of operators. Finally, it likewise defined the percentage of the State participation in the ownership that could be transferred to the private sector (a maximum of 65%, or more, in case the State does not participate in capital increase).

In the period between 1998 and 2001, the private sector participation took place in two distinct ways, namely by selling the assets of public companies up to the maximum limit mentioned above (65%) and by means of concession contracts. From 2001 onwards, the government favoured the second model, inhibiting the institution of the first model.

Subsequently, in mid 2002, the SISS developed norms whose main objective was to enhance transparency in the regulated companies. From March 2003 onwards, WWS operators were compelled to provide specific, detailed and uniform information on their management, their revenues and the costs of diverse activities. Thus, the SISS was able to characterize an 'efficient company' in an objective and credible manner. This was a fundamental concept to determine the tariff system in the regulatory sphere in Chile.

2.4.2 Legal and institutional framework

2.4.2.1 Responsibilities

Various governmental entities are responsible for the WWS sector in Chile. These include the MOP, the Ministry of Health, the Ministry of Economy, SISS, the Directorate-General for Water (*Dirección General de Aguas* – DGA), the National Directorate for Hydraulic Works (*Dirección Nacional de Obras Hidráulicas* – DOH), the National Commission for the Environment (*Comisión Nacional del Medio Ambiente* – CONAMA) and its Regional Departments (*Comisión Regionales del Medio Ambiente* – COREMAS), public and private companies and, finally, the Corporation to Promote Production (*Corporación de Fomento de la Producción* – CORFO).

The main function of MOP is to provide the nation with infrastructure services. It is responsible for constructing public works for rural services, including those associated with water resources and hydraulic infrastructures.

The responsibility for regulation in urban areas has been jointly attributed to the SISS (economic regulation) and the Ministry of Health (in terms of drinking water quality). Apart from this, along with the Ministry of Economy, the Ministry of Health is also in charge of regulating and supervising WWS operators in rural areas (cooperatives and committees). Additionally, the Ministry of Economy approves the tariff systems proposed by the SISS.

As aforementioned, SISS supervises and regulates the WWS sector in urban areas. In order to ensure its political independence, it was created as a decentralized institution, with its own patrimony and financial and administrative autonomy. In addition to the respective tasks of regulating prices, the quality of service and public service obligations, the SISS is responsible for imposing fines and penalties upon companies when they violate prevailing norms. SISS is also entrusted with the task of handling customer complaints.

Although no public authority exists for rural areas, as with the SISS for urban areas, the DOH, supervised by the MOP, has been responsible, since 1994, for implementing the National Rural Programme for Drinking Water Supply in rural areas and has regulatory functions in this scope.

Reporting to the MOP for all its activities, the DGA is responsible for planning, developing and managing water resources at a national level.

The environmental regulation of all sectors is carried out by the Secretary-General for the Presidency, through the CONAMA. The CONAMA is a state institution whose mission is to promote the environmental sustainability of the policies and strategies defined by the government. The CONAMA performs its activities by means of the COREMAS, whose functions include fostering consultations and coordinating the environmental aspects.

The CORFO was created in 1939 and is a Chilean state entity entrusted with the task of promoting the productivity and progress in diverse sectors, including the WWS sector.

Unlike the development that characterizes the WWS in urban areas, these services have still not been consolidated in rural areas. In this context, there is a public programme in Chile known as Rural Drinking Water Supply (Agua Potable Rural – APR), whose main objective is to provide technical, financial and legal assistance to operators in rural areas, namely the cooperatives and the committees. Figure 2.16 shows the main entities with responsibilities in the water sector in Chile (adapted from ADB, 2001).

2.4.2.2 Regulation

According to the present regulatory model, WWS operators are liable to SISS action. As already mentioned, this sector-specific national authority was created by Law No. 18,902, in 1990, and is only responsible for urban areas. SISS is an autonomous governmental agency that depends on the President, through the MOP, to whom it reports about all its activities. SISS was instituted as part of the process of legal, economic and institutional reforms for the WWS sector, which was begun at the end of 1980. The role of Superintendent, who is appointed by decision issued by the President of Chile, based on a previous public contest, is the most important in the SISS.

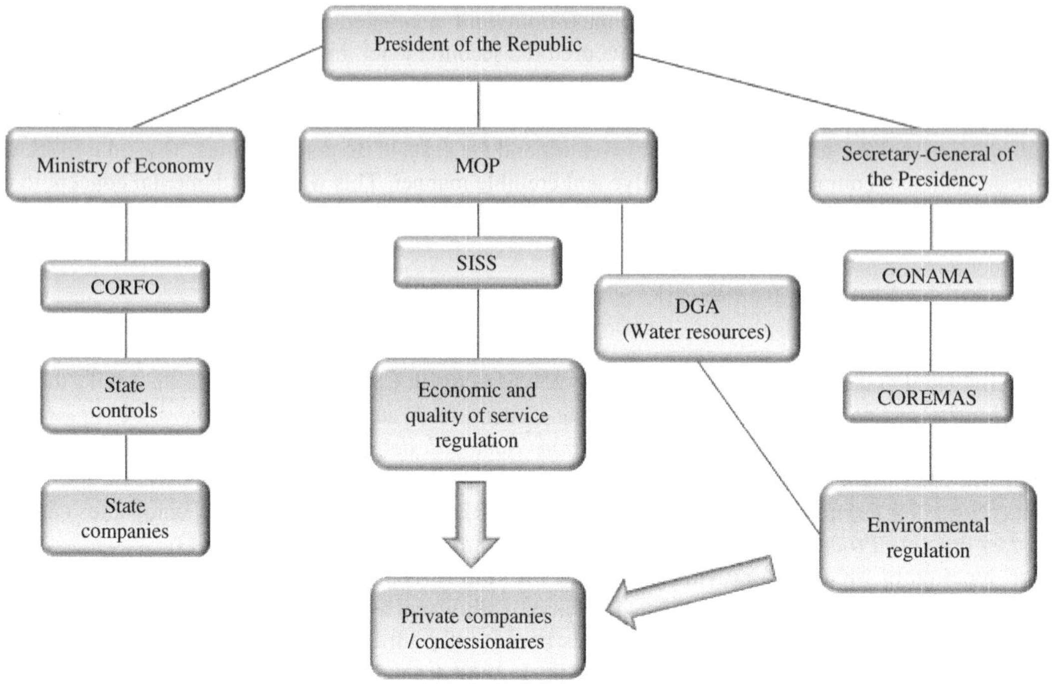

Figure 2.16 Institutional framework for the WWS in Chile

SISS is legally responsible for monitoring and supervising the water sector. The agency has normative responsibilities, which include reinforcing and ensuring compliance with the legislation associated with the WWS provision. It also plays a leading role in determining tariff systems. More specifically, SISS is responsible for the following: i) studying, proposing and reinforcing the fulfilment of technical norms concerning conception, construction and operation of WWS; ii) monitoring and reinforcing the tariff norms for the services being provided; iii) applying a BOT and other contracts regime, including awarding contracts, operations, transfers and the termination of contracts; iv) defining norms and controlling industrial effluents; and, finally, v) applying penalties and other types of sanctions to those which do not fulfil the legislation.

As a regulatory authority, the SISS issues instructions compulsory for all WWS operators and ensures compliance with the legal, regulatory and technical norms that rule different aspects of the services. The resolutions implemented by the SISS can be contested, not just before the SISS itself but also before Chilean courts of justice.

The SISS makes recommendations to the MOP with regard to the water sector. PPPs generally encompass the establishment, construction and operation of the water supply systems, from production to distribution, and the wastewater infrastructures, including collection, drainage, transport, treatment and discharge. These rights can be transferred, albeit conditioned by the competition, and are subject to the approval of the SISS.

The system of economic regulation based on the 'model' or 'efficient' company has been adopted in all sectors in Chile, which include public utility services. This method entails a comparison with a virtual, efficient company known as the 'model company', which is deemed to be the benchmark. Indeed, it is a fictitious and merely theoretical company created by the regulator for the sector with the aim of satisfying the projected demand in an optimal manner.

The SISS is overseen by the MOP and only has jurisdiction over the WWS sector and in urban areas. While performing its functions, the regulatory agency reported a sum of 7.04 million euros for its operating costs in 2007. At the end of that year it directly employed 157 people.

2.4.3 Market structure

2.4.3.1 Ownership and management of services

In Chile, WWS are almost totally provided by the private sector, although the public partner (CORFO) is present in the main operators. In any case, there are two main models of privatization in Chile. Up to 2001, there was above all, a full divesture of the operators by means of their assets sale. From 2001 onwards the model followed was that of concessions (privatization of management). Although the first model comprises only four companies, these entities are extremely large and correspond to 67% of the urban population (ESVAL, Aguas Andinas, ESSAL and ESSBIO). Private participation is always a majority holding.

With regard to the WWS in rural areas, there is a mixed scenario where both private and public operators co-exist. Although the respective proportions have not been identified, the public sector has a slight supremacy.

2.4.3.2 Players in the sector

In December 2007, the WWS for the urban areas were provided by 53 companies, comprising 52 operators dedicated to providing the water service and 50 operators which provide wastewater services. The 20 largest companies provide WWS to about 13.5 million inhabitants (approximately 4 million connections), where Aguas Andinas (Santiago) gains prominence due to its size, as shown in Table 2.16.

Table 2.16 Largest WWS operators in Chile

Company	Region	Customers (no.)	Localities (no.)
Aguas Andinas	Metropolitan	1,428,028	18
ESSBIO	Regions 6 and 8	610,842	87
ESVAL	Region 5	500,841	50
Aguas Nuevo Sur Maule	Region 7	195,011	30
SMPA	Metropolitan	175,848	3
Aguas Araucanía	Region 9	179,836	35
ESSAL	Region 10	167,429	33
Aguas del Valle	Region 4	165,780	21
Aguas Antofagas	Region 2	129,691	6
Aguas del Altiplano	Region 10	125,154	9
Aguas Cordillera	Metropolitan	113,253	1
Aguas Chañar	Region 3	73,517	10
Aguas Magallanes	Region 12	44,222	3
Aguas Décima	Region 10	36,608	1
Aguas Patagonia	Region 14	22,521	8
Servicomunal	Metropolitan	17,957	1
Servilampa	Metropolitan	4,308	1
Aguas Manquehue	Metropolitan	5,487	4
Aguas los Dominicos	Metropolitan	3,357	1
Coopagua	Region 5	3,273	1
TOTAL		**4,002,963**	**323**

With the aim of preventing a concentration of power in the market (and the possible abuse of a dominant position) at a national level, companies have been classified into three categories according to the total urban population to which they provide services. No individual or company should owned more than 49% of the companies within a specific category. Table 2.17 presents the different categories in the water sector.

The largest companies in the water sector in Chile are Aguas Andinas, which supplies 34% of the market, operating in the capital Santiago, whose majority shareholder (50.1%) is the Spanish company Aguas de Barcelona (and subsequently the French company Suez), and the *Empresa de Servicios Sanitarios del Bio-Bío* (ESSBIO), which operates in the sixth and eighth region around Concepción. The majority stake in this operator (50.97%) is held by the Argentinean fund Southern Cross. Finally, owing to its size, the *Empresa Sanitária de Valparaíso* (ESVAL) is also worthy of note. ESVAL operates in the region of Valparaíso and 47.92% of its shares are owned by the Hurtado Vicuña/Fernández León Group. These three companies are so important that together they supply water services to about 63% of Chilean customers.

Table 2.17 Categories based on the size of WWS companies in Chile

Category	Criteria	Companies (no.)	Urban population served
Large companies	15% of customers or more	2	50.5%
Medium companies	Between 4% and 15% of customers	6	34.3%
Small companies	Less than 4% of customers	45	15.2%

Among the minor companies, Aguas de Antofagasta, Aguas del Altiplano, Aguas Cordillera, Aguas Chañar, Aguas Magallanes, Aguas Décima, Aguas Patagonia de Aysén, Servicomunal, Aguas Manquehue, Servilampa, Aguas Los Dominicos, Coopagua can be highlighted and there are also 33 other smaller companies.

Although nearly all companies were privatized between 1998 and 2004, the Chilean state has maintained an important position in the water sector by means of CORFO. Table 2.18 below summarizes the main players in terms of operators in the water sector in Chile as of 2006 and their scope of operations (SISS, 2007a).

Table 2.18 Main corporate groups in the water sector in Chile

Economic group	Customers (%)	Country of origin
CORFO	35.7	Chile
Aguas de Barcelona	17.0	Spain
Southern Cross Group	10.7	Argentina
Grupo Hurtado Vicuña/Fernández León	8.8	Chile
Grupo Luksic	5.3	Chile
Grupo Solari	5.0	Chile
Municipality of Maipú	2.3	Chile
Iberdrola	2.0	Spain
Others	13.2	

2.4.3.3 Integration of the sector and services provided

In rural areas, rural cooperatives and committees for drinking water supply have been entrusted with the task of operating water supply systems. In rural neighbourhoods with a population density between 150 and 3,000 inhabitants and more than 15 households per kilometre of network, the National Rural Drinking Water Supply Programme has contributed towards achieving significant progress in terms of WWS provision. However, the small percentage of isolated dwellings that exist in Chile still lack the water service. Although a limited number of places have found solutions concerning the system of wastewater collection, most of them have numerous operational problems. Nevertheless, these small operators are mainly owned by the state.

In Chile, WWS are normally supplied by the same operator. In this context, the services are vertically integrated, i.e. the operators manage all the activities of the production chain, from the abstraction to the distribution of water. This is also the case with regard to wastewater services, i.e. services are integrated in this sector as well, including collection services and wastewater treatment. There is also a high degree of horizontal integration and only a limited number of operators are found for the entire urban population in Chile.

2.4.3.4 Numbers for the sector

In 2006, the production of water in Chile (excluding rural areas) reached about 1,451 million m^3, while consumption reached 958.3 million m^3, reflecting an increase of 2.7% as compared to 2005. Of this consumption, 54.7% is supplied by large companies, 27.5% by a set of medium companies and, finally, the remaining 17.8% are supplied by small companies. Of the total water abstracted, 51% is surface water and 49% is groundwater. In Chile, in 2006, the water sector recorded water losses of 37.6%. Figure 2.17 illustrates the annual evolution of water losses amongst regulated operators in the water sector in Chile (SISS, 2007b).

In Chile, the monthly per capita consumption of water diminished slightly between 2005 and 2006, reducing from 20 m^3 to 19.9 m^3 per customer per month, which amounts to 196 litres per inhabitant per day. The metropolitan region of Santiago has the highest monthly average of domestic consumption in the country, ranging from 125 m^3 in the case of Aguas de Manquehue, 112 m^3 in the case of Aguas los Dominicos to 44 m^3 in the case of Aguas Cordillera. In their turn, the medium-sized companies Aguas Nuevo Sur Maule, Aguas del Valle, Esval, Aguas Araucanía and ESSAL recorded average monthly consumptions of about 15 m^3 per customer.

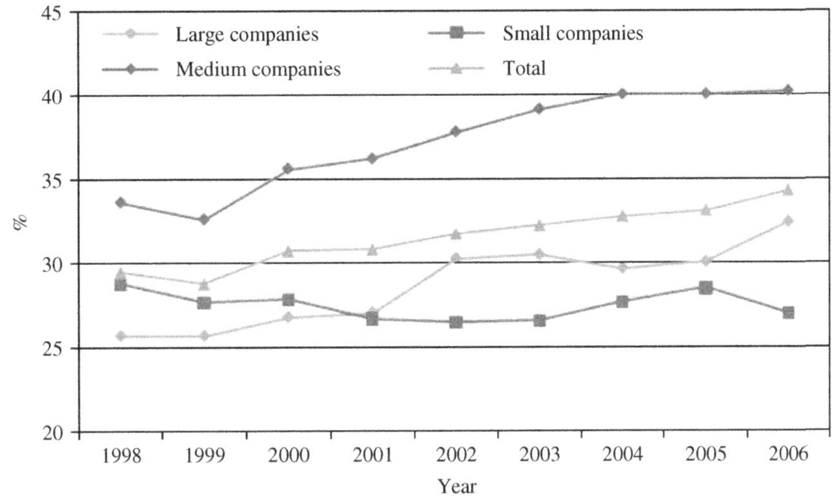

Figure 2.17 Annual evolution of the indicator water losses in Chile

According to the SISS, the level of water service coverage in urban areas at a national level is about 99.8%, which implies that 13.5 million inhabitants have access to the drinking water supply. With regard to the wastewater service, it covers about 12.9 million inhabitants corresponding to a level of coverage of 95.2%. Table 2.19 shows the levels of WWS coverage in urban and rural areas.

Table 2.19 Level of WWS coverage in 2006

Service	Coverage	Urban (87%)	Rural (13%)	Total
Water	Existence of network	100%	58%	95%
	Connection to network	99%	38%	91%
Wastewater	Existence of network	95%	62%	91%
	Connection to network	89%	5%	78%

The level of wastewater treatment service coverage has evolved considerably, having expanded from 20.9% at the end of 2001 to 81.9% in December 2006. In 2006, the volume of wastewater collected represented about 522 million m^3. Figure 2.18 illustrates the annual evolution of the WWS coverage (SISS, 2007b).

In Chile, in 2006, the revenues from the water service corresponded to 441.4 million euros while from the wastewater service they amounted to 315 million euros. In the same year, 339 million euros were invested in the WWS sector. Chilean operators invested about 283 million euros in the wastewater treatment service.

Water consumption is almost always measured by meters. In December 2006 94.4% of customers of the larger companies, 97% of customers of the medium companies and 98.8% of customers supplied by small companies had metering. The national average is about 96%.

In early 2008, the 20 largest companies in the water sector, supplying about 13.5 million inhabitants, had 9,566 employees.

2.4.4 Rules of the game

2.4.4.1 Tariff setting

The legal milestone that established economic regulation includes the tariff law and the respective regulation. This normative document takes into consideration the principles of economic efficiency, dynamic efficiency, intelligibility, equity and self-financing. In this context, tariffs are approved by the government, through the Ministry of Economy, according to the studies promoted by SISS. The tariff system is determined on the basis of the efficient or model company concept.

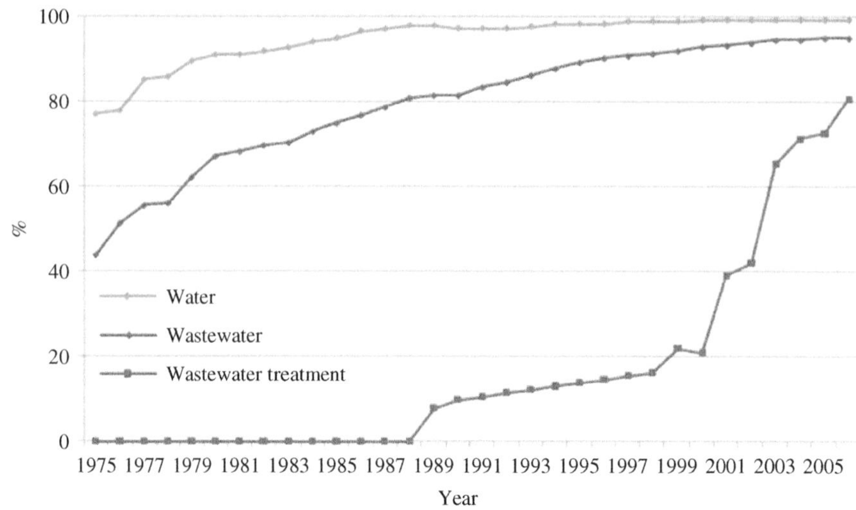

Figure 2.18 Annual evolution of the indicator level of WWS coverage in Chile

The efficient company model (articles 27 and 28 of the law), according to which tariffs are set, involves creating a hypothetical WWS company whose functions are performed in an efficient and optimized manner, considering prevailing norms and the geographical, demographic and technological restrictions that characterize the operation of the service (Chavez and Quiroga, 2002). Figure 2.19 shows the procedure for setting tariffs in Chile (Source: SISS).

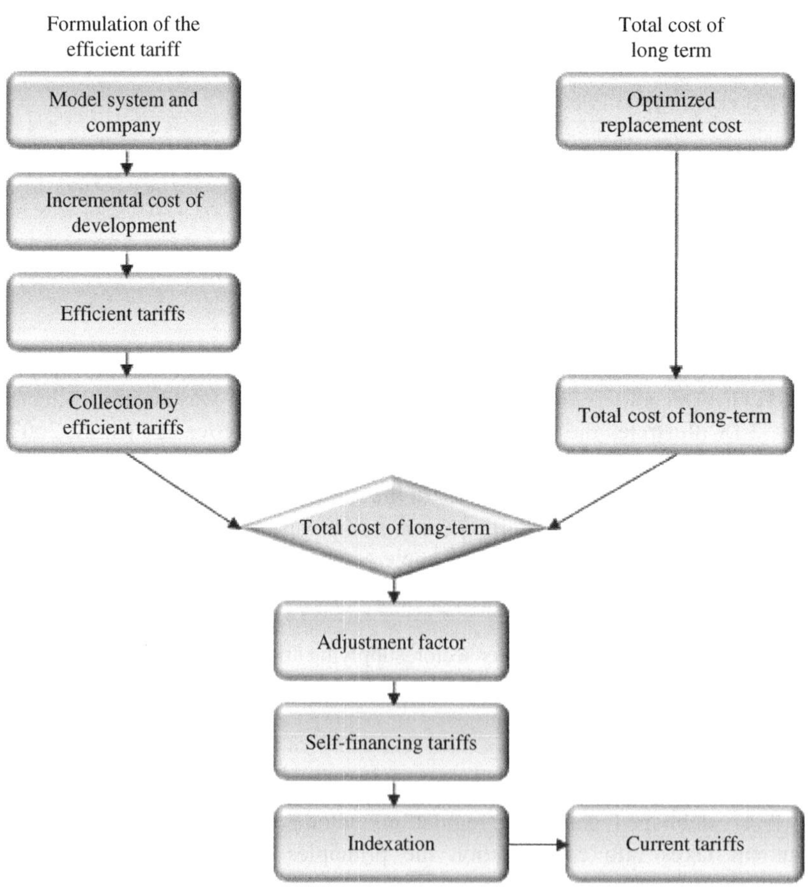

Figure 2.19 Technical procedure of tariff calculation

The principle of dynamic efficiency is reflected in the legal framework through the concept of efficient company model. Similarly to a situation of competition, this happens when the price established by a company is the result of a balance between supply and demand, and each company must adjust its costs to the efficient costs reflected in the price of the service. The concept of dynamic efficiency also implies that every time tariffs are established, gains in productivity are progressively incorporated.

Economic efficiency and, particularly, allocative efficiency imply that prices are defined on the basis of the concept of marginal cost. In a market without failures, the price reflects the cost of opportunity to produce an additional unit of a given good, reflecting the sacrifice in terms of resources that the production of this additional unit represents for society.

The process of setting tariffs follows a set of stages that have been defined by legislation. The specific procedures to be applied in the process are aimed at ensuring greater transparency. Figure 2.20 illustrates the regulatory procedure to set tariffs.

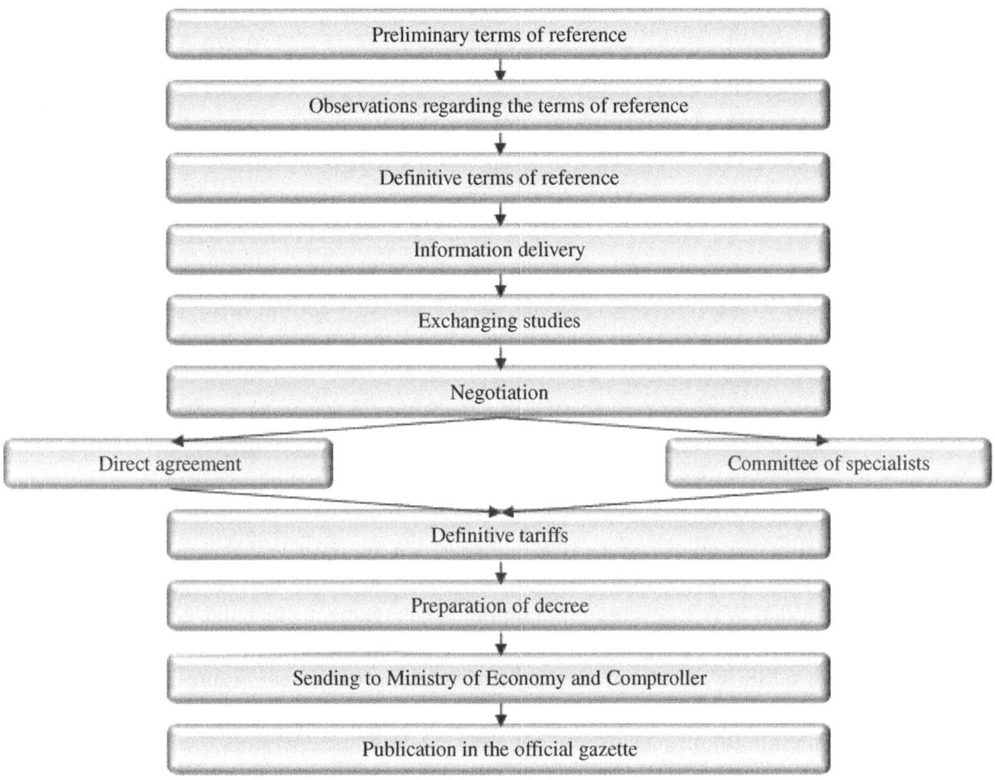

Figure 2.20 Administrative procedure diagram for the calculation of tariffs

In Chile, the main aspects involved in the process of setting tariffs include a forecast of demand, the design of a model company, the expansion project, the rehabilitation or maintenance of assets, value gains and indexation rules. Table 2.20 and Table 2.21 provide two examples of tariff systems implemented in Chile. The groups and zones reflect the different areas in which they operate and Group 1 represents the area of Greater Santiago.

Moreover, all Chilean customers are compelled to pay a tax for the fire-fighting service, consisting of 4.433 euros per month. VAT is applied at the prevailing rate to the charges, currently at the rate of 19%.

One particular aspect that is associated with the success of the regulatory model is the attribution of a consumption subsidy (20 m^3) for underprivileged families, which ensures a universal service in a nation where the poor domestic customers represent almost 15% of the total population.

The principle of cost recovery is applied and tariffs are based on marginal costs. The minimum cost of capital is defined by law.

2.4.4.2 Quality of service

Both the quality of drinking water, regularly monitored by the SISS, and the quality of service are characterized as being of a very high level. The regulatory agency acts so as to safeguard the implementation of the Chilean norm (NCh 409), which was later changed in 2005 and includes the standards of drinking water quality, water pressure and service continuity. Chile has also shown a significant evolution in this scope. During the early 1990s, the water sector faced problems with the treatment systems of some operators (e.g. chlorination). Consequently, in 1991, 20% of water services did not comply with bacteriological norms. In 2006, this figure was only 1%. During the same period compliance with disinfection norms increased from 89% to over 99%.

Table 2.20 Tariff system of Aguas Andinas for 2008

	Fixed component (€/month)	Variable components (€/m^3)						
		Water during normal season		Water during peak season		Excess of water consumption during peak hours*		Wastewater service
		Zone A	Zone B	Zone A	Zone B	Zone A	Zone B	
Group 1	0.78	0.404	0.400	0.388	0.384	0.893	0.887	0.399
Group 2	0.78	0.308	–	0.304	–	0.876	–	0.484
Group 3	0.72	0.350	–	0.351	–	0.981	–	0.138
Group 4	0.53	0.329	–	0.326	–	0.887	–	0.384
Group 5	0.86	–	0.250	–	0.242	–	0.624	0.376

* over 40 m^3

Table 2.21 Tariff system of Aguas Cordillera for 2008

	Fixed component (€/month)	Variable components (€/m^3)			
		Water during normal hours	Water during peak hours	Excess of water consumption during peak hours*	Wastewater service
Group 1	1.035	0.475	0.475	0.771	0.293

* over 40 m^3

In order to provide incentives for quality of service improvement amongst WWS operators, the SISS established a set of performance indicators. This group has been structured into diverse categories of indicators, namely accomplishing predefined works, WWS coverage, operational management, financial and commercial management, quality of service and environmental management. Table 2.22 shows the performance indicators designed by SISS. This tool is implemented by SISS by means of metric benchmarking exercises. Figure 2.21 shows an example of a benchmarking exercise developed by SISS related to water losses.

Finally, note that as an attempt to introduce some competition amongst operators the regulator prepares an annual ranking based on the quality of service and other information concerning their financial and operational aspects.

2.4.4.3 Public service obligations

In Chile, the legal framework unequivocally states that the tariff systems must reflect the principles of intelligibility, equity, self-financing and accessibility as priority values for the water sector. Moreover, prevailing legislation (Decrees No. 382 and No. 1199 and NCh 409) includes concepts such as universality, continuity and quality of service. In addition to this clear-cut definition in legal terms, SISS promotes specific benchmarking exercises aiming at encouraging improvements. The aspects considered concerning the WWS comprise continuity, pressure, customers' affordability and response to complaints. Figure 2.22 illustrates the evolution of the indicator regarding continuity of service.

As previously mentioned, governmental authorities are greatly concerned with underprivileged social classes. The first governmental action was to implement a subsidy, by means of Law No. 18,778 of 1989, which was subsequently changed in 1994, aimed at the WWS provided to poor customers. This subsidy, financed by the central government, is administered by the municipal authorities, which have also been entrusted with the process of selecting customers. The subsidies are directly deducted from the bill and the value is paid by the municipal authorities to the respective company.

The selection system is based on the socio-economic profile of customers, by means of indirect assessments which allow the State to focus its social action on the poorest sectors. These actions are performed by the Commissions for Social Community Assistance (*Comisión del Servicio de Asistencia Social*) through surveys.

Currently, the subsidy is applicable to the fixed and variable charges (only on the first 20 m^3, while additional consumption is billed normally), varying between 25% and 85% of the total value. The most common formula is to attribute 20 m^3 of water free of charge to poorer customers. Table 2.23 shows the conditions to apply for and exclusion from the subsidy for domestic customers with limited financial resources. However, when the subsidy ceases, customers can apply for the benefits again.

Table 2.22 Performance indicators used by SISS

Performance indicators	
a) Fulfilling predefined works	**e) Quality of service**
b) WWS coverage	Quality of drinking water
Water service coverage	Water service continuity
Wastewater service coverage	Drinking water service pressure
c) Operational management	Continuity of the wastewater service
Sources of supplies	Capacity to charge for services
Water production and losses	Response time to complaints
Distribution networks	**f) Environmental management**
Productivity indicators	Evolution of wastewater treatment
d) Financial and commercial management	Performance of the WWTP
Water billing	
Investments	
Operational costs	
Operational results	
Non-operational results	
Net profit	
Assets	
Profitability indicators	

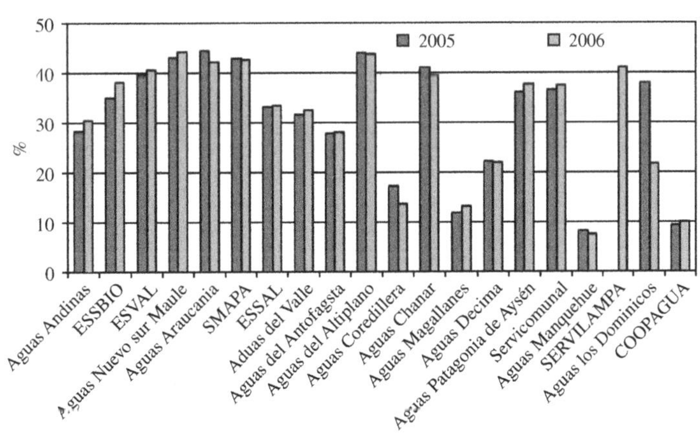

Figure 2.21 Indicator water losses per operator in Chile

Finally, another measure aimed at safeguarding social policies and strategies is related to the obligation (by means of Law No. 19,549 of 1998) of urban private companies to provide technical and administrative assistance to the rural committees and cooperatives for drinking water supply in their respective regions, as well as to perform the necessary activities to implement works for rehabilitation, improvements and the construction of new rural services.

2.4.5 Performance

The water sector in Chile is a success story in the context of Latin America, owing not just to the levels of coverage for the service (water supply and wastewater service) and the quality of service provided but also to the regulatory model that has been implemented.

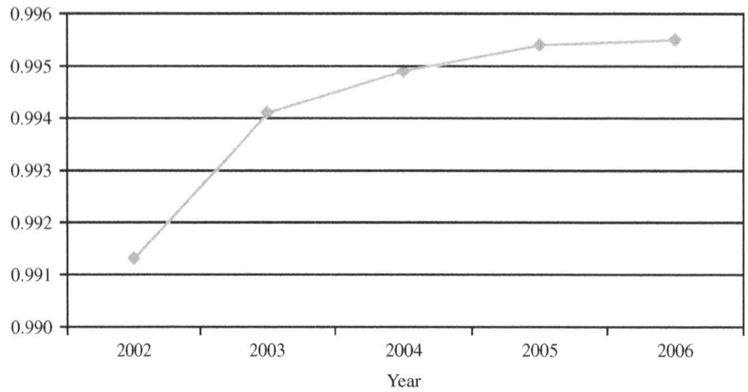

Figure 2.22 Annual evolution of the indicator continuity in the Chilean water sector

Table 2.23 Conditions for eligibility and exclusion from the subsidy

Social action subsidy
To be eligible for the subsidy it is necessary:
To be unable to pay for the service
To have no outstanding dues for services
To apply to the municipal authorities for this benefit in writing
Access to the subsidy ceases:
When some of the required conditions are no longer applicable
When the remainder of the bill is not paid on three consecutive occasions
When the municipal authorities are not provided with the required information to review the beneficiary's socio-economic conditions
When the period for which the subsidy was granted comes to an end (maximum of three years)

The stable legal, social, economic and regulatory scenario that characterizes the country is an asset that has attracted the private sector to the field of WWS activities, thus enabling large investments.

Another extremely important factor in the water sector in Chile is the financial assistance provided to domestic customers with scarce financial resources, which represents a fairly sustained policy of subsidization.

However, to ensure a greater development, the water sector in Chile needs to expand the competences of SISS to all WWS operators, i.e. to both urban and rural areas, and to make it more independent from the government.

2.5 COLOMBIA

2.5.1 Historical background

With a population of nearly 44 million inhabitants, spread over approximately 1,142,000 km^2, Colombia is organized administratively into 32 departments and a capital district (Bogotá). The departments represent the government in the different regions of the country and are administered by a governor. They are sub-divided into 1,119 municipalities, governed by mayors (*Alcaldes*).

The quality of the WWS in Colombia has increased significantly during the last decade. Despite the problems that exist, especially in rural areas, the main cities have an unusually high level of service in the context of Latin America. In certain aspects, Colombia is very different when compared to neighbouring countries (with the exception of Chile) in terms of the high levels of investment, the substantial cost recovery, the existence of relatively efficient public companies and, finally, a stable and healthy private sector participation.

Over the past two decades the water sector has undergone various reforms. The most significant development took place in 1994, when a new sector-specific policy was defined, characterized by a clear-cut strategy based on the growing importance of the municipalities, the participation of the private sector, WWS regulation, the full cost recovery principle and cross-subsidization.

During the first half of the twentieth century, the municipalities were responsible for providing WWS in Colombia. In 1950, as happened in many other countries, the Colombian government adopted a centralized scheme and created the Municipal Development Institute (*Instituto de Fomento Municipal* – INSFOPAL) and the Regional Water and Wastewater companies (*Empresas Departamentales de Acueducto y Alcantarillado* – ACUAS y EMPOS), which are responsible for funding, planning, designing and managing public services in the majority of municipalities. Nevertheless, some municipalities retained responsibility for supplying these services at a local level while others created municipal companies. The most illustrative example was the creation of the Medellin Public Company (*Empresas Públicas de Medellín* – EPM), in 1955, which had multisector responsibilities.

Subsequently, during the 1980s, the water sector faced an enormous crisis. INSFOPAL collapsed financially in 1989, thus returning the responsibility for supplying WWS to the municipal authorities. In addition, another institutional change in the water sector took place when the responsibility for this sector was transferred from the Ministry of Health to the Ministry for Economic Development. This represented a significant organizational change.

The National Constitution, dating from 1991, delegates the responsibility for supplying WWS to the municipalities, while also giving them the power to decide about the participation of the private sector in the management of this sector. The Constitution provides the legal framework and establishes a platform to introduce more reforms in the water sector, clearly separating the roles of operator and political decision-making.

In the aftermath of these reforms, the government produced Law No. 142 of 1994 (*Ley de Servicios Públicos Domiciliários*), which revolutionized the legal framework of the water sector. This law underwent some changes in 2001, while nevertheless maintaining its basic structure. The public services policies adopted by the liberal governments of César Gaviria (1990–1994) and Ernesto Samper (1994–1998) were characterized by modernization and openness to the corporate sector. Investments by the municipal authorities in the water sector increased considerably, thus improving the quality and efficiency of the services through private sector participation in the most underdeveloped areas of the country (where public companies had the greatest deficits), the establishment of a sector-specific regulator and, finally, by promoting cost recovery along with a policy of universal service, protecting customers from the poorest sections of society by means of cross-subsidization with customers who, *a priori*, had a greater capacity to pay for the services.

The first PPP was constituted in 1995 with the aim of modernizing the water sector, by creating a mixed capital company to supply WWS to the city of Cartagena. The PPP for the city of Barranquilla was formed in 1996 and other PPPs were subsequently established. In 2002, the government published the Programme for Corporate Modernisation with the purpose of introducing private sector participation in the water sector in small and medium-sized municipalities as well. This resulted in significant progress in terms of the quality and efficiency of WWS provision.

Law No. 142 regulates the WWS sector. This law represents the government's main legal tool in the water sector, aimed unequivocally at fostering efficiency and increasing competition, while simultaneously promoting private sector participation. Key elements of the law include the establishment of the Superintendence for Domestic Public Services (*Superintendencia de Servicios Públicos Domiciliarios* – SSPD) and the definition of the functions of the Commission for Regulation of Water, Wastewater and Waste (*Comisión de Regulación de Agua Potable y Saneamiento Básico* – CRA). Apart from promoting the cost recovery by means of tariffs and adjusting the limits of cross-subsidization between customers, the law also encourages the liberalization of all the operators of public services, including those that are traditionally considered to be natural monopolies.

The government of Alvaro Uribe sought to raise the level of the water sector in Colombia to provide a major impetus to increasing the coverage of the service, surpassing the Millennium Development Goals and improving the quality of service, especially in small cities and rural areas without these infrastructure services. To this end, the Colombian government created the Vice-Ministry for Water and Wastewater (*Viceministério da Água e Saneamento*) in 2006 and developed four new programmes: a programme for the organization of the WWS (*Planes Departamentales de Agua y Saneamiento*), another for wastewater collection (*Programa de Saneamento para Assentamientos* – SPA), a third for wastewater treatment (*Programa de Saneamiento de Vertimientos Municipales*) and a fourth programme for public health (*Programa de Lavado de Manos*).

The programme for the Regional Water and Wastewater Plans (*Planes Departamentales de Água e Saneamiento*) is aimed at planning and harmonizing the resources and regional schemes for supplying services at the level of each region in the country. The programme is first being implemented in four regions, which are facing particularly difficult circumstances, namely Sucre, Cordova, Magdalena and Guajira. This programme also intends to encourage automation in smaller companies in the sector, so as to benefit from economies of scale.

2.5.2 Legal and institutional framework

2.5.2.1 Responsibilities

Diverse entities are responsible for the water sector in Colombia, namely the Ministry for the Environment, Housing and Regional Development (*Ministerio de Ambiente, Vivienda y Desarrollo Territorial* – MAVDT), the Ministry for Social Protection (*Ministerio de la Protección Social* – MPS), the CRA, the SSPD, the Autonomous Regional Corporations (*Corporación Autónoma Regional* – CAR), the Directorate for the Budget and Territorial Development (*Financiera de Desarrollo Territorial S.A.* – FINDETER) and the municipal authorities.

In Colombia, the municipal authorities are responsible for ensuring that WWS are efficiently provided to local inhabitants by public service companies (article 5 of Law No. 142). Hence, the responsibility for WWS has been entrusted to public service companies, except in some specific cases defined by law, in which the municipal authorities can provide these services directly. In rural and suburban areas, municipal associations also provide water services.

The national water policy is a responsibility of MAVDT. Within this ministry, the Board for the Drinking Water Supply and Wastewater (*Dirección de Agua Potable y Saneamiento Ambiental* – DAPSBA) supervises the sector, defines regulatory policies and establishes guidelines for these services whereas the National Planning Department (*Departamento Nacional de Planificación* – DNP) conceives and monitors policies, by means of plans, programmes, studies and investment projects for the water and environmental sectors.

The Vice-Ministry for Water and Wastewater is responsible for establishing sector-specific policies, defining the framework for the national policies instituted by the DNP.

MPS has functions at the level of supervising the quality of water and public health.

CRA has responsibilities in terms of economic and quality of service regulation, establishing the tariff systems for operators by means of economic efficiency criteria and promoting competition between operators. It approves investment plans and defines and supervises the quality of service and the technical norms to be followed by WWS operators. Despite its varied and important functions, CRA is not responsible for monitoring many of its decisions, since this function has been assigned to SSPD.

SSPD is responsible for monitoring and supervising the public services (e.g. water, energy, gas and telecommunications). In this scope, it aims at assuring that operators provide these services in an adequate and efficient way, establishing uniform accounting criteria, supervising the administration of subsidies and monitoring the management of these subsidies.

CARs were created by Law No. 99 of 1993, and ensure environmental regulation. These entities are subject to the Ministry for the Environment and their functions include implementing policies, plans, programmes and projects in the area of renewable natural resources and the environment, as well as ensuring that the laws that have been established by the government with regard to attribution, administration, management and use of water resources are fulfilled.

FINDETER, created by Law No. 57, of 1989, is a commercial company overseen by the Ministry for the Treasury and Public Credit. It has been entrusted with the task of promoting regional and urban development, by financing and assisting with the design, implementation and administration of investment projects and programmes.

The municipalities are responsible for providing WWS. They can do it directly by delegating the mandate to municipal companies or by developing contractual or institutionalized PPPs. Figure 2.23 shows the institutional framework of the different entities involved in the water sector in Colombia.

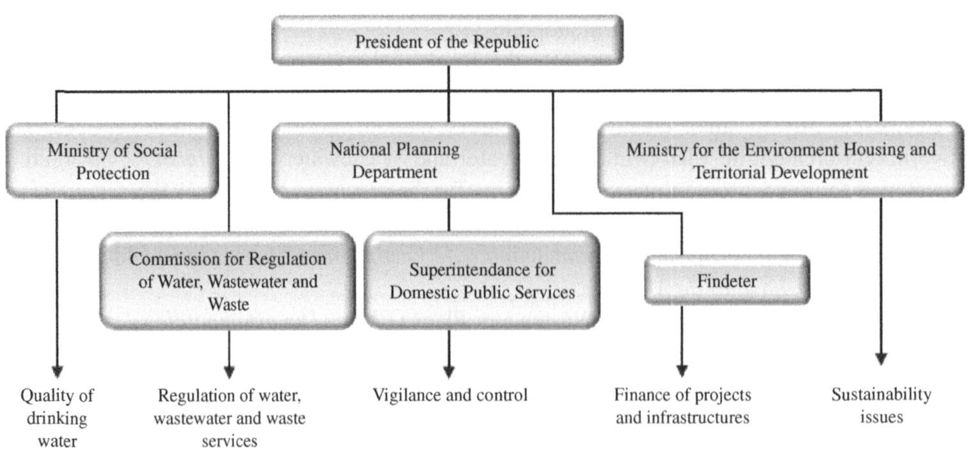

Figure 2.23 Institutional framework for the water sector in Colombia

2.5.2.2 Regulation

Colombia has an independent regulatory agency (CRA), created by Law No. 142 of 1994, which focuses on questions of economic regulation, quality of service and public service obligations for the WWS sector. CRA is a sector-specific regulator which also has functions with regard to solid waste services regulation.

CRA is a special administrative unit, overseen by the Ministry for the Environment. It is not a juridical entity but is independent in technical, administrative and patrimonial terms.

The economic regulation developed by CRA is, strictly speaking, based on yardstick competition, i.e. the tariff systems are determined according to the performance of operators in this sector which constitute best practices. The system of price caps, adopted for a regulatory period of five years, consists of three components respectively related to a) administrative costs, b) operational and maintenance costs and c) the costs of capital, which also includes environmental taxes. The first two components are based on efficient costs (best practices), while the third depends on the approval of the operators' development plans. This calculation is carried out separately for the water supply and for wastewater services. Best practices are determined on the basis of the non-parametric benchmarking technique of DEA. This daring model has, in fact, been strongly criticized by Colombian operators (see Marques & Contreras, 2007).

Decree No. 287 of 2004 defined the criteria and the methodology to be used to determine WWS tariffs, based on diverse principles, of which the most important is to provide incentives for economic efficiency. In fact, article 87 of Law No. 142 states that 'The tariff regime will be guided by the criteria of economic efficiency, neutrality, solidarity, redistribution, financial autonomy, simplicity and transparency'. Article 92 stipulates that, in tariff formulae, the CRA will ensure benefits by means of establishing a period of time for the average reduction of WWS costs, while operators are simultaneously provided incentives to become more efficient than the average and thus benefit from the advantages resulting from a greater efficiency.

CRA budget for 2006 was approximately 2.7 million euros. CRA is financially independent since it does not need to make use of the national budget. Its financial independence has been ensured by applying a surcharge on operators, known as the 'Special Contribution' (*Contribución Especial*), in compliance with Article 85 of Law No. 142. In this regard, all the operators of regulated services are compelled to pay this tax, which is calculated on the basis of the administrative costs of the companies. In 2007, CRA employed 105 people.

2.5.3 Market structure

2.5.3.1 Ownership and management of services

The WWS are publicly owned and the majority of these services are managed by the public sector. Although, in national terms, the private sector only represents a small percentage of WWS supplies, its participation has always been characterized by high levels of stability and sustainability and is considered to be a success story in the context of Latin America.

2.5.3.2 Players in the sector

According to CRA, in 2005 there were 2,886 operators providing water services and 1,071 operators providing wastewater services. This difference is basically due to the differences between the two services in terms of coverage.

The EPM is one of the largest public companies in Latin America. It is a municipal company that provides WWS, cleaning and urban waste services, electricity and local telecommunications and has a high level of performance. Public companies in the country's other large cities, such as the WWS company of Bogotá (*Empresa de Acueducto y Alcantarillado de Bogotá* – EAAB), also have good performance indicators when compared to many other WWS public companies in Latin America.

Unlike many other developing countries, where private sector participation in the water sector has been a very controversial issue, frequently resulting in an earlier termination of contracts, in Colombia private participation has taken place in a stable manner and is even considered by many to be a true success story. In 2004, Colombia had 125 private WWS operators and 48 mixed capital operators, which include large, medium and small companies.

Despite the good results (effectiveness) achieved by EAAB, this was due to a large 'injection' of public capital from the government. The situation became unsustainable, despite the substantial increase in tariffs. The solution found included keeping the statute of public operator for EAAB administration but in 2003 three private companies signed five management contracts for various phases of the value chain. EAAB also established a 20-year BOT contract, seeking the rehabilitation, operation and maintenance of one of its WWTPs. This contract was made with the Proactiva company, which provides wastewater treatment services to about 2.3 million inhabitants in Bogotá. This company, founded in 1997 with capitals equally contributed by the FCC (Spain) and Veolia (France), manages several water, wastewater and solid waste services throughout Latin America. In Colombia, Proactiva also operates, on an integral basis, the WWS for about 320,000 inhabitants in Monteria (20-year contract) and Tunja (30-year contract), where it serves approximately 150,000 inhabitants.

One can thus observe a proliferation of Spanish companies (private and public) in the water market in Latin America. The Spanish companies are present not just via FCC but also by Agbar, which provides WWS in Cartagena, supplying about 730,000 inhabitants, and by Canal de Isabel II Company and TECVASA.

The Triple A (AAA) Colombian Water Company is controlled by Canal de Isabel II Company, a municipal water company from Madrid and TECVASA (a private company from the Valencian Community). In Colombia, the AAA operates in Puerto Colombia, Soledad, Galapa, Sabanalarga, Sabanagrande, Santa Marta, Santo Tomás, Baranoa and Polonuevo, providing WWS to about 2 million customers. Table 2.24 shows the number of operators in the water sector in Colombia, according to the population served (Econometria, 2004).

Table 2.24 WWS and population served in the water sector in Colombia (2004)

Population	Municipalities (No.)	Type of operator		
		Public	Community	Private
		Total operators: 1,008		
>1 million	4	2	0	13
400,000–1 million	7	2	0	10
100,000–400,000	27	9	1	23
40,000–100,000	53	38	9	22
<40,000	1.007	799	32	48
Total	1.098	850	42	116
Percentage		**84.3%**	**4.2%**	**11.5%**

The participation of private initiative in the Colombian WWS sector began in Cartagena, in 1995, with the support of the World Bank. The most important examples are Cartagena, Barranquilla, Santa Marta, Tunja, Montería, Palmira, Girardot and Riohacha, whose operators are mainly Colombian and/or Spanish. The general performance of the WWS has improved where private participation occurred, in some cases far more than expected, for example, in Barranquilla, which achieved an impressive degree of success in the WWS coverage to the poor rural populations.

In 2006, amongst the 365 operators registered with SSPD, serving more than 2,500 customers each, 318 were public companies, 38 were municipalities and 9 were other kinds of entities. In recent years the municipalities have tended to withdraw from directly providing these activities and have created specialized companies for the purpose (*Empresas Industriales y Comerciales del Estado* – EICE), which, on account of their legal nature, have more flexibility to contract services under a private law regime, even though they are subject to administrative entities in terms of monitoring. Thus, in 2006, the majority of public service companies were of the EICE type, accounting for 53% of the total. Municipalities directly providing services represented 15% of cases, private companies 12%, official companies 13%, mixed companies 6% and authorized organizations corresponded to about 1%.

In some semi-urban zones there are successful examples of WWS being managed by community organizations, which are characterized by a high level of transparency and accountability. For example, *Junta Administradora del Aqueducto La Sirena en el Valle de Cauca*, on the outskirts of the city of Calí, manages 778 household connections, providing a continuous service to about 4,200 customers, a wastewater collection network and a treatment plant. Based on average consumption it bills an average price of 2.3 euros for 28 m^3 per month. There is no stratification of consumption and special tariffs are only applied to those who consume more than 28 m^3. The community has generated surplus funds that are invested in maintenance and are even used to build new WWS infrastructure. Another successful example is the case of the *Asociación Acueducto de Río Negro en Popayan*, formed ten years ago, which includes ten operators with 7,000 household connections.

2.5.3.3 Integration of the sector and services provided

Normally the WWS are characterized by vertical integration and by a limited degree of horizontal integration. As a general rule, the same entity provides both water and wastewater. However, some public operators who have been contracted externally, such as EAAB, are responsible for some of the stages of the production chain (water cycle). With a few exceptions (e.g. EPM), WWS operators in Colombia do not provide other services.

2.5.3.4 Numbers for the sector

In 2006, the water service in Colombia was characterized by a level of coverage of 88.3%. In that same year, the wastewater service had a level of coverage of 74.1%, although only 25% of the total wastewater collected was actually treated. In the same year, the water service had a level of water losses of approximately 49%. Table 2.25 shows the evolution of the levels of WWS coverage in Colombia (SSPD, 2002).

Table 2.25 Evolution of the level of WWS coverage (%)

	1990		1997		1999		2000	
Group	Water	Wastewater	Water	Wastewater	Water	Wastewater	Water	Wastewater
4 largest cities*	94.0	86.9	95.5	89.1	97.2	92.4	98.2	93.4
Cities >100,000 inhab.	80.0	74.8	94.2	77.7	96.1	84.2	96.4	85.5
Other municipalities	83.2	51.4	91.0	71.3	90.0	67.9	91.3	71.1
Urban zone average	**86.5**	**72.1**	**93.5**	**80.3**	**94.6**	**81.8**	**95.0**	**83.5**

* Bogotá, Medellín, Barranquilla and Cali

The average continuity of the water service in Colombia increased from 15.36 hours in 1993 to 19.82 hours in 2003. In 2007, the average continuity of the water service was slightly over 20 hours, varying between a 24-hour water supply (in the four largest cities) and an intermittent service in smaller cities.

In Colombia, the average consumption of water for human consumption in urban areas is about 59 litres per inhabitant per day.

The use of drinking water diminished by about 25% between 1996 and 2001, above all, due to a substantial increase in tariffs and the inclusion of low-income customers, as a result of the expansion of the service.

According to CRA, in 2006, the water services had revenues of 367.8 million euros and, as in the case of most developing countries, although the wastewater service is normally more expensive it generates a substantially lower level of revenue, in this case of 246 million euros. In terms of investments, 1.6 billion euros have been invested in WWS infrastructure in the past three years. Between 1995 and 2003, about 3 billion euros were invested, of which 16% were from private capital. Figure 2.24 shows the annual evolution of investments in the Colombian water sector.

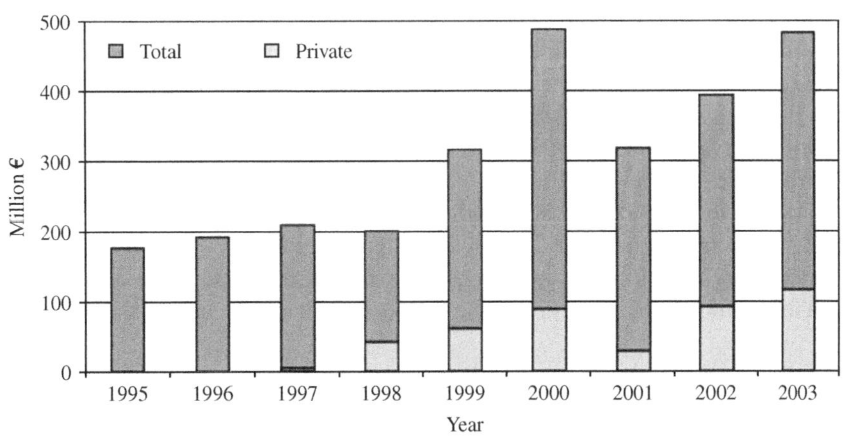

Figure 2.24 Annual evolution of investments in the water sector in Colombia

For the period between 2007 and 2017, SSPD has envisaged investments around 1.47 billion euros divided between water supply and wastewater collection and treatment.

2.5.4 Rules of the game

2.5.4.1 Tariff setting

The principles governing the determination of the tariff system in Colombia have been defined by Law. No. 142, irrespective of whether the companies are public, mixed or private entities. The tariff scheme that has been conceived, which is unique in Latin America, has leveraged a series of advances in the water sector, such as an important increase in water tariffs and the self-financing capacity of the WWS sector, an increase in transfers from the central government to municipalities for this sector, consequently resulting in a significant increase in investments and access to the service, and, finally, a greater protection of the poor socio-economic layers by means of tariff differentiation through blocks.

The tariff is calculated on the basis of efficient costs, according to the methodology defined at a national level by CRA (Decree No. 287 of 2004), by applying benchmarking (through the DEA technique). The tariffs are implemented by means of a fixed component and an increasing variable component, according to the consumption of water. In 2006, the average monthly bill for domestic customers was 4.51 euros for the water service and 3.18 euros for the wastewater service, for a total of 7.69 euros. The tariffs also differ according to the characteristics of the customer and are higher for commercial, industrial, official and special users.

In order to protect low-income citizens against tariffs that are too high, the regulatory agency applies factors that adjust the tariffs in each municipality to the estimated purchasing power of customers. Thus, CRA employs the concept of 'socio-economic layer', a tool used by the Colombian government to classify citizens on the basis of their respective income. Hence, each city has delimited geographical areas that correspond to each of the six socio-economic layers that have been defined, from layer 1 (very low) to layer 6 (high). The factors applied to water tariffs for customers range from between 0.3 and 1.2, as can be seen in Table 2.26.

Table 2.26 Socio-economic layers and corresponding adjustment factors

Block	Adjustment factor
1	0.30
2	0.60
3	0.85
4	1.00
5	1.20
6	1.20

The percentage of the Colombian population belonging to the social groups corresponding to layers 1–3 is much greater than the remaining, which corresponds to layers 5 and 6. For example, in Bogotá 73% of the population falls within layers 1–3 and only 10% falls within layers 5 and 6. In effect, the tariff system tends to generate permanent deficits, which require subsidies (transfers) from the national government or directly from the municipalities. Figure 2.25 illustrates the annual evolution of WWS tariffs according to customer blocks (SUI, 2008). Note that in Colombia the blocks correspond to social strata and not to consumption.

Table 2.27 shows the levels of subsidies that have been allocated over the course of time according to different blocks and with regard to the average WWS tariff.

Table 2.28 presents an overview of EAAB's tariff system for the WWS during 2007. The basic and non-basic consumption includes the environmental taxes, which are a significant element. The environmental taxes thus seek to discourage consumption, mitigating, to a certain extent, the environmental impact of water abstraction. These taxes are defined by means of the relationship between prevailing environmental regulation and the tariffs practised by the WWS operators.

2.5.4.2 Quality of service

Despite the great progress that has been made in the water sector at the level of WWS coverage, there are still serious problems in terms of the quality of water which contributes towards the high rates of mortality, especially amongst children. Although the general quality of service in Colombia has improved considerably during the last decade, there is still a very large gap between the largest cities and small cities and rural areas.

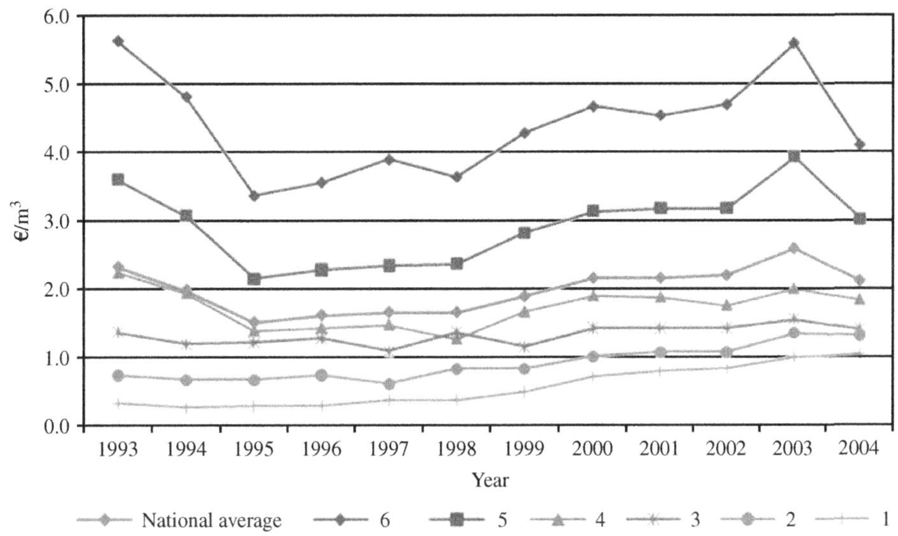

Figure 2.25 Annual evolution of the average tariff per customer block in Colombia

Table 2.27 Subsidy levels in average tariffs

Block	1	2	3	4	5	6
1996	–91%	–85%	–72%	–51%	–23%	1%
2000	–82%	–68%	–48%	–22%	13%	42%
2002	–73%	–55%	–30%	–12%	30%	61%
Maximum levels (Law No. 812 of 2002)	–70%	–40%	–15%	0%	–	–

Table 2.28 EAAB tariff system for 2007

Customer	Water supply (€/m³)			Wastewater service (€/m³)		
	Fixed component (2 months, €)	Basic consumption	Non-basic consumption	Fixed component (2 months, €)	Basic consumption	Non-basic consumption
Domestic						
Block 1	1.23	0.21	0.70	0.626	0.128	0.426
Block 2	2.46	0.42	0.70	1.252	0.256	0.426
Block 3	3.77	0.64	0.70	1.920	0.392	0.426
Block 4	4.10	0.70	0.70	2.087	0.426	0.426
Block 5	9.18	1.04	1.04	5.198	0.639	0.639
Block 6	11.22	1.11	1.11	7.222	0.682	0.682
Industrial	5.33	0.96		2.734	0.610	
Commercial	6.14	1.04		3.131	0.639	
Official	4.10	0.70		2.087	0.426	
Special	4.10	0.70		2.087	0.426	

The level of WWS coverage in rural areas is less than would be expected, keeping in mind the country economic development and, more specifically, the progress that has been achieved in the water sector. Although one of the reasons for this delay is the political conflict and the armed conflict, there are also other important underlying reasons, such as the fact that the municipalities use the transfers from the General Contributions System (Law No. 715 of 2001) for investments in urban areas. Apart from this, unlike some other countries in Latin America, Colombia does not have an institution or programme at a national level to provide technical support for the community organizations that serve rural areas. Law No. 142 has entrusted this task to governmental departments; however, this responsibility has not been implemented satisfactorily.

The Inventory of Rural Sanitation (*Inventario Sanitario Rural*), developed in 2002, has confirmed this situation and, amongst other aspects, has highlighted the fact that only 21% of rural community organizations had the necessary competences to develop WWS maintenance and 10% to administer the service. Only 32% issued bills and only 10% was billed on the basis of micro-measurement. With a view to increasing access and improving the quality of the WWS in this area, as stated above, in 2005 the government defined political guidelines for this service in rural areas in Colombia.

According to an analysis carried out by the SSPD in 2004, 72% of customers had access to quality water and 28% of customers received water that was unfit for consumption. In some cases, the water pressure was inadequate, significantly increasing the risks of contamination.

The wastewater services do not have a sufficient hydraulic capacity to meet needs, especially in poor areas, resulting in problems of overflowing and flooding. In 2006, about 25% of the wastewater produced in the country received some kind of treatment. The remaining 75% continued to be discharged without any kind of treatment, thus contaminating natural water resources.

Throughout the progress that has been achieved in the water sector in Colombia, the regulatory agency has performed a notable role in the fulfilment of its statutes, i.e. by defending the interests of customers and the public interest. However, in the course of its activities it does not prepare any kind of report to determine the operators' performance, which would thus encourage the use of benchmarking. It is also necessary to highlight, for example, that the level of water losses is almost 50%, which is an unacceptable figure, or even the high levels of a lack of continuity in service provision. In fact, the regulatory process that has been adopted by CRA provides incentives to enhance performance and develop the quality of service, since operators with negative results in this regard can suffer consequences in terms of a reduction in tariffs. However, this evaluation has not yet been carried out.

2.5.4.3 Public service obligations

As mentioned above, the regulation of the WWS sector in Colombia has been defined by means of Law No. 142 and subsequent changes. Apart from the principles associated with economic efficiency and the quality of service regulated by this legal framework, it also considers the universal nature of WWS and aims to ensure that these services are supplied to all Colombian customers.

With a view to promoting universal WWS and considering the low-income levels of the majority of Colombians, along with the widespread dispersal of customers, the prevailing policy in the water sector promotes cross-subsidization based on socio-economic strata and the characteristics of customers and between urban and rural areas. It has also implemented special tariffs for poor customers.

In order to promote a better interaction with customers, some operators, such as EAAB, have tried to optimize relations between the operator and customers through an entity known as the 'Consumer Defender' (*Defensor del usuário*). The main objective of this entity is to protect the interests of customers and it uses all available means to achieve this end, including handling complaints.

CRA Resolution No. 287 of 2004, regarding the tariff methodology, promotes a tariff system based on the principles of economic efficiency, neutrality, solidarity, redistribution, self-sufficiency, simplicity and transparency. All CRA's decisions, as well as the main facts pertaining to the water sector, are published in diverse forums, including the regulator's website.

2.5.5 Performance

Colombia has made tremendous efforts to achieve a high standard of performance in the water sector. Generally, and especially with regard to the water service in urban areas, this objective has been achieved and can even be considered to be a success story, keeping in mind the nation's geographical context. However, this has not yet been achieved with regard to WWS in rural areas. The creation and *modus operandis* of the regulatory model, centred on CRA, has been deemed to be an extremely positive measure. CRA has adopted a methodology for economic regulation based on productive efficiency comparisons between operators and performance incentives, bearing in mind aspects related to social regulation (e.g. a universal service) and environmental regulation (e.g. imposing taxes for this purpose). The regulatory procedures are transparent and neutral and reflect a very high level of know-how.

The main problems of the water sector in Colombia cannot be disassociated from the nation's problems. In addition to the political conflicts, which condition the development of the sector in a considerable part of the territory, the low per capita income and a low population density also affect the sustainability of this sector. The quality of service regulation can and must be improved, namely, by means of publishing and comparing performance indicators (e.g. benchmarking and sunshine regulation). Until very recently, CRA has, quite understandably, focused its activities on economic regulation and social regulation. An effective coordination between CRA and SSPD is of fundamental importance in terms of regulating the quality of service.

2.6 COSTA RICA

2.6.1 Historical background

Costa Rica is governed by a Political Constitution dating from 1949, which established a presidential system and a single state. The nation has about 4 million inhabitants distributed over 51,100 km^2 and is organized into seven provinces, whose governors are appointed by the President. These provinces are divided into 81 cantons, which, in their turn, are divided into 463 districts. The President of Costa Rica governs the country along with two vice-presidents, all of whom are elected by a popular vote every four years.

The water sector in Costa Rica has made great progress during recent decades, mainly in terms of the water service and especially as far as coverage is concerned. However, this is a sector which still faces enormous challenges, keeping in mind the limited number of wastewater connections, the poor quality of service provided and, finally, an insufficient cost recovery of the service by means of the prices charged to customers.

In order to improve this situation, the government has enacted several laws and carried out reforms that have regulated and introduced changes in the sector, in an effort to foster a positive evolution of the water sector along with a clear concern to safeguard the public health of customers.

Historically, the water sector in Costa Rica can be divided into two legislative periods, on account of two Water Acts, namely the XI Water Act, dating from 1884 (in effect 1884–1942) and Water Law No. 276, dating from 1942. During the first period, the water sector was characterized by a scenario in which the management of the water services was shared between the private and the public sector. The ministry that was mainly responsible for the water sector was the Ministry for Public Works. Water Law No. 276, which currently regulates the WWS, was implemented when responsibility for the sector was attributed to the Ministry for the Environment and Energy, mainly with regard to the water service.

Presently the water sector is primarily regulated by the 1942 Water Act, by the General Law for Drinking Water Supply (Law No. 1634 of 1953), which defines the responsibilities of the different governmental entities over the sector, as well as by the General Health Act (Law No. 5395 of 1973). Subsequently, the legislation that created the regulatory agency, Law No. 7593 of 1996, proved to be an important normative document in the promotion of efficiency and quality of service in this sector in Costa Rica, namely by establishing public service obligations, amongst other measures.

By means of its articles 267 and 268, the General Health Act (*Ley General de Salud*) establishes that '*The entire water supply system intended to be used and consumed by people must provide a continuous service of potable water, in a sufficient quantity to satisfy the needs of the people and with the necessary pressure to enable the correct functioning of the sanitary equipment in use*'.

It must be highlighted that all water supply systems are compulsorily subject to quality control by the Ministry of Health, which is the entity responsible for guaranteeing the drinking water supply in suitable conditions.

The government also sought to implement legislation that imposed more and better conditions for the operators, but the present situation of the country alongside the excessive costs of treatment plants resulted in the partial failure of this legislation and regulations.

The water sector in Costa Rica requires considerable investments in order to further develop, especially in terms of wastewater. In the past, more than 60% of investments in the sector were derived from government transfers, half of which were obtained from external loans. However, given the heavy investments that are necessary for this sector the government was unable to maintain the high levels of financing that it had provided until then. Consequently, the sector once again tried to increase revenues internally, by means of the tariff systems put into practice by the operators. Although there are some subsidies from the central authorities, the majority of investments in rural areas are supported by the Costa Rica Water and Wastewater Institute (*Instituto Costarricense de Acueductos y Alcantarillados* – AyA), the largest and most representative WWS operator in Costa Rica, as will be explained in further detail shortly.

The objective of AyA is to increase the level of coverage for the rural water service to 90% over the next three or four years through an agreement signed in March 2007 with Banco Popular and Central American Bank for Economic Integration (CABEI). The main aims of that agreement were related to promoting the sustainability of the *Asociaciones Administradoras de Acueductos y Alcantarillados Rurales* (ASADAS) to reaching a level of coverage of 100%, while simultaneously equipping the nation with an automated information system under ASADAS in order to improve records of operations and to set tariffs.

In 2002, AyA proposed a programme to modernize the sector (2001–2020), which promoted an increase in the water service coverage level in urban areas to 98.5% together with a better quality and a drastic improvement of the coverage of the wastewater service in urban areas to 89% in 2020. Apart from this, the programme challenged the sector to increase the level of coverage of the water service in rural areas to 90% by 2020. The total investment required to implement this programme was 1.1 billion euros, corresponding to approximately 55 million euros per year. This corresponds to four times the annual average investment between 1991 and 1998. In 2002, AyA estimated that, keeping in mind the increase in revenues and efficiency, only about 40% of the investments would have to be financed by the government. This situation reflects the many years of negligence in the maintenance of AyA assets.

2.6.2 Legal and institutional framework

2.6.2.1 Responsibilities

Responsibility for the water sector in Costa Rica has been shared between the Ministry of Health (*Ministerio de Salud* – MINSA), the Ministry for the Environment and Energy (*Ministerio del Ambiente y Energía* – MINAE), the regulatory agency for the sector – ARESEP (*Autoridad Reguladora de Servicios Públicos*), whose activities will be described next, the Competition Commission (*Comisíon de Defensa de la Competencia*) within the Ministry of Economy (*Ministerio de Economía*) and the cantons.

In compliance with the Drinking Water Supply Water Act, dating from 1953, the local administrative authorities, i.e. the municipalities, usually known as cantons, are ultimately responsible for the WWS provision in suitable conditions in terms of quality and reliability. However, the reality of the water sector in Costa Rica reveals that this service is mainly provided by other entities.

Apart from its role as an operator, AyA (the largest and main WWS operator in Costa Rica) has important responsibilities in the water sector with regard to the water supply to rural communities. It also monitors compliance with technical norms and guarantees the supply of a continuous service throughout the country, to the point of assuming the management of any WWS in areas where the venture by the respective operator might have failed. AyA serves as an example for other operators and is the main driver for the development of the water sector throughout the territory of Costa Rica.

In particular, the AyA's Department for Rural Works is responsible for planning, projects, financing and building water supply and wastewater service systems. Moreover, this body provides technical assistance to ASADAS and CAARs.

In fact, AyA plays an important political and regulatory role, albeit indirectly, since it monitors compliance with technical norms, can take over systems that have broken down and, finally, also advises the ministry responsible for developing the sector. In many situations this dual role results in a conflict of interests.

MINSA is responsible for supervising the water sector. However, the ministry has been entrusted with diverse aspects of governing the nation, namely to ensure the protection and improvement of the population's state of health, by effectively exercising its mandate and its institutional leadership. It focuses on promoting health and social participation, following the principles of transparency, equality, solidarity and universality. Although MINSA is keenly aware of its responsibilities in the water sector, namely safeguarding the water quality in the country, its performance has proved to be rather ineffective. There are instruments that require all operators to ensure the quality of the water being supplied, such as health regulations or closing water systems, in practice, the Ministry of Health has a merely representative role.

MINAE has functions at the level of managing water resources in Costa Rica, in compliance with the Environmental Act (Law No. 7554 of 1996), and is also responsible for environmental licensing. Other ministries also play a role in the water sector and there is sometimes a distinct overlapping of functions and responsibilities.

The Board for Water and Environmental Services is a department of ARESEP which has been entrusted with the task of carrying out the necessary studies to set prices and tariffs to guarantee the quality and supply of public services associated with the water sector, namely irrigation and drainage, the water supply and wastewater services, whenever these are not municipal services. Essentially, this entity's basic task is to recommend the approval of tariffs, according to prevailing technical rules and regulations, which include, among other factors, the principles of cost recovery, financial balance, economic efficiency, quality norms, quantity, reliability, opportunity, social equality and environmental sustainability. Figure 2.26 shows the institutional framework of the water sector in Costa Rica.

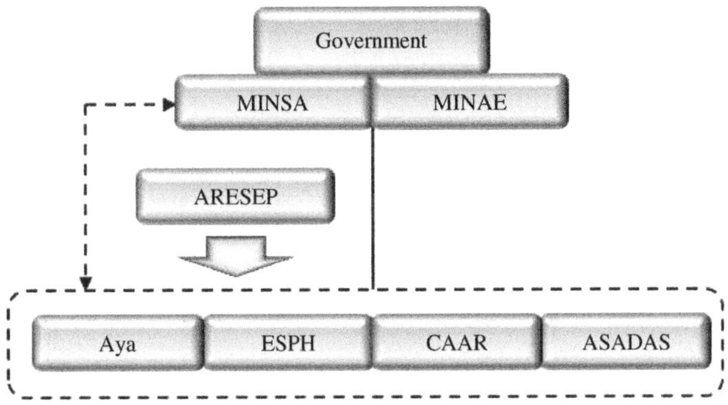

Figure 2.26 Institutional framework of the water sector in Costa Rica

2.6.2.2 Regulation

There is a sector-specific regulator for the water sector in Costa Rica. The ARESEP was created in 1996 by Law No. 7593. It is an independent, multisector regulator, encompassing the sectors of telecommunications, electricity, water, oil, irrigation and transport (various means of transportation). It plays the role of an economic regulator, which includes the task of setting tariffs for the operators, monitoring and controlling efficiency and quality of service and protecting customer interests. In 2006, the operating costs of ARESEP amounted to more than 3.65 million euros. This value, which includes the expenditure incurred with all the regulated sectors, is obtained entirely by means of tariffs charged to operators, which are calculated according to the volume of water supplied in the case of the water sector.

ARESEP has 175 staff but only 16 have been allocated to the water sector. The regulatory agency has the authority to establish regulations. The President (director-general) and the board of directors are appointed by Parliament for a period of four years. However, the regulator's independence statutes allow it to make decisions without needing the government's approval and, moreover, it is financially autonomous i.e. its budget does not depend on the government.

The tariff methodology applied to the water sector is based on a calculation of the operating, maintenance and capital costs, enabling a return on the capital invested, i.e. it is based on rate-of-return regulation. The regulatory interval is one year. The tariff-setting process involves public participation, where customers can express their opinions about the price changes.

In terms of customer protection, customers can file a complaint with the regulator, directly with the government or can even apply to the services of customer protection entities such as the *Defensoría de los Habitantes*. The regulator seeks to solve and mediate conflicts.

The regulator is compelled to prepare an annual report, which describes all its activities. In its turn, this report is submitted to Parliament.

2.6.3 Market structure

2.6.3.1 Ownership and management of services

As mentioned above, the water sector is mainly governed by public managing entities and AyA stands out owing to its role in directly and indirectly managing the WWS sector, within and outside its respective area of jurisdiction.

One can also note the participation of some private operators, who supply water to individual or smaller systems.

2.6.3.2 Players in the sector

The water sector in Costa Rica includes over 1,800 operators for the water service and only five for the wastewater service. However, AyA (especially) and *Empresa de Servicios Públicos de Heredia* (ESPH S.A.) have major importance because of their size and WWS coverage in the country.

WWS operators can be divided into public companies (AyA and ESPH), municipalities, committees managing water supplies in rural areas (*Comités Administradores de Acueductos Rurales* – CAAR), associations managing water and wastewater services (*Asociaciones Administradoras de Acueductos y Alcantarillados Rurales* – ASADAS) and, finally, private operators. Table 2.29 presents the numbers for each of these categories as well as the corresponding percentage of coverage in the country.

Table 2.29 Operators and their water service coverage in Costa Rica

Operator	Operators (No.)	Coverage of service (%)	
		2002	2007
AyA	1	43.2	46.5
ESPH	1	4.7	4.7
Municipalities	30	17.1	17.4
CAAR and ASADAS	1,800	24.4	25.5
Private	Unknown	9.0	4.0

AyA is a centralized public institution which reports to the Ministry of Health for all its activities. It is managed according to political criteria and this management frequently changes according to political cycles. AyA has been attributed responsibilities at the level of operating WWS, covering about 46% of the population of Costa Rica. The majority of these services are provided in urban areas and AyA supplies only 3% of the rural population.

ESPH is an autonomous multisector public company, governed by private law, which provides WWS and electricity services in 3 cantons located in the centre of the country and covers 5% of the total population of Costa Rica.

CAARs and ASADAS are small operators that supply water services in rural areas. Their activities depended, in the initial construction of the system, on subsidies from AyA, but not the services are operated without its help. Private operators represent isolated systems supplied by individual wells. Table 2.29 shows the different WWS players in Costa Rica according to their number and the corresponding level of coverage.

2.6.3.3 Integration of the sector and services provided

The water sector in Costa Rica is characterized by a horizontal and vertical integration of the respective services into a single operator, supplying the services for the whole water cycle, from water abstraction to distribution and from collection to wastewater treatment. It must be noted, however, that the wastewater service still represents a small percentage of coverage in terms of population. As such, in areas where this service is available, the operator that supplies the water service simultaneously manages the wastewater service.

Normally operators in the water sector only provide WWS, except for ESPH, which, as mentioned, is a multisector operator. Apart from these services, ESPH also provides electricity to about 100,000 inhabitants.

2.6.3.4 Numbers for the sector

In Costa Rica, the water sector has an annual production of about 350 million m^3, of which almost 53% is produced by AyA and ESPH (2006). The main source of water is obtained from groundwater, which accounts for about 60% of production. In this regard, it is possible to state that Costa Rica has an average daily consumption of 210 litres per inhabitant per day (including the volume of water which is not billed).

In 2006, the level of coverage of the water service was 98%. This high level of coverage is merely indicative, given that only about 50% of the water supplied is of a suitable quality, the majority of which is provided by AyA. There are also breaks in continuity, which jeopardise the reliability of the service.

In 2001, about 21% of the population was covered by the wastewater service. However, only 4.5% of wastewater was really treated. The largest share corresponded to coverage in the region of San José (the capital), which had a coverage of 50%, followed by the regions of Heredia and Cartago, both of which had about 15% coverage. Other regions had levels of wastewater service coverage of less than 10%.

In 2006, AyA produced about 310 million m^3 of drinking water and ESPH produced about 30 million m^3. Costa Rica was characterized by the fact that about 50% of the volume of water produced was not billed. In particular, AyA did not account for about 54% of the water it produced and ESPH did not bill about 48%. The level of micro-measurement for the water service is 96.4% in Costa Rica.

AyA collects about 50 million m^3 of wastewater and only treats 14 million m^3, while ESPH collects 5 million m^3 of wastewater and only treats 1 million m^3.

Water service operators in Costa Rica directly employ about 3,000 people. Table 2.30 summarizes the revenues earned and the investments carried out in the WWS sector in Costa Rica in 2006.

Table 2.30 Revenues and investments for the main operators in Costa Rica

	AyA		ESPH	
	Water	Wastewater	Water	Wastewater
Annual revenues (million €)	56.16	5.79	3.86	0.34
Investments between 2004 and 2006 (million €)	31.1		2.14	0.62

2.6.4 Rules of the game

2.6.4.1 Tariff setting

The WWS tariff setting needs to be approved by the Costa Rica regulatory agency, (ARESEP). The socio-cultural scenario, along with political instability, has had an impact on the financial fragility of operators. There is a large deficit between the revenues obtained and the costs incurred in providing WWS in this country.

The tariff systems do not allow the total costs of the service to be recovered. The tariffs are, above all, adjusted according to the need for short-term working capital, instead of being implemented on the basis of principles of economic and financial sustainability. AyA requests increases in the tariff system based on its short-term needs.

Furthermore, AyA has not been compensated by the government for assuming very relevant social functions of subsidization for water supplies in rural areas. This constitutes a heavy financial burden for AyA and is one of the main reasons for its financial difficulties. When the government abandoned subsidies for the rural sector, it was not aware of the financial impact that this measure would have on AyA and, consequently, on the whole water sector in Costa Rica.

The issue of cost recovery is clearly more serious in rural areas, where tariffs have no possibility of recovering the real cost of the services. In 2001, on average, rural tariffs represented the equivalent to 0.12 €/m^3, as compared to the 0.19 €/m^3 charged by AyA.

The tariff system that is usually applied by the operators comprises two components, namely a fixed component, which takes into account the fixed costs of the service, and a variable component, according to the volume of water consumed. Table 2.31 and Table 2.32 summarize the tariff systems established by the main WWS operators in Costa Rica, respectively, AyA and ESPH S.A.

Table 2.31 AyA tariff system for 2008

Tariff system		Variable component (€/m^3)			Fixed monthly component (€)
		0–15 m^3	16–40 m^3	>40 m^3	
Water					
Customer	Domestic	3.347	0.350	0.367	6.921
	Corporate	13.994	0.904	0.948	38.849
	Preferential	5.286	0.278	0.286	17.805
	Government	8.111	0.427	0.427	52.933
Wastewater					
Customer	Domestic	0.801	0.076	0.08	1.565
	Corporate	3.246	0.209	0.209	8.983
	Preferential	1.237	0.071	0.071	4.426
	Government	1.898	0.109	0.109	13.315

The tariff methodology applied in the water sector is based on rate-of-return regulation for regulatory periods of three to five years. The operating, maintenance and administrative costs are calculated along with the net assets, which allows for a given rate-of-return (*rédito de desarrollo*). In economic terms, the methodology adopted tries to balance total revenues with total costs, where the latter includes the return of the invested capital, since according to prevailing legislation tariffs must allow for a competitive return on capital and guarantee a suitable development of the activity.

Table 2.32 ESPH tariff system for 2008

Tariff system		Variable component (€/m^3)		Fixed monthly component (€)
		0–15 m^3	>15 m^3	
Water				
Customer	Household	2.000	0.243	4.218
	Commercial	9.595	0.641	15.535
	Institutional	4.278	0.484	7.485
Wastewater				
Customer	Household	0.911	0.067	1.790
	Commercial	1.300	0.095	2.557
	Institutional	1.040	0.077	2.046

2.6.4.2 *Quality of service*

Ever since it was established, ARESEP has collected information about the activities of the regulated entities. However, to date, this information has not been used in the regulatory process. Although the statutes of ARESEP state that it is to '*permanently evaluate the quality of the regulated public services*', this entity has not yet developed specific actions in order to provide incentives to improve WWS quality.

ARESEP does not play the role of promoting improvements in the quality of service of WWS operators. However, ADERASA (*Asociación de Entes Reguladores de Aguas de las Américas*) has developed a set of ten performance indicators, which seek to compare the performance of WWS operators in the Americas and thus, by means of benchmarking, encourage improvements in the quality of service. Table 2.33 shows the set of the ten performance indicators that have been adopted by ADERASA.

Table 2.33 Performance indicators adopted by the ADERASA

Indicator	Unit
Population served and connected to the water supply	%
Level of wastewater service coverage	%
Micro-measurement coverage	%
Importance of groundwater	%
Production per bill	m^3/bill/day
Consumption per inhabitant	l/inhabitant/day
Water losses	%
Bursts in the networks	no./km
Quality of drinking water (compliance)	%

To a certain extent, the reduced increase in water service prices throughout the country has concealed the poor quality of service in this sector. In 2005, only about 70% of the population was being supplied with water of a suitable quality. Moreover, 96% of all the wastewater that was collected was discharged into the environment without any kind of treatment, generating a high risk of contamination for public health and for water resources.

In most of the country's regions the capacity to produce water is very close to current demand. Thus, there is a very high risk of facing shortages in water production in the near future and, in fact, various cities have already been affected by this phenomenon and needed to draw on rationalization.

According to a news report dating from 2007, published in *La Nación*, about 17% of Costa Rican customers drink water that does not meet the minimum quality standards for human consumption and the programmes for improvement have stagnated. In terms of water that is not billed, the water sector in Costa Rica has a very high percentage of water losses compared to the water produced, revealing a serious level of inefficiency which greatly jeopardizes the continuity of the service and its economic and financial sustainability.

2.6.4.3 Public service obligations

In Costa Rica, prevailing legislation (Law No. 7593 of 1996) clearly defines the public service obligations that operators must respect, namely universality, continuity, quality of service and accessibility. However, the procedures to implement these obligations are not very transparent.

The legal framework makes no reference to establishing prices for customers with special needs, although some of the customers who live in more marginal areas get water services free of charge. It must be stressed that a high percentage of Costa Rican citizens live close to the poverty line.

2.6.5 Performance

Considering Costa Rica's socio-economic context, the water sector has demonstrated a reasonable level of performance. Some aspects of the institutional framework, especially the role of the regulator and the high level of coverage of the water service, are positive features to bear in mind.

However, the sector requires some reforms and large investments, particularly to meet the challenges of collecting and treating wastewater. The structure of the tariff system can and must be improved and there should be an unequivocal separation between the social functions of the state and the operators, especially as far as AyA is concerned. The universality of the sector should not be guaranteed at the cost of the economic and financial sustainability of the operators. The quality of service also calls for greater attention from the regulator.

2.7 FRANCE

2.7.1 Historical background

France is a democratic and semi-presidential republic, whose President is elected by direct and universal suffrage. It is the President who appoints the prime minister and presides over the council of ministers. The prime minister is responsible for coordinating the government. The French Parliament is bicameral, comprising the Senate (*sénat*), elected indirectly, and the National Assembly, elected by popular vote.

In administrative terms, France is organized into regions, which, in their turn, are divided into departments (*départements*) that are further subdivided into districts (*arrondissements*). With an area of about 547,000 km^2 and a population of about 64.4 million inhabitants, France is divided into 26 regions, consisting of 22 metropolitan regions, including Corsica, and 4 overseas regions, respectively Guadalupe, Martinique, French Guyana and Réunion. The departments comprise 96 metropolitan departments and 4 overseas departments (*départements d'outre-mer*). Each department is an administrative division as well as a state and a territorial collectivity (*collectivité territoriale*). The districts are divided into cantons, which subsequently are divided into municipalities (*commune*). Currently, France has 36,782 municipalities.

WWS in France are characterized by their universality and good quality. The large-scale private sector participation by means of concessions and leasing contracts (*gestion déléguée*) and the existence of river basins agencies (*agences de l'eau*), which charge taxes and seek to finance environmental investments, are some noteworthy features when compared to other developed countries.

As in most EU nations, the institutional framework governing WWS management was reorganized during the early 1990s and established by means of a vast number of legal diplomas.

The main normative documents include the Water Acts of 1964, 1992 and 2006, and all the legislation derived from them. Among the basic principles of this legislation it is possible to highlight the fact that water resources are not subject to any kind of public or private appropriation. From the legal point of view, water resources have the statute of *res nullius*; however, this is not the case with regard to the use of water at an individual level, where each property owner can use the water resources available in his property.

At an administrative level, the 1964 Water Act introduced the concept of the global management of the water cycle, developing a policy to control the quality of water and establishing the legal framework to implement the six water agencies, with a view to monitoring, promoting and financing the management of water resources. In financial terms, the law defined financial rules applicable to all the basins, based, at the time, on the principles of customer-pays and polluter-pays, imposing taxes on pollution and the consumption of water. In political terms, the law established the six agencies referred to, with the purpose of involving all the stakeholders in the decision-making process, while managing water resources on the scale of river basins.

In this context, the 1992 Water Act reinforced this structure, making it compulsory to prepare plans to manage water resources on the scale of river basins and on a local scale, implementing some essential principles regarding the water tariffs, the granting of licences to use water and public participation, among other measures. Moreover, this law also reinforced the penalties for environmental infractions and, last but not least, recognized water resources as a national patrimony (‘*patrimoine commun de la nation*’), explicitly acknowledging the economic value of water.

Diverse laws were enacted during the 1990s with the aim of ensuring an increased monitoring of WWS. The so-called ‘Sapin Law’, dating from 1993, had a great impact on the operators’ management. Although it does not refer directly to the water sector, it tries to prevent corruption and foster transparency in economic activities and government procedures. In fact, it precisely defined the legal framework to be applied in cases where responsibility for supplying local public services (water, wastewater, waste, etc.) was delegated to private operators, including rules regarding tender procedures.

Likewise, the 1995 Barnier Law, which limited the duration of contracts to 20 years, unless significant investments are required and prohibited an upfront payment by the private company, enabled greater competitiveness and consequently resulted in a reduction in WWS tariffs. It also made it compulsory for operators to publish an annual report about their activities, with information about quality and the prices charged for the services provided by the municipality, clearly reflecting the government’s concern for environmental issues and customer protection.

The 1995 Mazeaud Law, regarding public contracts and the delegation of public services, made it obligatory for the concession granting authority to publish an annual report and accounts, which also include the impact on the quality of service. It also implemented provisions that made it possible for competent and independent entities to carry out audits in these areas.

Concerning the environmental quality, wastewater began to be treated in the 1960s in France. Keeping in mind the growing environmental problems, the federal government created the Ministry for the Environment in 1971 and was one of the first countries in Europe to create such a ministry. Concerning the wastewater treatment and the quality of drinking water supply, French regulations have been greatly influenced by EU legislation, implemented by transposing and adapting diverse EU directives.

Finally, Law No. 1772 of 2006 defined the guidelines for the five-year plans for the water agencies (starting from the period between 2007 and 2012). This normative document transposed the Water Framework Directive to French national territory, addressing all the questions that arise while managing water on a river basin scale, such as support and actions for treating wastewater, the treatment and disposal of sludge, industrial effluents, solidarity for investments in rural areas, the protection of water abstractions, controlling water losses, preventing flooding, managing the aquatic environment and international cooperation.

Also in 2006, a Water Law was published, which regulated important issues such as the obligation to determine performance indicators, the creation of the national office ONEMA and the definition of rules to protect the water resources and the water cycle.

2.7.2 Legal and institutional framework

2.7.2.1 Responsibilities

Various public entities are responsible for the water sector. These include the Ministry of Health, the Ministry for Public Works, the Ministry for the Economy, the Ministry for the Environment and, especially, the Water Agencies, the National Agency for Water and Aquatic Environments (*Office National de l’Eau et des Milieux Aquatiques* – ONEMA), the French Institute for the Environment (*Institut Français de l’Environnement* – IFEN), the municipalities, the National Federation of Concession Granting Authorities and *Régies* (*Fédération Nationale des Collectivités Concédantes et Régies* – FNCCR), the French Court of Audit (*Cour des Comptes*), the French Standards Association (*Association Française de Normalisation* – AFNOR), the Operators Association (*Fédération Professionnelle des Entreprises de l’Eau* – FP2E) and the Consumer Associations (*Association de la Consommation éco-citoyenne* – CLCV and the *Coordination Nationale des Associations de Consommateurs d’Eau* – CACE).

Through its Department for Social and Sanitary Affairs (*Direction Départementale des Affaires Sanitaires et Sociales* – DDASS), the Ministry of Health has innumerable functions related to the water service, such as licensing abstractions and supervising the water quality. In its turn, the Ministry of Agriculture, through its Department for Agriculture and Forests (*Direction Départementale de l’Agriculture et de la Forêt* – DDAF), is responsible for wastewater treatment and monitoring of water resources quality. The Ministry for Public Works, through its Department for Equipment (*Direction Départementale de l’Equipement* – DDE), is responsible for assisting municipalities to manage WWS, constructing infrastructures and managing the public water domain. Finally, through its Department of Consumption, Competition and Fraud Prevention (*Direction Départementale de la Consommation, de la Concurrence et de la Répression des Fraudes* – DDCCRF) the Ministry of the Economy plays an important role in the area of public procurement and, especially, the delegation of services to private operators. The Regional Directorates (*Directions Régionales de l'Industrie, de la Recherche et de l'Environnement* – DRIRE) monitor the accounts and the activities of the operators that have been delegated responsibilities.

One of the main responsibilities of the Ministry for the Environment is to produce environmental regulations. Norms for wastewater discharge, norms regarding the quality of drinking water supply and the framework for water resources management are defined by EU directives.

The six French water agencies play an important role at the level of river basins, bringing together all the stakeholders in a sort of 'Water Parliament'. They have essentially financial functions, both in terms of obtaining revenues (e.g. taxes for water abstractions and the discharge of wastewater) and in terms of using these revenues in a suitable way (financing infrastructures). They are also responsible for preparing development plans, which are aimed, above all, at six priority areas, namely decontamination and purification in local communities, actions against industrial pollution, controlling pollution originating from the agricultural sector, improving the protection of water resources, drinking water supply and the preservation of natural aquatic environments. To a certain extent, the French water agencies apply the principles of the management model based on river basins.

ONEMA, which is supervised by the Ministry for the Environment, is a technical body that is a reference in terms of monitoring and knowledge about the state of water resources and the ecological functioning of aquatic environments. Its functions include disseminating knowledge about the WWS. It is involved in research programmes and stores the information obtained about the management of water resources electronically, assisting the National System for Information about Water (*Système d'Information sur l'Eau*). It also intervenes in the field of monitoring water use, supervising practices in this area. Finally, ONEMA has innumerable practical functions, using its knowledge and the specialized technical resources of its team while diagnosing the state of the environment and of French water resources.

IFEN's, now belonging to the Ministry for the Environment, mission is to produce and provide information about the environment in France. To this end, this institute organizes, collects and processes data that have been compiled about natural and technological risks, in association with the network of the main producers and users of environmental information. Moreover, the institute develops and harmonizes methods to process information, compiles statistics, promotes synergies between the statistics obtained and the geographical information that has been collected and prepares and produces environmental indicators and indicators for sustainable development in the areas that are under the supervision of the Ministry for the Environment. It works in close collaboration with the National Institute for Statistical Information (*Conseil National de l'Information Statistique*).

In France, the municipalities are responsible for the WWS, either individually or through various forms of municipal associations. Table 2.34 shows the respective proportions (IFEN, 2005). In particular, the mayor (*maire*) is responsible for all matters related to the WWS sector. The municipalities can operate WWS directly or contract a private operator to operate these services individually or jointly.

Table 2.34 Stages of the WWS production chain

Activities	Municipal (%)	Inter-municipal (%)
Production of water	28	72
Distribution of water	32	68
Wastewater collection	70	30
Wastewater treatment	61	39

FNCCR is a collective organization that was founded in 1934. It is formed by public and private operators of French local public services and defends the interests of its members at a national and European level. As the representative of its associated collectives, FNCCR has a great influence, especially concerning legislators.

Almost all private companies managing WWS in France are members of the FP2E (earlier known as the *Syndicat Professionnel des Entreprises de Service d'Eau et d'Assainissement* – SPDE). Its members, which include Alteau, Lyonnaise des Eaux, Saede, Saur, Société des Eaux de Fin d'Oise, Sogedo and Veolia Water and Nantaise des Eaux Services, employ 32,000 individuals and provide WWS to about 46 million inhabitants. In this context, the FP2E is the 'voice' of private operators to defend their interests.

The functions of the Court of Audit include monitoring the management of all public or private operators. Among its many activities, the Court of Audit supervises the efficiency, effectiveness and quality of these operators. It plays the role of auditor not just for accounts but also regarding performance and compliance with political objectives. In this context, the Court's role is not just limited to presenting criticisms but also includes a more in-depth approach, suggesting recommendations. The Court publicizes its conclusions, contacting directly all the stakeholders and after that its comments are made public. It has played an extremely important role in the water sector.

AFNOR has diverse functions in the water sector, such as standardization, publishing and giving out information, training and certification.

Finally, the CLCV, which was founded in 1952, and the CACE, which was created in 2001, bring together various regional associations. They are among the most important consumer associations in France and protect and defend customer interests. There are 14 consumers associations in France.

Figure 2.27 summarizes the institutional framework for the water sector in France (Marques, 2005).

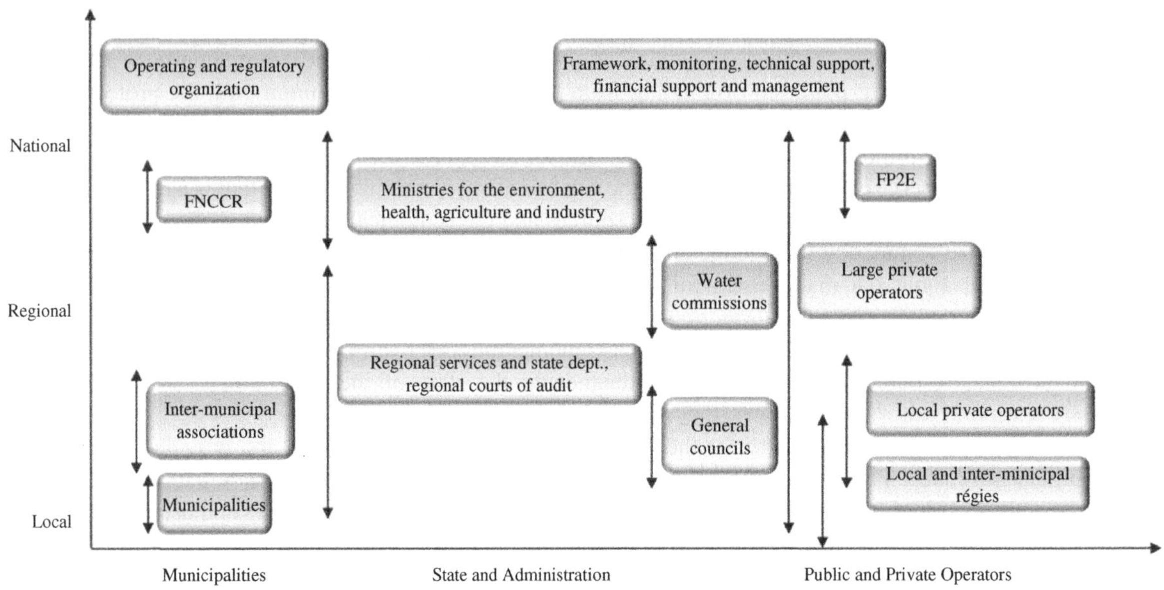

Figure 2.27 Institutional framework for the water sector in France

2.7.2.2 Regulation

In France, as is the case of most EU countries, there is no sector-specific regulatory agency for the WWS with functions such as setting and approving tariffs and defining norms for the quality of service control. In this field, the high number of systems and their limited size have also made it unfeasible to implement economic regulation or at least to regulate the quality of service to minimize the natural monopoly effects, the assets-specific features that characterize the WWS and the existence of incomplete contracts between municipalities and operators. In addition, the prevailing view in contractual regulation (regulation by contract) is that rights and obligations are already defined and that competition while accessing the market eliminates the failures of the water market.

In the past, the entities responsible for the sector agreed on the creation of a water regulatory agency (*Le Haut Conseil des Services Publics d'Eau et d'Assainissement*), which, owing to the difficulties of implementing economic regulation under the present framework of the WWS, was supposed to act as an observatory (*Observatoire de l'Eau*) and to put into practice sunshine regulation by means of publicizing the WWS performance. However, ONEMA will act as an observatory computing performance indicators of the sector.

From another perspective, owing to the public service aspects of the WWS, customers are relatively protected by French legislation and jurisprudence, particularly by means of the DDCCRF. Apart from this, in France, the administrative courts and all legislation are very effective. The Regional Treasury Directorates are also obviously involved in economic regulation and have the right to audit the balance sheets of the *régies* and, more importantly, the operators to whom responsibilities have been delegated. Moreover, these entities also play a central role in clarifying the notion of renewal and the provisions that were allocated for this purpose in the course of the delegation. The Court of Audit is another body that regulates and supervises the sector. It is a supra-governmental entity, which ensures greater respect and obedience.

Finally, it must be noted that, in the absence of a regulatory agency, the associations of the water sector, especially the FP2E, play an important role with regard to self-regulation, promoting the development of the sector. In this context, in 2005, the FP2E implemented a set of indicators for aspects related to the drinking water quality. In 2007, the FP2E expanded its set of indicators, encompassing the water sector performance as a whole, which proved to be a very useful tool to enhance the general quality of the information about the sector. As a nation that exports WWS technology and operators (which include the largest private groups in the world), France has an additional responsibility in the promotion of the quality of service.

2.7.3 Market structure

2.7.3.1 Ownership and management of services

In France, the municipalities (*communes*) are responsible for providing drinking water to the population and for collecting and treating wastewater. In case municipalities voluntarily decide to join forces, these activities are transferred to the intermunicipalities associations and are implemented by a commission to whom all the municipalities involved transfer responsibilities.

The majority of the population (75%) is supplied by private operators, who have had a presence in this sector since the eighteenth century. As mentioned above, the municipalities and associations are free to choose from organizational arrangements to manage and operate the WWS. The alternatives vary between direct management (*régie*) and delegated management (*gestion déléguée*). The latter includes leasing contracts (*affermage*) and concession arrangements. The operators that are constituted can have mixed capitals (public and private); however, ownership is always public.

2.7.3.2 Players in the sector

From the 36,569 municipalities in France, approximately 60% have a population of less than 500 inhabitants. About 29,300 out of that number have been organized into 2,000 intermunicipal systems, which supply WWS to about 66% of the French population. The municipalities comprise the remaining operators, which amount to more than 17,300. France has about 14,900 water service operators and about 14,400 wastewater service operators.

In France, models for WWS management vary between direct management, known as *régie*, and management that is delegated under a concession agreement. There is also the so-called *régie intéressée*, in which the entity to whom responsibility has been delegated (which only manages the system) receives a payment according to the volume billed, and the delegated management, or leasing, which is the most common organizational arrangement in France. In this model, the private operator manages services and charges tariffs to customers, but is not responsible for investments and as an alternative transfers the revenue to the municipalities in exchange for the amortization of assets. Normally, these contracts have durations of between 10 and 15 years. The average contract length is 11 years. More than a third of the operators have chosen to delegate their services, supplying water services to about 75% of the French population and wastewater services to about 52%. According to the *Cour des Comptes* (2003), leasing contracts (in which investments are financed by municipalities) represent about 85% of delegated management contracts. Concession arrangements (in which private operators finance investments) have longer durations than leasing. In both cases the municipality or the intermunicipal system establish the tariff system for the WWS and continue to own the infrastructures. Table 2.35 summarizes the organizational arrangements in the French WWS.

Table 2.35 WWS organizational arrangements in France

	Water service (%)	Wastewater service (%)
Public operators	27	45
Mixed companies	1	<1
Private companies	62	55
Veolia	39	28
Suez	19	18
SAUR	11	8
Other companies	3	<1

The largest French companies operating in the water sector are Lyonnaise des Eaux, part of the Suez Environment Group) and Veolia Water (Compagnie Générale des Eaux or Vivendi). Veolia Water provides water services to about 39% of the population, i.e. 24.5 million inhabitants, while Lyonnaise des Eaux supplies about 19%, i.e. approximately 14 million inhabitants, and SAUR, with a market share of about 11%, supplies about 6.4 million inhabitants.

The remaining private operators represent less than 3% of the water supplied to the French population. Table 2.36 summarizes the characteristics of the three main WWS operators in France (Barraqué & Bris, 2007). Table 2.37 depicts the evolution of the private sector concerning the water services in France (adapted from Reynaud, 2007).

Table 2.36 Statistics of the three main operators of water services in France

Company	Volume of business (million €)		Customers (millions)	Presence	Staff (No.)
	Total	Water			
Suez Environment (excl. energy)	11,100	5,900	80 for water; 50 for wastewater	70 countries	72,130
Lyonnaise des Eaux	1,700		14 for water; 9 for wastewater	2,600 municipalities	8,332
Veolia Environment (excl. energy)	25,240	8,800	108 for water and wastewater	57 countries (water)	271,153
Veolia Water France	4,400		24.5 for water; 16 for wastewater	8,000 municipalities	28,000
SAUR Group	1,400	1,100		8 countries	12,400
SAUR France	896		5.5 for water and wastewater	6,700 municipalities	6,800

Table 2.37 Evolution of the participation of the private sector in France

Year	1938	1964	1975	1979	2001
Public	83%	56%	50%	47%	30%
Private	17%	44%	50%	53%	70%
Veolia			42.3%	35.6%	51.1%
Suez			26.9%	24.6%	24.3%
SAUR			5.8%	10.4%	13.1%
Others			19.2%	16.4%	9.9%
SODEBA			5.8%	6.8%	–
SDEI			–	6.5%	–
Independent			–	–	1.9%
Population served (millions)	**42**		**52**	**53**	**60**

The city of Paris is a special case, even in international terms, since there are two different operators within the same city. A part of the service has been delegated to the Lyonnaise des Eaux (*'Paris rive gauche'*) and the remainder of the system has been entrusted to Veolia (*'Paris rive droite'*). The present situation, which, since 1985, has corresponded to two concession contracts, evolved from an earlier regime of leasing contracts. However, Paris decided to go back to the public management in 2010.

2.7.3.3 Integration of the sector and services provided

Operators in France, especially public operators, have very limited horizontal integration and a high level of vertical integration. The main private operators are characterized by vertical and horizontal WWS integration, i.e. in certain municipalities these operators encompass all the services related to the water cycle. Besides this, since they are conglomerates and specialize in providing public services, private operators integrate the provision of various public services, such as the water supply, wastewater, the distribution and production of energy, solid waste management and even public transportation, although not always altogether.

Wastewater services are sometimes provided by the operator responsible for the water services but in some cases are provided directly by the municipality, even if through an intermunicipal public utility within the scope of the water service.

2.7.3.4 Numbers for the sector

In 2006, the total production of water in France was 33.5 billion m^3. From this amount, about 6 billion m^3 were used for drinking water supply. The main source of water is groundwater, which accounts for about 60% of production (FP2E, 2008). The percentage of water losses is about 26%.

It has been estimated that the water sector in France directly employs approximately 60,000 individuals (FP2E, 2008).

In France, almost the whole population is connected to water distribution networks, i.e. more than 99%. With regard to wastewater, owing to the existence of local solutions, the level of wastewater service coverage is about 80%.

In 2004, the water average price for domestic was 3 €/m^3 (IFEN, 2007). Given that the average consumption per inhabitant and per day was about 165 litres, this results in an annual average bill of 177 euros per customer.

In 2004, the water agencies had a budget of 1.4 billion euros, primarily aimed at financing WWS infrastructures. Table 2.38 presents the costs in France for activities related to the WWS provision (CNE, 2006). Figure 2.28 presents an analysis of the financial scenario for 2006 for the water service, in millions of euros (FP2E, 2008).

Table 2.38 WWS costs in France in 2003

Costs	Water supply (10^6 €)	Wastewater (10^6 €)	On site sanitation (10^6 €)	Industrial effluents (10^6 €)
OPEX	5,850	5,299	165	1,095
Investments:	1,611	3,511	670	416
Network	1,200	2,325	–	–
Other assets	411	1,186	–	–

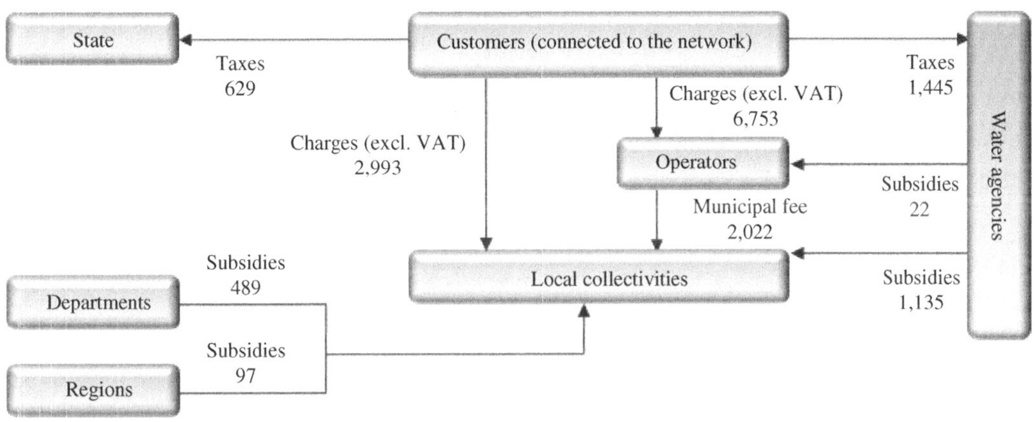

Figure 2.28 Funding for water services in 2006 (10^6 €)

According to the FP2E, private WWS had a turnover of 5.1 billion euros in 2004, which represents 42% of the total water and wastewater turnover.

In France, the WWS sector requires heavy investments. The main assets of the majority of the systems date from before 1960. The WWS patrimony includes more than 850,000 km of water mains and approximately 250,000 km of sewers. While the water supply systems have been evaluated at 130 billion euros, the wastewater systems represent approximately 75 billion euros. In this context, the costs for renovating these systems have been estimated at 1.5 billion euros and about 800 million euros per year, respectively. Currently, the rate of network replacement varies between 0.6% and 1% per year.

2.7.4 Rules of the game

2.7.4.1 Tariff setting

WWS prices vary according to the operational or organizational arrangement and can be mainly explained by the technical complexity of the large interconnected networks, which are preferably organized at an intermunicipal level and delegated to a private operator. In the case of the *régies*, the administrative authorities (municipalities) approve the prices every year, while in situations of delegated management the prices vary in line with the indexed formulae in the contracts, although they must always be approved by the municipal authorities. Table 2.39 shows the variation in the average price of water according to the operating model and form of organization (Barraqué & Bris, 2007).

Table 2.39 Variation of the water tariff according to organizational arrangement

Organization form	Organization			
	Municipal (€/m^3)	Intermunicipal (€/m^3)	Mixed (€/m^3)	Average (€/m^3)
Régie	2.19	2.85	2.48	2.54
Delegated	2.93	3.44	3.25	3.28
Mixed	2.60	3.04	2.97	2.97
Average	2.59	3.19	2.99	3.01

A significant part of investments in the WWS, about two-thirds, depends on funds collected by the Water Agencies, based on the principles of customer-pays (abstractions) and polluter-pays (discharge), and directly on the State (e.g. the Ministry of Agriculture). The remaining is obtained from customers through the tariff system. The tariff system generally comprises a fixed component and a variable component, which increases according to consumption. Drinking water supply is always metered. Table 2.40 summarizes the annual costs for domestic customers in four French cities for 2005 (IWA, 2006).

Table 2.40 Costs for domestic customers in four French cities (€)

	Water		Wastewater*	Other taxes
	Fixed component	Variable component*		
Bordeaux	43.75	154.33	195.69	61.56
Lille	23.02	156.07	178.27	85.03
Lyon	45.66	186.56	108.57	60.35
Paris	14.14	146.09	122.30	88.29

* Based on 200 m^3/year (€/200 m)^3

In France VAT is later applied to these costs, at the rate of 5.5%. Table 2.41 shows the WWS tariff system of five French cities (Veolia, 2005).

Table 2.41 Tariff system of five French cities (€/m^3)

	Fixed component	Variable component*	Wastewater treatment	Other taxes	VAT and/or other taxes	Total tariff
Paris	0.08	0.85	0.73	0.55	0.12	2.33
Syndicat d'Ile de France	0.11	1.28	0.98	0.56	0.16	3.09
Bordeaux (*Syndicat*)	0.26	0.91	1.17	0.37	0.15	2.86
Lille (*communauté urbaine*)	0.14	0.92	1.14	0.54	0.15	2.88
Lyon (*communauté urbaine*)	0.26	1.19	0.67	0.40	0.14	2.66

* Based on and divided by 200 m^3/year

The fixed component of the tariff system represents about 17% of the average bill and the variable component, which is based on consumption, corresponds to approximately 83% of it. About 90% of municipalities implement this two-part tariff system for the water service and about 38% for the wastewater service.

According to a study carried out by the NUS consulting firm (2008), in 2006 the average tariff for water and wastewater in France was 2.92 €/m^3, taking into account the largest cities. Based on the same study, it was possible to find out that when wastewater services were excluded from this tariff, the figure was 1.48 €/m^3. Furthermore, the average annual water bill in France is about 175 euros per inhabitant, of which 85 euros correspond to the water service and 90 euros to the wastewater services.

In 2005, costs for domestic customers for water bills could be divided in the following way: 46% corresponded to the water supply, 37% to wastewater services and, finally, the remaining 17% were related to taxes and charges. These taxes were for the water agencies and the *Fonds National pour le Developpement des Adductions d'Eau* (FNDAE), while the VAT was for the central government. The FNDAE was created in 1954 and is managed by the Ministry of Agriculture. It is specifically aimed at constructing water and wastewater infrastructure in rural and small municipalities. Table 2.42 shows the billing structure for water services in France for 2000.

Table 2.42 Billing structure for water services

Components	Percentage (%)
Production and distribution of water	42
Collection and treatment of wastewater	31
Water agency taxes	20
FNDAE tax	1
Tax for navigability (VNF tax)	0.5
VAT	5.5

It must be emphasized that the average cost of WWS almost doubled between 1990 and 2004. This is due to diverse reasons but one of the main causes is the substantial increase in the rates and taxes applied to billing.

Figure 2.29 illustrates this aspect (Barraqué & Bris, 2007). Figure 2.30 shows the costs incurred with various public utilities in the budget of French domestic customers.

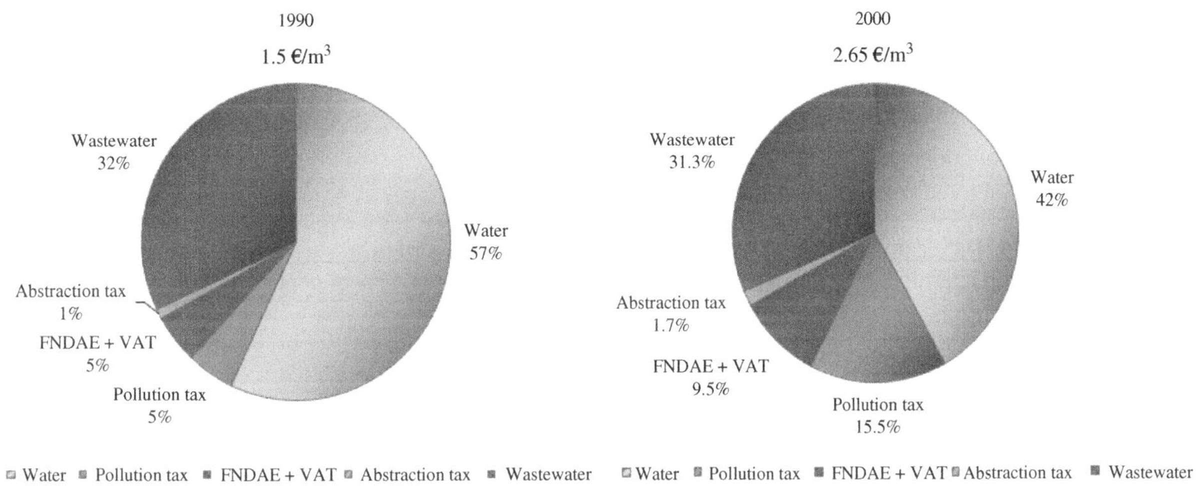

Figure 2.29 Evolution and breakdown of WWS billing in France

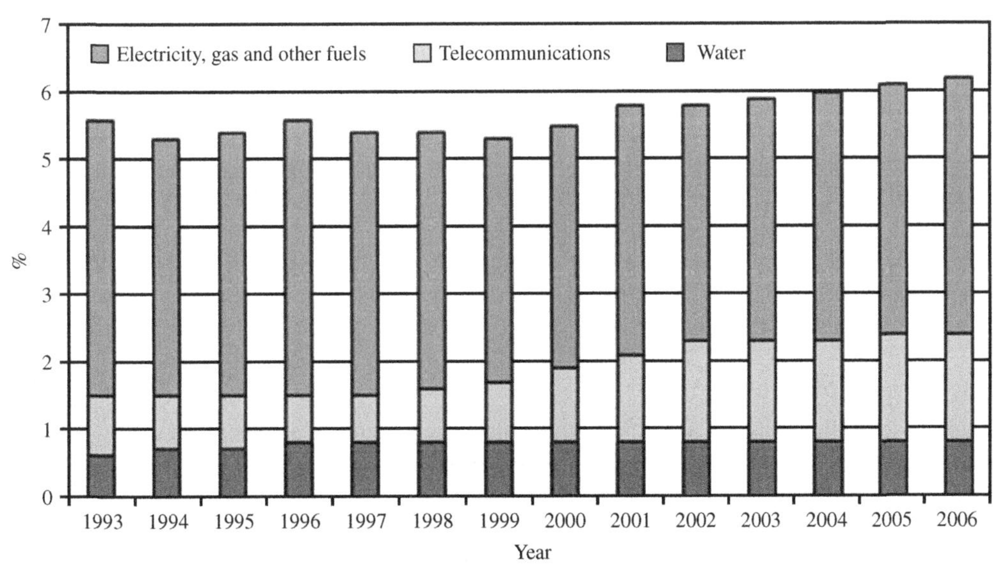

Figure 2.30 Annual evolution of costs of the French services of general interest

2.7.4.2 Quality of service

The French WWS sector is generally characterized by an excellent quality of service. The drinking water quality is high and in conformance with EU legislation. However, according to a survey by the Centre for Information about Water (*Centre d'Information sur l'Eau* – CIEAU), in 2004, only 71% of respondents felt that tap water was of a good quality. As in other EU nations, the drinking water quality is monitored at two levels, firstly by the service provider, on a permanent basis, and secondly by the competent authorities based on samples.

The quality of essential public services (price versus quality of service) is strongly debated in French society. If, on the one hand, existing rules ensure a minimum level of service, such as, for example, a continuous supply of water and mandatory information about the drinking water quality, which is to be given out to customers every year, on the other hand, there are no clear rules regarding the basic rights of customers. The 2006 Water Law established that customers have the right to access to water under socially and economically and acceptable prices. However, different voluntary benchmarking exercises have been tested or are currently being developed, by private and public operators as well as by diverse associations and government entities. Research bodies and/or universities have also played an important role in this area. Generally, private operators use their own sets of indicators as a tool for internal assessments.

In order to develop the water sector and consequently to promote the quality of service, the FP2E, in collaboration with the BIPE (a consultant) prepares an annual report about all the activities of the WWS in France. Apart from detailing all the parameters and elements that characterize the water sector, from 2008 onwards, it began to develop a set of individual performance indicators for the water and wastewater segments and common indicators. These indicators are presented below in Table 2.43, although they concern the private sector as a whole and not each operator individually (FP2E, 2008).

Table 2.43 Performance indicators carried out by the FP2E

Performance indicators	
Water supply	
Unscheduled interruptions of the service	<3% (93.18% of population served)
Compliance with bacteriological tests	99.69%
Compliance with physical-chemical tests	98.60%
Policy for patrimony	59.08%
Index for the protection of water resources evolution	59.85%
Wastewater	
Conformity of the wastewater discharge	82.31%
Compliance index with the wastewater discharge law	1.35
Degree of treatment of sludge	>95%
Common	
Rate of complaints	<1%
Debt of customers	<0.7%
Existence of an Advisory Committee for local services	7 committees for 10 private operators

In order to improve the quality of service and to consequently provide better services to customers it is relatively common for operators to develop a service charter (*charte de service*) where they promise to respect a set of minimum obligations. Table 2.44 presents the service charter of the operator for the city of Nantes (Nantes Metropole).

2.7.4.3 Public service obligations

As far as public service obligations are concerned, all the monitored results related to water quality are published at the Town Hall and are made available to customers for a period of three years.

A summary of the results of the drinking water quality tests must be sent to each customer once a year. This document explains the meaning of the main quality parameters and provides indications about the quality of water provided over the last 12 months. In case of emergency, particularly lack of compliance or abnormal pollution, customers must be informed as quickly and efficiently as possible according to the circumstances, using the radio, local newspapers, street notices and other means deemed to be suitable for this purpose.

Table 2.44 Service charter of Nantes Metropole WWS

Service charter
Permanent monitoring of the quality of water
Simplifying means of payment
Telephone assistance service 24 hours per day, 7 days per week
Systematic answers to mail
Intervention at households within 2 hours
Maximum delay of 2 hours for scheduled meetings
Subscriptions handled in 24 hours
New connections installed within 4 weeks
Meters installed within 48 hours
Prior notice of 24 hours if the service is to be interrupted
Information provided regarding unscheduled interruptions of water
Notification in cases of abnormal (excessive) consumption

The mayor is legally responsible for publishing, every year, a report about the price and quality of service (*Code Général des Collectivités Territoriales,* Article *L.2224-5*). The report must provide technical and financial information such as, for example, *inter alia*, a description of assets, compliance with specifications, water prices and evolution trends and outstanding loans. However, the municipalities do not always publish this document. Only 85% of municipalities fulfilled this obligation in 2000. Similarly, many municipalities do not circulate information about the quality of water, i.e. only 62% respected this requirement. In some situations (42% of cases) it was, in fact, the operator, as opposed to the mayor or a member of the municipal council, who prepared the reports.

With the purpose of improving transparency in the water sector, the French government has promoted access to delegation contracts, although without publishing or making them available to the public, by providing, for example, information about prices (calculations and variations) and the functions of the operator.

Municipalities with over 10,000 inhabitants are legally compelled to create local committees to inform customers about local public services, including WWS. Since 2002, according to the law regarding local democracy, the mayor must ask for the opinion of the *Commission Consultative des Services Publics Locaux* (CCSPL) before taking any decision about the management of the services, irrespective of whether it is a public or private operator. The majority of CCSPL members represent different interest groups.

There is no national social tariff for water services. However, there can be social support provided by the respective competent authorities and by some municipalities, which can subsidize a part of bills for low-income families in order to avoid the disconnection of services. This social support is partially financed by a tax deducted at source from operator revenues, which is about 0.1% of the water bill, created by the national convention for solidarity in issues related to water (*charte nationale 'solidarité eau'*).

The universal service is a principle that is currently in the phase of being studied, namely, the first article of the new Water Act of 2006 stipulates that '*Everybody has the right to access potable water, at prices that are acceptable to all'*. Some cities, such as Bobigny, on the outskirts of Paris, implemented measures, in March 2007, to prohibit the disconnection of water, wastewater and electricity services, based on an Environmental Charter and new housing rights.

Private operators founded the CIEAU with a view to providing a better information service to clarify doubts and questions by the public. This structure was created as a non-profit organization, governed by the 1901 law. The CIEAU compiles information, which it later makes available in a way that is accessible to all stakeholders. Its main function is to disseminate as much information as possible, which is later published, containing elements about the setting of WWS prices, quality of service, regulations and norms, drinking water and wastewater treatment and practical advice, among other issues.

2.7.5 Performance

The French model for the water sector is based on the public ownership of WWS, competition for the access to the market and regulation by contract. The various management models, subject to the local administrations, compete among themselves. The main problems are, above all, related to the existence of an oligopoly, constraining competitive access to the market, the fragmentation of the market, although private groups have significant economies of scale in each region, and sometimes a lack of transparency, despite the progress that has been achieved in the last decade. In any case, in the context of transparency, although the principles of best practices are enshrined in French law, neither are they frequently respected nor are infractions swiftly punished. Finally, it must be noted that the water sector in France is not yet self-sustaining. It is financed by diverse entities and the costs incurred on account of the services are significant.

In contrast, both the quality of service and the environmental standards are excellent and France is at the forefront of WWS in global terms, not only with regard to environmental aspects and technologies but also as far as operational and management aspects are concerned. The French water industry is strategically and historically important for the nation and plays a significant role in the economy.

2.8 INDONESIA

2.8.1 Historical background

Indonesia is governed by a presidential system, in which the President, who is simultaneously the head of state and head of government, is elected along with a vice-president. The main legislative body in Indonesia is the Popular Advisory Assembly (*Majelis Permusyawaratan Rakyat*), which consists of the People's Representative Council (*Dewan Perwakilan Rakyat*) and the Council of Regional Representatives (*Dewan Perwakilan Daerah*).

Indonesia has a population of about 235 million inhabitants and covers an area of approximately 1.9 million km^2. It is organized administratively into 33 provinces, managed by the respective governors, three of which are special territories and one is the territory of the capital city of Jakarta. The country is further divided, at a local level, into more than 450 districts and municipalities, governed by local authorities and mayors.

The Asian Crisis of 1997, which affected Indonesia in 1998, had serious consequences for the country, including in the water sector, particularly in Jakarta. In this context, the water service in Indonesia underwent several changes, especially from the 1990s onwards, with the entry of private operators who began to manage services and the subsequent associated institutional reforms.

This situation during the late 1990s resulted in political reforms for the water sector, characterized by a greater concern for increasing levels of access to the service, the sustainability of the service provided and the promotion of new investments in the water supply systems. In this regard, in 1999, the government enacted Law No. 22, which was aimed at Regional Autonomy, assigning to local governments and regional companies (*Perusahaan Daerah Air Minum* – PDAM) the responsibility for investments as well as for system improvement. The objective was to improve their effectiveness and service coverage.

The WWS sector is regulated by Law No. 7 of 2004, relating to Water Resources, by Law No. 8 of 1999, on Customer Protection, by Law No. 23 of 1997, concerning Environmental Management, by Government Regulation No. 16 of 2005, on Development on Water Supply Provision System, and, finally, by Law No. 23 of 1992, regarding Public Health.

Law No. 7 of 2004 and Government Regulation No. 16 of 2005 regulate the drinking water supply, making it necessary to supervise and monitor the quality of drinking water and ensure that it is supplied at an accessible price. These normative documents promote a balance between the efforts of customers and operators and encourage improvements in efficiency and an expansion of the water service. Moreover, this law defines the different players that depend on the Ministry for Public Works, with responsibilities related to water resources, and their functions, likewise enabling the participation of private companies and community organizations as operators of water services in the Indonesian water market.

By means of its article 3, Law No. 8 of 1999 established the legislative framework for protecting customer interests, simultaneously aiming at increasing social awareness in this area.

In terms of environmental legislation, the Environmental Management Act or Law No. 23 of 1997 provides the legal framework for the WWS sector, namely by regulating environmental quality, including norms for wastewater discharges.

Finally, the Health Act, or Law No. 23 of 1992, regulates the quality of water and, among other measures, defines the minimum requirements for providing drinking water supplies.

Apart from this legal framework for the national context, in Jakarta, with a view to stimulating the development of the water sector, the provincial government opted to issue concessions to two companies for the WWS sector, which, until then, had been provided by the PAM Jaya (the former public provincial operator). The first step towards this objective was the signing of a cooperation agreement (CA) in June 1997, which involved all the interested parties, namely the PAM Jaya and the two concessionaires.

However, owing to the Asian financial crisis in 1997, which affected the criteria defined in the CA, the contract proved to be unfair and relatively 'heavy' for customers. Keeping in mind the recommendations of the Regional Council, the CA inevitably needed to be renegotiated, a process that began in 2000.

After a long process of negotiation, which ended in October 2001, a new CA was signed (*Restated Cooperation Agreement* – RCA). With a view to primarily protect the interests of customers, as well as those of the other parties involved in the RCA (PAM Jaya and two concessionaires), a regulatory agency was established, known as the Jakarta Water Supply Regulatory Body (Badan Regulator PAM DKI-JWSRB).

2.8.2 Legal and institutional framework

2.8.2.1 Responsibilities

In Indonesia, responsibility for the WWS sector has been divided amongst diverse entities, both at a central and at a regional level, reflecting the high level of decentralization in this sector. At a central level, the main entities include the Ministry of Health (*Departemen Kesehatan*), the Ministry of Public Works (*Departemen Pekerjaan Umum*), the Ministry for Public Housing (*Departemen Dalam Negeri*), the Ministry for the Environment (*Badan Pengendalian Dampak Lingkungan*) and, at a regional level, the regional governments. The National Planning Development Agency (*Badan Perencanaan dan Pembangunan Nasional* – BAPPENAS), the Association of Operators (*Persatuan Perusahaan Air Minum di Seluruh Indonesia* – PERPAMSI) and the JWSRB are some other entities that also play an important role in this sector.

The Ministry of Health (MoH) is responsible for establishing guidelines, promoting public health and monitoring quality of drinking water regulations. It establishes directives and guidelines for the regional governments with regard to developing the wastewater service and implementing sanitary and environmental awareness programmes. The activities of the MoH in this scope are primarily aimed at rural areas.

The Ministry of Public Works is responsible for all works at a national level, so as to ensure that the regional public services are complying with technical norms. This ministry is responsible for providing due assistance to the water sector. Within this ministry, the Directorate-General for Human Facilities (*Cipta Karya*) promotes suitable assistance and technical guidance for the regional governments to conceive and build infrastructures. The Directorate-General for Water Resources is responsible for managing, allocating and protecting surface water resources and, keeping in mind that Indonesia is practically sustained by surface water, it thus plays a very important role in this sector. This agency is also responsible for implementing projects in the water sector, such as the Integrated Infrastructure Development Programme (IUIDP), established in 1990.

The Ministry of Home Affairs (MoHA) supervises decentralization and monitors the non-technical aspects of the performance of regional governments, including the PDAMs. This ministry has been entrusted with the task of promoting directives with regard to tariff setting and evaluating and monitoring the financial and operational performance of the PDAMs. The activities of the MoHA are primarily aimed at urban areas.

The main function of the Ministry for the Environment is to produce legislation and regulations regarding diverse aspects of environmental protection, which includes the water resources.

The BAPPENAS is responsible for planning and coordinating urban and rural infrastructure in the process of national reforms. Apart from this, the agency plays a fundamental role in formulating policies regarding water supply, wastewater services and waste management.

The regional governments enjoy regional autonomy and are directly responsible for the water and wastewater services. These governments also own WWS infrastructure and are responsible for regulating the PDAMs, safeguarding compliance with legislation in the WWS sector as well as WWS operation and maintenance.

PERPAMSI was founded in 1972 and is an association of local-government-owned water supply companies (water supply operators) in Indonesia. Presently, this association has about 300 water companies as its members, apart from 60 extraordinary members who play a role in the water sector (suppliers, manufacturers, contractors, etc.). In this scope, the main functions of this association include disseminating best practices in different water service areas and training water service staff.

Finally, notwithstanding its limited regulatory jurisdiction, considering Indonesia's dimensions, JWSRB, which was established by a Governmental Decree of the Province of Jakarta in 2001, plays a fairly important role in the water sector, functioning as an independent, impartial and professional body.

2.8.2.2 Regulation

The sector-specific regulator for the water service in Indonesia is the above-mentioned JWSRB. However, unlike what is usually the case outside Indonesia, this is a local entity which only functions within the jurisdiction of the city of Jakarta.

One of the functions of JWSRB is to ensure a balance of interests between customers and water operators, with an emphasis on protecting the rights of customers to obtain the water at a reasonable price and suitable quality. The JWSRB is also important for the credibility of the sector and thus to promote the participation of the private sector in water activities. Various committees have been established within JWSRB to facilitate its functions. These include the Tariff Committee and the Public Opinion Committee (KPPM), which, amongst other functions, carries out a survey regarding the quality of service, and the Specialized Committees.

In the course of its functions to protect customer interests, JWSRB is empowered to monitor the levels of quality of service, namely with regard to water supply, in terms of a suitable quantity, quality, continuity and pressure, with acceptable and fair tariff systems (protecting customers against monopolies). It improves transparency in the water sector and promotes the establishment of community water organizations, whether they are operators or not. Apart from this, JWSRB encourages and helps to set associations of water customers while likewise promoting the establishment of a water forum.

Organizations of water customers are entities that try to protect the interests of customers. These organizations can be residents' associations or entities that are registered as societies or with the local governments. In Jakarta, the organization known as the Water Supply Customers Committee (KPAM) plays an important role in the provision of the water service throughout this province. Its managerial members need to be approved or appointed by JWSRB.

JWSRB promotes the development of an annual water forum amongst the diverse stakeholders (operators, customers, government agencies, professional associations, universities, community leaders and NGOs). The main function of the forum is to discuss WWS matters and customer interests in this area. The forum is usually called the FKPM (*Customers Community Communication Forum*).

In 2006, JWSRB had operating costs of about 257,000 euros. Its budget is wholly guaranteed by payments made by customers, corresponding to 0.22% of the water bill. Presently, this regulatory body employs 30 staff, comprising 5 members with managerial functions, 15 assigned to the technical department and 10 to the administrative department. Figure 2.31 shows the organizational structure of JWSRB. Once again, it is important to stress that no regulatory bodies exist in other provinces and other government authorities directly perform these regulatory functions.

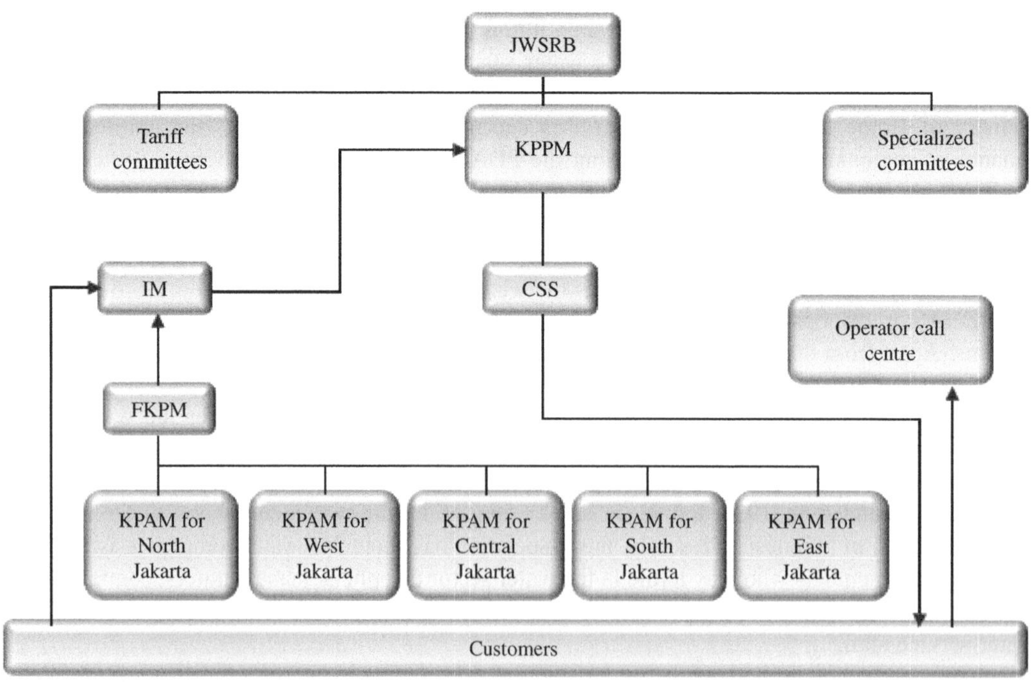

IM – Information management; CSS – Customer satisfaction survey

Figure 2.31 WWS regulatory framework in Jakarta

RCA stipulates that the concessionaire must be remunerated by means of a fee paid by the concession granting authority, according to the volume of water sold every semester, while customers pay the water tariff, approved by the governor, to the concession granting authority. This fee represents the payment of the concessionaires for their services and investments, by means of the volume of water sold. It is calculated on the basis of financial forecasts prepared by the concessionaires up to a predefined rate of 22% per year. It is indexed to the CPI each semester and is applied uniformly to all categories of customers. Thus, part of the business risk is transferred to the concession granting authority.

The water tariff, which represents the price paid by customers per m^3 of water consumed and which varies according to the kind of customer, is based on the affordability principle and cross-subsidization. Following article 26 of RCA, the authority that sets the tariffs is subject to the local government or city council of the province of Jakarta.

Ever since the regulatory body was established, tariff proposals are submitted through it for the governor's approval. In order to ensure the sustainability of the cooperation and to protect customer interests, automatic tariff adjustments were incorporated on the basis of the Governmental Decree No. 2459 of 2004. This was followed by a tariff system rebates, which considers the performance of the concessionaires while setting average tariffs.

Finally, the association for the water operators, PERPAMSI, also plays an important role with regard to the water sector self-regulation by means of its diverse functions, especially by disseminating best practices among its members. Moreover, this association has considerable power near government bodies and can influence the decision-making process.

2.8.3 Market structure

2.8.3.1 Ownership and management of services

As in most countries, the WWS in Indonesia are mainly provided by the public sector. Cultural reasons and the prevailing institutional structure in Indonesia have contributed to this situation.

Despite extensive government efforts to develop the participation of the private sector in the WWS sector, this only began to have an impact during the 1990s. The aim was to increase the availability of financial resources and to improve the efficiency and performance of WWS providers as well as their self-sustainability in terms of funding.

A significant number of initiatives to promote the participation of the private sector have been implemented but have generally been unsuccessful. As a consequence, the private sector has shown little interest in operating and maintaining water systems in secondary cities, due to the absence of a suitable regulatory framework and to the opposition of existing operators, PDAMs and the public in general.

In any case, the privatization arrangement that has been adopted in Indonesia only includes the management and operation of WWS and never the ownership.

2.8.3.2 Players in the sector

In Indonesia, water is supplied to the majority of the population through regional water companies, usually known as PDAMs. Until 1968, only some cities had public water supply systems (PDAMs), which grew as the country developed. In 1995, there were 306 PDAMs. However, with the general Asian crisis, they decreased in number, although they have once again resumed a positive growth. Presently, water is provided by 319 PDAMs throughout Indonesia.

About 245 PDAMs, i.e. 77% of all operators, supply water to fewer than 25,000 customers. This fact, along with the socio-economic situation of the country, indicates a clear weakness of the operators in terms of the self-sustainability of the water service.

Currently, the presence of the private sector in water supplies is essentially due to private participation in the two concessions for the city of Jakarta (PAM Lyonnaise Jaya/Palyja and Thames PAM Jaya/TPJ Thames/Aetra), the private operator in Batam, the Adhya Tirta Batam (ATB), which supplies about 900,000 customers, and a mixed company in Bali (Bali Province).

In 1997, the government of the province of Jakarta invited two multinational companies, Lyonnaise des Eaux (RAM Lyonnaise Jaya/Palyja) and Thames Water International (PAM Jaya/TPJ Thames), to submit proposals for managing and improving water services for Jakarta. These two companies were invited to become partners of the PAM Jaya to serve people in Jakarta and each of them was given responsibility for managing the service in one part of the city. The main objective of this project was to ensure the self-financing of the water service by means of the tariff system, even though these concessions were also meant to further social principles. This idea of dividing the city was, above all, an attempt to launch competition between the operators, resulting, at least in theory, in a quality of service improvement.

Palyja supplies water to the western part of Jakarta, serving about 52% of the 4.45 million inhabitants who live in this area, while the TPJ/Aetra (the new company replacing TPJ) provides water to the eastern part of Jakarta, supplying about 49.6% of the 4.43 million inhabitants of this zone. The two areas of cooperation are delimited by the Ciliwung River. Palyja is controlled by Suez, which holds a 51% stake in Palyja, while the remaining shares are divided between Astratel and Citigroup. Thames Water from the UK originally owned TPJ but today this operator belongs to Acuatico, a Singapore-based firm formed with Indonesian capital.

Before the participation of the private sector, the water service in Jakarta was managed and operated by PDAM DKI Jaya (PAM Jaya), which belonged to the government of the province of Jakarta (PEMDA DKI). As a public entity, the PAM Jaya has operated the water service in Jakarta since 1922.

In rural areas, most families have their own water sources, normally shallow groundwater, rainwater or surface water from nearby rivers and springs. Community systems and/or cooperatives provide water to about 30% of the rural population.

On the other hand, the wastewater service is provided in only ten cities, through seven PDAMs, one municipal company and two systems managed directly by the local government. The largest operator of this service is that of the city of Jakarta, through the municipal wastewater company PD PALJAYA. Only 3% of the population of Indonesia has access to wastewater services.

External assistance for developing the water supply system has proved to be extremely important for Indonesia. This assistance has been provided through international financial institutions, namely the World Bank and the Asian Development Bank, bilateral agreements with countries such as the United States of America (USA), Canada, Japan, Australia and the Netherlands and, finally, through international organizations such as UNICEF.

2.8.3.3 Integration of the sector and services offered

Some WWS are made available by multisector operators, which provide water, wastewater and urban waste disposal services, the PDAMs. As referred to above, the wastewater service in Jakarta is provided by different entities. Wastewater service is provided by PD PALJAYA, while water supply service is provided by two operators (Palyja and Aetra).

2.8.3.4 Numbers for the sector

About 83 billion m^3 of water are tapped every year in Indonesia, of which only 8% is used for human consumption. Agriculture accounts for the bulk of the water consumed, representing 91% of total consumption. The remaining 1% is consumed by the industrial sector.

About 90% of the population lives in areas under the jurisdiction of one of the 319 PDAMs, but only 39% of urban residents (18% of the Indonesian population) have access to the water service.

These shortcomings are even more visible in the wastewater services sector, where only 3% of the urban population has access to the wastewater service and out of this percentage only 2.25% receives some kind of treatment. In this scope, operators of wastewater services annually collect about 10.5 million m^3 of wastewater. The water services are characterized by fairly high water losses, varying between 30% and 50%, depending on the region.

Indonesia needs to increase its investments in the water sector by at least 1% of its Gross Domestic Product GDP, i.e. 2.6 billion euros every year, in order to achieve the Millennium Development Goals.

In 2005, Indonesia spent about 3.8 billion euros on infrastructure. This expenditure was mainly financed by the local governments (42%) and by the central government (42%), followed by the provincial governments, which contributed with 16% of the total expenditure. In this context, it has been estimated that expenditure with the water service represented about 10% of the total government outlay on infrastructure. Table 2.45 shows the distribution and percentages of expenditure and investments at the different administrative levels (World Bank, 2005).

2.8.4 Rules of the game

2.8.4.1 Tariff setting

In Indonesia, as in most developing countries, despite the enormous efforts of the authorities responsible for this sector, the tariffs of the water services do not even cover the operating and maintenance costs. Moreover, the sector suffers from a high percentage of non-payment of bills as well as water losses.

The tariff system practised in the water sector in Indonesia, especially by the PDAMs, according to the guidelines issued by the Ministry of Home Affairs (MoHA's regulation No. 23 of 2006), consists of the use of volumetric tariffs, with or without blocks. When blocks are used, they are normally progressive. The tariff structure is planned to enable cross-subsidization for low-income customers. The value of the tariffs is set by the local government, with the approval of the city council (municipal assembly), based on customers' affordability and their socio-economic situation. Naturally, affordability is also considered in conjunction with the cost of water production. The average costs that domestic customers bear range between 1% and 4% of the family income, although in some situations the cost of accessing the water service is much higher.

Table 2.45 Investments and expenditure at different administrative levels

	Central	Provincial	District	Total
Infrastructure (million €)	1,600	600	1,600	3,800
Total	24,800	2,600	9,800	37,200
% on infrastructure	6.4%	23.7%	16.3%	10.3%

In the Jakarta system, in particular, there are two kinds of payments: the fee or water charge and water tariffs. The water charge, which is paid to the concessionaire, is adjusted every semester, using an indexing formula, which considers factors such as foreign exchange indexes, the CPI, variations in the costs for labour, chemicals, electricity, and raw water/treated bulk water prices. The water charge WC is reformulated (rebased) every five years. On the other hand, the water tariffs are directly related to the service provided by the operator and seek to recover costs, without undermining the social principles mentioned above.

A part of the tariff system revenues is also retained by PAM Jaya for social purposes, namely to fight unemployment (*golden handshake*), although the real application of these funds is not known.

Although there are no differences between the two concessionaires in terms of the tariff system, the water charges are different. The tariff is obtained from customers, according to a tariff system, which is structured on the basis of a variable progressive component, varying with the consumption of water. The regulator calculates, reviews and proposes the tariff to the governor of the province of Jakarta. A Tariff Committee was created within the regulator with the specific objective of setting tariffs. The RCA, which was established between the public authority and the concessionaires, stipulates that the level of the tariffs must be adjusted to the financial capacity of customers in the province of Jakarta. Table 2.46 summarizes the tariff system implemented by the two concessionaires in Jakarta.

Table 2.46 WWS tariff system in Jakarta for 2008

Category		0–10 m^3	11–20 m^3	>20 m^3
I	Naval dockyards, fire-fighting services, social and charitable facilities and orphanages;	0.039	0.039	0.039
II	Public hospitals, low-income housing, washing stations, water trucks and very simple apartments;	0.039	0.039	0.071
IIIA	Simple housing units and apartments;	0.174	0.238	0.284
IIIB	Medium housing, medium apartments, kiosks, small repair workshops, small commercial establishments and non-commercial private institutions;	0.248	0.312	0.397
IVA	Luxury apartments, embassies/consulates, government offices, buildings of international entities, commercial buildings, tailors, military buildings, medium-sized commercial establishments, shops, restaurants, private hospitals, health clinics/laboratories, lawyers' offices, small industries and medical clinics;	0.362	0.440	0.532
IVB	Luxury hotels (1–5 stars)/motels, saunas/beauty salons, night clubs/cafés, banks, large repair workshops, companies, skyscrapers/condominiums, refrigeration companies, textile industries, warehouses/other industries, boats and PT Jaya Ancol;	0.692	0.692	0.692
V	BPP Tanjung Priok (Port).	0.816	0.816	0.816

With regard to the wastewater service, the prevailing tariff system normally uses the characteristics of the dwelling units as a reference, but there are also operators, such as PDAM Bandung (West Java Province) and PDAM Banjarmasin (South Kalimantan Province), which determine the prices for the wastewater service on the basis of the tariff system for the water supply.

For Jakarta, the PALJAYA defines the tariff system for the wastewater service according to a fixed rate, based on the size of the dwelling which, for domestic customers, represents an average of about 0.60 euros per month.

2.8.4.2 Quality of service

The water sector in Indonesia has diverse shortcomings. Although in general there tends to be a continuous service, in some areas the water service is available only intermittently and has inadequate pressure. In terms of the quality of drinking water, about 30% of the water distributed by the water companies is contaminated.

In order to monitor the performance of operators and assess customer satisfaction, JWSRB has developed a set of performance indicators, encompassing technical aspects, such as quantity, quality, continuity, pressure, water measured, water losses and new connections, and non-technical aspects, such as billing, service interruptions, complaints and the tariff system.

Performance indicators are an important tool for evaluating the performance of operators and are also used in the context of the Customer Satisfaction Survey (CSS). This survey is carried out to ascertain the degree of customer satisfaction with regard to the water service. The CSS was implemented according to the Law No. 8/1999 on customer protection. The CSS is one of the tools used to monitor the water services and provides a basis for future tariff regulation policies. More specifically, the CSS is an instrument that helps to achieve the short-term objectives that have been defined, such as identifying areas that have development problems and hurdles for expansion, thus supporting the decision-making process. Table 2.47 provides an overview of the performance indicators developed by JWSRB applied to the province of Jakarta.

Table 2.47 Performance indicators adopted by JWSRB

Number	Description	Performance indicators
1	Continuity	Available 24 hours a day Fewer than 3 interruptions of the service per week
2	Pressure	Reaching second floors
3	Quantity	Adaptability to daily needs
4	Quality	Compliance with quality norms and physical observations (turbidity, colour and smell)
5	Meter reading	Monthly reading (always on the same day) and accuracy of the reading
6	Water bills	Bills received monthly
7	Answers and complaints	Operator is capable of receiving complaints and requests 24 hours a day
8	Meter checking and calibration	Meter checked every 4 years or, at least, recalibrated
9	Communications with the operators	Availability of communications services on the part of the operator (at least one telephone number available 24 hours a day)

These indicators are later published in a report prepared every year by the JWSRB and are publicized through the media (sunshine regulation). However, imposing this system to ensure an efficient and effective water service, along with the strong pressure by government entities, will require heavy investments in the sector, which will thus result in substantial increases in the water tariff. This fundamental reality is based on the full cost recovery principle. Since the supply of drinking water is also viewed as a basic human need, the establishment of tariffs must consequently keep in mind the financial capacity of the customers being supplied. There have been some controversies in the water sector in Indonesia in this regard.

In this context, three pillars already mentioned (FKPM, KPAM and Information Management) form the so-called 'Water Voice System'. The Water Voice is an independent entity that is not part of the operator and stands for the interests of customers, in terms of both the tariff system and the quality of service being provided.

2.8.4.3 Public service obligations

Keeping in mind the socio-economic context of the majority of Indonesia's population, the geographical conditions and the population density that characterize the nation, there is no law in Indonesia that explicitly guarantees the universality of the service. However, access to the water services has been taken into account with the recent Water Act of 2004 as well as Government Regulation No. 16 of 2005, which simultaneously guarantees customer protection.

Operators normally have call centres and customer support offices, especially to handle complaints. For example, PAM Jaya classifies complaints into 8 categories: interruption of the service/low pressure, bursts, meter irregularities, low quality of service, works, inaccurate billing, absence of meter readings and others.

In Jakarta, the RCA stipulates that the concessionaires must achieve technical objectives and maintain a minimum quality, such as the aspects shown in Table 2.48. Amongst the stipulated parameters and indicators one can find parameters regarding the volume of water billed, the non-revenue water volume, the number of connections, the level of coverage of the service, the quality of the drinking water supplied to customers, the water pressure, client relations, interruptions in the water service and new connections.

Table 2.48 Minimum quality levels for the WWS sector in Jakarta

Pressure at customer sites
Maintaining pressure levels at 7.5 atm
Customer relations
All calls must be answered within 30 seconds
Answers to complaints regarding mains within 2 hours
Answers to complaints regarding the interruption of services within 4 hours
Answers to complaints regarding the quality of water within 6 hours
New connections
New connections must be established within 1 working day, after all payments and necessary paperwork have been resolved
Repairs for interruptions in the distribution service
Repairs will be carried out according to the following parameters:
Tertiary mains up to 100 mm under normal conditions within 6 hours, under difficult conditions within 24 hours
Secondary mains between 150 mm and 250 mm under normal conditions within 12 hours, under difficult conditions within 24 hours
Primary mains between 300 mm and 450 mm, under normal conditions within 24 hours, under difficult conditions within 48 hours
Primary mains larger than 450 mm within 72 hours

2.8.5 Performance

Currently, in Indonesia, the WWS sector is characterized by the existence of various problems. An inadequate management, insufficient political efforts and will, scarce government resources and an insufficient regulation are some of the factors that need urgent attention. The four main areas of intervention are related to the financial sustainability of the infrastructure for the services, the need to balance public and private interests by allocating risks between the stakeholders, promoting competition and adequate regulation and the need for restructuring, in order to improve and encourage transparency in the sector.

Other priorities are linked to the self-sustainability of the PDAMs, namely by means of the tariffs being charged, allowing for cost recovery and the management of the service without political interference. Owing to the limited dimensions that characterize PDAMs, incentives could be provided for some mergers, resulting in economies of scale. Along with these factors, the aspect of the non-payment of bills for the water service is also very important and is an area that requires intervention.

Despite the difficulties faced by the sector, the authorities have been implementing innumerable measures that intend to attenuate, or even eliminate, some of these shortcomings. In this context, the institution of a regulatory body in Jakarta, as well as the entire structure to protect the interests of customers, is a great step forward in terms of developing the quality of the water services being provided.

2.9 ITALY

2.9.1 Historical background

Italy is a democratic republic. The 1948 Italian Constitution established a bicameral Parliament, which consists of a Chamber of Deputies and a Senate, and an executive system comprising a Council of Ministers, headed by the prime minister. The President of the Republic is elected by Parliament for a seven-year term, along with a certain number of regional delegates. The President chooses the Prime Minister, who then proposes a cabinet of ministers subject to the President's approval.

In 2007, Italy had a population of about 59 million inhabitants, distributed over 301,333 km^2. It is administered at three organizational levels: 20 regions (*regioni*), about 100 provinces (*province*) and over 8,000 municipalities (*comuni*). However, five of the above-mentioned regions have a special status, (Friuli-Venezia Giulia, Sardinia, Sicily, Trentino-Alto Adige and Valle d'Aosta), which gives them greater autonomy to legislate about diverse matters that are not state monopoly. This special status is due to cultural, linguistic and geographical reasons.

Historically, as in most European countries, the municipalities are the main authorities responsible for WWS provision. However, in order to promote greater self-sustainability in the water sector, and consequently more robustness, the Italian government began a process of reorganization concerning the level of local responsibility in the sector. This reform began with the publication of Law No. 36 of 1994, usually known as the Galli Law (the name of the first member of the Italian Parliament who signed the legislative bill), and was based on the following pillars: a) the creation of integrated water services, encompassing the services of abstraction, treatment and distribution of water and the collection and treatment of wastewater, excluding stormwater networks; b) the definition of Optimal Territorial Areas (*Ambito Territoriale Ottimale* – ATO), based on hydrographical and politico-administrative criteria; c) the identification of ways of cooperation between local authorities, so as to form the body known as the Authority for Optimal Territorial Areas (*Autorità di Ambito Territoriale Ottimale* – AATO), along with its respective organizational arrangements; d) planning and monitoring the management of services in an integrated manner; e) the introduction of a tariff system which applies the user and polluter pay principles (transfers all the operating and maintenance costs and costs of capital to customers) and which provides incentives for improved performance, as well as the definition of a single tariff for each ATO and, finally, f) the creation of a governmental entity to oversee these reforms and their implementation, the COVIRI (*Comitato di Vigilanza sull'uso delle Risorse Idriche*), established within the Ministry for the Environment (*Ministero dell'Ambiente e della Tutela del Territorio e del Mare*).

However, over 13 years after the Galli Law was published, these reforms have still not been concluded. The processes of delegating management to a significant number of ATOs have not yet been completed.

In Italy, the first law developed in the context of the water sector was enacted in the mid-nineteenth century and was aimed at planning water usage (Law No. 2248 of 1865). This law defined the process of authorizing the use of water and sought to protect the land and the population by means of hydraulic works.

Subsequently, Law No. 129 of 1963 was considered to be the first attempt to regulate planning for water resources in Italy. It defined a national water plan and divided the nation into different water districts, on the basis of administrative levels and historical factors, and it was expected at that time that these districts would correspond to the future scale of the water services.

The second attempt began when state reforms were implemented and functions were transferred to the regions, paving the way for reinforcing local levels and the first policy to protect water resources. The municipalities continued to be the main entities responsible for organizing the service although there was a greater participation by the regions at the level of national legislation and local water management. This process began in 1972 but was implemented later and still continues to evolve (Law No. 112 of 1999). Smaller hydraulic works and permission for concessions were transferred to the jurisdiction of the regions, as was, in 1977, the management of water, by means of Law No. 616.

This transfer of jurisdiction also had a perceptible impact on tariffs. The price regime in Italy had been controlled by a single authority since 1944, when the *Comitato Interministeriale Prezzi* (CIP) was founded. In 1974, the role of CIP was transferred to the provincial level, to the Provincial Price Committee (*Comitato Provinciale Prezzi* – CPP). The objective of this body was to bring prices closer to real costs of service in each region. However, in 1984, tariff increases were limited as a measure to contain inflation. Between 1987 and 1990, the *Legge Finanziaria* allowed this rule to be disregarded, in order to recover at least 60% of costs by means of the tariff system. Nevertheless, WWS operators could not use profits and the tariff rules were neither simple nor transparent. The provincial committee came to an end in 1994 and the temporary measure became the official policy until the implementation of the reform that begun the same year.

In 1976 significant changes were implemented in terms of water policies in Italy concerning the protection of water resources. Earlier, water was considered to be a resource to be exploited but from 1976 onwards it was deemed to be a protected resource. The Merli Law (Law No. 319 of 1976) was the first important law concerning water pollution and introduced norms for the discharge of wastewater and a plan to use and protect water resources. The law was also conceived to introduce general WWS criteria, with the aim of managing water usage, and to establish indicators regarding the quantity and quality of water resources. In this context, the regions and the municipalities are responsible for monitoring, planning and organizing activities concerning the water sector at a local level.

Law No. 183 of 1989 further developed the regulatory framework for the water sector, defining the river basin as the unit for the integrated management of water, under the leadership of water authorities, with the participation of the state and the respective regions. It also introduced the river basin plan, which, amongst other activities, regulates water abstraction. Italy has been divided into 11 national river basins, managed by 6 national basin authorities, 18 inter-regional basins (involving regions and the state administration) and diverse regional basins, whose management has been delegated to the regions. This law introduced the River Basin Plan (*Piano di Bacino*), a legal instrument that defines guidelines aimed at enhancing, protecting and ensuring a balanced management of water within Italy, on the scale of river basins.

Law No. 142 of 1990 assigned to the municipalities (and provinces) the responsibility of managing local public services. The WWS sector has consequently been reorganized since 1994 and water policies have been reinforced. In 1994, the overall reorganization of these services began with the aforesaid Galli Law. This law introduced a new level of WWS management, known as ATO. Within the ATO, operators manage the whole WWS production chain. This is known as the Integrated Water Service (*Servizio Idrico Integrato* – SII).

Finally, concerning the WWS delegation that was introduced by the Galli Law, the Decree No. 280 of 2001 established possible management alternatives. In particular, it specifies that the procedures of the AATO while delegating management to operators that involve a private entity must be done by means of a public tender, which is to award the contract to the proposal that offers the greatest economic benefits. However, Law No. 448 of 2001 (*Legge Finanziaria, 2002*) made it compulsory to have public tenders for local public services, but also defined the so-called 'safeguard period', in which these tenders could also not be held.

Law No. 326 of 2003 (*Legge Finanziaria, 2004*) modified Law No. 448 of 2001, allowing for the following organizational arrangements for WWS management: delegation to private entities, delegation to mixed capital public companies or delegation to public companies formed by capital derived from participating municipalities within the ATO (in-house managing). Public procurement is used in the first two cases while in the third case the company has to exclusively provide WWS and cannot participate in managing other services. It is important to note that in the past some contracts were awarded without public tenders, which induced the EU to accuse Italy of disrespecting public contracts directives.

The primary objective of recent legislation that has been carried out by the government, namely Legislative Decree No. 152 of 2006, is to enhance the quality of human life. It should be implemented by preserving and developing conditions and promoting a prudent and rational use of natural resources. For this purpose, this decree reorganized the coordination and integration of these matters, in compliance with prevailing legislative criteria and principles, and the functions of the regions and local authorities, according to EU directives. The norms of this decree are applicable to the area of the human, instrumental and financial resources outlined in existing legislation, without additional burden on public finances. This normative document formally altered the Galli Law and introduced changes at the level of the COVIRI responsibilities, expanding its functions to encompass urban solid waste.

Diverse regulations have been produced regarding the environment. One of the most important normative documents, Law No. 152 of 1999, combined different policies concerning the environment, health and the economy, seeking to protect water resources. This normative document constitutes the legal framework to protect surface water and groundwater, describing the general environmental objectives to prevent and reduce pollution. It also reorganized the entities responsible for monitoring pollution, aiming at improving the quality of water, adapting protection for special uses, achieving a sustainable and durable use of water resources, giving priority to drinking water, and maintaining the natural capacity of water bodies for regenerating water, so as to protect wildlife. In order to fulfil these objectives the decree declared that regions should implement a set of measures to be established in the plan for water resources protection (*Piano di Tutela delle Acque*).

Apart from this federal legislation, the regions have also developed their own respective laws, in diverse areas. Finally, it can be noted that during the past two decades Italian regulations in this sector have been shaped by EU legislation.

2.9.2 Legal and institutional framework

2.9.2.1 Responsibilities

Responsibility for the water sector has been divided amongst the Ministry for the Environment (*Ministero dell'Ambiente e della Tutela del Territorio e del Mare*), the Ministry of Health (*Ministero della Salute*), the Ministry for Infrastructure

(*Ministero delle Infrastrutture*), the different regions, the Federation of Energy and Water Operators (*Federazione delle Imprese Energetiche e Idriche* – Federutility), the National Agency for Environmental Protection (*Agenzia Nazionale por la Protezione dell'Ambiente* – ANPA), which is subject to the Ministry for the Environment and defines, for example, rules, norms and service levels for the sub-sector of wastewater services, the National ATO Association (*Associazione Nazionale Autorità e Enti di Ambito* – ANEA) and COVIRI, an entity that plays an important regulatory role at a national level and whose functions will be described in further detail in the following point.

The Ministry for the Environment is responsible for the WWS, providing a general framework for service levels (e.g. continuity) and tariff rules (standard calculation method). The Ministry of Infrastructure plays an important role in managing hydraulic works and also administers EU funds for the sector. The Ministry of Health is responsible for monitoring the water quality.

The different regions have had a great deal of political influence in Italy since the 1970s. Apart from the intervention while implementing the Galli Law, with regard to the creation of ATOs (which must respect the borders of river basins or administrative areas) and methods of cooperation between municipalities, they provide and determine the institutional framework of contracts between the AATOs and operators. The regions also have numerous functions in the water sector, since they are responsible for regional river basins (monitoring the quality and quantity of the water that is available), and they approve and license the respective uses of water, collecting taxes on water abstraction (e.g. groundwater).

At a regional level, after ANPA was created, in 1994, each region in Italy was provided with a branch of this agency, known as the Regional Agency for Environmental Protection (*Agenzia Regionale per la Protezione dell'Ambiente* – ARPA), which substituted the existing local sanitation units (*Unità Sanitaria Locale* – USL) in matters related to environmental issues. ARPA helps operators monitoring the pollution generated by the various agents in the water sector, such as the discharge of industrial wastewater and other environmental situations which subsequently may arise.

The regions are responsible for implementing the Galli Law, defining the borders of the diverse ATOs and the framework for intermunicipal cooperation. They also create the structures and areas of jurisdiction in which WWS operations will be contracted. In most cases, the ATO borders coincide with those of the provinces, which effectively coordinate the municipalities involved. In some cases, only a single ATO was defined within a region (e.g. Puglia, Basilicata, Valle d'Aosta and Sardinia). The main ways of cooperation between municipal authorities are called *convenzioni*, agreements between municipalities, which are frequently coordinated by a large municipality or by the province, or *consortium*, i.e. the creation of a new local authority dedicated to the WWS sector. The regions are responsible for preparing regional plans (*Piani stralcio*) to protect water resources and establish the limits for the discharge of wastewater on the basis of these plans.

Once this stage of a regional framework has been concluded, the reforms can be implemented on a local scale. AATO is the local authority responsible for organizing WWS activities within the ATO. The AATO must develop a technical, financial and economic plan (*Piano d'Ambito*), which includes an analysis of the present state of the infrastructure and services and future requirements, the new service levels that are to be adopted and the investments that are necessary to achieve them, minimum productivity growth, the tariff system and the method of adjustment with regard to the progress attained in the service. After this has been developed, the AATO can delegate WWS management and operation and is then responsible for its monitoring and supervising.

ANEA is a non-profit national AATO association whose objectives are to study and discuss issues to promote uniform WWS regulations and activities in coordination with the Ministry for the Environment. Members of this association collaborate at the level of sharing experiences and debate key issues related to their development. ANEA has played an increasingly important role corresponding to the growth of the number of AATOs in Italy and, consequently, of its members.

As afore mentioned, the operator is chosen by the AATO for the geographical area of the ATO and is exclusively responsible for the WWS in this area (SII). This operator is a company governed by private law, but in terms of ownership it can be based on private, public or mixed capital. With the pressure and compulsory nature of the water sector reforms, many of the operators were transformed into private capital companies. In fact, the capital of these companies is increasingly derived from private operators. Apart from this, even the budgetary law for 2002 defined that the structural arrangement for WWS management would be, in the future, constituted by private capital companies and that other kinds of companies would be an exception. Figure 2.32 shows the organization of the water sector in Italy after the implementation of the Galli Law.

Relations between the AATO and the operator are regulated by an administrative contract, known as the '*Convenzione di affidamento*', and the nature of these contracts is determined by the regional governments.

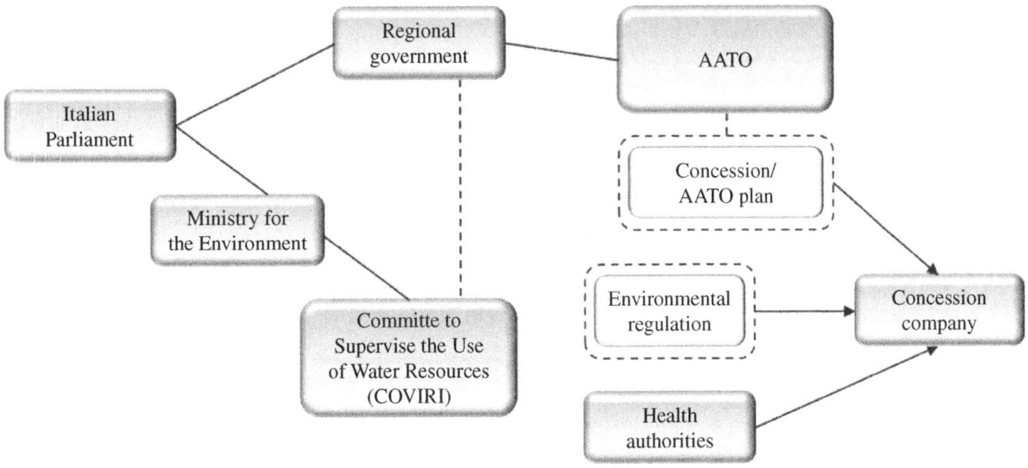

Figure 2.32 Institutional framework of the water sector in Italy

Federutility is an association of operators of public services, including WWS operators. It aims to ensure quality in the sector, sharing information and contributing towards developing public services.

The *Federconsumatori* plays an important role with regard to customer protection. Founded in 1988, it is a non-profit organization whose main objective is to protect the interests of customers.

Finally, the Competition Authority (*Autorita Garante della Concorrenza e del Mercato*), which monitors the concentration of power in the sector and the possible abuse of the dominant position, has also issued opinions in order to ensure that the rules for public contracting are upheld, especially the requirement to hold a public tender while choosing private partners in mixed companies.

2.9.2.2 Regulation

In 1994, the Galli Law created a committee to monitor the water sector (COVIRI), an autonomous body that is nonetheless supervised by the Ministry for the Environment. The role of COVIRI has never been clear and therefore has always been subject to controversy. Apart from its functions of observing whether the principles of the legislation regarding the reforms for the water sector were being fulfilled, the committee must regulate the setting and adaptation of tariffs and to protect the interests of customers. The most problematic issue has been the formula used to set tariffs based on a price cap methodology. COVIRI was recently restructured by means of Law No. 152 of 2006, which also renamed it the Authority for Monitoring Water Resources and Urban Waste (*Autorità per la Vigilanza sulle Risorse Idriche e sui Rifuiti*), since its functions now also encompassed the urban solid waste that was earlier the responsibility of the Waste National Observatory (*Osservatorio Nazionalle sui Rifuiti*), even though this has not yet become operational.

As defined by Law No. 34 of 1994, the main objective of COVIRI is to ensure that WWS operators are efficient, effective and self-sustaining. The committee can suggest actions against operators if any of the principles mentioned in the Galli Law are violated or even against their administrators, in order to defend the customers' rights and interests.

In Italy, the regulatory process for the water sector involves different levels: the Ministry for the Environment, through the proposals issued by COVIRI, especially regarding the national tariff system; the regions, which prepare drafts of contracts; and, finally, the AATOs, which are empowered to ensure, monitor and define tariffs within their jurisdictions, as well as to monitor and supervise the quality of service and the commitments assumed by means of contracts. Figure 2.33 shows the regulatory structure that has been developed for the water sector in Italy.

In terms of economic regulation, the functions of the committee include proposing criteria to determine tariffs using the standardized price cap method, which are then approved by the government, specifically by the Ministry for the Environment. As long as they comply with prevailing legislation, the WWS operators can decide and approve tariffs in the various areas of the country.

As an association for the sector, Federutility also plays a regulatory role, albeit at the level of self-regulation. One of its important functions is to disseminate best practices among its members, given the growing concerns regarding the improvement of the quality of service and the economic and environmental sustainability of WWS operators. ANEA also has similar functions.

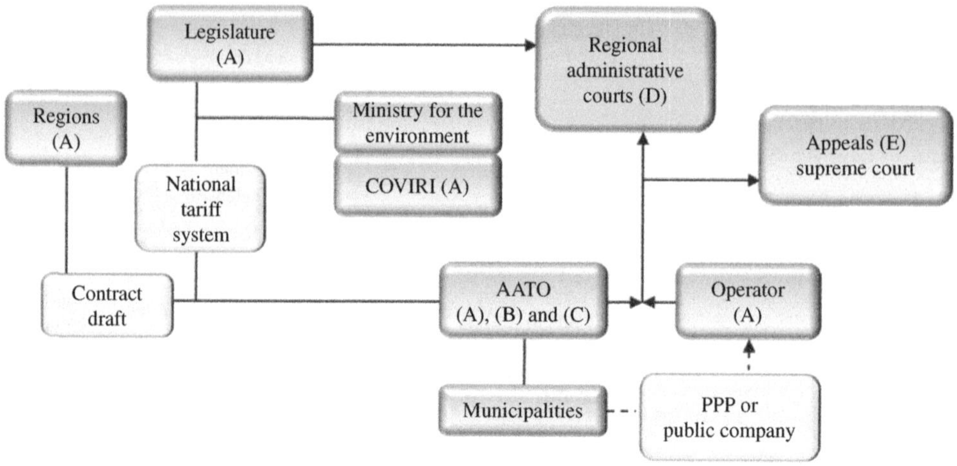

(A) Defining the draft of contracts and regulatory rules
(B) Monitoring, i.e. collecting and analysing information about performance
(C) Verification of the framework of prevailing rules
(D) Interpreting and applying rules
(E) Judging appeals against decisions made in the course of (D)

Figure 2.33 WWS regulatory framework in Italy

2.9.3 Market structure

2.9.3.1 Ownership and management of services

In the past, WWS in Italy were mainly provided by public operators. With the implementation of the Galli Law and the creation of 91 ATOs, the private sector increasingly began to provide these services. Despite the contracts established with private operators, the infrastructures are always owned publicly. From the 57 out of 91 planned ATOs that were operational in 2006, 13 were public (in-house), 11 were private and 33 were mixed capital companies, although with a majority public stake.

So far, there is no accurate information about private sector participation in Italy. It has been estimated that private companies are directly responsible for just over 10% of the sector but if one considers their indirect participation (through mixed capital companies) private participation is more than 50%. However, the private sector in Italy includes the presence of large multinational companies, namely Veolia (in the regions of Sicily, Calabria, Latina, Piemonte, Liguria and Tuscany), the Suez Group (Tuscany), Aqualia owned by the FCC Group (in the regions of Sicily and Calabria) and Saur, which, in the meanwhile, sold its position to ACEA (Sicily and Campania), as well as large Italian companies, both from the water sector (ACEA) and other utilities (IREDO and HERA) and even companies from associated or distinct areas (e.g. construction).

2.9.3.2 Players in the sector

Until the ATOs were implemented, the water sector in Italy comprised 3,704 water services operators and 4,278 operators of wastewater services. The largest percentage of these operators represented direct management by municipalities. It must be noted that these data are not strictly disaggregated, since there were operators who provided both services. However, with the restructuring and after the reforms defined by Law No. 36 of 1994, and Law No. 152 of 2006, are completely implemented, WWS will be managed by 91 operators, corresponding to the number of ATOs in the country. The majority of these operators are mixed capital companies (where the public sector has a majority shareholding), while the presence of totally public or totally private operators is relatively balanced but fewer in number than the mixed capital companies.

For example, in Rome, ACEA (where Suez has 8.6% of shares, 51% is held by the municipality of Rome and the remainder is traded on the stock market) was awarded a contract for 30 years. ACEA provides WWS to about 7 million inhabitants, i.e. 14% of the Italian population, and is the largest operator in the water sector in Italy. Companies of the ACEA Group directly or indirectly operate integrated WWS management in the regions of Liguria, Tuscany and Lazio (Rome and Frosinone).

Apart from its presence in the shareholding structure of ACEA, Suez has invested capital in Acque Spa, in Acque Toscane in Fiesole, in Montecatini Terme and Ponte Buggianese, in Acquedotto del Fiora, in Nuove Acque in Arezzo and, finally, in Publiacqua, all of them operating in the region of Tuscany and providing WWS to about 2.7 million inhabitants i.e. to 75% of the region's population.

Other large multinationals, such as Veolia and Aqualia (FCC), also have a significant presence in the water sector in Italy. Veolia has capital in companies operating in 38 municipalities in the ATO of Latina, supplying services to about 600,000 inhabitants, as well as in regions such as Liguria, Veneto, Piemonte, Emilia-Romagna and Sicily, through Acqualatina, Genova Acque, SIBA and Siciliacque companies. Siciliacque alone provides WWS to 1.5 million inhabitants.

In its turn, Aqualia is present in companies that supply services in Sicily and Calabria, such as the operator in Caltanisseta, which provides WWS to 700,000 inhabitants.

The HERA Group is another important player, with activities in the fields of energy, water and solid waste, managing WWS in about 400 municipalities in the Emilia-Romagna region, through seven territorial companies (Hera Bologna, Hera Ferrara, Hera Ravenna, Hera Rimini, Hera Modena, Hera Forlì-Cesena and Hera Imola-Faenza). In this context, the company provides WWS to about 1 million inhabitants.

The multi-utility operator IRIDE, which recently acquired AMGA Genova, provides WWS to about 915,000 inhabitants, through Acque, AMTER and IDRO-TIGULLIO companies, the latter two in the region of Liguria.

There are other players in the water sector with a noteworthy presence in the AATOs, including financial entities, construction companies and other service companies.

2.9.3.3 Integration of the sector and services provided

As previously described, in the past WWS were mainly provided by the public sector and especially by municipalities. With the restructuring of the water sector, operators provide integrated services, both vertically and horizontally, within the jurisdiction of the ATO, in which a single operator is responsible for the whole water cycle.

The public or private operators present in the water sector sometimes integrate other services, such as electricity or gas. A particularly noteworthy case is that of the operator for the city of Rome, ACEA, which is responsible for supplying water, wastewater services, electricity, public lighting and heating. Another example is that of AMGA Genoa, which was acquired by the IRIDE Group in 2006 and provides water, electricity and gas services.

2.9.3.4 Numbers for the sector

In Italy, the water sector is characterized by an annual production of about 7.6 billion m^3 of water for public consumption. This is equally derived from groundwater and surface water, 35% of the latter being obtained from springs. Domestic consumption represents approximately 65% of all the water that is produced for public consumption.

It has been estimated that the per capita amount of water supplied to Italians corresponds to about 230 litres per inhabitant per day.

It has been estimated that about 40% of the volume of water produced is not billed.

In 2007, the level of coverage of the water service was about 96%, while approximately 84% of the population was covered by wastewater services. Wastewater systems tended to be separate and there was also a significant percentage of stand-alone systems. In 2002, it was estimated that stand-alone systems represented about 70% of total systems. However, from the wastewater that is collected and drained only 74.8% is treated in wastewater treatment plants.

In 2004, WWS operators obtained total revenues of about 4.2 billion euros. In the same year, 1.5 billion euros were invested in the water sector. Figure 2.34 shows the decline in government investments in the water sector over the course of the years (Danesi *et al.*, 2007).

In 2001, the water sector in Italy had about 64,000 employees (Danesi *et al.*, 2007), with an increase of 8,000 staff in the past five years. A study with more recent data (2005), although based on a sample of 25% of all ATOs, revealed a very significant reduction in the number of employees after the Italian water sector was restructured. This study estimated that each ATO had an average of 378 staff members (Fraquelli & Moiso, 2005), which would correspond to about 35,000 employees for the whole water sector in Italy.

2.9.4 Rules of the game

2.9.4.1 Tariff setting

The tariff system usually applied by operators is a two-part tariff system for the water service, based on a fixed component, according to the meter size or the property size, charged to all customers, and a variable component, according to consumption and divided into blocks. This variable component usually increases with consumption. There are normally five blocks for domestic customers, namely the low, the basic and other three blocks for higher levels of consumption. Sometimes these blocks also depend on the size of the family in question. In terms of wastewater services, there is normally only one variable component for collection and another for treatment. Tariffs for this service frequently comprise a single block. Table 2.49 shows the WWS charges for domestic customers in three Italian cities (2005).

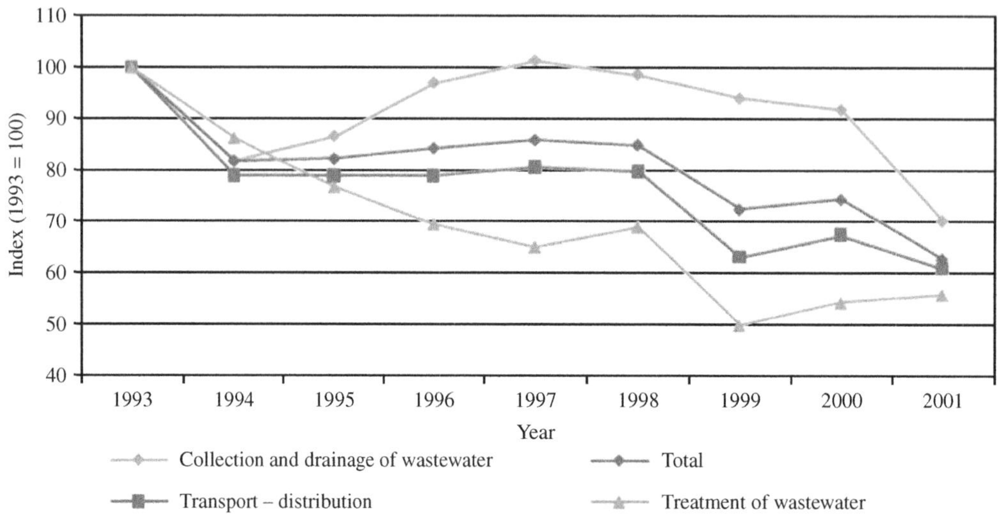

Figure 2.34 Investments in the WWS sector in Italy

Table 2.49 WWS tariff systems for three Italian cities

	Water (€/200 m³)		Wastewater (€/200 m³)	
	Fixed component	Variable component	Collection	Treatment
Bologna	6.2	128.3	17.8	50.3
Palermo	25.1	91.5	15.0	41.8
Rome	7.7	49.0	17.8	51.0

Over and above the tariff systems, customers also pay VAT, which is 10% for the water sector in Italy. It must be noted that other components are also charged for services provided, some of which are fixed annual sums, such as the charge for maintaining fire hydrants. Reclaimed water is also subject to a special tariff.

Tariff systems differ according to the nature of the customers. Amongst domestic customers a distinction is often made between resident and non-resident domestic customers. In this regard, it is important to note that cross-subsidization exists between users of the water sector in Italy, since larger customers are almost always charged considerably higher tariffs, which also vary according to the nature of the consumption. Table 2.50 shows the tariff system implemented by L'Acquedotto Pugliese, the company that provides services to the region of Bari.

According to COVIRI, WWS represent, on average, a cost of about 250 euros annually for domestic customers, corresponding to an average tariff of 1.23 €/m^3. Table 2.51 shows the variation of tariffs in Italy with the WWS reforms (Danesi *et al.*, 2007).

The use of the tariff system is considered to be the main instrument to ensure the economic and financial viability of WWS operators, which envisages a full cost recovery, as stipulated by article 13 of Law No. 36 of 1994. Another objective associated with the implementation of the Galli Law, which is simultaneously related to the creation of ATOs, consists of defining criteria to establish tariff systems according to very clear rules, based on the investment plan for each ATO.

Over the course of time the WWS sector has been heavily subsidized by various levels of the government. Until the Galli Law, the municipal authorities proposed tariffs, which were approved by the CPP and later by the National Price Commission (NPC). The main concern was to curb the effect of inflation rather than to recover costs, which became increasingly larger, thus also requiring higher subsidies. In 1991, the NPC, influenced by the regulatory process of OFWAT, decided to impose a price regime that provided other kinds of incentives for operators, based on price cap regulation. The Galli Law made it possible for this reform to attribute this function to the regulator. In 1996, COVIRI established (by Decree) a new system to set tariffs, according to the formula below, based on the Average Real Tariff (ART) determined by revenue cap regulation. In this system, the operating cost is limited to a maximum ceiling of 30%, unless otherwise authorized by COVIRI, beyond the theoretical value obtained by the following formula:

$$T_n = (C + A + R)_{n-1} \times (1 + \Pi + K)$$

where:

T is the tariff to be determined for the year n;

C corresponds to the operating costs of the previous year (obtained by means of regressive statistical formulae defined in the Decree);

A represents the depreciation of the previous year;

R refers to the return on invested capital (fixed regulated value of 7%) of the previous year;

Π is the expected rate of inflation for the year being regulated;

K is the price cap.

Table 2.50 Tariff system of L'Acquedotto Pugliese for 2008

Service	Tariff (€/m^3)
Water	
Fixed component (according to the meter)	11.11–121.62 (€)
Low tariff (0–73 m^3)	0.419
Basic tariff (73–101 m^3)	0.499
1st Block (101–146 m^3)	0.873
2nd Block (146–256 m^3)	1.346
3rd Block (>256 m^3)	1.669
Wastewater	
Wastewater collection	0.091
Wastewater treatment	0.264

Table 2.51 Tariffs and billing for the water service in Italy

	Average	Minimum	Maximum
Tariff (€/m^3)			
Before the reforms (2002)	1.01	0.58	1.52
After the reforms (2003)	1.23	0.85	1.83
Water bill (€/year)			
Before the reforms (2002)	202.59	115.84	304.66
After the reforms (2003)	245.51	169.27	365.52

The formulae to calculate all the parameters have been stipulated in the Decree mentioned. The price cap K was defined as a percentage index, which the local authorities can use while adapting levels of average tariffs from year to year, in which K cannot exceed certain values, depending on the tariff level of the previous operating year. In this method, the ART is established by means of the formula ART × VW = T, where the ART is the value of the average real tariff, VW is the volume of water supplied and T corresponds to the total revenues obtained by means of this formula. The tariff system is reviewed every five years. In 2002, COVIRI proposed the revision of this regulatory model, although maintaining its entire structure and methodology (price cap formula). This revision was due to some problems that were detected during its application, such as the failure to include fees in the cost structure, the failure to update some values for the current year, tariff articulation and the reviewing and monitoring mechanism (Alpaos & Valbonesi, 2008).

2.9.4.2 Quality of service

In terms of the quality of water, the situation in Italy is variable but has improved considerably in recent years. Given the geographical structure of the country there are many areas that are more vulnerable to pollution (e.g. nutrients), especially the large lakes and the upper Adriatic region, which include the most industrialized and populous regions in Italy.

In order to more effectively achieve its main objective, i.e. consumer protection, in 2006 COVIRI suggested to the Ministry for the Environment that benchmarking should be carried out between WWS operators. This proposal comprised the development of a set of 55 performance indicators.

These WWS indicators are divided into three groups: 19 technical indicators, 18 management indicators and 18 financial and economic indicators. In its turn, the technical indicators group is further subdivided into indicators regarding the service coverage (3), WWS characteristics (1), the supply and quality of services (10) and the characteristics of the infrastructure (5). The group of management indicators is subdivided into administrative indicators (2), relations with customers (4), service guarantees (2) and maintenance of services (10). Finally, the remaining 18 indicators form a single group of economic and financial indicators.

Other entities such as Federutility and the regions and operators themselves have developed benchmarking activities, especially based on performance indicators and metric benchmarking. Figure 2.35 shows the water losses indicator for the different regions, as part of a benchmarking exercise carried out by the operator HERA for 2004 (HERA, 2005).

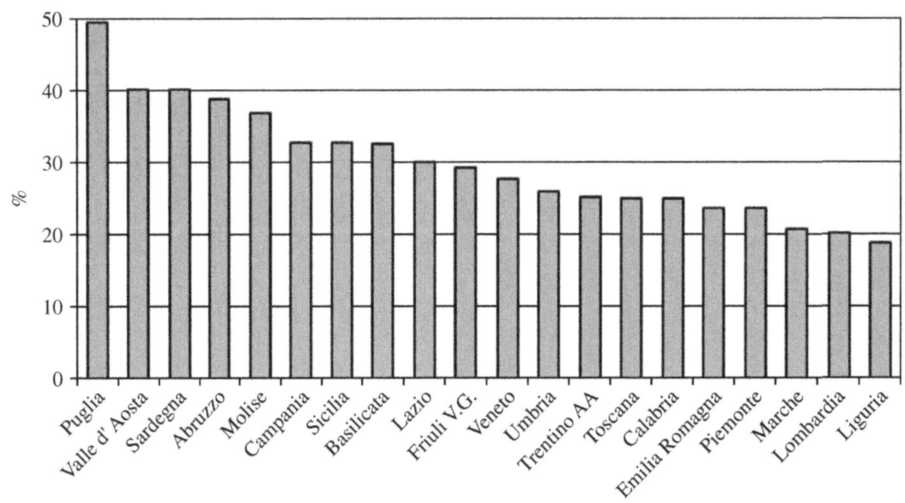

Figure 2.35 Water losses indicator in the different regions of Italy

One of the objectives of the reforms that were implemented in the water sector was to improve the quality of service. In this scope, each AATO would have to define a service charter establishing, on the one hand, the public service obligations, which will be analysed in the next point, and, on the other hand, a commitment regarding the service levels to be implemented. The investment plan must consider the planned levels of service, which should be included in the economic and financial study. This provision of the Galli Law has resulted in a substantial improvement of the quality of service for almost all the indicators with greater and better accountability and greater transparency. Table 2.52 shows the service charter of ACEA, the company operating services in Rome.

2.9.4.3 Public service obligations

Prevailing legislation, namely Law No. 36/1994, makes it compulsory for WWS operators to guarantee a service that is adequate for the needs of customers (adaptability) at a reasonable cost (accessibility), in compliance with Italian and EU environmental protection norms. Apart from this, regulations have also sought to introduce the concept of serviceability amongst operators, so as to encourage them not to allow their infrastructure to deteriorate and to manage assets suitably.

The regulations in effect (D.P.C.M., dated 4 March 1996) have also defined more specific public service obligations, namely regarding minimum levels of service (*Livelli Minimi di Servizio*). These obligations are applicable to water abstraction and volume supplied, the quality of drinking water provided, the operational quality control and the quality of customer service. Penalties have also been prescribed for situations in which operators do not uphold some of their commitments.

The water sector in Italy reflects a great concern with social issues at the level of tariff setting. In cases of large families, some local operators have modified the tariff structure that increases in line with the level of consumption.

This payment is applied according to the basic or low rate. Other operators establish percentages for reducing tariffs in the case of poor customers. Such families have to submit a series of proof to become eligible for and benefit from these advantages. Delays and non-payment, as well as the disconnection of services, are regulated.

Table 2.52 Service charter of ACEA

Water service	
Meter readings	At least once a year
Bills	At least twice a year
Urgent situations	24 hours
Restoration of the service in case of scheduled maintenance	Within 24 hours
Prior notice regarding interventions in cases of scheduled maintenance	Within 24 hours
Failures in meters or other surface equipment: Response time for the first intervention	72 hours
Failures in underground infrastructure or equipment: Response time for the first intervention	72 hours
Restoring the service: Maximum time after being notified	72 hours
Connections or changing meters for existing networks Activation	9 days
Connections or changing meters where networks do not exist	
Estimates	30 days
Implementing the task	45 days
Authorized consumption	9 days
Verification of the meter	20 days
Calibrating faulty meters	20 days
Verification of contradictory meters	30 days
Verifying the level of pressure	20 days
Termination of contracts by customers	15 days
Situations of delayed payments Reconnecting services after payment is received	The following day
Resolving inaccurate bills	30 days
Waiting time	
Average time	30 minutes
Maximum time	90 minutes
Answers to written complaints	30 days
Wastewater service	
Continuity and regularity of the service Urgent situations	24 hours
Failures or obstructions of sewers or canals Maximum time	72 hours
Cleaning and sweeping after flooding Maximum time	72 hours

Within the regulatory system, operators are obliged to regularly provide information to the AATO, within the scope of the system and process for compiling information and data regarding the water sector.

All operators have service charters, which stipulate their obligations with regard to customers. Examples include Acque Toscane, in the municipality of Montecatini Terme (region of Tuscany), or Publiacqua (also in Tuscany).

2.9.5 Performance

Notwithstanding the delays and the controversies that have occurred, the restructuring model for the water sector in Italy has produced good results.

Although it has been heavily criticized, mainly by professionals from the water sector, the results have been clearly evident, both at the level of economies of scale (going from several thousand operators to 91 AATO) and improvements in the quality of service. Economic, financial and even environmental sustainability have become far more viable, with one of the best price-versus-quality ratios in the water sector in Europe.

In terms of less successful aspects, apart from the time taken to implement the reforms, which have still not yet been concluded, it is possible to highlight contracts being awarded to private operators without public tenders, the lack of penalties for failures to comply with the service quality norms that have been adopted, cross-subsidization between services and customers in the tariff system and the national regulatory authority's lack of independence and tools for its activities. The price cap regulation formula is not perfect and may cause some unfair situations. It can therefore be improved, although it is an important and noteworthy step.

2.10 MOZAMBIQUE

2.10.1 Historical background

Mozambique is a presidential republic, where the political party with a parliamentary majority appoints the government. In administrative terms, Mozambique is divided into provinces, which are further organized into districts. In their turn, the districts are divided into administrative posts, which are finally subdivided into localities, the lowest level of representation of the central State.

Mozambique has an area of approximately 8,016,000 km^2 and a population of 21.4 million inhabitants. It is divided into 11 provinces, including the city of Maputo, which has the status of a province and a provincial governor. However, it is very common, even at an official level, to say that Mozambique only has ten provinces, since often no distinction is made between the city and province of Maputo. The provinces are administered by governors, who are appointed by the President of the Republic. Since 1998, these administrative divisions have been complemented by 33 local authorities, known as municipalities [the 23 cities plus a town in each province, except Maputo (city), which only has one administrative unit, the municipality and the city with the same name]. Apart from this, Mozambique is also divided into 128 districts.

After the country became independent, some changes were introduced to the administrative divisions, in terms of the nomenclature of both the administrative levels and units. Many of these names were changed to their original nomenclature with local roots to remove the colonial implications. There was, however, a level of continuity with regard to the territorial division itself, since the colonial districts were called provinces, the council districts and surrounding areas were named districts and the posts were known as localities, although the latter were reformulated in 1986. In this context, there has been a significant evolution since 1995, with the creation of new districts and municipalities, in addition to the reintroduction of the administrative posts, which had been eliminated in 1975.

Given Mozambique's rather precarious socio-economic scenario, coupled with the constant droughts that occur in the country and the negative impact of natural disasters (e.g. floods), the Mozambican government has implemented, since 1986, a programme to reform the water sector, with the purpose of obtaining funding, especially from the World Bank. As a consequence, privatization turned itself into a necessity and initial measures were introduced in 1989. The government's decision to go ahead with privatization was strengthened in 1999, with the entry of SAUR and Águas de Portugal (AdP) into the Mozambican water market.

In this context, the government enacted a set of laws, mainly during the late 1990s, which sought not only to protect water resources but also to assist the management and sustainable use of water. Thus, the Water Act or Law No. 16, dated 1991, was a fundamental instrument to promote the interests of Mozambique's population. This law establishes that water resources belong to the public domain, and also defines the general principles for water management, the need to create an inventory of all the water resources in the country, the general regime for water use, the priorities to be kept in mind, the general rights of customers and the corresponding obligations, among other aspects.

Despite the considerable progress that had been made in the water sector since Mozambique became independent, there was still a long way to go with regard to service and coverage levels. This included the poor state of the infrastructures whose life cycle was about to come to the end, the high percentage of population without access to a regular water service and the urban population with access to piped water but being provided with irregular and poor quality of service. In rural areas and in cases where hand pumps were installed, it was even more difficult to maintain water services operational. The situation was far more critical in the context of the wastewater service.

In this scenario, the Mozambican government introduced reforms for the water sector by means of a National Water Policy in 1995, which focused on different conditions and areas of activities, namely the water supply in urban, suburban and rural areas, the integrated management of water resources and wastewater services, as well as the respective goals and corresponding costs. The National Water Policy also established guidelines for privatizing the urban water services. Two years later, the Mozambican government and the World Bank jointly established the first National Water Development Programme (*Programa Nacional de Desenvolvimento de Águas* – PNDA I), which included arrangements for private sector participation in the WWS, among other measures. The government established an operating contract for 15 years for Maputo and 5 year contracts for the cities of Beira, Quelimane, Nampula and Pemba. As events unfolded, it became necessary to implement a new plan, which happened in 1999 with the PNDA II, which had a budget of 79 million euros for investments in infrastructures. The PNDA II was specifically designed to implement private sector participation by means of concession, leasing or management contracts.

The Framework for Delegated Management (*Quadro da Gestão Delegada* – QGD) was established in 1998, with the approval of Decree No. 72. This law defined the legal framework for delegated management and created new bodies in the sector, although it was only effectively implemented about a year later. The objective of the QGD was to ensure the efficiency of the WWS management and respond to the sector's need for planning and development, as well as to implement policies and achieve the goals and parameters that had been defined in the National Water Policy.

The QGD encourages a separation of regulatory and government functions from the operational functions. Certain activities are carried out in conformance with economic criteria, such as the tasks performed by the FIPAG, while some decisions are more technical, impartial and standardized, like the ones related to the Water Supply Regulatory Council (*Conselho de Regulação do Abastecimento de Água* – CRA) competencies. The QGD also considers the growing involvement of the local authorities and that institutional support should accommodate this gradual transition, especially while making decisions related to the development of infrastructure and public investments.

It must also be stressed that although the CRA is an independent regulatory authority, it does not act alone, but instead interacts with the internal and external entities that belong to the QGD and which influence and determine the success of its respective activities, contributing towards the desired final result, i.e. achieving the goals that were defined by the QGD and their integration as a whole.

The creation of the CRA, by means of Decree No. 74 of 1998, simultaneously resulted in the creation of specific regulations for the WWS, namely the Drinking Water Quality Regulations in 2004, which standardized the minimum levels required for the drinking water quality.

Finally, the development of the 2007 Water Policy was aimed at promoting a suitable use of water resources, so as to ensure a sustainable use. Moreover, this document advocated safeguarding the quality of the environment, providing incentives for preventing water resources pollution and aiming to achieve a balance for customers as a whole, in cases of multiple and conflicting use of water within the public domain.

2.10.2 Legal and institutional framework

2.10.2.1 Responsibilities

Responsibility for the water sector has been delegated to diverse entities, namely the Ministry for Public Works and Housing (*Ministério das Obras Públicas e Habitação* – MOPH), through the National Directorate for Water (*Direcção Nacional de Águas* – DNA), the CRA, the Water Supply Investment and Patrimony Fund (*Fundo de Investimento e Património do Abastecimento de Água* – FIPAG), the Ministry of Health, through the National Directorate for Health (*Direcção Nacional de Saúde* – DNS) and the Regional Water Administrations (*Administrações Regionais de Águas* – ARA).

The ARAs are public bodies with legal status and are supervised by the MOPH, through the DNA. In this regard, the ARAs are responsible for the operational management of water resources, organized on the basis of river basins. The DNA functions include defining policies, strategic planning for the water sector, fostering the general improvement of the WWS, coordinating different players in the sector, regulating and producing technical norms and, finally, promoting decentralization and preventing concentrations.

According to the Water Law No. 16 of 1991, the functions of the Ministry of Health comprise establishing parameters to be used for monitoring the drinking water quality so that it is deemed fit for human consumption. In this context, the DNS, under the Ministerial Law No. 180 of 2004 is the authority responsible for ensuring that the drinking water quality is monitored, as well as for supervising the implementation of legislative provisions. The water supply systems must be examined by a Water Inspector, appointed by the DNS, or a representative, who can freely access any point of the drinking water supply system.

The CRA was created as a public corporation under the terms of Decree No. 74, dated 1998. This institution consists of three public members proposed and appointed by the MOPH. The CRA is responsible for regulating the water service and for the issues related to the FIPAG and Private Operator Companies (Águas de Moçambique and Vitens), especially concerning the tariffs adopted, the quality of service, and programmes to expand the network. The new tariffs proposed by the FIPAG are only really implemented after they have been approved by the CRA. The regulator also collects opinions and complaints from customers and municipalities, and has powers of arbitration, conciliation and mediation in conflicts between the FIPAG and the operator.

The government body that signed the contract with the consortium that won the public tender for water services management in Mozambique was the FIPAG. This body is composed by a managing board and an executive committee. The board has a representative of the MOPH, who is its President, and includes four other members. The Executive Director is supported by a vice-president, assisted by a group of managers specialized in public service companies, investment programmes or financial management. The authority and responsibilities of the FIPAG are related to investments and budget management, efficiency maximization and cost recovery, contracts management and monitoring and also the reinforcement of contractual obligations of the selected private operator (World Bank, 1999).

In short, the functions of FIPAG include implementing investment programmes and their management, contracting private operators, delegation to operators, managing debts and, finally, ensuring the sustainability of the services concerning the operating and maintenance costs.

NGOs are also an important tool providing support for customers. Figure 2.36 shows the different entities with relevant responsibilities in the water sector in Mozambique (adapted from Alvarinho, 2002).

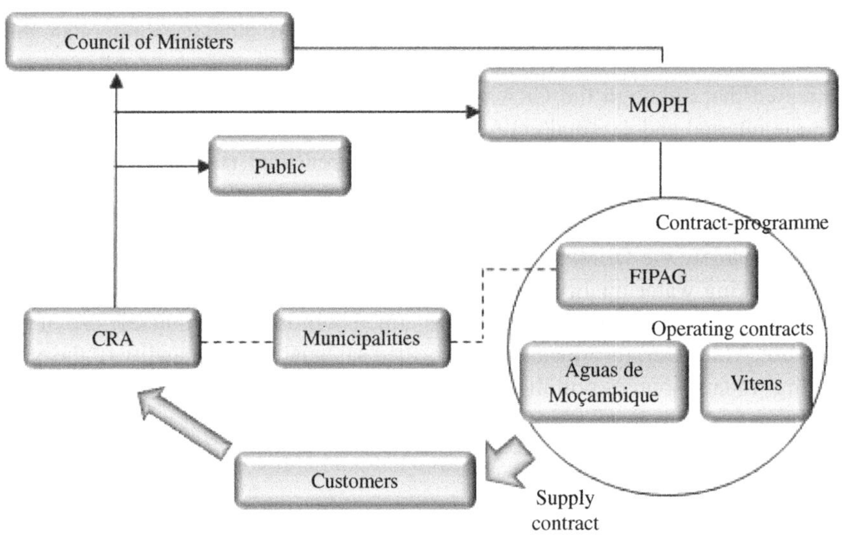

Figure 2.36 Institutional framework of the water sector in Mozambique

2.10.2.2 Regulation

The water sector in Mozambique has a sector-specific regulatory agency. The CRA is a body that, owing to its independence and technical competence, ensures a balance of interests between private and public sectors. Moreover, the CRA responsibilities include protecting customer interests, economic regulation, reconciling interests between the concession granting authority and the operator, monitoring and assisting in processes of delegated management, identifying the needs of current and future customers and, finally, promoting the QGD improvement. Thus, the scope of the CRA's actions ranges from the government, at one end, and local authorities, at the other end, so as to defend customer interests.

The CRA is a body established under the public law. It has a legal status and administrative and financial autonomy and is governed by the dispositions of its statutes and internal regulations, by the specific norms that are applicable to public services and administrative and financial autonomy and other applicable legislation. The regulatory agency is a collegial body consisting of three members appointed by the Council of Ministers, after a joint proposal presented by the MOPH, the Ministry of Finance and the Ministry for State Administration. While exercising its functions, the CRA has the same rights and obligations as those of the State concerning the protection of its facilities and staff. However, it is accountable to the Council of Ministers for its actions.

The CRA only regulates the services provided by the delegated systems, i.e. 14 systems. In order to carry out its activities, in 2007 the CRA spent about 1.68 million euros, including the remuneration of its 26 staff. Despite its independence, the CRA still depends on the national budget, since the taxes that have been imposed on operators are not enough to ensure its self-sustainability.

2.10.3 Market structure

2.10.3.1 Ownership and management of services

The water sector in Mozambique is mainly dominated by private operators, through the Águas de Moçambique and Vitens companies, which provide water services to 2.5 million inhabitants. The remaining the population is supplied by the public sector.

The first large consortium present in the Mozambican water supply sector was formed in 1999 during the programme aimed at privatizing public companies. This consortium was known as 'Águas de Moçambique' and was headed by SAUR International. Two years later, SAUR sold its entire stake in the consortium to Águas de Portugal, which is still active to date (beginning of 2008).

2.10.3.2 Players in the sector

The first concession contract with a private operator in Mozambique was signed in September 1999 in Maputo. Later, other contracts were signed with the private sector in four other cities. Águas de Moçambique is the operator responsible for operating the water service in the city of Maputo, in the municipalities of Matola and Boane and in the cities of Beira, Quelimane, Nampula and Pemba, encompassing a total population of more than 2 million inhabitants. Strictly speaking, the PPP that was formed was not a concession contract but a lease contract since the investment continues to be under the responsibility of the FIPAG.

The duration of the leasing contract in Maputo is 15 years. With regard to the other cities, management contracts were signed for a period of five years. Under the terms of these contracts, the operator is only responsible for operating and maintaining the FIPAG infrastructures. Customers have a commercial relationship with the FIPAG rather than with the private operator. The FIPAG periodically pays the private operator a rent for the service provided, part of which is related to the private operator's performance regarding the goals that were established.

In 2005, a new private operator, the Dutch company Vitens, entered the water supply market in Mozambique. Currently, this company operates in the cities of Inhambane, Maxixe, Xai-Xai and Chókwè and in 11 suburban towns, providing water services to about 550,000 inhabitants. However, the service continues to be provided in poor conditions, with only 30% of the population having access to drinking water supplies and with only two cities having a continuous supply of water (24 hours).

The drinking water supply sector comprises 7 public municipal operators, 11 public district operators and about 200 small private operators. In the scope of wastewater services, only two municipalities have such a service (Maputo and Beira). Another 22 municipalities have wastewater collecting systems, even though they are rather precarious.

In short, the water service is available in 23 cities and in 68 towns. Within the group of cities, 14 are provided by means of systems with delegated management and 5 are managed through state companies. The remaining services correspond to 70 systems managed locally. Figure 2.37 shows the institutional framework of the water sector in Mozambique in 2007 (adapted from DNA, 2007).

2.10.3.3 Integration of the sector and services provided

The water sector, and especially the water service, is vertically integrated in Mozambique, since the same operator (e.g. Águas de Moçambique) is responsible for providing the entire production chain of the water cycle, from abstraction to the distribution, including treatment and storage.

In Mozambican cities that have wastewater services, these systems are managed by the operator that is simultaneously responsible for the water service.

2.10.3.4 Numbers for the sector

There is very little information available about the volume of water produced in Mozambique. However, the delegated systems annually produce about 85 million m^3 of water. In 2006, water services covered 26% of the population and wastewater services reached 4.6% of inhabitants, as illustrated by Table 2.53 (DNA, 2007). In places where the water service exists, generally it is not provided continuously and, on average, in 2006, it was available for 15 hours per day.

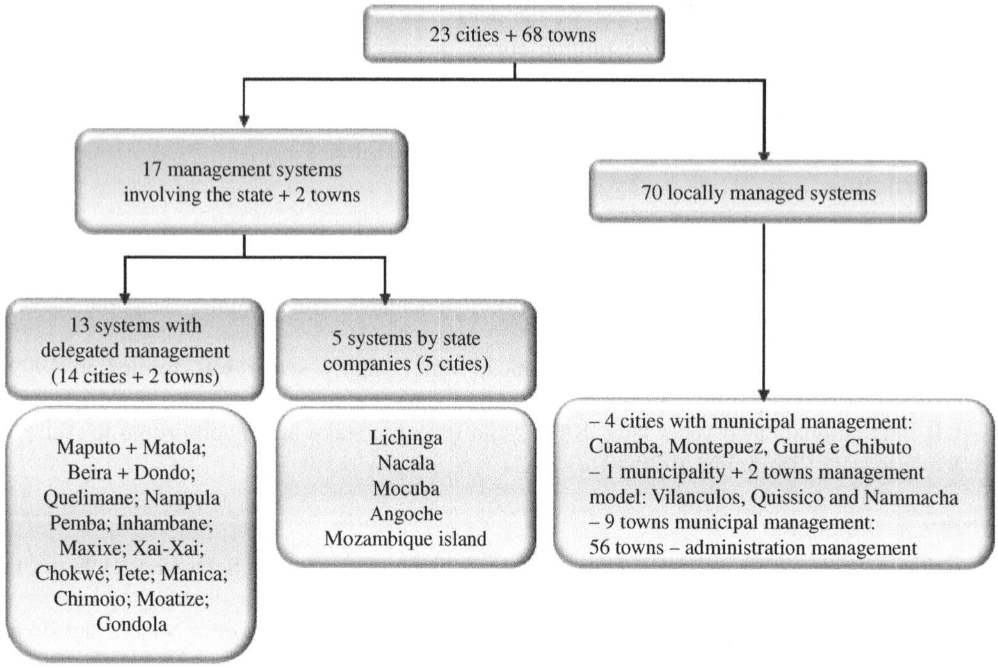

Figure 2.37 WWS in Mozambique

Table 2.53 Annual evolution of the WWS coverage level in Mozambique

Year	Water supply				Wastewater services			
	Rural		Urban		Rural		Urban	
Variable	Coverage (%)	Pop. served (millions)	Coverage (%)	Pop. served (millions)	Coverage (%)	Pop. served (millions)	Coverage (%)	Pop. served (millions)
2002	35	4.7	33	1.2	27	3.6	42	2.0
2003	36.3	4.9	35	1.4	30	4.0	42	2.1
2004	40	5.5	36	1.5	33	4.5	46	2.9
2005	41.2	6.2	36.5	1.5	35	4.7	52	3.1

In Mozambique, more than 50% of the volume of water produced is not billed (it has been estimated at 51% of total production). In 2005, the government announced objectives to be achieved during the period 2005–2009, which included increasing the water service level of coverage to 60% in 2009 that would correspond to about 4 million inhabitants living in urban areas. In order to achieve the Millennium Development Goals, the government has established a target of 70% coverage for the water service by 2015.

The investment programme implemented in the five systems under delegated management involved about 57 million euros in 2005. This amount was divided into 17.8 million euros for Maputo/Matola, 11.7 million euros for Beira/Dondo, 15.1 million euros for Quelimane, 5.5 million euros for Nampula and 2.7 million euros for Pemba. About 23% of the total value invested is allocated for infrastructure rehabilitation and emergency repairs, which, when combined with the works to expand the systems, are contributing towards increasing the quantity of water available to customers. The supply has already tripled in Pemba and doubled in Nampula. It is also expected that, when compared to the situation in 2003, investments will double in Beira and in Quelimane, and increase by 14% in Maputo.

In the four cities referred to above, investments of about 18.5 million euros have been planned for repair and expansion works and for projects to improve management and institutional capacity-building. In comparison to the capacity available in 2004, these projects will triple the existing capacity in Maxixe, double the capacity available in Xai-Xai and increase capacity one and a half times in Chókwè and Inhambane. Coverage indicators are expected to evolve considerably, as observed in Table 2.54 (DNA, 2007).

Table 2.54 Coverage level evolution of 9 cities under delegated management

Year	9 systems under delegated management								
	Maputo %	Beira %	Quelimane %	Nampula %	Pemba %	Xai-Xai %	Chókwè %	Inhambane %	Maxixe %
2002	38	24	13	19	52	–	–	–	–
2005	43	23	13	24	60	40	55	57	8
2009	Mozambique hopes to achieve an overall coverage of at least 60%								

2.10.4 Rules of the game

2.10.4.1 Tariff setting

The CRA is responsible for defining and setting tariff systems for the water services integrated into the QGD, under the terms of the Water Tariff Policy, dated 1998, and in conformance with Decree No. 74, dated 1998. The Tariff Policy defines water as being an economic good and says that tariffs should reflect the need to recover costs. The tariff systems are proposed by the FIPAG and are approved by the CRA.

The variation of the tariff system in the cities of Maputo, Matola, Beira, Dondo, Quelimane, Nampula and Pemba is an integral element of a policy that seeks a full cost recovery of the service, through payments to customers, aiming at ensuring the self-sustainability of the systems while always preserving their social viability. These investments had a significant and visible impact in a short period of time and, consequently, have resulted in a substantial improvement of the water service.

The WWS tariffs of the cities of Xai-Xai, Chókwé, Inhambane and Maxixe, which were recently integrated into the QGD, continue to follow a process of greater adjustment, since they are at a considerably lower level than the tariffs of other cities. It is expected that by 2009 it will be possible to achieve a level of prices that enables self-sustainability.

An analysis of the tariff systems among the different water services reveals that there is a regular policy in this area, i.e. the CRA acts in order to make the methodology used to implement prices among the operators uniform. Currently, the prevailing tariff systems comprise a fixed component, related to the meter rental and maintenance, and a variable component with progressive blocks, according to the consumption of the final customer (domestic or industrial).

The two-part tariff system that has been implemented, approved by the CRA and proposed by the FIPAG has been summarized in Table 2.55. In this regard, minimum revenue is guaranteed by means of a minimum consumption of 10 m^3 every month and by the fixed component, related to the meter size installed. VAT (17%) is applicable on these prices, on 75% of the water bill referring to the total amount, encompassing both the fixed and variable components.

Table 2.55 Tariff system in Maputo and Matola for 2007

Consumption blocks	Tariff (€/m^3)
Average tariff	0.432
Public taps	0.257
Household	
Minimum consumption of 10 m^3 (month)	3.429
Consumption between 10 m^3 and 20 m^3	0.529
Consumption between 20 m^3 and 30 m^3	0.537
Higher levels of consumption	0.543
Industry, commerce and public consumption	
Consumption up to 25 m^3/month (commerce and public consumption)	13.214
Consumption up to 50 m^3/month (industry)	26.429
Additional consumption (€/m^3)	0.529

Part of the price paid by customers is used to pay a tax to the FIPAG and another tax to the CRA. The tariff system is fixed for five years and is adjusted regularly according to a formula defined in the contract, i.e. a cost index during the respective period. There is no tariff system for the wastewater service.

2.10.4.2 Quality of service

Under the terms of the contracts, operators are responsible for continually monitoring the drinking water quality, in conformance with the requirements that have been established. The concession granting authority is responsible for auditing and monitoring compliance with drinking water supply norms. In this context, the contract can stipulate that tests must be conducted regularly to determine whether the operator is providing a service according to the water quality levels that have been agreed upon. The quality must be monitored at the point of treatment and at strategic sites along the distribution networks, as agreed in the contract.

An important development is the increased reliability of the service and the drinking water quality in all the systems as a result of the investments that are underway. These developments are enabling an expansion of the service, reversing the tendency towards reductions or stagnation, which was visible during the 1990s.

Despite this progress, the water sector still lacks self-sustainability. Mozambique's circumstances, where about 70% of the population lives below the poverty line (Water Aid, 2005), and the need for heavy investments to counter the persistent droughts that affect the country are some of the aspects that constrain a stronger improvement in the quality of service provided. However, it is also important to note the efforts that have been made by entities that are directly or indirectly responsible for improving the WWS.

2.10.4.3 Public service obligations

Within the scope of serving one of its main functions, i.e. protecting customer interests, the CRA defined the 'General Conditions for Drinking Water Supply Contracts', which regulates the main rights and obligations of the water services customers that are a part of the QGD. Table 2.56 shows the main clauses contained in this document.

Table 2.56 Contractual principles established between operators and customers

Principles	
• Providing good quality of drinking water in compliance with the Regulations for Public Water Distribution and Drainage Systems;	• Bill average consumptions if the connection does not have a meter;
• Maintaining and repairing distribution networks;	• Notifying customers and immediately take measures to repair situations of unscheduled interruptions in the service;
• Monthly reading of meters;	• Giving customers information about the tariffs being practised and establish water connections;
• Sending debit notes to customers before the payment of the bills is due; • Inspecting the infrastructures of customers; • Responding to requests and complaints within a maximum period of fifteen (15) days, except for complaints regarding the values of bills, which must be answered within ten (10) days;	• After ascertaining that conditions for implementing a connection are viable and the connection fee and other costs have been paid, if the connection is within 25 metres of a network main, the company must activate the connection within a maximum period of twenty (20) days.

2.10.5 Performance

The water sector in Mozambique has been characterized by significant progress in recent years. Although its performance is still far from acceptable, let alone what is desirable, the problems that the country faces, with an extreme level of poverty, make a greater development of the water sector unfeasible.

Keeping in mind the socio-economic context, strong points of the water sector in Mozambique include political and institutional stability, namely of the entities associated with the water sector, the significant number of technical staff in the country and a private sector that has already gathered some experience (almost ten years).

As for negative aspects, it is possible to highlight the fragmentation of water management into many sectors and institutions, dependence on the central administration for developing, operating and maintaining the water systems and water resource management systems, difficulties in viewing water as a good with an economic value and the failure to articulate the management of water resources with the quality of water, health, the environment and economic development. Although investments have increased significantly, the contribution of customers is still limited, the resources subsidize or benefit primarily the middle and upper classes and there are few resources, or even no resources, for the wastewater services. The WWS needs heavy investments to increase the level of coverage. At the same time, social policies must be created, to allow for the universal service.

2.11 PORTUGAL

2.11.1 Historical background

In constitutional terms, Portugal has a semi-presidential system, comprising four organs of sovereignty, namely the President of the Republic, who is the head of the state (regulatory power), the Assembly of the Republic (legislative power), the government (executive power) and the courts (judicial power).

Portugal has a population of about 10.3 million inhabitants (2001 census) and an area of about 93,400 km^2. It is administratively organized into districts, municipalities (the basis of the country's administrative structure), also known as council districts managed by town halls, and into parishes, which are administered by parish councils. Currently, Portugal has 18 districts, 308 municipalities and 4,257 parishes.

The WWS in Portugal has witnessed significant developments, especially during the last two decades concerning the access and quality of service, and even the market structure. The water sector first began to be restructured in 1993, with the introduction of deverticalization, separating the 'wholesale' segment from 'end-users' (retail segment). Then, private capital was allowed and, later, a sector-specific regulatory agency was instituted. These reforms had an unequivocal impact on the progress and quality of drinking water sector and the success of the regulatory model has been widely acknowledged, both in Portugal and abroad.

In Portugal, the WWS are regulated by various legislative documents, among which it is possible to highlight the legal framework that set the attributions of the sectors, established by Law No. 46/77, later changed by Law No. 372/93 and Law No. 88-A/97; the authorization of private capital participation and the deverticalization of the sector, by means of Law No. 379/93; and the basis for concessions, both for wholesale systems through Law No. 319/94 and Law No. 162/96 for water and wastewater, respectively, and for retail systems, by means of Law No. 147/95, which, in its turn, also created the Observatory for these services. Other important measures concern the rules for the regulation of the water sector. Law No. 230/97 created the Institute for the Regulation of Water and Solid Waste (IRAR). Law No. 362/98 established its statutes, and Law No. 151/2002 expanded its functions. It is equally important to note the regulation of the local corporate sector, and the implementation of municipal companies, by means of Law No. 58/98, later replaced by Law No. 53-F/2006, as well as the legislation protecting customers of essential public services, namely Law No. 23/96, recently altered by Law No. 12/2008.

Finally, with regard to environmental aspects, the Portuguese legislation has incorporated all EU regulations, especially the Water Framework Directive by means of Law No. 58 of 2005, and defined the water resources ownership under the Law No. 54/2005.

In Portugal, until the mid 1990s, the local administration was exclusively responsible for the WWS. In 1993, Law No. 372 (revoked by Law No. 88-A/97), which altered Art. 4 of Law No. 46/77, made it possible for WWS to be operated by private companies and other similar entities, by means of concession contracts. The legal diploma also made a distinction between regional systems and single municipality systems. The former serve at least two municipalities as a 'wholesale' provider and require substantial investments by the State for reasons of national interest. All the other systems are single municipality systems, even if they are managed by associations of municipalities.

Once the law about the sector's attributions was altered, thus allowing private capital to invest in some activities within the water sector, Law No. 379/93 defined the legal regime for operating and managing regional and municipal systems to produce, treat and distribute drinking water and to collect, treat and discharge wastewater. Under the terms of Art. 3 of this law, the State directly operates and manages regional systems, or attributes these functions by means of concessions to public corporate entities or to companies resulting from the association of public and private entities, where the public sector must compulsorily hold the majority stake. This normative document was amended by Law No. 103/2003, which clarified the public interest assignments of the regional systems.

Law No. 319/94, altered by Law No. 222/2003, regulates the legal regime for constructing, operating and managing regional systems for abstraction and drinking water treatment, when granted by means of concessions arrangements. The juridical regime for building, operating and managing multimunicipal systems to collect, treat and discharge wastewater, when granted by means of concessions arrangements, are regulated by Law No. 162/96. On the other hand, Law No. 147/95 governs the legal regime for concessions for single municipality systems and created the National Observatory for Regional and Municipal Systems. However, this diploma was partially revoked by Law No. 362/98, which approved the statutes of IRAR.

The creation of IRAR in 1997, by means of Art. 21 of Law No. 230, proved to be an important measure both to safeguard the rights of customers and the economic sustainability of the systems. Law No. 362/98, which was altered by Law No. 151/2002, as already mentioned, approved the statutes of IRAR and put an end to the Observatory. These changes in this diploma gave IRAR the necessary powers to supervise and monitor the drinking water quality.

In 1998, Law No. 58 regulated the creation of municipal, inter-municipal and regional companies by municipalities or associations of municipalities. These could be established under three arrangements: public companies (in which the municipalities, associations of municipalities or administrative regions hold all the capital), public capital companies (in which the municipalities, associations of municipalities or administrative regions have invested capital along with other public entities) and companies with a majority public stake (in which the municipalities, associations of municipalities or administrative regions hold the majority of the capital in association with private entities), which represent institutionalized PPPs. Later, in 2006, Law No. 53-F/2006 was published, which established the new juridical regime for the local corporate sector, revoking Law No. 58/98. Among the new elements of this normative document is the material contained in Art. No. 11 which states that local corporate sector entities engaging in activities within the regulated sectors are subject to the authority of the respective regulatory agency and that it is essential to hold public tenders to choose private partners for institutionalized PPPs.

In terms of public service obligations, it is important to note the recent approval of Law No. 12/2008, which altered Law No. 23/96 and created some juridical mechanisms aimed at protecting customers of essential public services. This normative document is a significant landmark for customer protection with the aim to assure a minimum set of services, deemed to be indispensable for the quality of life in modern society, which includes the WWS.

Finally, it is important to note regulations related to the environment and water, which are, naturally, very vast. Law No. 58/2005 is especially worthy of note. It approved the Water Act and transposed the Water Framework Directive to Portuguese national legislation, establishing the basis and the institutional framework for the sustainable management of water resources. Regarding the institutional framework, the Water Act has implemented the principle of the river basins region, which serves as the main unit for planning and management, as stipulated by the Directive. Five Regional Administrations for River Basins (*Administrações de Região Hidrográfica* – ARH) were created, whose headquarters are the same as those of the current Commissions for Coordination and Regional Development (*Comissões de Coordenação e Desenvolvimento Regional* – CCDR).

2.11.2 Legal and institutional framework

2.11.2.1 Responsibilities

Numerous and diverse kinds of entities are jointly responsible for the water sector. In administrative terms, it is important to note the roles of the regulatory agency (IRAR) and of other bodies of the Public Administration, namely the Ministry for the Environment, Territorial Planning and Regional Development (*Ministério do Ambiente, do Ordenamento do Território e do Desenvolvimento Regional* – MAOTDR), the Water Institute (*Instituto da Água* – INAG), the Portuguese Environmental Agency (*Agência Portuguesa do Ambiente* – APA), the CCDRs, the Directorate-General for Consumer Affairs (*Direcção-Geral do Consumidor* – DGC), the Competition Authority (AC), the Portuguese Association for Water Distribution and Wastewater (*Associação Portuguesa de Distribuição e Drenagem de Águas* – APDA), the National Laboratory for Civil Engineering (*Laboratório Nacional de Engenharia Civil* – LNEC) and the municipalities. Finally, apart from being involved in operating WWS, the Águas de Portugal (*Águas de Portugal* – AdP) company plays a very relevant and structural role in the sector's development.

IRAR is the sector-specific regulator for the WWS (as well as for urban waste). As part of its functions, IRAR supervises and regulates WWS concession arrangements. Its objectives include safeguarding the quality of service delivered and supervising and ensuring the equilibrium and sustainability of the sector, under the terms of its statutes and the law. It is also the competent authority for monitoring the quality of drinking water, encompassing all Portuguese WWS.

The MAOTDR is the highest entity responsible for the diverse policies of the environmental sector, which includes the water policy in the scope of the water sector.

The management of water resources has been entrusted to INAG, which was created in 1993 and has its own patrimony and administrative autonomy, although it is supervised by the MAOTDR. INAG's mission is to propose, follow closely and ensure the implementation of national policies in the field of water resources, so as to safeguard the sustainable management of this sector as well as to guarantee the effective application of the Water Act.

The APA was created in 2007, as a result of a merger of the Institute for the Environment and the Institute for Waste, with the purpose of creating conditions for a greater effectiveness in the management of environmental policies and sustainable development. This agency is also responsible for proposing, developing and monitoring the implementation of policies in the areas of climate change and emissions of atmospheric pollutants, environmental impact assessments, prevention of serious risks, integrated pollution monitoring and environmental education, ensuring the participation of the public and environmental NGOs and the circulation of information.

The CCDRs are decentralized bodies of the central administration (MAOTDR), whose role is related to environmental regulation, general coordination, planning, licensing and inspections, namely in the case of water abstractions and wastewater discharge. The CCDRs are also responsible for managing EU funds allocated for financing investments in the water sector. However, the Water Act created the ARHs, which cover part of the CCDRs' functions.

Law No. 58/2005 (Water Act) created the ARHs, to serve functions with regard to water management, including planning, licensing and monitoring. The ARHs for northern Portugal, Central Portugal, the Tagus Valley, the Alentejo region and the Algarve region were established in mainland Portugal.

The DGC is a public institute created to promote the policy of customer rights, as well as to coordinate and implement measures aimed at customer protection, information and education and to support customer organizations. In cooperation with IRAR, this body plays a significant role in protecting WWS customers.

The AC, which was created in 2003, has transversal powers over the Portuguese economy to apply rules for competition, in cooperation with the sector-specific regulatory bodies. The AC's mission is to ensure that the rules of competition are implemented in Portugal, according to the principles of a market economy and free competition, with a view to ensuring that markets work efficiently. It seeks to achieve a high level of technical progress and secure greater benefits for customers. The AC collaborates with IRAR in all matters related to competition in the WWS market.

The APDA was created in 1986 and is an entity that represents and defends the interests of WWS and all the other players in this sector. Simultaneously, this association encourages the handling, research and development of matters related to drinking water quantity and quality, collection, treatment and the final discharge of wastewater. It is a forum where professionals from diverse backgrounds and areas can intervene in the water sector.

The LNEC is a research laboratory that is a reference at a national and international level in the field of civil engineering, including the area of hydraulics and, especially, sanitary engineering.

The municipalities are responsible for providing WWS. In this context, they can delegate these services to private entities by means of concessions arrangements or to other entities, such as parishes and customer associations.

The AdP company, which belongs to the State corporate sector, is the main corporate group in the environmental sector in Portugal. Its mission is to contribute towards resolving national problems in the areas of the water supply, wastewater services and the treatment and collection of waste, within a framework of economic, financial, technical, social and environmental sustainability. Presently, it encompasses more than 50 companies in the scope of its activities, including some international activity.

2.11.2.2 Regulation

There is a sector-specific regulatory agency in Portugal, the IRAR, which, according to its statutes, has important regulatory functions with regard to regional and municipal WWS and urban waste systems that have been delegated as concession arrangements. It seeks, above all, to protect customer rights as well as to safeguard the sustainability and economic viability of the systems. IRAR is also responsible for monitoring and controlling the quality of drinking water, encompassing all WWS in Portugal, with the exception of Madeira, which has its own competent regional authority.

IRAR is an entity governed by public law, with legal personality, administrative and financial autonomy and its own patrimony. It is subject to the supervision of the MAOTDR and therefore is not independent. The board of directors is also politically named and have three-year mandates. However, it is financially independent since it is entirely funded by the concessionaires of the regional and municipal systems, by means of a tax created to support its activities.

The process of consolidating the regulatory model began in 2003. This model is now fully operational and its strategy is based on two broad planes of intervention: one at the level of the structural regulation of the water sector and the second at the level of regulating the actions of operators (conduct regulation).

The structural regulation of the water sector in horizontal terms seeks a suitable level of separation of the operators by geographical units and by types of services (markets), without undermining desirable and necessary economies of scale and scope economies, creating better conditions for competition and allowing a more effective regulation. Vertically, it intends to ensure an appropriate level of aggregation of the operators, when they develop their activities in more than one of the successive stages of the production process necessary to provide the service.

At the same time, IRAR's strategy also includes regulating the actions of the operators concerning economic aspects and the quality of service provided. Economic regulation is the most important form of regulating the behaviour that is allowed for operators, insofar as monopolistic prices tend to be higher than the prices practised in competitive markets. Achieving lower prices, which simultaneously allow for the economic and financial viability of the operators, corresponding to a fairer situation for customers, requires a strong intervention of the regulatory agency. Economic regulation also includes evaluating the investments of the operators.

Regulating the quality of service is a way of supervising actions that are inextricably intertwined with economic regulation, constraining the operators' permissible behaviour in the context of the quality of service they provide to customers. The functions of IRAR with regard to the quality of drinking water corresponds to a particular case in terms of the quality of service since the institute goes beyond traditional regulatory functions. Figure 2.38 shows IRAR's regulatory model (IRAR, 2003).

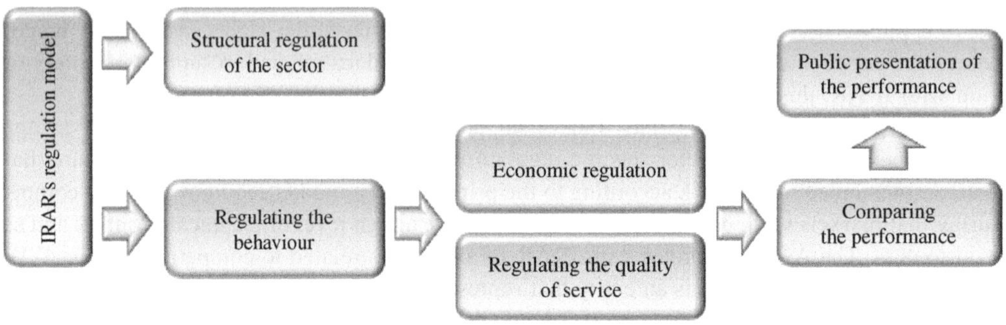

Figure 2.38 Regulatory model of the IRAR

According to the new organic law of the MAOTDR (Law No. 207/2006), IRAR will continue under the indirect administration of the State, with the status of a Public Institute, and will be renamed as The Water and Waste Services Regulation Authority (*Entidade Reguladora dos Serviços de Águas e Resíduos* ERSAR). Apart from its functions as the regulator for the economic aspects and the quality of service in the water and waste sectors, the new organic law also outlines other functions of the ERSAR, which continues to be the competent authority to monitor the quality of drinking water. It encompasses all WWS and also guarantees the structural regulation of the sector, while promoting comparisons and raising public awareness about the activities of the operators.

As already mentioned, the regulatory model defined for IRAR is somewhat limited in terms of its functions, especially concerning tariff regulation. With regard to municipal concessions, IRAR only issues opinions on the tender process for municipal systems concessions and the drafts of contracts. It does not intervene in the process of setting tariffs, which is a decisive element while choosing concessionaires, except in situations of economic and financial rebalancing. The tariffs are regulated by the terms of the contracts (as are tariff revisions). With regard to concessions for multimunicipal systems, IRAR pronounces itself upon the tariffs that the concessionaire proposes every year to the concession granting authority.

No regulatory activities have as yet been envisaged for IRAR in the context of regulation of municipal systems, other than concessions, apart from questions related to the drinking water quality. However, it must be highlighted that the expected and desirable reform of IRAR's statutes will expand its scope of action to cover entities that are part of the local administration in terms of customer protection issues. It is also important to note that the local corporate sector Law envisages that IRAR will be involved in their regulation, although the way in which this will be implemented has not yet been clarified.

IRAR has adopted sunshine regulation as its main regulatory method for the quality of service, which is based on publicizing the results of the performances of regulated entities. To this end, IRAR compiles an annual report with performance results, obtained on the basis of a set of performance indicators applied to the operators. The preparation of this document, which is based on benchmarking, includes a joint assessment of performance, with comparisons between the operators, and an individual assessment of the performance of each operator in qualitative and quantitative terms (Marques, 2006b). In this regard, the awareness that IRAR seeks to achieve concerning the activities of the operators is developed by means of pressure by customers and citizens in general, through their protection groups, the media, the political classes (government/parties) and NGOs, the so-called sunshine regulation.

The objective of this regulatory approach is to 'embarrass' the operators that perform poorly, so that they will more likely correct their weaknesses (name-and-shame policy). Although this method does not set tariffs and its coercive power is limited, the display and public discussion of the behaviour of the regulated operator has very positive effects, inducing competition between the regulated operators and resulting in progressive performance improvements in the whole market (Marques, 2005).

In the course of its activities, in 2007 IRAR operating costs were about 5 million euros, being self-financed by means of regulatory taxes and water quality taxes. At the end of 2007, IRAR directly employed 62 staff.

2.11.3 Market structure

2.11.3.1 Ownership and management of services

WWS ownership and management has mainly been associated with the public sector. Although in the past the largest operating companies were private operators (e.g. CAL in Lisbon), the assets still belonged to the State or to the municipalities. There was no private management in the water sector in Portugal between 1974 and 1993. From 1993 onwards there was a proliferation of private sector participation, which, in 2007, covered almost 20% of the total population. Apart from this, there has been an accentuated growth in juridical-formal privatization (public utilities governed by private law).

2.11.3.2 Players in the sector

The municipalities are responsible for providing WWS. They can do so directly, by means of municipal services (which are directly managed by municipal bodies and do not have administrative or financial autonomy), or through semi-autonomous utilities (which are administratively and financially autonomous and are managed by their own Boards of Directors, although they do not have legal personality). Municipalities can also delegate their responsibilities to parishes or to municipal companies, seeking, in the case of the latter, a greater level of entrepreneurship and a clear option for private law. WWS activities can also be attributed to private operators, by means of concession arrangements. An analysis of operators at the level of parishes, which are quite significant numerically, reveals that presently in Portugal water services are supplied by 405 and wastewater services by 315 operators.

In Portugal, many municipalities do not supply 'wholesale' water services, which are provided by regional systems controlled by the AdP Group, although they do participate in those companies as minority partners. In 2006, the AdP supplied 'wholesale' water services to 200 municipalities, albeit only partially in many of these, through 14 companies (including EPAL). In terms of 'wholesale' wastewater systems, this number amounted to 186 municipalities and the services were provided through 16 companies. There are also two companies that were concessionaires of municipal systems (inter-municipal), namely Águas do Vouga and Tratave, which also belonged to the AdP through the participation of Aquapor/Luságua and two associations of municipalities which manage the systems directly (e.g. AMCAL).

In terms of 'end-user' services, the dominant organizational arrangement continues to be the municipal service, although the number of municipal companies and concessionaires has grown in recent years and this trend is expected to continue in the near future. Without including parish operators, which represent a very small part of the population, Table 2.57 summarizes the different forms of 'end-user' systems operating in Portugal. The public companies shown herein represent the atypical situations of EPAL (Lisbon), Águas de Santo André (Santo André) and IGA (Porto Santo).

EPAL is a special case in Portugal due to its history, size and importance. This water company, which serves the city of Lisbon and surrounding municipalities, is a company with entirely public capitals, where 100% of shares are held by AdP. EPAL provides retail services only in the municipality of Lisbon. This company directly supplies 564,000 inhabitants (Lisbon) and indirectly supplies 2.6 million inhabitants in 26 municipalities that buy 'wholesale' water from EPAL.

It is also important to note the relevance of the management model based on municipal companies. At the beginning of 2008 there were 18 municipal companies, representing 1.6 million inhabitants (15% of the Portuguese population). Of these, 3 were institutionalized PPPs, since they involve private partners, and 3 others were in the midst of the tendering process.

As mentioned above, AdP is the main player in Portugal. Apart from its dominant presence in the 'wholesale' segment, it is also a significant player in 'end-user' systems, through Aquapor and Luságua, competing with other private players. The remaining private players include Compagnie Générale des Eaux (CGE/Veolia), which is responsible for 4 companies, potentially serving about 270,000 inhabitants, AGS (Somague/Sacyr Group) with 8 participations and 670,000 inhabitants and Indaqua (Mota-Engil, Soares da Costa and Hidrante), with 5 companies and 540,000 inhabitants. Aqualia (FCC Group) also has 3 small systems. In 2008, about 25 systems (29 municipalities) privatized WWS, as showed in Table 2.58, corresponding to about 1.8 million inhabitants (17% of the Portuguese population).

Table 2.57 'End-user' WWS organizational arrangements

Type of management	Water		Wastewater	
	Number*	Population**	Number*	Population**
Municipal services	229	4,192,375 (40.6%)	244	5,066,832 (49.1%)
Semi-autonomous services	26	2,228,223 (21.4%)	24	2,206,585 (21.4%)
Municipal companies	18	1,567,049 (16.0%)	20	1,650,826 (16.0%)
Concessionaire companies	24	1,737,033 (16.8%)	17	1,393,841 (13.5%)
Public companies	3	593,404 (5.7%)	–	–
Total	**300**	**10,318,084**	**305**	**10,318,084**

* Year 2008; ** Population as per the 2001 census.

Table 2.58 Retail WWS concessionaire companies

Location	Scope*	Population**	Company	Start	Duration (years)	Shareholders
Mafra	W	54 358	CGE de Mafra	1994	25	CGE/Veolia
Fafe	W	52 757	Indaqua Fafe	1996	25	Indaqua
Ourém	W	46 216	CGE Ourém	1996	25	CGE/Veolia
Setúbal	W and WW	113 937	Águas do Sado	1997	25	Aquapor/Luságua
Planalto Beirão	W	77 017	Águas do Planalto	1997	15	Aquapor/Luságua
Batalha	W	15 002	Águas do Lena	1997	15	Aquapor/Luságua
Trancoso	W and WW	10 889	Águas da Teja	1997	25	Aquapor/Luságua
Figueira da Foz	W and WW	62 601	Águas da Figueira	1999	25	Aquapor/AGS
S. Tirso/Trofa	W	109 977	Indaqua S. Tirso/Trofa	1999	35	Indaqua
S. M. da Feira	W and WW	135 964	Indaqua Feira	1999	35	Indaqua
Valongo	W and WW	86 005	Águas de Valongo	2000	30	CGE/Veolia
Cascais	W and WW	170 683	Águas de Cascais	2000	25	Aquapor/AGS
Paredes	W and WW	83 377	Águas de Paredes	2001	35	CGE/Veolia
Alcanena	W	14 600	Luságua Alcanena	2001	15	Aquapor/Luságua
Carrazeda	W and WW	7 642	Águas de Carrazeda	2001	30	AGS
Gondomar	W and WW	164 096	Águas de Gondomar	2001	25	Aquapor/AGS
Alenquer	W and WW	39 108	Águas de Alenquer	2003	30	Aquapor/AGS
P. de Ferreira	W and WW	52 985	AGS Paços de Ferreira	2004	35	AGS
Barcelos	W and WW	122 096	Águas de Barcelos	2004	30	AGS
M. Canavezes	W and WW	52 419	Águas do Marco	2005	35	AGS
Matosinhos	W and WW	167 026	Indaqua Matosinhos	2007	25	Indaqua
Vila do Conde	W and WW	74 391	Indaqua V. do Conde	2007	40	Indaqua
Abrantes	WW	42 235	Abrantagua	2007	25	Aqualia/ LenaAmbiente
Elvas	W and WW	15 500	Aqualia	2008	30	Aqualia
Campo Maior	W and WW	8.387	Aqualia	2008	30	Aqualia

* W – water, WW – wastewater; ** 2001 census

After including the institutionalized PPPs corresponding to the municipal companies, the presence of the private sector reaches 19%. If one includes the systems that are currently under tendering processes, it has been estimated that the total population served by private operators is more than 2.2 million inhabitants (21% of the population).

2.11.3.3 Integration of the sector and services provided

In Portugal, the WWS are mainly deverticalized, i.e. the 'wholesale' and 'end-user' systems, both for the water supply as well as for the wastewater services, are provided by different entities. In horizontal terms, the operators have a limited degree of integration and (even after excluding the parishes) there are 300 operators for almost 10.4 million inhabitants.

Operators almost always provide water and wastewater services together. Sometimes, they also provide urban waste services and, less frequently, transport or other utilities. There is only a limited number of operators providing a water supply or wastewater services separately.

2.11.3.4 Numbers for the sector

In 2007, 862 million m^3 of water were produced in Portugal. The main source of water is surface water (more than 65%), and the production of drinking water was about 7% of the total consumption of water. Likewise in 2007, about 35% of the volume of water produced for the water services in Portugal was not billed, of which the vast majority corresponded to water losses.

In Portugal, the average per capita consumption of water diminished between 2002 and 2005, reducing from 169 litres to 153 litres per inhabitant and per day. Figure 2.39 shows the evolution of the average annual national per capita consumption (IRAR, 2007).

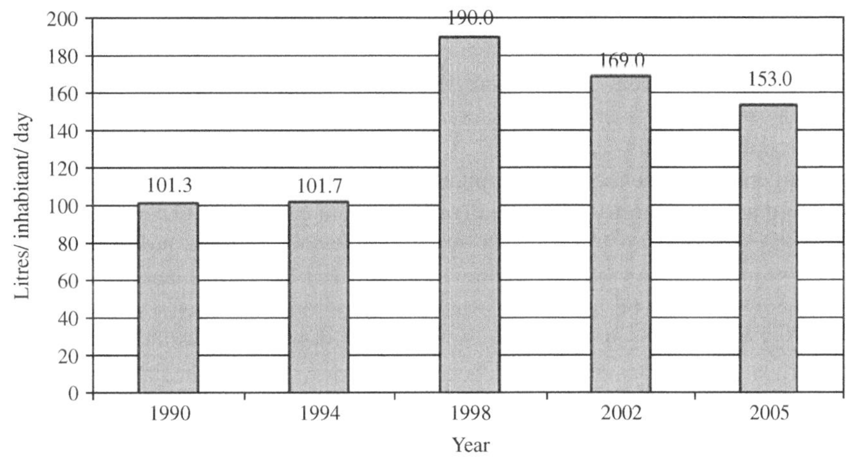

Figure 2.39 Annual evolution of the average national per capita consumption in Portugal

According to IRAR, the level of coverage of the water service in urban areas at the end of 2005, at a national level, was about 91.1%, which means that about 9.4 million inhabitants had access to water supply. In terms of the wastewater service, this still needs to be further developed, in that period covering about 7.7 million inhabitants, which represents a level of coverage of 75.1%.

Figure 2.40 shows the levels of WWS coverage. It must be emphasized that about 2% of the population is isolated and, in principle, will never be served by public networks (IRAR, 2007). It is also important to note that the goals established in the PEAASAR I (Strategic Plan for Supplying Water and Wastewater Services) for 2006 envisaged levels of coverage of 95% for the water service and 90% for the wastewater service.

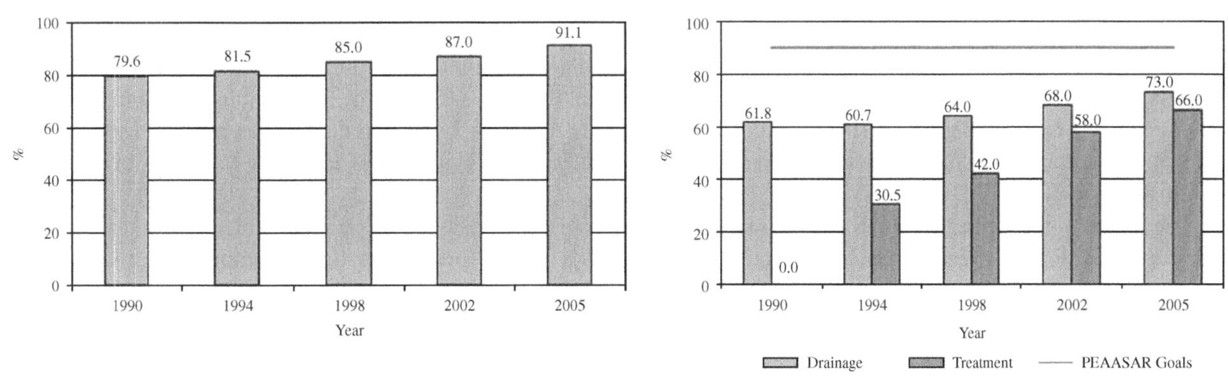

Figure 2.40 Annual evolution of WWS coverage in Portugal

The level of coverage of the wastewater treatment service has evolved considerably and, in December 2006, represented 66%. At the end of 1994, this figure was 30.5%. In 2006, the volume of wastewater collected represented about 315 million m^3.

In 2006, the water service in Portugal had revenues of 1.2 thousand million euros, while the wastewater service had total revenues of 450 million euros. In the same year, the WWS were characterized by a volume of annual investments of 900 million euros, a substantial part of which was allocated to wastewater treatment.

In 2006, the WWS directly employed about 17,500 individuals.

The water services include about 75,000 km of mains while the wastewater service includes approximately 40,000 km of sewers.

2.11.4 Rules of the game

2.11.4.1 Tariff setting

In Portugal, the establishment of prices and tariffs depends on the management model. IRAR does not have any kind of functions regarding municipal services, semi-autonomous services and the different models of municipal companies. Thus, it is the responsibility of the local administration, in this case the Municipal Assembly, to approve the tariff systems proposed by the Municipal Executive or by the board of directors of the semi-autonomous services. In the case of inter-municipal or inter-municipal semi-autonomous services, the inter-municipal assembly is responsible for approving tariffs. In the case of municipal (or inter-municipal) companies, it is possible to find different situations, according to the nature of the company in question. If it is an institutionalized PPP, then the General Assembly approves the tariffs, proposed by the Board of Directors. In the case of other municipal companies, the Town Hall or the Board of Directors approves tariffs.

In municipal systems involving concessions arrangements, tariffs are defined in the concession contract signed between the concession granting authority (Town Hall) and the concessionaire. The setting of tariffs is established in the tender proposal, under the terms of Law No. 147/95. IRAR cannot directly interfere in tariff setting, unless the economic and financial equilibrium of the concession is jeopardized by unpredictable reasons when the contract is signed. In such situations, IRAR can be invited to issue its opinion on the matter.

In regional systems, IRAR has other kinds of powers, being requested to issue opinions about the proposed tariff systems, based on an investment programme that has been defined previously to the date of the concession contract and revised annually as well as on the annual report of the companies and the respective budget. The formula for defining the tariff system proposed in the concession contracts of multimunicipal systems consists of a hybrid methodology, based on the rate-of-return regulation method established contractually and introducing a mechanism to share gains in productivity.

Although it varies from contract to contract, the rate-of-return is fixed by the concession granting authority and consists of a base rate (Treasury Bonds or Euribor) plus a risk premium of 3%, applied to the capital stock achieved and the legal reserve.

As already stated, EPAL is a special case in the entire Portuguese panorama, likewise with regard to tariff setting. The tariff system is the result of an agreement that is periodically signed between the State and EPAL, where the State is represented by the Directorate-General for Companies. Law No. 230/91 establishes the directives for setting the prices to be applied by EPAL. The tariff system consists of a fixed component and a variable component, which depends on the volumes of water supplied, as well as the respective principles for application. Customers are classified into domestic and non-domestic categories and EPAL is authorized to charge, as part of the price, an additional component, whose value is fixed in the agreement, per m^3 of water billed to all its customers in the city of Lisbon, with the exception of the Town Hall. In the case of the Town Hall, the amount is allocated to paying for the value of municipal consumption. The normative document assumes that the prices for the sale of water and the remuneration of the services provided ensure revenues that allow operating costs to be covered and guarantee adequate levels of self-financing and return on the capital invested, i.e. the economic regulation of EPAL is based on the method of rate-of-return.

It must be noted that each WWS has its own classification of customers and its own tariff blocks, which sometimes vary a great deal between operators, although prices always rise in proportion to consumption. It must also be highlighted that there is always a fixed component for the water supply, corresponding to the meter size, which increases according to the type of meter, and a variable component associated with blocks of consumption. Metering exists in 100% of the water sector in Portugal. Concerning wastewater services, the tariff systems vary even more between operators, but normally also include a fixed component (for making the service available) and a component that varies with the consumption of water. A component specifically related to wastewater treatment can also be included. In the past, charges for wastewater services were frequently based on the areas of the property. In general, the WWS, especially in inland areas of Portugal, is heavily subsidized by the general budget of the municipality. The drinking water supply almost always subsidizes the wastewater service and, within the water service, industrial and commercial customers and customers with high levels of consumption subsidize other (domestic) customers. It must also be noted that there are never any charges for the stormwater service and it is common to find social tariffs and special tariff systems for large families.

The following tables (Table 2.59, Table 2.60 and Table 2.61) show the tariff systems of different models of WWS management in Portugal, namely those of EPAL (state company), the SMAS of Oeiras and Amadora (semi-autonomous services), of Águas de Cascais (Concessionaire) and of the council district of Porto de Mós (municipal service) for domestic and non-domestic/industrial customers. In addition, VAT at the rate of 5% is applied to these values.

Table 2.59 Tariff system of EPAL for 2008

EPAL		
Water	Domestic	Non-domestic
Fixed component (€/month)	3.90–2,129.85	9.36–2,854.80
Variable component (€/m^3)		
0–5 m^3	0.1530	
5–20 m^3	0.5754	1.3460/1.0199*
>20 m^3	1.3460	

* Private Institutions of Public Interest

Table 2.60 Tariff system for domestic customers for 2008

Services	Semi-autonomous services (Oeiras and Amadora)	Private company (Águas de Cascais)	Municipal service (Porto de Mós)
Water			
Fixed component (€/month)	2.95–216.11	3.42–133.59	2.05–8.10
Variable component (€/m^3)			
0–5 m^3	0.492	0.400	0.510
5–10 m^3	1.037	0.729	0.860
10–15 m^3			1.020
15–25 m^3	1.989	1.211	1.530
>25 m^3	2.076	1.951	2.650
Wastewater			
Fixed component (€/month)	–	3.12	1.160
Variable component (€/m^3)			
0–5 m^3	0.268	30% of the water bill	–
>5 m^3			20% of the water bill

2.11.4.2 Quality of service

The drinking water quality is regularly monitored by IRAR and is characterized by a very high level. The regulatory agency acts so as to safeguard compliance with water quality standards, according to the Portuguese and EU legislation. Portugal has evolved considerably in recent years also in this area.

In order to encourage the quality of service amongst WWS, IRAR carries out an annual benchmarking exercise, in which it establishes, compares and publicizes the performance of regulated operators (concessionaires). To this end, it has defined a set of 20 performance indicators, which have been divided, keeping in mind the WWS and the 'wholesale' and 'end-user' aspects of these services. These groups have been structured into different categories of performance indicators, namely indicators aimed at protecting customer interests, the sustainability of the operator and the environmental sustainability. Table 2.62 summarizes the performance indicators defined by IRAR for the water supply and for wastewater services. Table 2.63 shows the average national results for the first three years in which the benchmarking scheme was implemented by IRAR. The results are generally very positive, although there are some indicators that have worsened. However, this was due to changes in the definitions of these indicators and a greater rigour in implementing them. Figure 2.41 shows, for example, the evolution of the indicator related to the use of WWTPs in the context of the wastewater services (IRAR, 2007).

Table 2.61 Tariff system for non-domestic customers for 2008

Services	Autonomous services (Oeiras and Amadora)	Private company (Águas de Cascais)	Municipal service (Porto de Mós)
Water supply			
Fixed component (€/month)	2.95–216.11	4.75–166.09	2.05–8.10
Variable component (€/m^3)	1.7511 (0–150 m^3)	0.400 (0–5 m^3)	0.82 (0–50 m^3)
		1.258 (5–120 m^3)	
	2.3952 (>150 m^3)	1.727 (120–400 m^3)	1.13 (>50 m^3)
		2.080 (>400 m^3)	
Wastewater service			
Fixed component (€/month)	–	3.12	1.160
Variable component (€/m^3)			
0–5 m^3	0.268	30% of the water bill	–
>5 m^3			20% of the water bill

Moreover, IRAR issues comments with regard to the results of each indicator for every regulated operator. This assessment is carried out on the basis of a classification system, which compares the performance obtained with reference values (optimum values, or close to them, which the regulator considers to be reasonable and achievable by the operators), keeping in mind the factors of the context in which each operator acts.

In this process, the IRAR classifies the performance of each operator in qualitative terms, considering the quality of service provided, whether it is poor, average or good, by means of the differences between the results obtained by the indicator and the interval of reference values. IRAR publishes the results, classifying them according to the performance indicator.

IRAR explains, for each operator, the results that have been obtained by calculating the performance indicators, quantifying and classifying the quality of service as unsatisfactory, average or good, attributing it a red, yellow or green ball, respectively.

The annual regulatory process carried out by the IRAR results in the publication of an annual report (Annual Report for the Water and Waste Sectors in Portugal), which presents the results that have been obtained. This document is a crucial element of the regulatory model that is currently in effect, not just because it plays an important role in promoting a greater effectiveness and efficiency of the operators, but also because it provides reliable and easy to interpret information to customers and other stakeholders regarding the services provided.

2.11.4.3 Public service obligations

In Portugal, public service obligations, which seek to protect the interests of customers, are considered in the Water Act (Law No. 58/2005) and the Essential Public Services Act (Law No. 23/96, subsequently altered by Law No. 12/2008). Within this legal framework, the universality and the continuity of the WWS are guaranteed by the Water Act and by the Essential Public Services Act, respectively.

The Essential Public Services Act considers diverse aspects to safeguard the interests of customers: the right to participation (organizations representing customers have the right to be consulted while defining regulations for the legal framework governing public services), the obligation to provide information when supplies for pubic services are suspended (the service cannot be suspended without suitable notice, except for fortuitous cases or *force majeure*) and the right to partial quittance (the payment of a public service cannot be refused). In addition, it includes the assurance of quality standards (any service being provided must conform to high quality standards), minimum consumption (imposing and charging for minimum consumption is prohibited), billing (customers are entitled to receive a bill that duly specifies the sums that are presented therein), lapse by prescription (the right to require payment for the price of the service lapses after a period of six months after it has been provided), the injunctive nature of rights (any agreements or dispositions that exclude or limit the rights attributed to customers by this law are considered to be null and void) and safeguards entitlements (all legal dispositions that specifically prove to be more favourable to customers are safeguarded).

Table 2.62 Performance indicators adopted by the IRAR

Performance indicators	
Water service	**Wastewater service**
Protection of the customer interests	**Protection of the customer interests**
Customer service accessibility	Customer service accessibility
AA 01 – Service coverage	AR 01 – Service coverage
AA 02 – Average water charges	AR 02 – Average water charges
Quality of service supplied to customers	Quality of service supplied to customers
AA 03 – Service interruptions	AR 03 – Flood occurrence
AA 04 – Water tests performed	AR 04 – Answers to written complaints
AA 05 – Quality of water supplied	**Sustainability of the operator**
AA 06 – Answers to written complaints	Operator's economical and financial sustainability
Sustainability of the operator	AR 05 – Operating cost coverage ratio
Operator's economical and financial sustainability	AR 06 – Unit running costs
AA 07 – Operating cost coverage ratio	AR 07 – Solvency ratio
AA 08 – Unit running costs	Operator's infrastructural sustainability
AA 09 – Solvency ratio	AR 08 – Treatment utilization
AA 10 – Non-invoiced water	AR 09 – Treatment of collected wastewater
Operator's infrastructural sustainability	AR 10 – Wastewater pumping capacity
AA 11 – Fulfilment of the water intake licensing	AR 11 – Sewer rehabilitation
AA 12 – Treatment utilization	AR 12 – Service connection rehabilitation
AA 13 – Transmission and distribution storage capacity	Operator's operational sustainability
AA 14 – Mains rehabilitation	AR 13 – Sewer clogging
AA 15 – Service connection rehabilitation	AR 14 – Pumping station clogging
Operator's operational sustainability	AR 15 – Sewer structural breakdowns
AA 16 – Mains failures	Operator's human resource sustainability
Operator's human resource sustainability	AR 16 – Employees
AA 17 – Employees	**Environmental sustainability**
Environmental sustainability	AR 17 – Wastewater tests performed
AA 18 – Utilization efficiency of water resources	AR 18 – Fulfilment of the wastewater discharge parameters
AA 19 – Utilization efficiency of energy resources	AR 19 – Utilization efficiency of energy resources
AA 20 – Final destination of sludge from the water treatment	AR 20 – Final destination of sludge from the water treatment

So as to increase economic accessibility to the WWS, diverse municipalities all over Portugal have implemented a Family Water Tariff, based on a proposal by the Portuguese Association for Large Families, which considers the number of people in the family unit. This measure has been widely adopted by operators. Many other WWS have implemented other kinds of social tariffs, defined on the basis of criteria related to age, income, illnesses or other factors.

Table 2.63 Performance indicators applied to Portuguese retail service

Performance indicators for the 'end-user' water services			
	2004	**2005**	**2006**
AA 01 – Service coverage	86	86	84
AA 02 – Average water charges	1.06	1.13	1.24
AA 03 – Service interruptions	2.24	0.97	0.66
AA 04 – Water tests performed	97.95	99.93	99.46
AA 05 – Quality of water supplied	99.12	99.21	98.95
AA 06 – Answers to written complaints	80	66	73
AA 07 – Operating cost coverage ratio	1.17	1.22	1.24
AA 08 – Unit running costs	0.88	0.92	0.90
AA 09 – Solvency ratio	0.28	0.35	0.29
AA 10 – Water which is not billed	26.1	25.1	23.6
AA 11 – Fulfilment of the water intake licensing	23	1	3
AA 12 – Treatment utilization	69	62	62
AA 13 – Transmission and distribution storage capacity	1.3	1.4	1.4
AA 14 – Mains rehabilitation	0.9	1.6	1.3
AA 15 – Service connection rehabilitation	2.2	2.9	2.6
AA 16 – Mains failures	99	81	67
AA 17 – Employees	3.5	3.4	3.8
AA 18 – Utilization efficiency of water resources	18.6	16.9	15.8
AA 19 – Utilization efficiency of energy resources	0.5	0.5	0.5
AA 20 – Final destination of sludge from the water treatment			

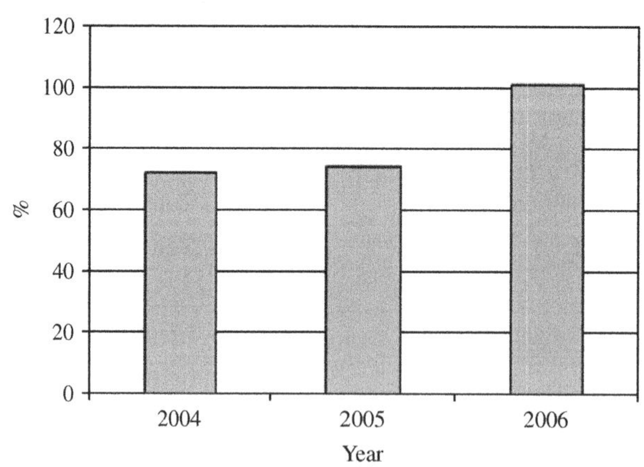

Figure 2.41 Annual evolution of the indicator reflecting the use of WWTPs

In an attempt to improve transparency in the water sector and customer awareness about costs in this sector, the Local Finance Act (Law No. 2/2007) established that tariff systems should be made available on the website of the respective municipality, along with a complete breakdown of the costs of the service.

Moreover, so as to ensure a greater efficiency while handling complaints by customers, operators are obliged to have a complaint book and send all complaints received to IRAR within a period of ten working days.

Finally, it must be mentioned that the drinking water quality has improved significantly in recent years and nowadays generally it has excellent levels of quality.

2.11.5 Performance

In Portugal, the WWS has witnessed extremely important developments in the past two decades. The diverse legal landmarks that have been established, the organizational changes and, above all, the heavy investments that have been made have resulted in a significant increase in WWS coverage, so much so that, in general, these services are currently supplied with a high level of quality. The excellent drinking water quality supplied to customers likewise reflects the success of this evolution. Apart from this, the large participation of stakeholders in recommendations and the technical credibility and reputation of operators for being free from political factors, especially considering their institutional vulnerability, are also factors that have contributed towards the national and international recognition of the vast operational and regulatory progress that has characterized the water sector in Portugal.

It must be mentioned, however, that Portugal still faces important challenges in this area, namely the need to expand the regulatory model to all the WWS in Portugal, the fact that, despite significant efforts, operators are frequently not self-sustaining, the need for operators to improve their productive efficiency, the clarification of the role of the State in the water sector and the separation of regulatory functions from operational functions and, finally, an improvement of the existing legislation governing this sector, which, in some cases, is outdated and, in other instances, is still limited in the context of the new requirements.

2.12 ROMANIA

2.12.1 Historical background

Romania is a democratic republic, in which legislative power is shared by two chambers, the Senate and the Chamber of Deputies. The President holds executive power and appoints the Prime Minister. The Prime Minister leads the government whose members he chooses. The government is subject to a parliamentary vote of approval. The country extends over an area of 238,391 km^2 and has approximately 22 million inhabitants. In administrative terms, it is organized into 4 macro-regions (NUTS I), 8 regions (NUTS II), 42 counties (*județe*), 319 cities (*oras*) and 2,686 parishes (*minicipiu*). Each of the parishes has its own local council and is governed by a mayor (*primar*). In Romania, only 103 of the largest cities have the status of municipality, which gives them greater autonomy and administrative power regarding local matters.

In the context of the legislative framework of Romania, the first normative document regarding the protection of water was published in 1891 and concerned the preservation of water resources. The Water Regime Law (*Legea Apelor*) was subsequently promulgated, in 1927. Later, in 1953, legislation was produced concerning the rational use and protection of water and a new law was published in 1974, which intended to protect the environment with regard to the rational management of water resources and protecting and monitoring the water quality. This law was later changed, in 1989, by Law No. 5. In 1976, Law No. 1 approved a national programme aimed at the future development of Romania's river basins. The programme was coordinated by the National Institute for Hydrology and Water Management (*Institutul National de Hidrologie si Gospodarire a Apelor* – INHGA). This institution has been in existence for over 100 years and plays a very significant role in the water sector. The INHGA is currently coordinated by the National Water Administration (*'Apele Romane'*).

The legal framework for the WWS sector in Romania is essentially based on two laws, the 1995 Environmental Protection Law and the 1996 Water Law, which govern the decisions and regulations produced by governmental authorities.

The Water Law is aimed at the control of water resources and activities carried out in the area of water management. This law has delineated a broad approach to govern the sector at the level of river basins and to manage water resources, associating social and economic development with the pollution of natural eco-systems. Moreover, this law also stipulates that domestic use has priority over all other kinds of use of water resources.

Apart from promoting the conservation and protection of water resources, maintaining an ecological balance, the Environmental Protection Law (Law No. 137) of 1995 (*Legea Protectiei Mediului*) incorporates key economic principles which seek to ensure the participation of all stakeholders in terms of decisions and management.

For the first time in Romania, the legal system was established with public participation, with reviews and comments regarding licensing, policies and legislation. Simultaneously, the ministries responsible for supervising the sector, especially the Ministry for the Environment and Water Management (*Ministerul Mediului și Gospodăririi Apelor* – MMGA), changed some of the norms that established the parameters for the quality of drinking water, and the criteria for classifying and regulating the use of water, in order to comply with EU legislation.

In 2002, the government appointed the *Apele Romane* as the authority responsible for implementing the Water Framework Directive, integrating the INHGA into this organization. Over the course of 2001, 11 River Basin Committees were created in Romania, organized according to the decentralized bodies of the *Apele Romane*. These committees joined the main players responsible for water management and environmental protection in Romania, i.e. the MMGA, the Ministry for Public Health (*Ministerul Sănătăii Publice* – MSP), WWS operators, mayors, the Environmental Protection Inspectorate (IPA), water customers and NGOs.

In 2006, by means of Law No. 51, rules were created for the services of general interest (*Lege servicilor Comunitare de Utilitâti Publice*), defining a set of norms keeping in mind their essential role.

The implementation of Law No. 219 of 1998, gave municipalities the freedom to delegate their services to the private sector by means of concession arrangement.

2.12.2 Legal and institutional framework

2.12.2.1 Responsibilities

There are several entities responsible for the WWS management. These include the MMGA, the MSP, the *Apele Romane*, the local IPA, the local administration, the regulator (*Autoritatea Nationala de Reglementare pentru Serviciile Publice de Gospodarie Comunala* – ANRSC), the Romanian Water Association (*Asociatia Romana a Apei* – ARA), the Competition Authority (*Consiliul Concurenteie*) and NGOs.

The MMGA develops national policies and strategies at the level of managing and protecting water resources. This entity has diverse functions in this area, *inter alia*, strategic planning, including the definition of the water management strategy at a national level and development programmes, producing legislation and policies for the water sector, preparing its budget, establishing norms as well as monitoring and controlling them, dealing with administrative procedures that seek to regulate water resources by means of licensing systems and, finally, it also plays a role in international cooperation in this area.

The MSP also plays an important role in the water sector and is the main entity responsible for monitoring the drinking water supply.

Under the tutelage of the MMGA, the IPAs are the authorities responsible for inspecting and monitoring the implementation of legal norms. The local IPAs in Romania are responsible for issuing licences and authorizations and for inspecting and monitoring the quality of drinking water and the discharge of effluents. Currently, there are 42 IPAs, corresponding to each county, including the city of Bucharest.

The *Apele Romane*, created in 2002, performs important functions concerning the national strategy for water management. The *Apele Romane* is a public company controlled 100% by the state and overseen by the MMGA. The company has diverse competencies associated with the management of 11 river basins, such as managing water abstraction in Romania. The *Apele Romane* is a self-sustainable company, recovering all its costs by means of the prices charged for water sold and other services provided. In addition to its other functions, the *Apele Romane* manages the watercourses, such as rivers and lakes, as well as groundwater, taking care to ensure that this water is used in a rational manner. This authority also plays a role in protecting water resources against degradation and pollution of water quality, preventing and monitoring accidental pollution, as well as preventing against floods and droughts, coordinating operations involving water dams and other reservoirs. At the same time, the *Apele Romane* monitors the quality and quantity of water resources, along with the use of water, and designs the management of budgets for river basins. Finally, this entity is also responsible for applying strategies aimed at the water sector and the environment and implementing policies and rules carried out by the MMGA.

The Romanian constitution stipulates that the local public administration is the main authority responsible for providing WWS. In effect, local bodies are also empowered to decide how to manage these services, providing them directly, delegating it to a public enterprise or delegating it to a private operator via a concession arrangement.

The ANSRC, the regulatory agency for the water sector, was created recently and has functions in terms of licensing, monitoring and accompanying concession contracts.

The ARA is an association of WWS operators and plays an important role in the sector, namely by safeguarding the interests of its members with regard to the policies adopted at a state level, by disseminating best practices and by supporting, encouraging and funding research in the water sector, in order to focus on innovation and technological progress.

The Competition Authority supervises the prices/tariffs of specific services, such as the sale of water by the *Apele Romane*. This authority was created in September 1996 and is a state organization specialized in the area of competition. It implements policies and strategies aimed at promoting competition.

Environmental NGOs in Romania are in a phase of strong evolution and represent a valuable source of information, which can be compiled, processed and used to develop the sector. The mechanisms used by NGOs to collect information are extremely efficient. Owing to the nature of their activities and the number of their members, their intervention frequently has a significant effect on the decision-making process. The general role of environmental NGOs includes discussing current practices and the behaviour of communities. Their activities intend to raise social awareness, promote ecological education and pressure decision-makers with the aim of protecting the environment effectively. Figure 2.42 shows the institutional organization of the water sector in Romania (WaterTime, 2004).

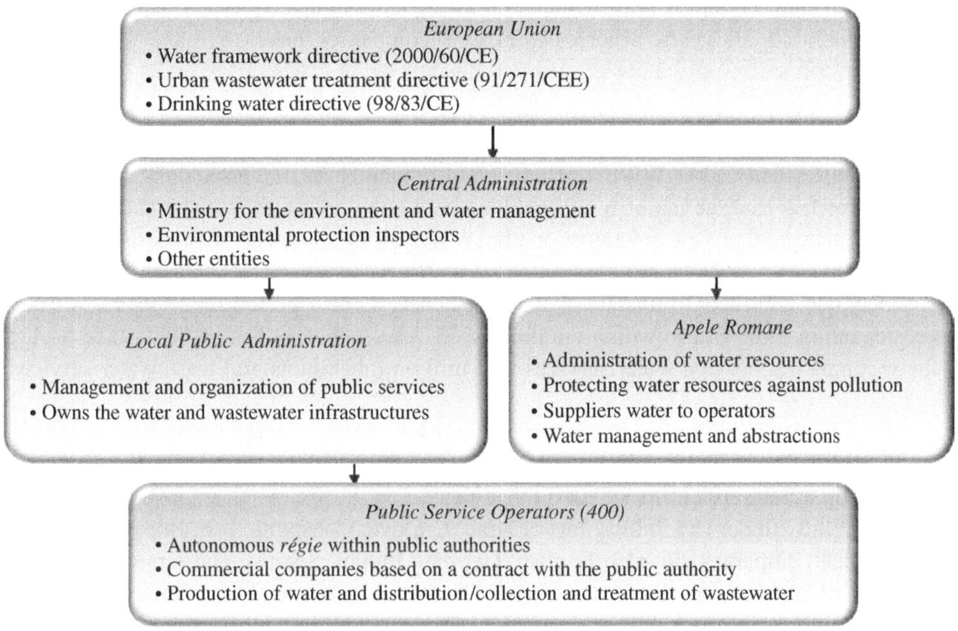

Figure 2.42 Institutional organization of the water sector in Romania

2.12.2.2 Regulation

Romania has a multisector regulatory agency with functions in the water sector, namely the ANRSC, which is overseen by the Ministry for the Public Administration. This body is a public institution of national interest, with a juridical personality. Its main functions are to regulate, monitor and control, at a national level, activities in the area of services of general interest. The ANRSC, created in 2004, is a multisector regulatory agency, which supervises the electricity, heating, transport, public lighting, water, wastewater and stormwater sectors. While carrying out its functions, the ANRSC is an independent regulatory agency with vast powers, including the power to issue secondary legislation. It is directly funded by operators and has an advisory board, which issues its opinion beforehand regarding more important decisions.

The main functions of the ANRSC while regulating operators of public services are as follows: a) to prepare the regulatory framework, keeping in mind the organization, coordination and functioning of public services; b) to certify/authorize operators, irrespective of the nature of their capital; c) to approve WWS prices and tariffs according to established parameters; d) to supervise tenders for delegating services; e) to safeguard the preparation and harmonization of standards, norms and technical recommendations according to prevailing national and EU legislation; and f) to arbitrate, suspend or change licences and permits.

Concerning the water sector, the ANRSC has, to date, performed functions at the level of licensing WWS operators and approving the prices proposed by these operators. In the meanwhile, this agency has produced the regulatory framework for the water sector, by means of Law No. 241 of 2006. This law defines the main procedures related both to tariff regulation, depending on the kind of operator, and to the quality of service supervision and the main public service obligations.

As an association of operators, the ARA also plays a role in regulatory activities in the WWS sector, although in terms of self-regulation. The dissemination of best practices among its members is an important activity, especially in the context of growing concerns regarding improvements in the quality of service and the economic, financial and environmental sustainability of the WWS sector.

2.12.3 Market structure

2.12.3.1 Ownership and management of services

WWS operators are generally organized as autonomous *régies*, within the local administration, and as commercial companies, by means of delegation by the local administration. These commercial companies may or may not involve private capital. In any case, despite the recent increase of private sector participation in Romania, particularly in the largest cities such as Bucharest, Ploiesti, Falticeni and Timisoara, in numerical terms WWS operators still continue to be mainly public enterprises.

Both these organizational frameworks entail the management and operation of the WWS, since their infrastructures always have public ownership. Private sector participation is usually assigned by means of a concession contract with a duration ranging from 20 to 30 years, where the private companies have to make significant investments in infrastructures.

2.12.3.2 Players in the sector

In 2006, there were more than 2,000 WWS operators in Romania but, of these, only about 400 operators were autonomous. Despite the fact that the majority of operators continue under public ownership, recently there has been an increase of the presence of private operators in the WWS sector, namely Vivendi (Veolia Water) and Suez-Lyonnaise des Eaux (Ondeu).

Vivendi has concession contracts to operate WWS in Bucharest for a period of 25 years, with an 83% stake in ApaNova Bucareste, and in Ploiesti also with a duration of 25 years, where it has an 80% stake in ApaNova Ploiesti. In total, both these companies provide water services to 2 million inhabitants and wastewater services to 1.8 million inhabitants.

Suez-Lyonnaise des Eaux established a concession contract for 25 years in Timisoara, in partnership with the local water company, Aquatim. Suez holds a 51% stake in the venture while the remaining capital belongs to the municipality. It provides water services to 400,000 inhabitants.

PPP projects are currently underway in the cities of Arad, Craiova, Constanta, Iasi, Brasov and Falticeni, which are attracting not just French companies, but also English, German, Italian, Spanish and Israeli enterprises.

The vast majority of autonomous operators, over 80%, are commercial companies, with public or private capital, while the remaining operators are autonomous, although without legal personality. Private participation already involves more than 30 operators. The reduction in the number of players has taken place gradually over the course of time, although there are incentives aimed at encouraging associations. This trend is expected to grow in the future.

2.12.3.3 Integration of the sector and services provided

In Romania, the WWS sector is vertically integrated and services are always jointly managed, although they might also occasionally provide other activities (e.g. urban waste service).

There is a limited level of horizontal integration and each operator serves an average population of 55,000 inhabitants.

Some of the operators (e.g. ApaNova Bucareste) buy water from the *Apele Romane* company but have their own water treatment infrastructures.

2.12.3.4 Numbers for the sector

In Romania, the water sector is characterized by an annual production of about 9 billion m^3 of water (data from 2006). The main source is surface water, which represents about 66% of total production. Domestic consumption accounts for the second largest use of water, corresponding to 30% of the total, while industrial consumption is the sector with the highest level of consumption, i.e. 50%.

It has been estimated that the consumption per capita for Romanians corresponds to about 200 litres per inhabitant per day. This figure has been steadily declining since 1999.

There are no accurate data regarding the non-revenue water volume in Romania. However, it has been calculated that this varies between 20% and 50% of the volume of water produced, resulting in an average of 35%.

In 2007, the level of water service coverage was about 70%, consisting of 89% in urban areas and 18% in rural areas. In the same year, the wastewater service covered approximately 50% of the population.

The water sector in Romania employs about 150,000 workers (UNDP, 2004) and its patrimony includes more than 42,000 km of mains and about 17,000 km of sewers.

In 2006, about 600 million euros were invested in the WWS sector. It has been estimated that it will be necessary to invest 18 billion euros in the sector by 2018 in order to comply with EU legislation (water, wastewater and nitrates directives).

2.12.4 Rules of the game

2.12.4.1 Tariff setting

In Romania, the process of WWS tariff setting is mainly regulated by Decree No. 32, 2002, which was later changed in 2003 by Decree No. 35, with regard to the governance of these services.

Operators can opt for a two-part system, i.e. comprising a fixed component and a variable component, according to the consumption of water. However, WWS operators are normally characterized by implementing a tariff system that only considers the variable component, usually based on consumption and irrespective of the type of customer. It must be noted that in the last 14 years, the price of water has increased 16,000 times. In 2003, the average tariff for domestic customers represented 0.23 €/m^3 for the water service and 0.05 €/m^3 for the wastewater service (Speck, 2006). These values, which are the most recent figures available in literature in this area, are certainly outdated. For example, the tariff system of the operator in Bucharest tripled between 2003 and 2007. Figure 2.43 shows the evolution of the tariff systems in Bucharest and in Romania.

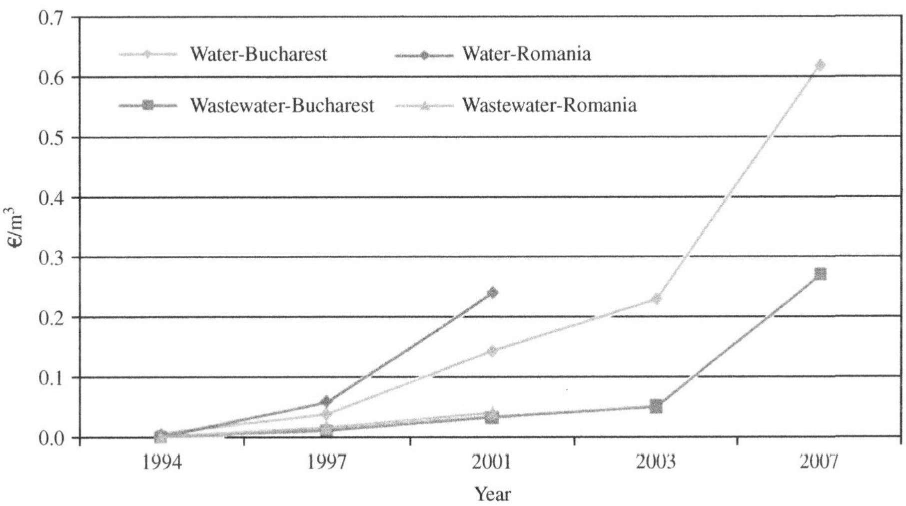

Figure 2.43 Evolution of the WWS average in Bucharest and in Romania

The tariff systems for public operators are revised every 3 months, if the CPI shifts more than 3% or if there is a significant increase in the costs of production. This review takes always place at the request of the operator, keeping in mind increases in the cost of service, profit margins, the development index and the quantity of water supplied.

The municipalities pay for water abstraction under the jurisdiction of the *Apele Romane*, which they later treat, and set their own prices, albeit subject to approval. The cost of water purchased from *Apele Romane* varies, depending on the source and infrastructures used.

Despite the existence of a high percentage of water losses in the service and usual delays in paying WWS bills, the prices practised by operators are low and are not enough to guarantee the cost recovery of service provision and the heavy investments that need to be made in this sector.

As mentioned, tariff systems are almost always defined based on a proportional part, according to consumption, although the metering index is less than 90%. Table 2.64 shows the tariff systems for the cities of Bucharest and Ploiesti. Table 2.65 shows the tariff systems implemented by three public operators chosen randomly. These prices do not include VAT, which is 19% for both services.

The *Apele Romane* implements taxes on water abstraction, which vary according to its purpose (producing electricity, municipal distribution and groundwater abstraction for industrial customers).

2.12.4.2 Quality of service

Since the early 1960s, Romania has developed diverse norms pertaining to water management in general and to the quality of drinking water in particular. The most relevant norms are the National Standards (*Standarde de*

Stat – STAS), such as STAS 1.342, regarding the quality of drinking water supply, STAS 1.343, related to the water use, STAS 1.846, which controls the discharge of wastewater, STAS 4.706, concerning the categories of quality and conditions for surface water, and STAS 9.450, for water used for irrigation and agriculture.

Table 2.64 Tariff systems practised by Vivendi for 2007

	Water (€/m^3)	Wastewater (€/m^3)
Bucharest	0.62	0.27
Ploiesti	0.53	0.12

Table 2.65 Tariff systems of three public operators for 2005

	Water (€/m^3)	Wastewater (€/m^3)
Braila	0.297	0.042
Oradea	0.365	0.204
Sibiu	0.213	0.098

With regard to the supervision of the quality of service, this is still implemented by self-regulation on the part of operators. The ARA has taken an interest in implementing norms and technological progress but has not yet developed any specific measures in this area, for example, by means of statistical information, calculating performance indicators and implementing benchmarking.

The ANRSC has also not implemented any measures to this end. Despite the fact that this is one of its functions, its lack of development has meant that it has still not been able to perform this statutory obligation.

Thus, there is very little information available about the quality of service being provided. Some individual studies contain some statistical information about the water sector, but it is almost always rather scanty. Figure 2.44 shows elements for two indicators (in 2001), one referring to water losses and the other to micro-measurement according to the different Romanian regions (UNDP, 2004).

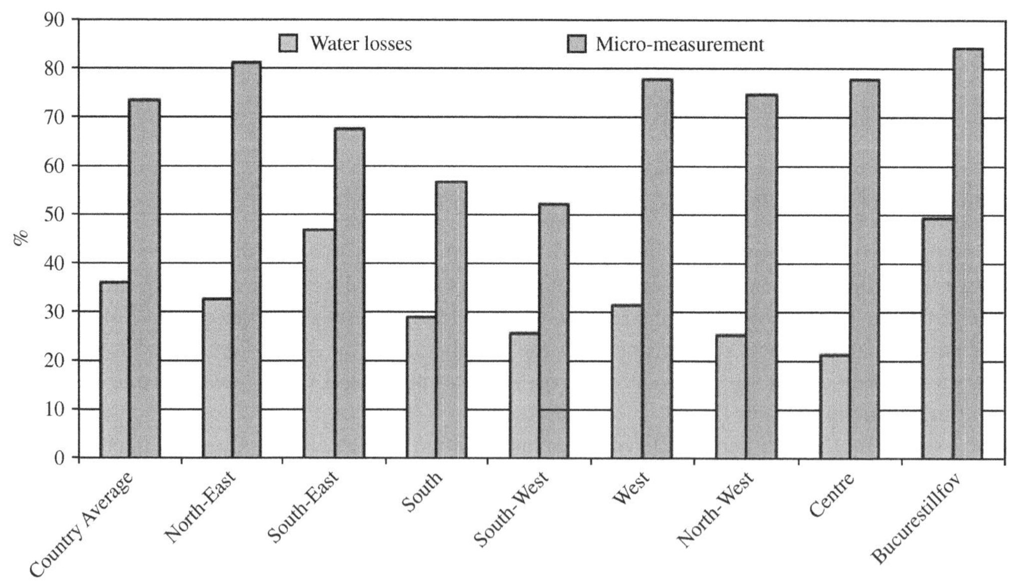

Figure 2.44 Indicators for water losses and metering according to the different regions

2.12.4.3 Public service obligations

Currently, services of general interest are regulated by the ANRSC, including the WWS. This agency has the capacity to promote regulation at various levels, including imposing public service obligations. Law No. 241, 2006, produced by the ANRSC and exclusively dedicated to the water sector, has imposed a set of public service obligations, such as the quality of service, equity of tariff systems, transparency and accountability, self-sufficiency, continuity, adaptability, accessibility and compliance with general legislation.

It must be noted that, prior to this, Law No. 51 of 2006 comprised strong measures with regard to customer protection, although this legislation was common for all essential public services. For example, it included the obligation to provide a universal service, which was not considered by Law No. 241, exclusively focused on the water sector. Even before this, Decree No. 35 of 2003 required WWS operators to guarantee the service continuity, in both quantitative and qualitative terms, adaptability to customer demand, non-discriminatory access and, finally, compliance with prevailing norms and regulations.

2.12.5 Performance

The water sector in Romania has made significant progress in recent years. The heavy investments that have been made in this sector and the increase in the level of coverage reflect this positive evolution. The creation of a regulatory agency for services of general interest, which include WWS, is another very positive aspect and will have a relevant impact in the near future.

Some causes for concern include, without doubt, the low economic and financial sustainability of the WWS sector and the prevailing strategy for the participation of the private sector, which has, above all, focused on the main urban areas, ignoring rural areas, with no solution in sight. A larger scale of operations would also enable further benefits, as would the availability of more and better information about the water sector, which is essential to promote a more effective and sustained decision-making process.

2.13 SINGAPORE

2.13.1 Historical background

Singapore is a parliamentary republic, where the President of the Republic is the head of State. The Prime Minister is the head of the government and has executive power. In its turn, legislative power is divided between the government and a unicameral parliament.

The nation has about 4.7 million inhabitants over an area of about 704 km^2. The country is severely limited in terms of water resources. Despite the heavy rainfall that characterizes the country, its capacity to store rainwater is insufficient and it has to import water from neighbouring countries as the most suitable solution.

In this regard, acknowledging that this extensive scarcity of water resources could constrain the island's development, in 1927 the government established the first agreement for water supply with Johor (Malaysia). Presently, Singapore's drinking water supply is guaranteed by two agreements signed in 1961 and 1962 with Malaysia, which terminate in 2011 and 2061, respectively

During the 1980s and 1990s, Singapore made enormous efforts in order to establish a suitable legal and institutional framework for environmental matters. These were vanguard initiatives in international terms and included water services, controlling pollution and building a modern wastewater and stormwater systems. With all these institutional tools, Singapore is known as the 'Garden City'.

From the 1990s to date, the government has been focusing on measures that guarantee the sustainability of the demand for water. As a form of overcoming these deficits in natural resources and preserving existing resources, the island city-state of Singapore has made a great deal of financial resources available for research and technology in recent years and has consequently developed an impressive know-how and capacity to manage water resources. Since the available reservoirs only guarantee 50% of the demand for water, Singapore has opted to develop other sources of water, namely, choosing to use reclaimed water for non-drinking water purposes (NEWater) and implementing desalination systems since September 2005 (Cheng, 2007).

Moreover, the government has developed a 'Singapore Green Plan 2012', which ensures the sustainability of the water supply, as the main focus of the water policy in Singapore. This policy consists of increasing the robustness and resilience of the water sources, developing the principles of managing demand, encouraging private sector participation and studying and implementing new technologies.

Following a similar line of orientation and keeping in mind the fragility that characterizes the sustainability of the drinking water supply, the government of Singapore has adopted a new strategy, known as the 'Four Taps Strategy', whose four sources of water are the stormwater reservoirs, water received from Johor in Malaysia, water obtained from recycling processes (NEWater) and water obtained by means of desalination. This strategy, associated with other large projects, such as the Deep Tunnel Sewerage System (DTSS) and the Marina Barrage, has made the country increasingly robust and independent in this area.

Currently, the WWS sector is essentially regulated by the Wastewater and Drainage Act. In this regard, codes of conduct have been instituted, such as the Code of Practice on Wastewater and Sanitary Works (CPSSW), Code of Practice on Surface Water Drainage (CPSWD) and the Code of Practice for Water Services (CPWS). The codes of conduct contain information related to the minimum requirements for planning, building and operating WWS infrastructures.

The CPSSW establishes guidelines regarding suitable planning and projects for the wastewater system, only stipulating the minimum and imperative requirements that must be upheld for projects. Apart from the minimum requirements, this code has also prescribed engineering best practices for planning, preparing projects and building wastewater systems.

The CPSWD specifies the minimum engineering requirements for draining surface water. Duly qualified technicians and entities must ensure that all the aspects regarding the drainage of surface water are effectively considered while planning, preparing projects for and implementing new proposals to develop and build new infrastructures.

The CPWS defines directives regarding the planning, construction and repair of sources of water supply for all domestic, commercial and industrial customers. The code also establishes the minimum parameters for providing the water service.

Finally, the government is greatly concerned about the environment. In this regard, there are numerous norms and other legal documents regulating the discharge of wastewater and the inherent pollution. In particular, the 2002 Environmental Pollution Control Act can be highlighted, which limits the temperature, the CQO, the CBO, the suspended solids, the dissolved solids, the pH values and 28 other different parameters for effluents.

2.13.2 Legal and institutional framework

2.13.2.1 Responsibilities

In Singapore, responsibility for the WWS has been divided among various government bodies, namely through the Ministry for the Environment and Water Resources (MEWR) and, under its supervision, through the National Environment Agency (NEA), the Environment and Water Industry Development Council (EWI) and the Public Utilities Board (PUB). The Singapore Water Association (SWA) also has important functions.

The MEWR began as the Ministry of the Environment in 1972 and its main mission is to promote an environment that ensures a high quality of life and high levels of public health. It is also responsible for managing the systems for collecting and treating wastewater. Reforms were introduced in the sector in September 2002, which changed the functions of the MEWR. It was now responsible for strategically managing national water resources and was entrusted with overseeing the wastewater service. Through its two statutory boards, the NEA and the PUB, the government has sought to achieve its targets, by means of continuous innovation and research.

The PUB is the water authority responsible for providing a reliable water service in suitable conditions, keeping in mind the economic principles of fairness and efficiency, so as to sustain Singapore's economic growth. The PUB also has functions regarding the collection, drainage and treatment systems of domestic wastewater and stormwater.

The NEA was created in 2002 and its main function is to implement the environmental policies established by the MEWR. Under the supervision of the NEA, the Departments for Environmental Protection and for Environmental Public Health and the Meteorological Services work together to ensure a high-quality and sustainable environment for present and future generations of Singaporeans.

The SWA aims to play an important role to promote Singapore and reinforce its profile as a point of reference for all services and technologies related to the water sector. This association seeks to foster mutually beneficial synergies with local companies in order to develop a vibrant and dynamic local water industry. It periodically holds a forum for cooperation and for exchanging ideas and knowledge between member companies, both within Singapore and internationally. The forum is a platform for networking between members and to disseminate strategic information in terms of emerging business opportunities and new technologies.

The EWI was established in 2006 by the MEWR and is responsible for developing the environmental market, which also includes the water sector. The EWI's functions are implemented by an advisory board and an executive committee.

2.13.2.2 Regulation

Despite the enormous development that characterizes the water sector in Singapore, there is no sector-specific regulatory agency explicitly responsible for economic regulation and for ensuring the quality of service and public service obligations. However, the transversal regulation implemented guarantees the public service obligations, while self-regulation contributes towards an excellent quality of service.

Even concerning economic regulation, the absence of a regulator is not very evident, since the services are not very high-priced and are provided with an excellent quality of service. Apart from this, the contextual framework of the WWS sector dissuades possible controversies with customers. Thus, the regulation of prices and tariffs is superfluous and unnecessary.

In the meanwhile, there is a professional association in the water sector, which, among its various functions, plays an important role concerning self-regulation and disseminating best practices to be implemented in Singapore, by keeping side by side with other international experiences and practices.

2.13.3 Market structure

2.13.3.1 Ownership and management of services

WWS are mainly provided by a single operator that manages all the services. However, there has recently been a proliferation of private activity in water services management, namely concerning the DTSS, the Marina Barrage, the new infrastructures of NEWater and the desalination plant.

In any case, the Singapore operator (PUB) is a public entity and WWS assets are always publicly owned.

2.13.3.2 Players in the sector

In Singapore, the WWS are provided by a state public entity, i.e. the PUB. In 2003, the PUB developed a treatment technology which allows it to collect all the water used in the country (re-using wastewater) to produce NEWater, which is currently a popular brand of bottled drinking water. However, this service began with the aim of supplying non-domestic customers, such as industries and commercial establishments, but after seeing that the results achieved concerning the quality of water were better than the quality of water obtained from Malaysia, this water began to be used for domestic purposes, although in a bottled form.

As stated above, one of the PUB's main objectives being the water sector development, in 2005 it delegated the management of the desalination plant to Hyflux, a Singaporean group that operates internationally in diverse sectors, under a BOOT contract valid for 25 years. Hyflux also produces bottled water, called Desal H2O, at this station. Hyflux has also won the tender for building and operating the new NEWater plant.

2.13.3.3 Integration of the sector and services provided

In Singapore, the PUB, as the main operator in the water sector, integrates WWS horizontally and vertically. In this sense, it supplies the whole value chain of the water service, from the production to the water distribution. This is also the case in the wastewater segment, being responsible for all the aspects ranging from collection to wastewater treatment, including recycling.

2.13.3.4 Numbers for the sector

About 500 million m^3 of water are consumed every year in Singapore. Domestic consumption accounts for 53% of the drinking water supply, commerce/industry consumes 40%, government authorities consume 6.7% and the remaining water is used in shipyards.

The country is characterized by the WWS universality, since the entire population of Singapore has adequate access to WWS (100% of the population). Apart from this, an especially noteworthy feature concerning the development of the water sector in Singapore is that water losses were only 4.5% in 2006, possibly one of the most remarkable value in the world.

In Singapore, the main assets of the water sector comprise 14 reservoirs, with three further reservoirs currently under construction, one desalination plant and three NEWater plants. In this context, in 2006, 1,230,000 m^3 of drinking water were made available every day, as were 81,000 m^3 of NEWater and 112,000 m^3 of industrial water. The consumption of water per capita in Singapore in 2006 was 158 litres per inhabitant per day.

As far as costs of production are concerned, NEWater costs about 50% to 60% less to produce when compared to desalinized water. In 2005, NEWater sold water at a tariff of 0.53 €/m^3.

There are currently three NEWater plants producing approximately 0.9 m^3 per second. About 6% of this figure is used indirectly for drinking water purposes, covering 1% of Singapore's drinking water requirements, which are close to 13 m^3 per second.

Singapore's desalination plant is one of the largest in the world. It has been operational since 2005 and produces about 136,000 m^3 of water per day, thus satisfying 10% of the demand for water in this island nation.

2.13.4 Rules of the game

2.13.4.1 Tariff setting

In Singapore, customers use two types of water services, differing with regard to the treatment and the purpose for which the water will be used, namely drinking water and non-drinking water and domestic and non-domestic uses. Thus, contrary to what usually occurs internationally, customers in Singapore are endowed by drinking water infrastructures and are also provided with a water service that is unfit for human consumption.

In this context, keeping in mind the different services provided, tariffs vary with the costs of production and the guidelines of the two services. The tariff structure of the drinking water service is implemented by means of tariff blocks, which increase in proportion to consumption, while the non-domestic water service is only charged according to a block based on consumption, as illustrated in Table 2.66.

Table 2.66 WWS tariff system of the PUB for 2008

Type	Consumption Block (m^3/month)	Tariff (€/m^3)	Water Conservation Tax (% of the tariff)	Waterborne Fee (€/m^3)	Sanitary Appliance Fee (€)
Domestic	1–40	0.54	30	0.14	1.37 per connection
	>40	0.65	45	0.14	
Non-domestic	–	0.54	30	0.28	
Shipyards	–	0.89	30	–	–

The main purpose of this tariff system is to discourage the consumption of water, reflected in the implementation of diverse taxes, namely the water conservation tax, the waterborne fee and the sanitary appliance fee. Prices are also subject to VAT, at 3%.

As mentioned earlier, the non-domestic water service (non-drinking recycled water) and NEWater have a significantly lower cost than the costs involved in providing drinking water. The tariff structure for this service comprises a unit tariff based on consumption, which is about 0.20 €/m^3 and 0.53 €/m^3, respectively.

Therefore, this is an additional incentive to use these water services as compared to the other available water services. These water services cost less because they are exempt from taxes.

2.13.4.2 Quality of service

The PUB develops its activities with the purpose of ensuring the welfare of customers. The company has established a set of performance indicators, which reflect the guarantees and commitments made to customers. In this context, the PUB has defined a series of internal performance indicators, which can be classified into three groups of indicators: general feedback, feedback via PUB-One and performance indicators regarding the quality of service. Table 2.67 shows the indicators related to general feedback.

Table 2.67 Group of indicators regarding general feedback

General feedback	Response time
Customer Service Feedback Form	5 business days
Quality Service Helpline	Immediately
Email	3 business days
Mail and fax	5 business days

The second group of indicators, i.e. feedback via PUB-One, is related to operational problems, encompassing aspects such as bursts, inadequate pressure, sewer blockages, sewer collapses and flooding. In such situations, the PUB has committed to provide answers to requests within 2 hours.

Finally, the PUB has developed a set of specific indicators to encourage and assess the quality of service. These performance indicators adopted by the PUB are summarized in Table 2.68.

Additionally, having the best international practices as a reference, the government has established standard values for the quality of drinking water. These values are merely guidelines and can vary in relation to the quality of 'wholesale' water supply or changes in the source of water supply, owing to the operational requirements of the PUB. However, these standards conform to WHO recommendations.

Table 2.68 Performance indicators regarding the quality of service

Quality of service indicators	Response time
Processing Licensed Plumbers' application for water supply	12 business days
Time for completion of new water connections after receipt of road opening approval and site is available for laying of connection	6 business days
Duration to turn on water supply after completion of piping works	1 business day
Processing Drainage Development Control Plans	12 business days
Processing architect's application of Drainage Interpretation Plans	7 business days
Processing of Certificate of Statutory Completion (CSC)	7 business days

In Singapore, according to the WHO, the drinking water supply is of an exceptionally high quality. In this context, in a an attempt to maintain these quality standards, the government has created a system of (heavy) fines to encourage compliance with norms regarding the discharge of effluents, thus ensuring that the current quality will not be affected in the future.

2.13.4.3 Public service obligations

As stated in the previous point, with regard to the quality of service the operator focuses a great deal on customer service. The PUB has developed an interface between customers and the operator based on the so-called CARE model, more precisely, CALL, ACTION, RESPONSE and EVALUATION. Figure 2.45 illustrates the CARE model, showing the cycle of activities between the PUB and its customers.

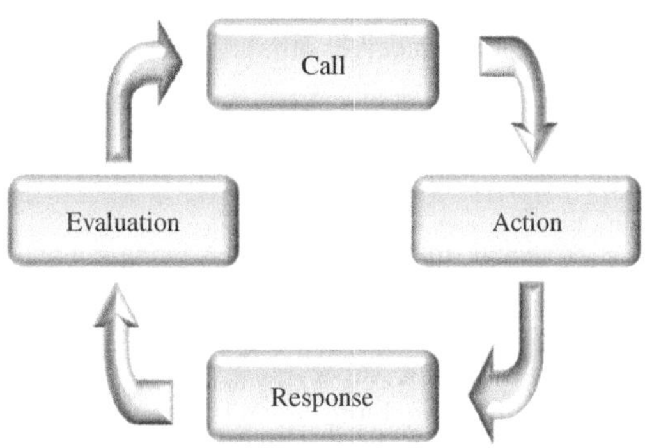

Figure 2.45 CARE model implemented by the PUB

When the operator receives feedback from the public (CALL), an immediate action (ACTION) is implemented, collecting all the necessary information and/or sending this information to the concerned department to take due measures (RESPONSE).

Feedback must be monitored and recorded. After taking measures and resolving the matter, the department must assess (EVALUATE) the causes and take steps to ensure that the problem does not occur again. Finally, the customer is informed about the results of his/her call.

In Singapore, in order to further protect the interests of domestic customers and to facilitate communications by customers regarding problems, especially in terms of service interruption, the operator (PUB) makes available services by means of the following channels: telephone, email, fax, SMS, Webchat and VoIP.

The Singaporean government has a high level of environmental consciousness and, as such, tries to transmit this philosophy to the citizens. Public awareness about different environmental aspects is developed by means of three main measures: specific campaigns, the educational system and the Clean and Green Week, which was begun in 1990. The first specific campaign, called 'Keep Singapore Clean', focused on environmental issues and water management.

2.13.5 Performance

The water sector in Singapore is known for its high level of development, not only due to the excellence of quality of service provided with, for example, extremely low water losses, but also due to the excellent drinking water quality supplied, irrespective of the kind of treatment, customer and the purpose for which the water is to be used.

Other noteworthy features include the great efforts of governmental authorities to make the country self-sufficient concerning water production, by means of some pioneering technological processes. It is expected that this objective will have been achieved by 2011. It is also important to mention the high level of environmental awareness of the government and of the Singaporean population in general.

2.14 SWEDEN

2.14.1 Historical background

Sweden is one of the richest countries in the world in terms of water resources availability. The country has an area of about 450,000 km^2 and approximately 9 million inhabitants. It is structured into three regions (*landsdelar*), namely *Götaland* (southern Sweden), *Svealand* (Central Sweden) and *Norrland* (northern Sweden), 21 counties (*län*) and 290 municipalities (*kommuner*).

Since WWS are considered to be natural monopolies, they are subject to stringent regulation by the competent entities. Sweden is a reference in this area, controlling the *modus operandis* and the performance of operators. This administrative approach, even in the absence of a sector-specific regulatory agency for this sector, is an excellent example at the level of customer protection.

In this scope, Sweden has an encompassing and robust legislation concerning the environment and the water sector. Three fundamental laws provide the framework for the regulation of the sector: the Water and Wastewater Act (*Lag – 1970: 244 – om allmänna vatten – och avloppsanläggningar*), which was replaced on the 1st January 2007 by The Public Waterservices Act (*Lag om allmänna vattentjänster, 2006: 412*), the Environmental Code (*Miljöbalk*, 1998: 808) and the Food Safety Act (*Livsmedelslagen, SFS 1971: 511*).

The Water and Wastewater Act is an essential legal tool for the regulation of the water sector in Sweden in the context of its liberalization and modernization. Amongst other dispositions, this law stipulates that the municipalities are responsible for guaranteeing that customers receive suitably high-quality WWS or for ensuring that they are supplied by means of delegation. In the Swedish water sector, the public or private operator directly responsible for operating the service is commonly known as the 'principal' (*huvudman*). According to this diploma, the National Water and Wastewater Tribunal (*Statens VA – nämnd*) is responsible for solving conflicts between operators and their respective customers at a national level.

The main competencies in terms of water management have been delegated to local or regional authorities. The only ones that can own and be the principal of a public water and wastewater utility are a municipality, a municipality-owned company or a federation of municipalities. At a regional level, the County Administration Council (*Länstyrelsen*) is responsible for examining, supervising and coordinating the water sector. At a local level, there are 290 municipalities responsible for planning, building and operating WWS infrastructure which also have their ownership.

In order to consolidate legislation dispersed over various areas, in 1999 the government implemented the Environmental Code, merging into a single document 15 environmental laws, including the Water Act (protection of national water), the Environmental Protection Act (monitoring and licensing the discharge of wastewater), the Health Protection Act and the Public Hygiene Act. Before this normative document was implemented, environmental protection was regulated by about 200 norms.

This code started to regulate environmental standards and defined the measures to be developed in this sector, so as to prevent and minimize the environmental impact of water abstraction and the discharge of effluents.

Finally, in the water sector, the main change introduced with the implementation of the Food Safety Act was the fact that it established drinking water as a food product and it consequently began to be subject to the quality standards equivalent to any other food product, assigning legal responsibility for this area to the Ministries for the Environment and Agriculture.

2.14.2 Legal and institutional framework

2.14.2.1 Responsibilities

In Sweden, as in neighbouring countries, the municipalities are responsible for the WWS. However, under the terms of the law, the municipalities can choose to manage the services directly or to delegate them to a private operator. In addition, the municipalities can form federations (under public law). The National Water and Wastewater Tribunal (*Statens va-nämnd*) is the authority responsible for licensing water abstraction and the Regional Environmental Tribunals (*Regionala miljödomstolarna*) are empowered to authorize the discharge of treated wastewater by large WWTPs, while licensing at the level of small WWTPs is the responsibility of the County Administration Council (*Länstyrelsen*).

Apart from the responsibilities of service provision, there are other entities that play an important role in the water sector, namely in terms of supervising, regulating and protecting water resources. These organizations include the Municipal Committees for Environmental Protection and Health (*Miljö- och hälsoskyddsnämnden*), the Ministry for the Environment (*Miljödepartementet*), the Ministry of Agriculture (*Jordbruksdepartementet*), the Environmental Protection Agency (*Naturvårdsverket*), the County Administration Council, the Swedish Water and Wastewater Association (*Svenskt Vatten*), the Competition Authority (*Konkurrensverket*) and the National Water and Wastewater Tribunal (*Statens VA – nämnd*). Figure 2.46 summarizes the institutional framework of the water sector in Sweden as well as the relationships between the entities referred to above (adapted from Euromarket, 2003).

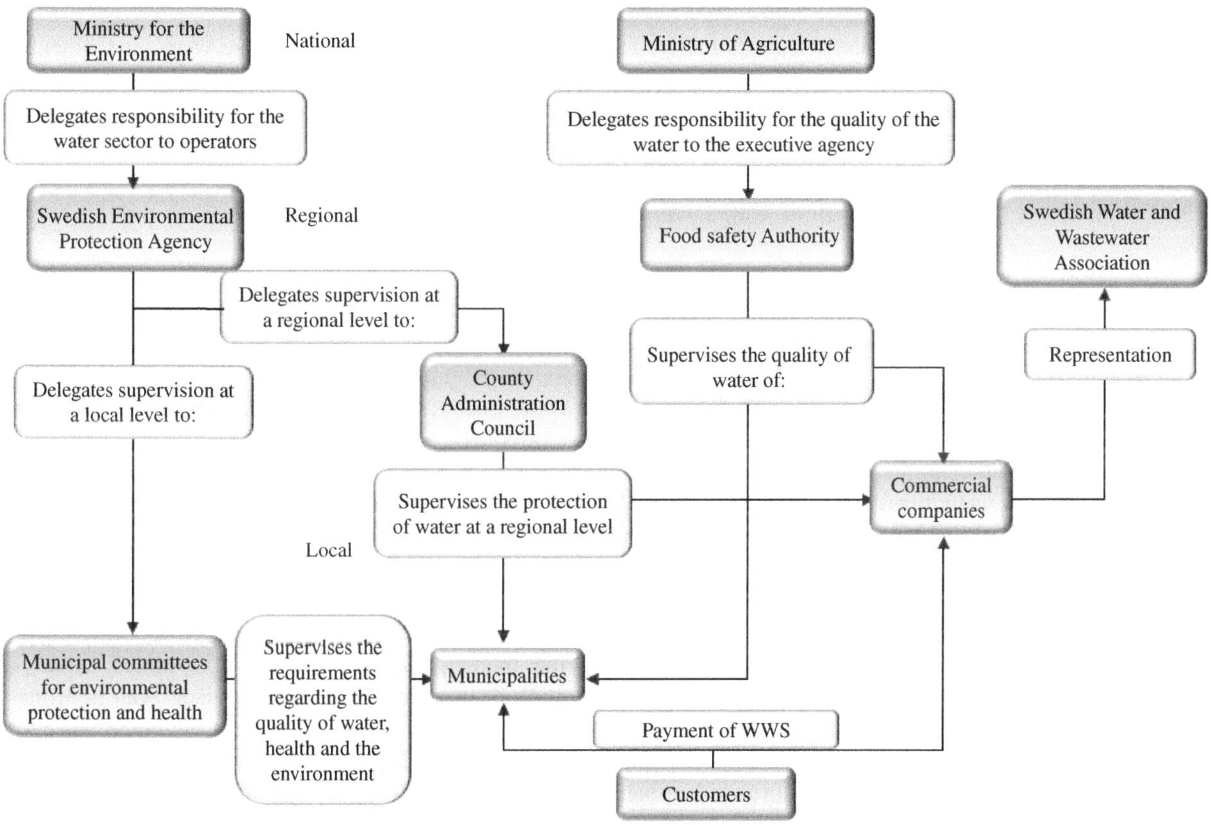

Figure 2.46 Institutional framework of the water sector in Sweden

The Ministry for the Environment is responsible for protecting water resources in Sweden and for the investments in the WWS that are considered to be of general interest. Diverse bodies supervise the quality of water resources, depending on the level in question, such as the EPA, which has jurisdiction at a national level, the County Administration Council, which monitors the sector at a regional level and, finally, at a local level the water sector is supervised by the Municipal Committees for Environmental Protection and Health.

The Ministry for Agriculture has overall responsibility for the quality of drinking water, since it is considered to be a food product, according to prevailing legislation. The Food Safety Authority (*Livsmedelsverket*) is the central authority responsible for supervising the quality of drinking water. This body receives and reviews the annual reports prepared by operators regarding the quality of drinking water, the developing programmes for improvements and the several audits made per year.

The EPA is the main body responsible for supervising the protection of water in Sweden. The function of this agency is to promote and coordinate ecological development and to act as a proactive source for policy changes, at both national and international levels.

The chief mission of the County Administration Council is to take steps to preserve environmental resources at a regional level, collaborating with all the municipalities to formulate objectives and prepare action plans for more under-developed areas. In cases where the environmental impact is limited, this body is also duly empowered to issue licences.

The Water and Wastewater Association was created by the municipalities in 1962 and includes all Swedish municipalities as its members. This association assists the WWS at a technical level as well as at an administrative and economic level. It also protects the interests of the municipalities in negotiations with other authorities and organizations, especially at the level of preparing legislation in this scope.

The Competition Authority aims to ensure that agreements do not undermine competition, intending to '*eliminate and counteract obstacles to effective competition in the field of production of and trade in goods, services and other products*' (Swedish Competition Act, 2002). It is also responsible for promoting proposals that intend to change legislation and other measures aimed at eliminating obstacles for effective competition, such as the continuous search for the perfect know-how regarding all aspects concerning the competition.

In 1996 the Water and Wastewater Association, Stockholm Vatten and the Water and Sewage Works in Gothenburg and Malmoe established the Swedish Water Development (SWD) AB, with a view to assisting municipal and international public organizations, by circulating and communicating Swedish knowledge and experience in the fields of ownership, operation and management of WWS. However, this company does not exist any more.

The National Water and Wastewater Tribunal is responsible for solving conflicts between operators and customers at a national level.

2.14.2.2 Regulation

In Sweden there is no sector-specific regulatory agency with explicit functions of economic regulation, monitoring the quality of service and public service obligations. However, the water sector in Sweden is characterized by a very effective transversal regulation, which guarantees public service obligations such as universality, and by self-regulation, essentially performed by the association for the water sector, which encourages the development and progress of the sector and, consequently, the quality of service provided.

In this sense, this association plays an important role in promoting a benchmarking system (DRIVA in the past and now VASS) amongst water operators (about 20 municipalities). It focuses, primarily, on reducing costs and improving operational efficiency. Stockholm, Gothenburg and Malmoe are also members of the 6 cities Performance Indicators project, which supports the benchmarking programme for the largest cities in Nordic countries.

The National Water and Wastewater Tribunal is responsible for defending and protecting customers, especially in terms of conflicts with operators, according to the article 36 of the Water and Wastewater Act.

In case the WWS are provided by the private sector, the municipality can appoint a so-called supervisor (*Tillsynsman*) to inspect and supervise the provision and operation of the services. This supervisor, who is in fact a regulator, seeks to ensure that citizens receive a high-quality water service at an appropriate cost, in compliance with the Water and Wastewater Act. The supervisor is accountable to the municipality with regard to all activities. To date, there has only been a single *Tillsynsman* in Sweden, namely, in the municipality of Norrköping.

2.14.3 Market structure

2.14.3.1 Ownership and management of services

In Sweden the WWS sector is traditionally managed directly by the 290 municipalities, with the exception of the largest urban centres, such as Stockholm, Gothenburg and Malmoe, whose operators are constituted by regional associations, which mainly provide wholesale services. The private sector also participates in the management of these services and is responsible for approximately 10% of the sector.

In 2002, 243 municipalities reported that they were organized as units to directly manage the WWS, 34 were municipal companies, 30 were intermunicipal systems, 6 operators had management contracts and 2 municipalities were totally private (Norrköping and Karlskoga). The structure of the water sector market in Sweden is summarized in Table 2.69 (Aqualibrium, 2003).

The participation of private companies in the water sector in Sweden has always been very controversial. The number of management contracts is quite limited and has been stable since 1995. As the sector works very well and has moderate costs, there is no need to implement reforms, which, whenever they exist, are normally ideological in nature. In 2002, there were only six management contracts (PPPs) with the private sector, all of which were of a short duration (four to seven years). However, the private sector has a strong participation in service provision.

Table 2.69 Institutional arrangements in the water sector in Sweden

	1995	2002
Present institutional arrangement		
Municipal administration	244	243
Municipal company	17	34
Intermunicipal organizations	44	30
Management contracts	7	6
Other	7	8
Unanswered	1	–
Future institutional arrangements		
Plans for changes	20	18
No changes planned	207	236
Unanswered	60	35
Total number of municipalities	287	289

2.14.3.2 Players in the sector

The number of municipal operators is fairly similar to the number of municipalities. The SWD counted 294 WWS operators in Sweden, for the 289 municipalities (in 2002). The presence of some supra-municipal entities reduced the number of operators, although some of them only sell wholesale water.

The majority of these operators, about 243, are municipalities which manage WWS directly, including Gothenburg, Malmoe and Uppsala.

At the end of 2002 there were 34 municipal companies. Despite their limited number, it is important to note the significant role they play in Sweden, since such organizations supply water to large cities, namely to Stockholm and Gothenburg and surrounding areas. In fact, Stockholm Vatten provides water services to about 1 million inhabitants, corresponding to 11% of the Swedish population, and also provides the wastewater service. Especially in the area of Stockholm there is a great deal of cooperation between municipalities, by means of formal or informal associations. In this regard, the water sector in Sweden reveals a clear tendency towards regional and bilateral cooperation between municipalities.

2.14.3.3 Integration and the scope of operator activities

In Sweden, mainly due to the fact that municipalities directly manage these services, the majority of operators provide all WWS together. If one were to consider only the municipal companies, it is possible to observe that only a small number supply the services simultaneously, namely Stockholm Vatten, Solna Vatten AB and Roslagsvatten AB.

The vertical integration of these services is common amongst operators, although there are exceptions to the rule, which promote both the wholesale water service and the wholesale wastewater services. For example, Sydvatten provides wholesale water to 12 municipalities, including Malmoe, whereas Syvab provides the wastewater service to 6 municipalities on the outskirts of Stockholm.

In the segment of the municipal companies there are five operators (multi-utilities) that provide other network services, such as electricity, in addition to WWS.

2.14.3.4 Numbers for the sector

The water sector in Sweden is characterized, above all, by the great abundance of water resources. The percentage of tap water for human consumption is extremely low (less than 2%), keeping in mind Sweden's European context. However, this abundance has also proved to be a factor that has served to discourage an efficient use of this resource, which is reflected in the level of the per capita consumption of 188 litres per inhabitant per day, corresponding to one of the highest levels within the EU.

Sweden produces about 900 million m^3 of drinking water every year obtained from three different sources: groundwater (24%), springs (25%) and surface water (51%).

The water sector in Sweden is characterized by a coverage level of about 100%, for both the water service and the wastewater service. This situation is even more remarkable considering the fact that only a small percentage of operators have relevant economies of scale, since the population is very small and is widely scattered in much of the country. The sector is also characterized by moderate levels of water losses, namely of about 22% (2004), which is worthy of note for a nation with such an abundance of water resources (IWA, 2006).

In Sweden, the WWS count about 6,000 employees (about one-third of which are administrative staff, one-third are operational and maintenance technicians and another one-third are treatment plant staff). In this regard, it must be noted that this figure, which represents a high level of productivity, has been achieved after cutting back about 4,000 employees over the last ten years. According to the water association, the objective is to attain a level of 4,000 staff for the water sector in Sweden over the next few years.

2.14.4 Rules of the game

2.14.4.1 Tariff setting

There are no regulations or a national structure in Sweden to set WWS tariffs and prices. However, under the terms of prevailing legislation, the municipalities must implement a tariff system capable of recovering from customers all the costs associated with the services, including a rate-of-return of their assets (capital cost) but without allowing for unreasonable profits.

According to current legislation, the WWS are disaggregated in administrative terms and, despite the aspects of a natural monopoly, operators are not allowed to derive extra revenues from the services, apart from the costs of service provision, so as to subsidize other services that are not self-sustainable (e.g. urban waste). The costs of providing the service include all operating and maintenance costs and the costs of capital, which in turn comprise the depreciation of investments, interests and a return over the equity.

The large-scale involvement of municipalities in the WWS sector could prove to be, *a priori*, a predominant factor for funding services provided to customers. However, an analysis carried out in 1998 regarding the sector revealed that, on average, 99% (in 2001) of the operating costs and costs of capital were covered by tariffs implemented by operators and by the taxes for connections to the system, therefore leading to a total cost recovery (Watertime, 2004).

In Sweden, prices and tariffs for the water service are set by means of a system with two components, a fixed component and a variable component, depending on the volume of water consumed or other factors. The fixed component covers, on average, 32% of the total cost of the water service.

The prices for the wastewater services are set similarly, i.e. prices are divided into a fixed component, which likewise encompasses about one-third of the cost of service, and a variable component. However, there is a growing tendency to increase the fixed component by the operators. For example, in Stockholm, the variable component was totally replaced by a fixed component, which includes the total cost of the wastewater services.

In 2000, the government introduced a component in the tariff system, as a kind of environmental tax, for the discharge of stormwater into the wastewater system. This tax is charged on an annual basis according to the area of the property. However, customers who have the capacity to avoid depositing stormwater in the public network (e.g. soil soakage) are not subject to this tax. Additionally, VAT is also charged on these prices, which, in Sweden, is 25%.

However, currently, the government does not introduce any components in the tariff systems. This is a decision of the municipalities in their territory.

Table 2.70 shows the tariff systems of three operators for 2005, based on a consumption of 200 m^3 (IWA, 2006).

Table 2.70 Tariff systems of three WWS operators in Sweden for 2005

	Water (€/m^3)			Wastewater (€/m^3)
	Fixed component	Variable component	Other component*	Variable component
Stockholm	0.245	0.238	0.483	0.678
Gothenburg	0.259	0.287	0.550	0.755
Malmoe	0.251	0.268	0.519	0.662

* referring to stormwater

Industrial customers pay for WWS provision at production cost and benefit from special conditions. Possibly, owing to the abundance of water resources, large-scale customers can benefit from some discounts. In terms of agriculture, no prices have been established, since owners are entitled to use the groundwater resources within their properties and this is the main source of production for agriculture.

2.14.4.2 Quality of service

The water sector in Sweden is known not just for the reliability of its services but also for the high quality of service provided. The role of the operators' association of the operators has contributed to a certain extent towards this situation, namely through the implementation of a benchmarking system (DRIVA and now the VASS) among water operators (about 20 municipalities), focusing, in particular, on costs and the operational efficiency of WWS. This system not only directly benefits the participants of the project but also the whole industry indirectly.

The main Swedish cities, such as Stockholm, Gothenburg and Malmoe, are voluntary members of a benchmarking system which involves the six largest Scandinavian cities, with a view to optimizing results, since the respective countries have very few operators with these operational dimensions. This benchmarking scheme consists of developing 28 performance indicators, which cover the areas of economic sustainability, customer satisfaction, quality of service and reliability, environmental quality and human resources.

A considerable number of studies have shown that, despite an unfavourable spatial distribution of the population, Sweden has managed to achieve a very competitive service in comparison to other European countries, with a better performance than many countries which have more favourable scales of operations.

2.14.4.3 Public service obligations

In Sweden, the public service obligations of universality and equity have been enshrined in the law.

By means of the Principle of Public Access to Administrative Documents (*offentlighetsprincipen*), Swedish citizens have a long and firm tradition of the right to access all documents regarding public services and to participate in their administration, which has had a notable impact on the WWS.

In situations where WWS do not comply with the quality requirements, the County Administration Council applies penalties, imposing precautionary and corrective measures, so as to safeguard the customers' health.

Under prevailing legislation, the operator of the wastewater service is compelled to monitor the parameters regarding the quality of the effluents discharged. Apart from this, operators must also submit an annual environmental report with these analyses to the supervisory authority, normally the Municipal Committees for the Environment and Health.

2.14.5 Performance

Sweden is one of the richest countries in the world in terms of water resources. It has a high level of performance, with an excellent quality of service/price relationship. 99% of WWS costs are recovered by means of revenues obtained from customers.

Sweden has a relatively small population for its land area and hence does not benefit at all from economies of scale. Consequently, a large part of the Swedish population is spread throughout the country, served by different players, leading to higher costs for providing the service. The government granted some subsidies for smaller municipalities in terms of water tariffs in the past but now there are no subsidies in Sweden concerning water tariffs.

2.15 THE NETHERLANDS

2.15.1 Historical background

The Netherlands, officially known as the Kingdom of the Netherlands, is a constitutional monarchy. The monarch is considered to be the head of state, but the government institutes legislation based on a formally declared constitution. Government officials, such as the Prime Minister, are elected through a democratic process. The Parliament is constituted by two chambers (the Upper House or Senate and the Lower House). The country has an area of 41,526 km^2 and a population of about 16.5 million inhabitants. It is organized into 12 provinces, each of which is administered by a governor (*Commissaris van de Koningin*). In their turn, the provinces are divided into 443 municipalities (*Gemeenten*). The Netherlands is presently described as a decentralized unitary state, characterized by a well-known and large-scale participation of the Dutch citizens in decision-making processes and consensual approaches, known as the '*Polder model*'. This participatory democracy is also evident in the water sector, where there are water boards (*waterschappen*), which are democratically elected governmental units with functions in the area of water management. The water boards are often described as the oldest democratic form of government in the Netherlands. Although they are based on a tradition of local, cooperative and participatory governance, they are characterized by a basic principle according to which the individuals with a greater interest in the functions of the water board are those who influence it the most. Moreover, owing to the size of its territory and its relation with the sea, the Netherlands has always paid due attention to matters related to water.

The WWS has undergone several progresses, both at a legislative and at an organizational level. However, it is still possible to recognize some signs of the sector's early institutional roots. This can be seen with regard to wastewater treatment, where, ever since they were first founded in the Middle Ages (sometime between the twelfth and the fourteenth centuries) the water boards have continued to be responsible for managing water resources and the wastewater treatment service. They still co-exist in the same organizational arrangement, although they are more closely grouped and benefit from greater efficiencies of scale than in the past.

Within the organizational and institutional evolution that took place, it is possible to highlight the great concern and focus of the central authorities on the providing integration of the services and consequently economies of scale. The large number of water companies (229 in 1938), municipalities and water boards during the early twentieth century led the government to develop a specific legal framework, namely the Water Supply Act of 1957, which was later altered in 1975 and 2000, promoting a fusion between operators in order to grant the respective authorities the power to introduce changes in their organization and consequently achieve economies of scale.

In the Netherlands, water services are always provided by public operators, although they are governed by private law and operate, in fact, as private operators. In reality, the Dutch Public Limited Companies (PLCs) are characterized by an exceptional quality of service, corresponding to a high level of customer satisfaction, which, among other factors, has deterred the private sector participation in these services and their regulation by an independent regulatory agency. Moreover, the nation is known for having a central government that takes an active interest in environmental issues and particularly in questions related to water, developing a very robust legal framework, which, in many situations, acts as a model for European directives. Prevailing legislation regarding the water sector is varied. Among other normative documents are the Water Supply Act of 1957 (changed in 1975 and 2000), the 1992 Water Services Act and the 2004 Water Services Ownership Act.

The administrative and financial structure of the water boards is based on the Water Services Act (*Waterschapswet*), which came into effect in 1992, establishing general rules for the composition of their governing bodies. Large landholders and other owners elect representatives directly, while the industry representatives are chosen by the chamber of commerce. The proportion of seats available for each interest group is determined by the respective province. Local and regional responsibility is entrusted to the water boards, which are subsequently supervised by the provinces. Matters regarding water management at a national level are governed by an executive body of the Ministry for Transport, Public Works and Water Resource Management, namely the Directorate-General for Public Works and Water Resource Management (*Rijkswaterstaat*).

In September 2000, the Dutch Minister for the Environment, Jan Pronk, announced the intention to produce legislation that would prevent the privatization of PLCs in the Netherlands. The PLCs would continue to have exclusive rights over the services to produce and distribute drinking water in their jurisdictions, with regard to domestic customers, and would allow competition in the context of industrial customers, possibly to be served by private companies. A law to this effect was being drafted in 2001, but after the Dutch government resigned in 2002 the issue was left behind by the following government. The issue was later raised again and the government created a new water law (*Waterleidingwet*), which was finally approved by the Dutch parliament in December 2003 with the support of the main parties, forbidding the private sector participation in the water sector concerning its ownership and management.

In the context of preventing and monitoring the pollution of water resources, the Dutch government produced a set of normative documents, namely the 1969 Surface Water Pollution Act (*Wet Verontreiniging Oppervlaktewateren*), which introduced the principle of polluter-pays (*wie het water deert die het water keert*). Under this act, agents who produce pollution equivalent to 100 inhabitants must be licensed.

This law attributed the responsibility for implementing these policies to the provincial government; however, in most cases this responsibility was delegated to the water boards. As a result, the water quality management became the most important activity of these bodies, overshadowing the traditional priority given to managing the quantity of water.

The 1981 Groundwater Act (*Grondwaterwet*) regulated the management of groundwater. Other related laws, such as the 1986 Soil Protection Act (*Wet Bodenbescherming*) and the 1989 Water Management Act (*Wet op de Waterhuishouding*), were also important legislative landmarks. In 1995, the Dutch government introduced a tax on groundwater abstractions, reflecting existing concerns about the negative impact of excessive consumption.

2.15.2 Legal and institutional framework

2.15.2.1 Responsibilities

In the Netherlands the WWS are characterized by the existence of diverse players, who are duly decentralized and are responsible for the different phases of the water cycle in a way that is quite unique among EU countries. The PLCs are responsible for the water service, namely for activities related to the abstraction, treatment and distribution of water. The municipalities are the main entities at the level of managing the wastewater collection services and, finally, at the end of the cycle, the water boards have functions in terms of wastewater treatment. Figure 2.47 illustrates the role of the main players in the water cycle (Marques, 2005).

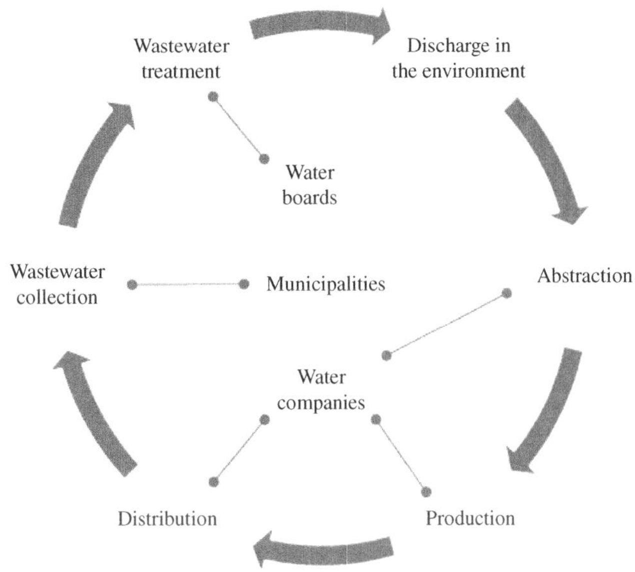

Figure 2.47 Water cycle and main players

As a consequence of the unbundling of activities within the water cycle, their coordination is a fundamental requirement within the water sector. For this purpose, it is essential that there are strong ties between the main players.

In this context, it is possible to identify two distinct groups that intervene in the water management system in the Netherlands. The first group comprises the protagonists which are directly involved in water management, concerning both political development and the provision of the service, i.e. the State, provinces, municipalities, PLCs and water boards. The other group includes the remaining stakeholders which, in spite of not playing a direct role in the sector, have a great influence regarding different aspects of its regulation, management and funding, such as the Association of Water Companies of the Netherlands (*Vereniging van waterbedrijven in Nederland* – VEWIN), the Association of Water Boards (*Unie Van Waterschappen*) and two banks, namely the Water Boards Bank (*Nederlandse Waterschapsbank*) and the Municipality Bank. Figure 2.48 shows the institutional organization of the main players in the water sector in the Netherlands (adapted from Watertime, 2004).

At a national level, the Ministry for Transport, Public Works and Water Resources (*Ministerie van Verkeer en Waterstaat* – V&W) and the Ministry for Housing, Territorial Planning and the Environment (*Ministerie van Volkshuisvesting, Ruimtelijke Ordening en Milieubeheer* – VROM) are the two most important ministries directly involved in the water sector. The V&W is responsible for areas related to the wastewater treatment services and the quality of surface water, while the VROM is responsible for the drinking water supply and for safeguarding its quality. VROM inspectors monitor the drinking water quality supplied to customers. The V&W and the Institute for Water Management and Wastewater Treatment (*Rijksinstituut voor Integraal Zoetwaterbeheer en Afvalwaterbehandeling* – RIZA) are the entities that prepare new legislation/regulations. The RIZA is a research and advisory institute aimed at assisting the functions of the *Rijkswaterstaat.* This institute compiles data and carries out research about the quality and quantity of water resources and, based on these data, issues recommendations.

The provinces are primarily responsible for managing groundwater as well as for supervising the performance, activities and accounts of the water boards. They also have the power to create or abolish water boards, and to determine their management functions, areas of operation and the respective governing structure. Moreover, the provinces also have various powers under the terms of the Water Supply Act, especially in order to promote efficiency in the sector, which includes the capacity to provide incentives for mergers between water companies. In brief, the provinces receive guidelines from the federal government, which they later transfer to the municipalities and to the water boards, although they are governed by their respective policies. Consequently, the provinces are also responsible for representing the municipalities and water boards before the national government.

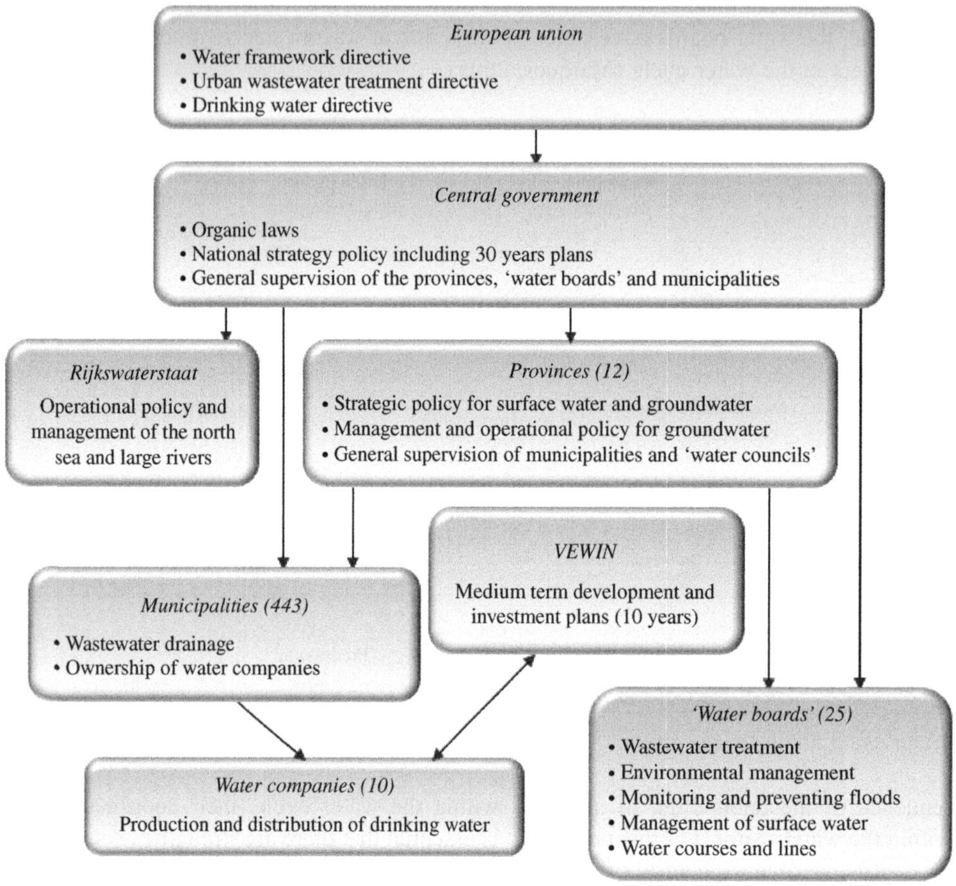

Figure 2.48 Main players in the water sector in the Netherlands

The municipalities are responsible, through their departments, for the drainage networks within their jurisdiction, including domestic wastewater and stormwater. However, as mentioned above, they are not responsible for wastewater treatment. The municipalities are represented by the Association of Dutch Municipalities (*Vereniging van Nederlandse Gemeenten* – VNG), whose main function is to advise its members in diverse fields whenever requested to do so. Furthermore, the municipalities are also shareholders in the PLCs and have some influence in this area. Finally, the Water Boards Bank (*Nederlandse Waterschapsbank*) is dedicated to funding the water sector.

The VEWIN was founded in 1952 on the basis of the former VWN. The VWN, which was originally established in 1899, sought to foster cooperation between water companies and was mainly founded with the aim of increasing knowledge among its members. The VEWIN is a representative association of water companies, whose main mission is 'to promote the healthy development of water supplies in the Netherlands'. The association intends to promote the interests of water companies, as long as these interests are not incompatible with its primary mission.

The VEWIN is responsible for diverse activities, such as the dissemination of best practices and producing publications about various aspects of the water sector. For example, the VEWIN prepares medium-term development plans for the water services (10-year plans). Simultaneously, the VEWIN is also responsible for the benchmarking system, which encompasses four aspects: water quality, quality of service, environmental quality and economic and financial performance. This benchmarking system has been a voluntary exercise but there is a proposal to make it compulsory.

The Association of Water Boards protects the interests of the water boards at a national level, especially before parliament. All the water boards are members of this association, which seeks, above all, to find solutions for common problems among its members.

The *Rijkswaterstaat* was created in 1798 and later restructured in 1971. It is a national government organization, overseen by the V&W, which is responsible for supervising the water sector and its strategic policies. The national government also has the power to formulate long-term development plans (30-year plans), which define the general policy regarding the sector's future, as well as future needs at the level of infrastructures. These documents complement the 10-year plans, which are prepared and revised by the VEWIN every five years.

The RIONED Foundation is a research centre dedicated to wastewater services in the Netherlands, which brings together municipalities, water boards, provinces, ministries, operators, consultants and builders. The RIONED Foundation is interested in all matters related to the management and operation of wastewater infrastructures and facilities, contributing towards innovation and efficiency improvement in the different activities in this field. The resulting information and knowledge is made available and shared with municipalities, government entities and other bodies that support and constitute the Foundation (RIONED, 2006).

The Dutch scientific community (e.g. universities) also provides an important contribution towards the performance of the Dutch water sector since it has always been at the forefront of worldwide research related to the water sector.

Finally, it is important to mention another entity that despite not being a part of this structure plays an important role in the water sector. It is a customer protection NGO (*Consumentenbond*), which represents 600,000 members and publicizes statistical information about the Dutch water sector. For example, it has been publishing the price/quality ratio of all Dutch water supply companies since the early 1990s.

2.15.2.2 Regulation

Although the subject of privatization and, consequently, WWS regulation is periodically discussed in the Netherlands, there is no explicit regulation of these services in the country. The Ministry for the Economy and Internal Affairs (*Ministerie van Economische Zaken*) has promoted this debate and made efforts to establish a sector-specific regulator, motivated, above all, by the positive impact of the regulatory agency for the energy sector (DTe) in the improvement of the economic efficiency of this sector. In fact, irrespective of the fact that the PLCs will continue to be publicly owned, there is a project for the creation of a regulatory agency for the water sector in the Netherlands. The delays in activating this entity, which has been planned since 1997, are related to the extremely positive role of the VEWIN in the self-regulation of the sector. Indeed, the VEWIN has behaved, strictly speaking, as a regulator, encouraging improvement in the operators' performance, concerning both the quality of service and the main economic and financial indicators. With its important and proactive role, and using benchmarking as its main tool, the VEWIN has postponed not just explicit economic regulation in this sector, but also privatization, since it has managed to maintain the Dutch water companies at the forefront of excellence at a global level, which limits the potential efficiency gains offered by privatization and regulation.

In the wastewater segment, the water boards, RIONED and VNG likewise have functions of self-regulation, encouraging the use of benchmarking and improved performance in wastewater services. The results of these initiatives are, however, less visible.

The Consumer Association (*Consumentenbond*) also plays an important role in terms of regulation, namely by protecting customer interests, in all the activities related to the WWS but, above all, with regard to the quality of service.

2.15.3 Market structure

2.15.3.1 Ownership and management of services

WWS are mainly characterized by the overwhelming participation of the public sector in terms of ownership and management, with one of the highest levels of performance in the world in different aspects. Although the private sector played a relevant role in the water sector in the past, especially during the 1980s, the need for heavy investments and the pressure of various interest groups (e.g. environmental organizations) diminished its importance in the subsequent decade. For instance, the 2004 Water Services Act prohibited the participation of the private sector in terms of drinking water supply. However, water for other purposes (e.g. industrial use) can be supplied by private operators.

The scenario is quite different in the segment of wastewater treatment. In 2002, the Delfland Water Board established a concession contract to build a WWTP in The Hague. This plant became one of the largest in Europe, treating effluents corresponding to about 1.7 million inhabitants. The public tender was won by a consortium, which included the Rotterdam Water Company and the multinational Vivendi, presently Veolia Water. The PPP contract, functioning under a BOT scheme, represents about 177 million euros in construction costs, with duration of 30 years.

2.15.3.2 Players in the sector

Without considering the limited participation of the private sector, the Dutch water sector comprises, as mentioned above, three types of public operators, according to the respective activity concerning the water cycle.

The water service is provided by PLCs. The framework and organizational arrangements of these operators are some of the main reasons for the success of Dutch water services, enabling, on the one hand, the efectiveness and efficiency of the private sector and, on the other hand, the accountability, legitimacy and transparency that are required to the public authorities, periodically reinforced at the time of the elections. The size of the PLCs has grown year after year. Since 1970 the central government has encouraged mergers of water services, so that they have a minimum economic dimension of 100,000 connections. At the end of 2007 there were only 10 operators: WBGR, Brabant Water, WMD, Vitens, Waternet, WML, DZH, PWN, Evidese and Oasen.

Diverse attempts have been made in the past to create multiutilities, namely by energy companies such as Nuon and Delta. Some of these companies even participated in the water sector in the Netherlands and expanded internationally in this market. The political measure to ban private initiative from this sector compelled them to withdraw, at least for the time being, from the Dutch market, although they maintain their interest in this sector abroad.

The water boards, of which there are currently 25, are responsible for managing and operating WWTPs and for managing river basins along with territorial planning related to water resources, nature conservation and environmental protection. The number of water boards, which are elected government bodies, has been diminishing over time as a result of mergers. Their revenues are derived from a tax on pollution and an operating tax.

The municipalities provide the wastewater collection (and stormwater services). Thus, there are 443 wastewater service operators in the Netherlands.

2.15.3.3 Integration of the sector and services provided

Water services in the Netherlands are vertically integrated. The PLCs supply all the stages of the production chain, from abstractions to distribution. The PLCs are usually fairly large since there are 10 operators for 16.5 million inhabitants. Although this was not always the case, the PLCs only provide the water service, with the exception of Waternet, which is also responsible for the wastewater service and other activities related to water resources.

The wastewater service is characterized by deverticalization. The end-user services, corresponding to the wastewater collection, are distinct from the wholesale services, which comprise the wastewater treatment. There is also a lower level of horizontal integration, especially with regard to the collection of domestic wastewater.

2.15.3.4 Numbers for the sector

In the Netherlands, the volume of water produced in 2006 was about 1,300 million m^3, of which 1,150 million m^3 correspond to the drinking water supply. Water losses in the same year represented nearly 10% of the total volume of water produced.

In 2006, WWS patrimony included 116,000 km of mains, which supplied water to the Dutch population, and 101,000 km of wastewater sewers.

The main source of water is groundwater, which represents about 60% of the overall production. The Dutch have an average daily per capita consumption of 123.8 litres.

The level of coverage of the water service is 99% and the figure for the coverage of the wastewater service is 98%.

In the Netherlands, the water services operators directly provide work for about 4,900 employees (VEWIN, 2005). Table 2.71 shows the breakdown of the water service costs in the Netherlands, by operator, for 2006.

2.15.4 Rules of the game

2.15.4.1 Tariff setting

Although Dutch WWS prices are generally considered to be high, this is not really the case after considering the principles of cost recovery and purchasing power parity, i.e. the burden that these services represent for customers, when compared to other European nations. Even in real terms, the average values are not very significant when compared to other European countries.

Table 2.71 Cost composition for the water service by operator in the Netherlands (€/m^3)

Operator	Total costs	Taxes	Capital costs	Depreciation	Operating costs
WBGR	1.06	0.18	0.07	0.13	0.68
Brabant Water	1.10	0.21	0.21	0.16	0.51
WMD	1.23	0.19	0.21	0.21	0.63
Evides	1.26	0.06	0.50	0.27	0.44
Vitens	1.31	0.22	0.24	0.23	0.62
Waternet	1.48	0.04	0.15	0.34	0.96
WML	1.49	0.17	0.43	0.28	0.61
PWN	1.54	0.03	0.23	0.35	0.93
Oasen	1.68	0.19	0.22	0.46	0.81
DZH	1.72	0.15	0.40	0.40	0.77
Sector average	**1.34**	**0.15**	**0.28**	**0.26**	**0.65**

The water sector in the Netherlands can be proud of its own financial institution, the Water Boards Bank (*Nederlandse Waterschapsbank*), which was created in 1954 by the water boards. This banking institution, which has a very high level of credibility (it has obtained an AAA rate), finances the entire water sector, granting long-term loans to water boards, municipalities and other public institutions that are involved in WWS management. It also finances these activities in international markets. Given their legal structure and the requirements to maintain a balanced budget, the water boards have a credit risk classification of 0, which likewise represents savings for the bank since no credit analysts or project evaluators are needed.

The water sector in the Netherlands is characterized by clearly separated and decentralized operators for the respective services. In this context, it can be noted that each operator implements its own tariff system.

The water boards impose two kinds of taxes, which fund more than 95% of the services provided. This tariff system is characterized by the implementation of a fixed component, called the 'water board fixed charge' and a variable component, which depends on the pollution produced. The rare deficits are subsidized by the central and local governments.

The charges related to pollution are based on an environmental pollution unit (EPU) (*vervuilingseenheid*), which represents the level of pollution of the wastewater discharged. A single domestic customer is normally charged one EPU and connections with more than one customer are charged for three EPUs. However, in some cases, this tax is implemented according to the meter size. In 2002, the average cost per EPU was 46.24 euros per year. A household pays, on average, about 240 euros per year.

The water boards spent about 2.3 billion euros in 2004, of which 27% related to investments and the remaining represented operating costs. From the tariff of 46.24 euros per EPU, about 75% represents the costs of treating wastewater and the remainder comprises the costs associated with expenditure for managing basins and the quality of water resources.

The majority of Dutch municipalities, about 90%, charge a tariff for the wastewater service, so as to fund activities in this sector. The charges reflect about 85% of the real costs of the service. In other municipalities, this funding is obtained by means of general taxes collected by the municipalities. Figure 2.49 shows the cost recovery evolution of the wastewater collection and drainage service over the course of time in the Netherlands (RIONED, 2006).

In terms of the water supply, the service provided by the PLCs represented, in 2006, an average cost of 1.69 euros per m^3 for domestic customers, which can be divided into three components. The first component corresponds to the PLC costs for providing the service (operating/maintenance costs and cost of capital), which represent 76.5% of the total cost. The second component is related to the abstraction taxes (groundwater), which represent 9.1%. Finally, the third component corresponds to state taxes (14.4% of the total cost of the service), including VAT (6%) and the tax on tap water (0.141 €/m^3 on the first 300 m^3 consumed). If one considers non-domestic customers, the average cost per m^3 of water, in 2006, in the Netherlands would reduce to 1.51 €/m^3 of water. It can be noted that among the PLCs, the tariff system varies according to the place being supplied.

The taxes related to water abstractions only apply to groundwater. This tax began to be instituted in 1995 and encompasses two components. The first one is charged by the provinces, as the contribution of operators for research regarding groundwater and to develop strategic plans. The second is set by the Ministry of Finance and is charged without being assigned to any specific purpose. The tax on tap water was implemented in 2000, to the detriment of a VAT increase, to leverage the water sector revenues. Table 2.72 (VEWIN, 2006) and Table 2.73 (IWA, 2006) show the composition of the tariff system for the water supply, both in average terms for the entire country and for some of the main cities. In the latter, the elements regarding the wastewater service have also been shown.

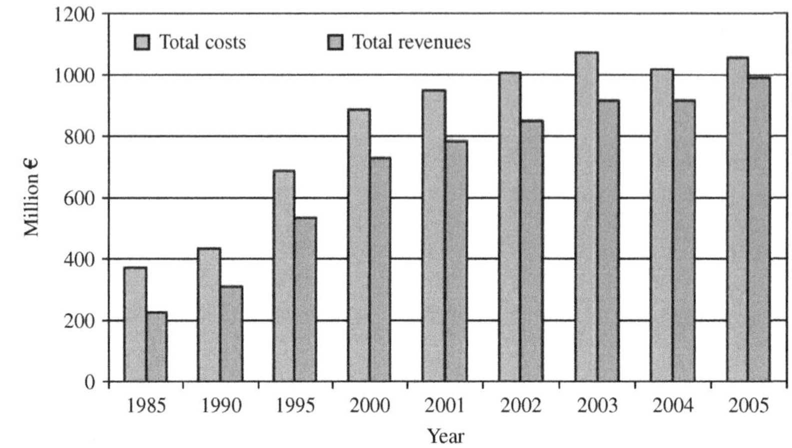

Figure 2.49 Costs and revenues in the wastewater sector

Table 2.72 Average tariff of the water service in the Netherlands

	€/m^3	%
Payment to the water company	1.29	76.5
Tax on groundwater abstractions	0.15	9.1
Taxes on drinking water supply and VAT (6%)	0.24	14.4
TOTAL	**1.69**	

Table 2.73 Summary of the prices charged to domestic customers in three cities

	Water (€/200 m^3)		Wastewater (€/200 m^3)	
	Fixed part	Variable part	Collection	Treatment
Amsterdam	31.31	206.13	88.57	140.90
Rotterdam	50.63	161.19	105.48	140.90
Eindhoven	31.40	137.69	100.65	108.70

The industrial customers are supplied by means of individual contracts. It must be noted that, for them, as well as for other non-domestic customers, the prices are established on the basis of the quantity of water consumed and tariffs diminish with an increase in consumption and according to the type of water consumed (Swami, 2004). Table 2.74 compares the customer charges for the various public services in the Netherlands.

2.15.4.2 Quality of service

The VEWIN has implemented a voluntary benchmarking system for its members (PLCs) since 1997, which encompasses almost all water services. The use of this system sought to achieve three objectives: increased transparency, accountability and the production of information to improve the operators' performance. The benchmarking exercises compare performance in four areas, as stated above: water quality, quality of service, environmental quality and economic and financial performance. In the last area, the benchmarking exercise is carried out annually while in the other categories they are carried out every three years.

These benchmarking exercises were implemented on a voluntary basis but plans to make them compulsory have already been announced. This study is promoted by developing a set of performance indicators in the areas referred to. The indicators are based on about 3,300 elements of data supplied by companies. In order to audit the information received, the VEWIN surveys customers directly so as to evaluate customer satisfaction and expectations with regard to the water companies (e.g. continuity of service).

Table 2.74 Comparison of charges for public services

Service	Year	Individual domestic	Family domestic
Drinking water	2004	89	179
Heating and gas	2004	972	972
Electricity	2004	519	711
Solid waste	2005	202	253
Pollution tax	2005	52	156
Water board charges	2005	38	38
Sanitation tax	2005	116	125
Municipal taxes	2005	669	730

For example, Figure 2.50 shows results for the indicator concerning the water service disruptions and for the indicator drinking water quality.

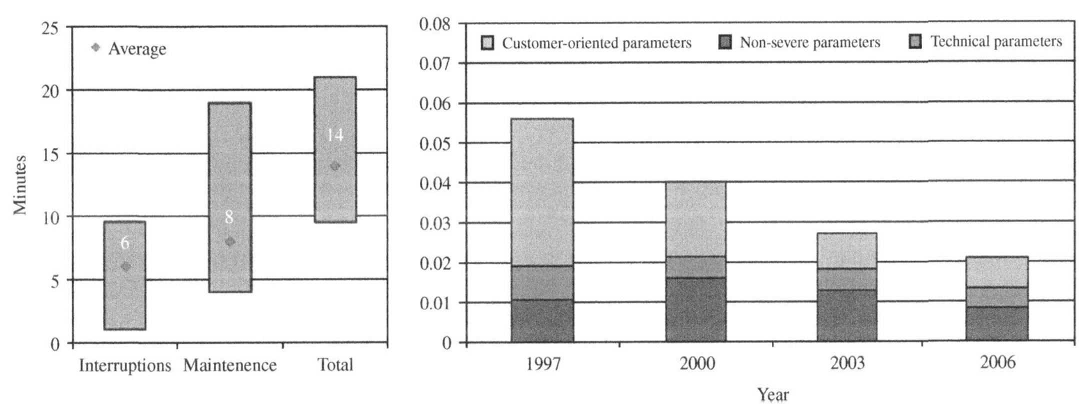

Figure 2.50 Indicators of water service disruptions and drinking water quality

The environmental aspect is accounted for and analysed by means of the energy consumption, the waste production and a dehydration index. These factors are later used to determine an environmental impact index.

The indicator drinking water quality is obtained by means of an aggregated index, called the water quality index, which reveals the level of compliance with legislation. A value of '0' is the best possible value, while '1' is the minimum acceptable level. Dutch water operators have very positive results in this area, as the figure above proves. Table 2.75 shows a summary of the evolution of some of the performance indicators calculated by the VEWIN (VEWIN, 2006).

Benchmarking is also applied to the wastewater service, with regard to wastewater treatment (water boards), as well as to the wastewater collection provided by the municipalities. RIONED has developed various initiatives within this scope. Its first study was carried out in 2004, encompassing 39 municipalities. Presently, the Foundation is doing a similar study that involves the participation of 42 municipalities. The benchmarking exercises carried out by the RIONED Foundation include not only the wastewater treatment but also the wastewater collection.

2.15.4.3 Public service obligations

According to the Water Supply Act, water companies are compelled to inform customers about their performance and the quality of water supply. Water companies must also collect and process the information requested by VROM inspectors, who monitor the water service being provided and publicize it.

The main public service obligations such as universality and continuity have been considered in legislation. However, quality of service levels for customers has not been stipulated.

Table 2.75 Performance indicators adopted by the VEWIN

Indicator	2000	2003	2006
Water quality (average index)	0.06	0.05	0.03
Non-compliance with norms	–	0.11	0.11
Water quality (by customers)	–	7.7	8.0
Customer satisfaction	7.6	7.6	7.6
Calls answered within 20 seconds (%)	–	51	38
Water supply disruptions (min)	–	–	14
Energy consumption (kWh/m^3)	0.47	0.48	0.47
Recycled waste (%)	90	94	94
Service costs (€/m^3)	1.28	1.31	1.34

According to prevailing legislation, i.e. the Environmental Management Act, the municipalities are compelled to prepare annual environmental and wastewater plans, whose drafts must be publicly evaluated and the water boards have the capacity to change them by means of petitions.

In terms of the quality of water, the government has developed legal norms (which, in many cases, are far more stringent than European norms), which aim to guarantee an exceptional quality of drinking water supply.

2.15.5 Performance

Generally, the Dutch experience is considered to be an outstanding case of excellence in the water industry, acting as a benchmark for countries around the world. There are various reasons for the success of the Dutch organizational and regulatory model in the area of water services. These include the private management model implemented by public operators, the representation and accountability of all stakeholders and the self-regulation of operators, developed by the VEWIN. The benchmarking scheme that has been introduced by this body has proved to be very effective, substantially improving the quality of service and resulting in lower costs. The reduction of the number of players was another aspect that contributed towards this scenario of excellence, since it enabled significant economies of scale.

Likewise, with regard to the collection and wastewater treatment service, the Netherlands has a high level of performance, albeit less noteworthy than in the water services sector.

There are practically no unsuccessful aspects in the Dutch WWS sector. Questions that could give rise to criticism are linked to the importance and possible dominance of the sector by the VEWIN and to the fact that this success is due to the size of the operators and corresponding efficiency of scale and not to the prevailing self-regulation model. Another issue is the wastewater collection service which could be operating at an inefficient scale. Furthermore, part of the revenues obtained from taxes imposed to the water sector, which are quite substantial, are not applied in the sector.

2.16 UNITED KINGDOM

2.16.1 Historical background

The UK is, formally, a parliamentary monarchy, where legislative power is associated with the government and executive power is represented by the council of ministers, which is headed by the prime minister. The UK is a highly centralized state, with the Parliament at Westminster comprising the House of Lords and the House of Commons. The tendency towards decentralization began in 1999, with the institutionalization of the Scottish Parliament and the National Assembly for Wales. Northern Ireland also has self-government, by means of the Northern Ireland Assembly.

The UK has a population of about 60.5 million inhabitants and an area of about 245,000 km^2. Although it is administered by a central government, the UK is made up of four countries: England (83% of the population), Scotland (9%), Wales (5%) and Northern Ireland (3%). The four countries have been administratively organized into regions, counties, districts and parishes.

In the UK, the WWS are generally known for their excellence and success and are an international benchmark. The universality, reliability and high quality of service, where the diversity of institutional ties between the countries that comprise the UK is especially worthy of note, are some best practices in this sector worldwide. Furthermore, the complete privatization of the WWS in E&W, the independent economic regulation (pioneering in the European context) and the introduction of competition in the water sector are other factors that have contributed towards this success.

The WWS are regulated by diverse normative documents in E&W. Although much of the regulatory activity took place during the late 1980s and the early 1990s, the initial phase began with the implementation of legislation in 1973, by means of the Water Act, which was later changed in 1983, 1989 and 2003. Likewise, the publication of the Water Industry Act 1991, which was altered in 1999, introduced important changes in the water sector.

The Water Act 1973 had an enormous impact on the water sector, owing to the organizational revolution it brought to the WWS with the creation of ten regional operators, thus merging thousands of entities that had been operating those services until then. However, it was only with the enactment of the Water Act 1983 that these authorities were effectively empowered to terminate the agreements that had been in effect in the past.

Subsequently, the main consequence of the implementation of the Water Act 1989 was the water sector privatization, with the creation of regional Water-and-Sewerage Companies (WaSC), which were issued operating licences for 25 years, and the Water-only Companies (WoC), which were smaller in size and operated on a more local scale.

The Water Industry Act 1991 introduced diverse changes in the functioning of the water market in E&W. In particular, it defined the functions and the *modus operandi* for the regulator (Director-General of Water Services), as well as the rights and obligations of the WWS concerning both the environment and customers in E&W. This 1991 law was amended in 1999 and at that moment included a new section regarding social aspects, encompassing issues such as a ban on disconnecting services and another section that created the Scottish regulatory agency (The Water Industry Commissioner for Scotland – WICS).

The main changes in the Water Act 2003 were related to the introduction of competition for large customers, which had not existed until then, with the development of a new abstraction licensing system and with the restructuring of the regulatory agency, replacing the office of director-general by a board of directors and likewise changing its name from Office of Water Services (OFWAT) to Water Services Regulation Authority (maintaining the OFWAT abbreviation).

In Scotland, the Water Act 1980 was one of the most important normative landmarks, which, among other aspects, regulated the quality of drinking water. The Water Industry Act 1999 created the WICS, the regulatory agency for the sector, whose statutes were later changed by the Water Services Act 2005 that also changed its name to the Water Industry Commission.

Furthermore, the Water Industry Act 2002 created Scottish Water, by merging the three existing public operators providing WWS to the population of Scotland. It also regulated the role of the economic regulator (Water Industry Commissioner), the Customer Panels and the regulator for the drinking water quality (Drinking Water Quality Regulator).

In Northern Ireland, a normative document from 1999 (Water Order) changed the existing organizational framework of the water sector in many aspects, which, until then, had been based on the Water Act 1972. This law would, for example, regulate the minimum levels for the drinking water quality as well as for the discharge of effluents. Later, in 2003, other environmental norms (Water Environment Regulations) implemented the Water Framework Directive, establishing even more stringent levels of quality.

Encompassing the entire UK, the Competition Act 1998 clarified the regulation of competition in the country, including liberalization and opening up the water sector to competition. The law identifies and addresses restrictive market practices and abuse of a dominant position in the market. In addition to these impositions, the law introduced new penalties for anti-competition acts.

2.16.2 Legal and institutional framework

2.16.2.1 Responsibilities

As in the case of legislation, responsibility for the WWS is similar among the different countries that constitute the UK. In E&W, the main organizations in the water sector include the government, through the Secretary of State for Environment, Food and Rural Affairs (SSEFRA), the Department for Environment, Food and Rural Affairs (DEFRA), the regulatory agency (Water Services Regulation Authority – OFWAT), the Environment Agency (EA), the Consumer Council for Water (CCW), the Drinking Water Inspectorate (DWI), the Competition Commission (CC) and, finally, the Association of Operators (WaterUK).

SSEFRA is the government body with the greatest responsibilities in the water sector and in the development of the water policy. SSEFRA is the top organization in this field, which includes licensing WWS and appointing the various regulators.

DEFRA is ultimately responsible for all aspects associated with the water policy in England, which includes water supply, management and quality of water resources and the supervision of all environmental aspects related to water and the water market. The National Assembly for Wales (NAW, the Department for the Environment in Northern Ireland and the Scottish Executive in Scotland) carry out similar functions to DEFRA.

OFWAT is an independent sector-specific regulatory agency responsible for the economic regulation of the WWS in E&W. This body was created as a decentralized institution, with its own patrimony and administrative and financial autonomy. Along with its functions in the scope of economic regulation (tariff system), OFWAT also regulates the quality of service and was recently given the power to impose penalties in the course of its functions. The Water Industry Commission for Scotland and the Northern Ireland Authority for Utility Regulation were created in Scotland and Northern Ireland, respectively, with regulatory functions and responsibilities similar to those of OFWAT.

The EA was created in 1996 and replaced the former National Rivers Authority (NRA). It plays an important role in environmental regulation, being responsible for preserving, developing, redistributing and ensuring the correct utilization and the quality of water resources in E&W. In this regard, the agency is the central body responsible for water resource planning in these two nations. The Scottish Environment Protection Agency (SEPA) and the Environment and Heritage Service (EHS) in Northern Ireland have similar functions.

The CCW was created by OFWAT, substituting the body known as the Water Voice, which existed until 2005. It is also an independent entity, with the statutory duty to represent customer interests, which also includes the functions of investigating and monitoring customer complaints regarding the WWS. There are currently six regional committees of this body, as opposed to the ten that were created with the water sector privatization. In Scotland, there is Waterwatch Scotland, which recently replaced the five committees that existed there and has functions similar to those of the CCW. In Northern Ireland, these competencies are the responsibility of the Consumer Council for Northern Ireland.

The DWI is an independent body, which, since 1990, has been responsible for monitoring the drinking water quality in E&W, taking coercive action in situations where norms are not being upheld. A similar entity in Northern Ireland has the same name as its counterparts in E&W. In Scotland, these functions are under the responsibility of the Drinking Water Quality Regulator.

The CC was established under the 1998 Competition Act and replaced the Monopolies and Mergers Commission (MMC) to promote and guarantee competition in the UK. It issues opinions about market concentration operations, market abuse and the existence of deviant behaviour, as well as about regulated sectors. In particular, WWS can appeal to this body concerning the OFWAT tariff decisions, which in the context of mergers and acquisitions also issues its own opinions. Figure 2.51 shows the main players in the water industry in E&W (Marques, 2005). Table 2.76 summarizes the main players and responsibilities in the water sector in the UK.

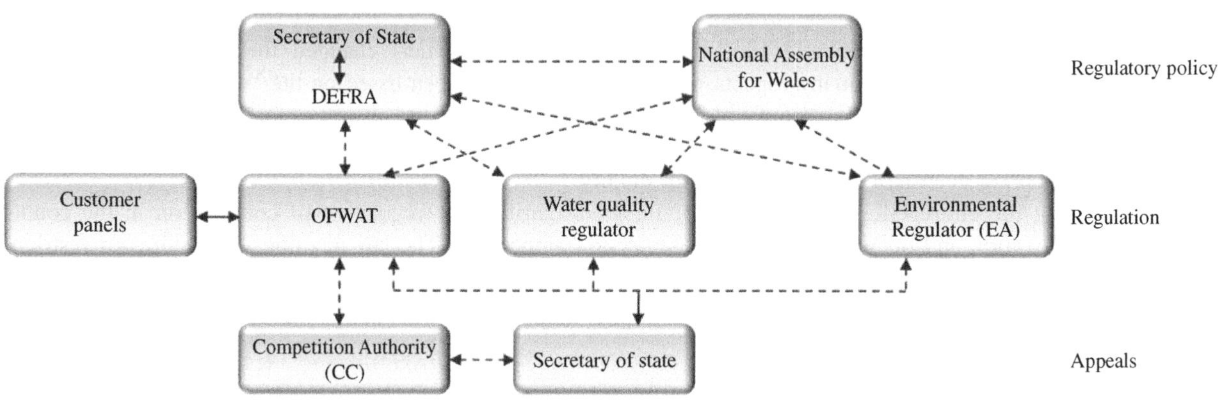

Figure 2.51 Institutional framework of the water sector in E&W

Finally, the main objective of the Association of Operators (WaterUK) is to protect the interests of its members as well as to represent them at a national and European level. It also disseminates best practices among its members. At the same time, the association plays an important role as a mediator between operators, regulators, governments and civil society.

2.16.2.2 Regulation

In the UK, the prevailing regulatory models have proved to be quite effective. They are based on three independent regulators (three-dimensional regulation): the economic regulator, the environmental regulator and the drinking water quality regulator. These three regulatory bodies are complemented by the existence of other entities that also have important regulatory functions, such as DEFRA, the CC and the CCW. In E&W, OFWAT, as the economic regulator, is responsible for safeguarding WWS activities, protecting customer interests, guaranteeing the financial viability of companies, promoting economic efficiency and introducing competition. The EA plays the role of an environmental regulator, which includes licensing abstraction and discharges and, finally, the DWI is responsible for regulating the drinking water quality.

Table 2.76 Main bodies responsible for the water sector in the UK

State	Water resources management	WWS	Environmental regulation	Water quality	Economic regulation
E&W	DEFRA	10 WaSC and 12 WoC (and new entrants)	EA	DWI	OFWAT
Scotland	SEPA	Scottish Water	SEPA	DWQ	WICS
Northern Ireland	EHS	Northern Ireland Water	EHS	DWI	NIAUR

In E&W, the privatization of the water sector consisted of the sale of company assets. At the same time, a sector-specific economic regulator was created for the water sector, the Director-General of Water Services (DG), responsible for the Office of Water Services (OFWAT), which played a role in monitoring the quality of service and customer protection. The assets were sold for about 5% of their estimated value (taking into account the payment of debts and the supply of a green dowry). Incentives were provided to encourage the private English companies that were formed at the time to expand internationally, an objective that has been successfully implemented. The English model of WWS organization and regulation was established based on private ownership and management and on centralized control by the State, by means of stringent regulation whose function is to substitute the market and, consequently, to protect customers, who are at the epicentre of the model.

With the privatization process, an operating licence was issued to WWS for 25 years, by the responsible government entity (Secretary of State for the Environment of E&W – SSE), which established a series of principles to be upheld by WWS, under the supervision of OFWAT. The task of environmental regulation was entrusted to the NRA, which was absorbed, in 1996, by the EA, while the drinking water quality was, and continues to be, regulated by the DWI. These entities work in close cooperation with OFWAT, especially to define quality goals which have an impact both on investment and economic regulation.

In 1989, the period for revising the tariff system (regulatory interval) was established at ten years, by the Department of the Environment, with the possibility of a mid-term review after five years and occasional intermediate reviews (interim determinations). After the initial determination, OFWAT was made responsible for reviews, which began in 1994. The fourth review has now almost been completed. The draft determinations, setting prices for the period 2010–2015 were announced to the public on July 2009. There has followed a short period in which OFWAT met with all water and sewerage companies. These meetings allowed the companies to challenge the review and suggest amendments. This review will be completed on November 2009 when OFWAT will publish its final determinations. Companies who are still not satisfied at this stage have the right to contest the decisions at the Competition Commission. However, while these cases are heard the new price limits will come into force from 1 April 2010. Any changes made by the Competition Commission will be applied retrospectively. It is yet to be determined whether a fifth review will take place in the same way in 2014. The calculation methodology is based on price cap regulation. The formula used consists of establishing the RPI $\pm$ K limit, where the RPI is the Retail Price Index and K represents the difference between Q and X, where Q corresponds to the impact of the new investments to be made and X is a factor that estimates the expected productivity gains. The value of the formula RPI $\pm$ K constitutes the upper limit for increases in the WWS weighted average charges. Figure 2.52 illustrates the regulatory procedure for determining the variation of the price caps, which guarantees the additional revenues to fund the investment programme.

The revenues obtained (tariff basket) are constituted by various items, namely from the volume of unmeasured water, the volume of measured water, the volume of measured wastewater, the volume of unmeasured wastewater and the volume of wastewater received, treated and discharged. They also include revenues derived from the stormwater service, which may or may not be included in the revenues related to wastewater.

The price cap determination for a WWS is largely based on the performance of other operators and hence employs the model of yardstick competition. In the first regulatory period, established by the government, the X factor was determined on the basis of the total factor productivity (TFP), estimated by means of index numbers. During this period, the government's forecasts proved to be very benevolent, resulting in exceptional profits in the WWS, which led to the subsequent application of claw-backs.

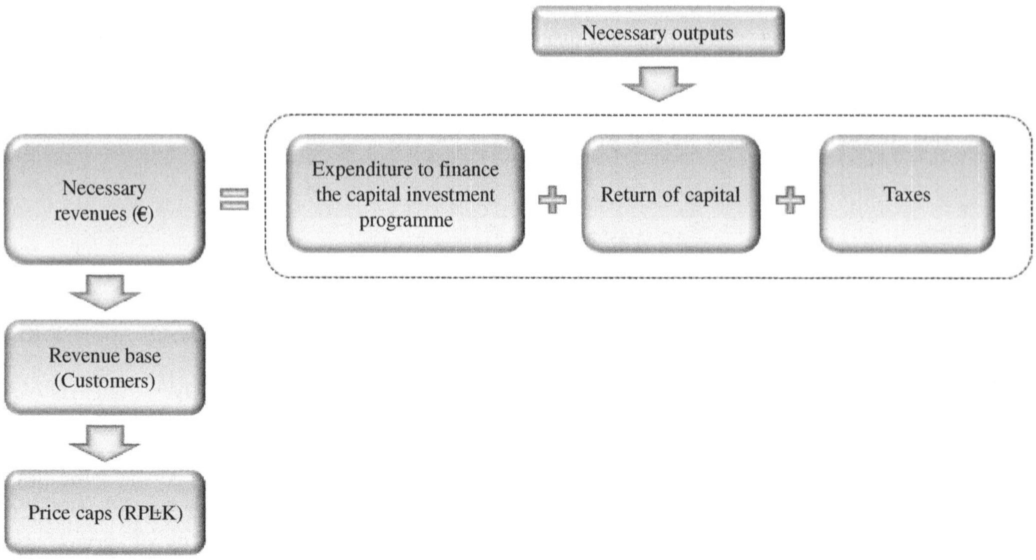

Figure 2.52 Calculating price caps

For the second regulatory period, from 1994 to 1999, OFWAT determined the X factor based on more refined benchmarking studies, using econometric methods (statistical regressions) and mathematical programming techniques (like DEA), which made it possible to determine the relative efficiency of the WWS and, thus, to estimate the expected productivity gains in the future.

These methodologies were only applied to the OPEX and were widely contested by operators owing to their shortcomings and the fact that they could not access the calculation models. Nevertheless, excessive profits were again recorded during the second period, which induced OFWAT to change the calculation formula. From this point onwards (third regulatory period), WWS began to pass all profits above those that were considered to be 'fair and reasonable' in one go to customers instead of in phases (glide path). The benchmarking methodologies continued to be used during this period and their application was extended to also include the CAPEX.

There were essentially three main aspects while calculating price revisions. The first concerned estimating gains in efficiency and productivity using benchmarking techniques. Table 2.77 illustrates the philosophy adopted, based on the creation of matrices for efficiency bands for the year 2006/2007 (OFWAT, 2007b). The second aspect consisted of accomplishing an investment plan (asset management plan), proposed by the WWS and accepted by the regulator. It would have to encompass the different components enabling maintenance and improvements in serviceability, i.e. the level of investment in maintaining infrastructure could not result in a quality of service deterioration. OFWAT supervises the quality of assets by systematically evaluating certain indicators. For example, for the water distribution networks, this analysis considers the number of bursts, low pressure, interruptions and abnormalities in the drinking water quality. Finally, the third aspect concerns the return of WWS, especially considering the calculation of the costs of capital and the assets to be remunerated. These costs encompass the regulated capital value, which results from OFWAT's assessment of the regulated asset base of the WWS. The cost of capital was obtained by the CAPM.

In the last two tariff revision periods (price review), from 2000 to 2005 and from 2005 to 2010, while maintaining the same principles, the following formula was adopted to determine the K factor:

$$K = -P_o - X + Q \pm V \pm S$$

where:

P_o – Adjustment related to the excess gains obtained in the past;
X – Future gains in productivity;
Q – Desired quality standard;
V – Improvement in the security and reliability of the supply;
S – Improvement in service levels.

Thus, the process of determining the K factor considers the recovery of a fair and normal return (P_o), gains in efficiency and productivity expected for the next regulatory period (X), the maintenance and serviceability of operators (S), establishing a balance between demand and supply concerning water resources (V) and improvements in the quality of water, environmental aspects and service levels (Q). These matters, as well as financial aspects, are reflected in an asset development and management plan (master plan), which is analysed, discussed and approved by OFWAT. It must also be noted that a carrot-and-stick policy has been instituted to determine the K factor, which allows best practices to retain part of the gains achieved, by reducing the P_0 limit (carrot), and penalizing worst practices by not allowing this reduction (stick).

Table 2.77 Relative efficiency of operators in E&W

OPEX \ CAPEX (only maintenance)	E	D	C	B	A
A	Northumbria		Southern, Thames, Yorkshire, Dee Valley, Portsmouth	Cambridge, North Surrey	
B	Anglian, Bournemouth and West Hampshire	Severn Trent, Wessex, Sutton and East Surrey	North West, Essex and Suffolk		York
C		South Staffordshire	Bristol, Tendering Hundred	Three Valleys	South West
D			Dwr Cymru, Folkestone and Dover, Mid Kent		
E			South East		

A – Most efficient, B – Above average, C – Average, D – Below average, E – Least efficient

It must be noted that, in accordance with its functions, OFWAT has played a very important role in promoting competition in the water sector in E&W. From the range of incentives provided, emphasis can be given to the inset appointments, where, for a given geographical area, the incumbent operator is substituted by another to ensure activities of drinking water supply and wastewater, access to the network by third parties (common carriage), competition on the frontier of the areas being supplied, competition in terms of commercialization and the use of private and unregulated systems. All these infrastructures have been developed to some extent in these nations and, although their market share is still very small, they have contributed significantly towards the dynamics and competitiveness of the water sector, reducing prices substantially for large customers.

In Scotland, the economic regulation implemented by the WICS, in practice, has proved to be fairly similar to the regulation of E&W. It is characterized by a system of maximum prices and the use of benchmarking as an essential tool. The first regulatory period extended between 2002 and 2006. The second period is currently underway and will conclude in 2010. At present, the WICS regulates a single operator, i.e. the Scottish Water.

In Northern Ireland, the multisector regulatory agency (Northern Ireland Authority for Utility Regulation – NIAUR) has only recently (2007) expanded its functions to include the WWS. The tariff system for the first regulatory period (2010–2015) is currently being determined, likewise based on a system of maximum prices. Just like in Scotland, this regulatory agency only regulates a single operator (Northern Ireland Water). Table 2.78 summarizes the main characteristics of the economic regulators in the UK.

Customer panels and committees also have important regulatory functions, not just in the scope of customer protection with regard to answers to complaints and resolving conflicts but also concerning price reviews. They participate in the regulatory process and give their opinion about the investment plans and the proposals of both OFWAT and operators to calculate prices.

In addition to the specific activities of the regulatory agencies, the operators' associations (especially WaterUK) also play an important role in terms of self-regulation. Besides trying to ensure discipline among operators, the dissemination of best practices has proved to be a primary activity for these organizations, with a view to achieving progress in the water sector and consequently protecting the interests of customers.

Similarly, by investigating whether the prices adopted are reasonable, the Competition Authority also plays a relevant role in regulating the water sector. Despite its vast multisector scope, this organization analyses and compares

WWS prices with the average price level, issues opinions on operations involving mergers and acquisitions in the water sector and is the body for appeals concerning decisions by OFWAT related to tariff setting.

Table 2.78 Main characteristics of the regulators in the UK

	OFWAT	WICS	NIAUR
Commencement of activities	1989	2005	2007
Field of regulation	Water and wastewater	Water and wastewater	Electricity, gas, water and wastewater
Total expenditure (million €)	14.5	6.3	1.5 (only in the water sector)
Source of revenues	98.5% from licences and 1.5% others	64% from Scottish Water, 35% from the Central Market Agency and 1% others	100% from the operator
Staff (No.)	190	18	65, but only 12 for the water sector
Economic regulation		Price cap regulation	
Regulatory period (years)	5	4	5
Capital cost	CAPM	WACC	CAPM after 2009/10

2.16.3 Market structure

2.16.3.1 Ownership and management of services

WWS ownership differs amongst the four nations that comprise the UK. In E&W, all WWS have been, since 1989, completely in the hands of the private sector.

In contrast, WWS in Scotland and Northern Ireland are provided by public operators, albeit governed by private law. There are also various PPPs in these two nations for diverse activities of the WWS production chain (e.g. WWTP).

2.16.3.2 Players in the sector

The ownership of privatized WWS in E&W reveals a mixed (private) model, which has undergone significant changes in the past five years, with a considerable growth in the role played by financial investors and private funds. This situation has been accompanied by a reduction in WWS own funds and their substitution by debt funds.

Until 1974, in E&W, WWS were provided by a significant set of companies and local services. In that year, the sector was reorganized, by de-municipalizing these activities and creating the ten regional companies, i.e. Regional Water Authorities (RWA), according to the principle of management based on river basins. In 1989, the Thatcher government, faced with the water industry's need for heavy investments, and after several years of study, decided to implement an absolute revolution in this sector. It decided on privatization in the strictest sense of the term, i.e. by selling assets.

In 1989, there were 39 private companies in E&W, covering more than 90% of the population. Of these, 10 companies, corresponding to the former RWAs, were simultaneously responsible for water and wastewater services (called WaSC), while the remaining 29 only provided the water service (WoC). The number of WoCs diminished in the years after the privatization as a result of various mergers and acquisitions. At the end of 2007, there were 10 WaSCs and 12 WoCs (although there were other licensed companies, which were new entrants).

Presently, in E&W there are 12 companies providing WWS and 13 companies that only supply water services. Hence, two more companies were created, in order to provide water and wastewater services, even though only active at a local level to provide WWS to three agglomerations that did not have access to water. Also in the context of the new entrants, the other new entities correspond to one company that supplies a single customer, another company that provides water services to an area with about 2,000 inhabitants and the remaining 21 companies are large operators, supplying more than 150,000 customers.

On the other hand, except for the two companies created in 2007, the wastewater services are operated by large companies, supplying more than a million inhabitants.

Five of the ten WaSCs are also listed on the stock exchange. Three investors control 45% of the shares of one of them, Northumbrian. 30% of another, South West Water, is owned by the Pennon Group. Severn Trent Water is dominated by the Severn Trent Group. North West Water is under the United Utilities Group. Finally, Yorkshire is under the control of the Kelda Group. Of the remaining five WaSC, Anglian, Southern and Thames are owned by private groups or financial entities. Wessex is owned by the Malaysian multinational YTL and the fifth unlisted WaSC is a non-profit entity (Glas Cymru).

The ten WaSCs were protected against acquisition for five years by the government's 'golden share'. As stated above, the number of WoCs declined from 29, in 1989, to 12, at the beginning of 2006, mainly due to mergers between WoCs and acquisitions by WaSCs. During the 1990s acquisitions were dominated by French multinationals operating in the utilities sector (Veolia, Suez and SAUR), as well as by companies linked to the energy sector that wanted to expand into the water sector (e.g. Enron, Scottish Power and RWE). However, since 2000, water multinationals and energy companies have been withdrawing from this market and in some cases new management arrangements have been sought, such as the non-existence of shareholders (e.g. Welsh Water) and the involvement of banks and other financial entities, as well as listings on the stock exchange. Of the WoCs, three are still owned by Veolia, one company is owned by Suez/Agbar, one belongs to a group from Hong Kong, another to a British group (Biwater) and the remainder is owned by private equity. Table 2.79 shows the main owners of the 10 WaSCs and the 12 WoCs in E&W, the country of origin and their nature.

Table 2.79 WWS in E&W

Company	Company	Owner	Country	Type
WaSC	Anglian Water	Osprey/AWG	UK	PE
	Northumbrian Water	–	UK	SEC
	North West Water	United Utilities	UK	SEC
	Severn Trent Water	Severn Trent	UK	SEC
	Southern Water	Royal Bank of Scotland	UK	PE
	South West Water	Pennon Group	UK	SEC
	Thames Water	Macquarie	Australia	PE
	Welsh Water	Glas Cymru	UK	NPC
	Wessex Water	YTL	Malaysia	M
	Yorkshire Water	Kelda	UK	SEC
WoC	Bournemouth/West H. Water	Biwater	UK	P
	Bristol Water	Agbar/Suez	Spain/France	M
	Cambridge Water	Cheung Kong Infrastructure	Hong Kong	M
	Cholderton Water	Cholderton Estate	UK	P
	Dee Valley	–	UK	SEC
	Folkestone and Dover	Veolia	France	M
	Mid Kent Water	UTA and HDF	Australia	PE
	Portsmouth Water	South Downs Capital	UK	PE
	South East Water	UTA and HDF	Australia	PE
	South Staffordshire Water	Arcapita Bank	Bahrain	PE
	Sutton and East Surrey Water	Aqueduct Capital	Germany	PE
	Tendring Hundred	Veolia	France	M
	Three Valleys	Veolia	France	M

SEC – Stock Exchange Quoted; M – Multinational; PE – Private Equity; NPC – Not-for-profit company; P – Private company

In Scotland, WWS are currently provided by a single operator, the Scottish Water, a state company that supplies the entire population of Scotland. However, with a view to introducing more competition in the market, the government has liberalized the water sector, allowing the appearance of new entrants such as Osprey, Satec and Aquavitae. As in Scotland, Northern Ireland Water is a state company that operates WWS throughout the territory of Northern Ireland.

2.16.3.3 Integration of the sector and services provided

Generally, WWS are integrated into a single operator in the UK. In this context, the services are vertically integrated, i.e. the operators perform all the services in the production chain, from abstraction to the distribution of water. This is also the case as far as the wastewater service is concerned, where services are also integrated, both in terms of collection and wastewater treatment. Thus, there is an equally high degree of horizontal integration, with a limited number of operators for the entire urban population in the UK.

There are rare exceptions in E&W, where about a dozen companies only provide services related to the water production chain, albeit on a large scale. Apart from this, there are individual systems, which also encompass a significant population.

2.16.3.4 Numbers for the sector

In the UK, more precisely in E&W, about 16 billion m^3 of water are produced every year. Surface water is the main source and accounts for 70% of production. From this volume of water produced, the electricity sector is the largest customer, followed by drinking water supply, accounting for 42% and 37.5% of consumption, respectively. Since water consumption is rarely metered, it is difficult to determine precisely the value of water losses; however, it has been estimated that losses are approximately 23.4% in the UK.

Although more than 75% of the population does not have a meter on their property, OFWAT estimates that the per capita consumption of water in 2006 was about 147 litres per inhabitant per day. Table 2.80 shows the evolution of the consumption per capita according to type of company. Table 2.81 displays the number of inhabitants with meters, according to the type of company. Table 2.82 summarizes the main characteristics of the water supply and wastewater services per country within the UK.

Table 2.80 Annual evolution of per capita consumption of water in E&W

	1995–96	1996–97	1997–98	1998–99	1999–2000	2000–01
WaSC	149	145	147	145	148	149
WoC	171	163	160	159	161	161
Total	**154**	**149**	**150**	**148**	**151**	**152**

Table 2.81 Population with and without meters for the water service

	Water		Wastewater	
	Non-measured	Measured	Non-measured	Measured
WaSC	34,434,000	8,765,000	40,414,000	10,919,000
WoC	7,866,000	1,933,000	–	–
Total	**42,300,000**	**10,698,000**	**40,414,000**	**10,919,000**

Table 2.82 Main characteristics of the water and wastewater networks

WWS	England	Wales	Scotland	Northern Ireland
Water				
Population supplied (millions)	50.7	3	5	1.7
Water produced (millions of m^3)	5,469		838	226
Network (km)	310,000	27,000	47,000	26,500
Water losses (%)	22		37	27
Wastewater				
Population supplied (millions)	48.7	2.9	5	1.4
Network (km)	285,000	19,000	49,000	14,500
Population connected to the network	96%		100%	83%

The WWS in the UK directly employs about 35,000 individuals, of whom 3,700 are in Scotland, 1,800 are in Northern Ireland and the remaining are in E&W.

Keeping in mind the characteristics of the water sector in the different countries, the need for investments and the revenues required are likewise different. Table 2.83 provides a summary of the investments and revenues per service and per country.

Table 2.83 Summary of investments and revenues

WWS (millions of €)	E&W	Scotland	Northern Ireland
Water			
Total revenues	5,452.6	554.5	188.6
Investments	7,153.8	982.8	321.4
Wastewater			
Total revenues	5,534.9	676.8	165.8
Investments	7,648.8	1,072.7	489.0

2.16.4 Rules of the game

2.16.4.1 Tariff setting

OFWAT is an added value of the English model. Its institutional structure gives it independence, powers and tools that allow this body to act in an efficient and competent way. OFWAT has provided an impetus to the water industry in E&W, which, in a few years, has become one of the most important and best performing WWS sectors in the world. In the past 15 years, prices have not been lowered in E&W, at least not in real terms. However, there has been an extremely high level of investment in the industry which led to a significant progress in quality standards, reflected in the improvement of the environmental quality and the quality of service delivered. In addition, there have been important gains in efficiency and productivity, ensuring value for money in these services. Table 2.84 shows these results.

Table 2.84 Variation of the average annual bill for customers in E&W

Variation in performance	Prices (€)			
	1990–1995	1995–2000	2000–2005	1990–2005
Improvement in environmental quality	58	54	46	158
Improvement in the quality of service	29	5	2	35
Improvement in efficiency	−8	−29	−96	−133
Total balance	**78**	**30**	**−48**	**61**

The price caps are fixed for a period of five years by OFWAT, thus enabling price stability for companies without greatly increasing risk. A series of mergers and acquisitions during the 1990s (frequently between a WaSC and a WoC operating in the same region) often included price reduction agreements as part of the conditions for mergers, outside the usual period of price cap reviews.

The determination of the price cap, and consequently the tariff system, is based on a process of analysis known as a price review. The review entails a process of consultation and negotiation with operators. The variations in percentage that are established are known as the K factor and are linked to the CPI each year, which, in the UK, is measured by the Retail Price Index (RPI). The formula is thus known as RPI-K. The process is based on estimates for expenditure (operations and capital) and the variation of productivity expected for the following regulatory period. Table 2.85 shows the average price caps established in E&W, according to the type of company.

In E&W, the tariffs implemented for domestic customers are fixed annually by the WaSCs based on the variation of the RPI and the K factor, determined by the regulator. Although the price caps are defined, operators have some freedom to set the tariff system, namely concerning the various components that are part of the price. However, the tariff systems must avoid discrimination among services and customers and allow the recovery of the real costs associated with each component included in the tariff.

The majority of the WWS in E&W continue to define tariffs without considering the water consumption. Thus, the tariff systems of domestic customers are mainly indexed to the value of the properties (rateable value – RV). Most of the WWS apply a permanent fixed rate. However, it is also common to implement a component related to the property itself, which dilutes the impact of the RV (two companies charge for their services based on a fixed component).

Table 2.85 Price caps (K) established in E&W

Average price cap	1990–95 K defined in the course of privatization	1995–2000 OFWAT decision	2000–05 OFWAT decision	2005–10 OFWAT decision
WaSC	5.0%	1.4%	−1.5%	4.3%
WoC	6.1%	0.4%	−2.4%	3.1%
Total	**5.2%**	**1.3%**	**−1.6%**	**4.2%**

Customers are charged for WWS via the application of a fixed component and a variable component, irrespective of whether or not a meter exists. The sole variation concerns the variable component, which, in cases where water is metered, is established according to the consumption of water, while in other situations this is implemented according to the characteristics of the property (RV). Table 2.86 shows the WWS charges for domestic customers (OFWAT, 2007a).

Table 2.86 Average bills for domestic customers

Unmeasured volumes	Charges change 1989–2007 (%)*	Average bill in 2007 (€)	Measured volumes	Charges change 1989–2007 (%)*	Average bill in 2007 (€)
Water	53.7	196.3	Water	−8.1	167.5
Wastewater	44.4	208.8	Wastewater	−3.3	188.8
Combined	48.8	406.3	Combined	−5.6	356.3

* Excluding the CPI

In E&W, the average WWS bill for domestic customers in 2007 was 390.3 euros (187.6 euros for the water service and 202.7 euros for the wastewater service). These values reflect an increase of 42% as compared to 1989.

The prices established by the WaSC for the treatment of effluents by non-domestic customers are calculated on the basis of the Mogden formula, which considers the polluting load of the effluents, which is especially important for some kinds of industrial wastewater. Table 2.87 shows some examples of the tariff systems adopted in E&W.

In Scotland and Northern Ireland, the tariff systems aim at totally recovering costs as well, including an adequate return on the capital invested. Just like in E&W, there are domestic customers who do not have meters and are subject to a single tariff component, which reflects all the service costs.

Concerning other customers, the WWS are charged on the basis of a fixed component, according to the meter, and a variable component, based on the volume of water consumed. Table 2.88 and Table 2.89 show the tariff systems adopted by the Scottish Water (2008) for customers with and without meters for 2008.

The first and second block diverge among the WWS, since consumption up to 25 m^3 is the limit of the first block for the water service while the limit for wastewater services is 23.75 m^3 (95% of access to the network). The second blocks relate to values higher than the consumption referred to, for both services, and have lower tariffs.

Apart from this, customers are obliged to pay a municipal tax for the drainage service, according to the municipal tax block (Council Tax band). This tax comprises two components, one related to the property (Property Drainage), with values ranging between 33.26 and 99.63 euros, and the other related to roads (Roads Drainage), with values between 33.26 and 99.63 euros.

Customers who do not have access to wastewater services are not obliged to pay for them. Table 2.90 and Table 2.91 show the tariff systems implemented by Northern Ireland Water for 2008.

2.16.4.2 Quality of service

In general, the WWS sector throughout the UK (in all the four countries) has a high level of quality. In this regard, the reforms instituted in the water sector (privatization in E&W and mergers in Scotland and Northern Ireland), associated with the implementation of regulatory agencies for economic aspects, the quality of service and customer protection have contributed decisively towards the progress achieved in the water sector in the UK.

Concerning the quality of service, the regulator also publishes an annual report comprising the levels of service (DG2–DG9) practised. This report includes aspects related to inadequate pressure, interruptions, restrictions on use, flooding, contacts concerning billing, written complaints, measured billing and the ease of telephone contacts.

Table 2.87 Tariff systems of Anglian Water and Folk and Dover WS for 2008

WWS	Water	Wastewater	
		Stormwater and wastewater	Wastewater
Anglian Water (WaSC)			
Unmeasured volume			
Fixed component (€/year)	164.13	198.68	156.22
Variable component (€/RV)	0.525	0.565	0.565
Measured volume			
Fixed component (€/year)	32.47	89.92	47.46
Variable component (€/m^3)	1.509	1.648	1.648
Seasonal customers*			
Fixed component (€/year)	–	–	–
Variable component (€/m^3)	1.942	2.980	2.351
Social tariffs			
Option I			
Fixed component (€/year)	211.06	261.02	218.55
Variable component (€/m^3)	–	–	–
Option II			
Fixed component (€/year)	84.92	147.37	104.91
Variable component (€/m^3)	0.810	0.797	0.797
Folkestone and Dover Water Services (WoC)		Water service	
Measured volume			
Fixed component (€/year)		36.56	
Variable component (€/m^3)		1.63	
Social tariff**			
Fixed component (€/year)		234.79	
Unmeasured volume		89.92	
Fixed component (€/year)		57.00	
Variable component (€/RV)		1.25	

* For domestic customers with an annual consumption of water below 75 m^3; ** In case consumption is less than 122 m^3

Table 2.88 Scottish Water tariff system for customers with meters

Components	Water	Wastewater
Fixed component (€/meter/year)	165.3–3,091.2	165.2–3,090.6
Variable component		
1st block (€/m^3)*	2.59	3.26
2nd block (€/m^3)*	0.94	1.54
Unit tariff (€/m^3)	0.94	1.54

* Meters up to 20 mm

The public display and media publicity of these elements (sunshine regulation) dissuade the WWS with the worst performances and encourage those who represent best practices. With the same objective, i.e. to increase the quality of service, OFWAT has explicitly considered performance regarding levels of service during the price review for the period 2000–2005, rewarding the five best WWS (+0.5%) and penalizing the five worst WWS (−0.5%).

Table 2.89 Scottish Water tariff system for customers without meters

Block of municipal tax	A	...	H
Water (€)	141.8	...	425.36
Wastewater (€)	160.3	...	480.87
Combined (€)	302.1	...	906.23

Table 2.90 Northern Ireland Water tariff system for customers with meters

Components	Water	Wastewater
Fixed component (€/meter/year)	77.99–1,980.6	38.99–990.3
Variable component (€/m^3)	1.18	0.83

Table 2.91 Northern Ireland Water tariff system for customers without meters

Components	Water	Wastewater
Unmeasured volume	Max. 493.2	Max. 493.2
Fixed component (€/year)	32.3	35.2
Variable component (€/1238 € of NAV)	11.0	11.8

NAV – Net Annual Value (equivalent to the RV)

In this scope, the benchmarking developed by the regulatory agencies has played a prominent role in stimulating the development of the water sector and thus the quality of service provided. Table 2.92 shows the evolution of the quality of service in the WWS in E&W since 1990.

Table 2.92 Levels of service evolution in the water sector in E&W

Description	1990–95 (%)	2003–04 (%)	2006–07 (%)
DG2 – Properties at risk of low pressure	1.33	0.04	0.02
DG3 – Properties subject to unplanned supply interruptions of 12 hours or more	0.33	0.14	0.15
DG4 – Population subject to hosepipe bans	14	0	30
DG5 – Properties subject to sewer flooding incidents (overloaded sewers and other causes)	0.03	0.01	0.02
DG5 – Properties at risk of sewer flooding incidents (once in ten years)	–	0.03	0.02
DG5 – Properties at risk of sewer flooding incidents (twice in ten years)	0.08	0.01	0.01
DG6 – Billing contacts not responded to (within five working days)	21.78	0.47	4.79
DG7 – Written complaints not responded to (within ten working days)	21.42	0.14	1.85
DG8 – Bills not based on meter readings	–	0.15	0.86
DG9 – Received telephone calls not answered within 30 seconds	–	5.85	–
DG9 – Telephone call handling			
– Calls abandoned		2.89	9.76
– All lines busy		5.44	5.66
– Call handling satisfaction			4.47

Figure 2.53 shows an assessment of overall performance, considering all the levels of service (Overall Performance Assessment – OPA) mentioned above. Thus, it is possible to determine the overall degree of satisfaction with the WWS quality of service (OFWAT, 2006).

So as to maintain and develop the water sector in the UK, OFWAT also promotes periodic benchmarking exercises, involving over 40 WWS internationally, encompassing countries such as Scotland (1), Northern Ireland (1), Canada (4),

Scandinavia (10), the Netherlands (6), Australia (6), USA (9) and Portugal (IRAR – 15% of the population). This project developed 23 performance indicators, which evaluate the areas of Bills to customers (2), Customer service levels (1), Water quality and environmental performance (2), Water delivered, leakage and water efficiency (4), Unit costs and relative efficiency (3), Network activity (2) and, finally, Financial performance (10). Apart from this, OFWAT periodically publishes reports about diverse aspects related to the functioning of the WWS, which make it possible to evaluate issues over the course of time and compare the performance of operators (e.g. water losses).

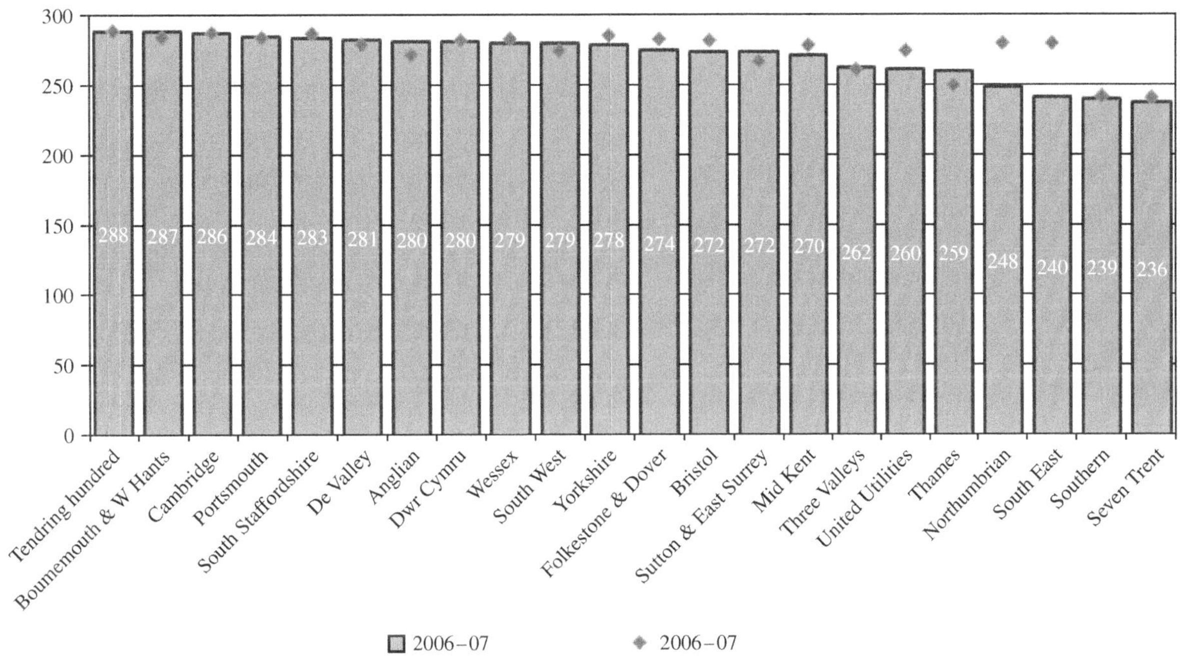

Figure 2.53 OPA indicator for the water services in E&W

In Scotland, the main function of the WICS is to promote the WWS quality of service. Given the limitations arising from only regulating a single operator, the regulatory agency has adopted WWS in E&W as benchmarks, apart from using the results achieved in past years as a reference, namely with regard to the main performance indicators.

2.16.4.3 Public service obligations

One of the essential characteristics of the regulatory model adopted in E&W consists of placing customers at the heart of regulation. In this regard, prevailing legislation clearly considers public service obligations such as universality, continuity, equity, the quality of service and accessibility. Likewise, in an innovative way, OFWAT introduced the minimum parameters for the quality of service, after which the WWS will have to compensate customers (GSS). For example, if a complaint is not answered within a specified time frame or if the water service is interrupted without prior notice and owing to the operator, customers receive a monetary compensation.

In April 2008, OFWAT reviewed the procedures for implementing the quality of service levels. One of the most important changes involved making these levels uniform within the territories of E&W, as well as changes in the guarantees related to the wastewater service.

It must also be highlighted that low-income customers and customers with special needs are protected by a specific regime, established by the Water Industry Act 1999, and services cannot be disconnected easily. Similarly, in 1989, during the institutional design of regulation, customers were provided with representation and their interests also became protected through the Customer Panels. The existence of these entities, especially aimed at handling complaints, reflects an important guideline regarding the rights of citizens. In addition to the policy designed to protect the interests of customers, operators have also defined special tariffs for customers with limited financial resources, such as customers benefiting from state subsidies and those who have three or more children, as well as special cases associated with chronic illnesses. Table 2.93 shows some examples of the indicators reflecting the quality of service adopted by OFWAT. Table 2.94 presents the quality of service levels implemented by OFWAT in E&W.

Table 2.93 Examples of indicators reflecting the quality of service

Service to customers	Period of privatization	2005–06
Percentage of written complaints answered within 10 working days	81.9% (1992/93)	96.9%
Percentage of billing contacts answered within 5 working days	79.8% (1992/93)	95.6%
Number of compensation payments (GSS)	11,388	34,040

Table 2.94 Quality of service levels established in E&W (€)

Quality of service levels regulation	Levels of service payments		Penalties for delayed payments	
	Domestic customers	Non-domestic customers	Domestic customers	Non-domestic customers
Appointments not made properly	24.75	24.75	12.38	12.38
Appointments not kept	24.75	24.75	12.38	12.38
Incidences of low water pressure	30.92	30.92	–	–
Incorrect notice of planned interruptions to supply	24.75	61.85	24.75	61.85
Service not restored* – initial period	24.75	61.85	24.75	61.85
Service not restored* – each additional 24 hours	12.38	30.92		
Written account queries and requests to change payment arrangements not actioned on time	24.75	24.75	12.38	12.38
Written complaints not actioned on time	24.75	24.75	12.38	12.38
Property's sewer flooded internally	Payment equivalent to the annual wastewater services charges (Min. payment 185.5; max. 1236.5)		24.75	61.85
Property's materially affected sewer flooded externally	Payment equivalent to 50% of the annual wastewater service charges (Min. payment. 92.75; max. 618)		24.75	24.75

* Supply not restored within the notified time (planned works) or when the supply is interrupted for an unscheduled period of time/ emergency situations.

Under the Water Act 2003, OFWAT acquired new powers, namely the capacity to apply penalties to the WaSCs and WoCs up to a maximum of 10% of their turnover, in cases of serious violations of customer service norms, legal obligations or licensing conditions.

Operators are legally obliged to develop and maintain an economical and efficient system for the water service in their respective areas. The WWS are responsible for supplying a clean and reliable water service, developing plans for water resources, duly submitted to the EA, and defining their point of view about water resources management over the next 25 years.

The WWS must also prepare drought contingency plans, in which they define solutions for different types of drought and propose and justify investments related to the water resources management to be incorporated into tariff reviews, promote the efficient use of water, in the name of customers, and, finally, maintain the water supply system in an economical and efficient way.

In Northern Ireland and Scotland, despite being less visible than in E&W, the regulatory authorities are also known for their great dedication to safeguarding customer interests and ensuring that public services obligations are fulfilled. Moreover, in Scotland, the Water Act 1980 made it mandatory for the local authorities to report all information concerning the WWS to the Council of Ministers.

2.16.5 Performance

The UK has a well-designed, modern and very robust regulatory model. Regulation is primarily concerned with safeguarding customer interests, without neglecting the sustainability of the operators. The different regulators have been able to successfully supervise and monitor the quality of service, as the improvements in this sector have proved.

The regulators have ensured that the different public service obligations are fulfilled through the establishment of contracts between the various stakeholders, and have maintained operator remuneration by means of suitable economic regulation, i.e. guaranteeing the economic and financial viability of the operators without allowing profits above a fair and reasonable level. Finally, it is important to note that in all its diverse economic, social and environmental aspects, regulation in the UK has been developed from a long-term perspective and with the purpose of ensuring sustainability.

The less successful aspects are related to an absence of universal metering to measure the volumes consumed and the prevailing high regulatory risk, which makes the British WWS less attractive to the private sector.

2.17 UNITED STATES OF AMERICA

2.17.1 Historical background

The USA is a presidential federal republic. Executive power is exercised at a federal level by the President and is independent of the legislative power, exercised by the Congress, which, in turn, consists of the House of Representatives and the Senate.

The USA has an area of 9.8 million km^2 and a population of about 303 million inhabitants. It is divided into 50 states and one federal district, the District of Columbia. Each state is organized into counties, with the exception of Louisiana, in which the subdivisions are called parishes, and Alaska, where the state subdivisions are called boroughs. The counties can also be further classified into smaller administrative units, known as municipalities, townships or towns (depending on the state) and boroughs. The USA presently has about 30,000 cities.

Among many unique features, the WWS in the USA is characterized by a very high level of water consumption. Over the course of the water sector history, this scenario has naturally given rise to many problems, namely concerning the treatment and discharge of wastewater and, consequently, the protection of water resources. In this regard, demonstrating a keen awareness of environmental questions, the federal government has designed a robust legal framework, which not only addresses the issue of effluents but also safeguards the quality of federal water resources.

In the USA, the water sector is mainly regulated by the 1974 Safe Drinking Water Act (SDWA) and by the 1972 Federal Water Pollution Control Act (P.L. 92-500), which with subsequent changes was then known as the 1977 Clean Water Act (CWA, P.L. 95-217). These laws provided the legal framework for this sector currently in effect in the USA and later resulted in the development of more restrictive laws, such as the 1988 Water Quality Act (P.L. 100-4), which led to the National Pollutant Discharge Elimination System whose requisites are also applicable to stormwater.

Along with subsequent changes, the CWA constitutes the basic federal law to prevent and mitigate water pollution in the USA. This diploma establishes objectives with regard to the quality of water resources, eliminating the presence of toxic substances and additional water pollution, ensuring that the quality of surface water is suitable for sports and recreation.

The SDWA, which was subsequently altered in 1986 and 1996, constitutes the main federal law that ensures the drinking water quality for all Americans, from large metropolitan cities to rural areas. By means of this law, the Environment Protection Agency (EPA) was given the power to establish norms concerning the drinking water quality and to supervise the implementation of these norms in all states, municipalities, boroughs and among WWS operators.

As the name itself suggests, the SDWA is aimed at guaranteeing the drinking water supply quality in the USA. However, this legal document assigns to EPA the task of establishing norms to limit the effect of pollution and other harmful elements for public health (National Primary Drinking Water Regulations). The norms include maximum levels for each element (Maximum Contaminant Levels) as well as the respective objectives (Maximum Contaminant Level Goals).

Finally, it is important to note that each state government has the power to establish its own legislation, as long as it is in conformance with the federal law, a situation which can be found in the majority of the states.

2.17.2 Legal and institutional framework

2.17.2.1 Responsibilities

Responsibility for the water sector has been divided among various entities, which include the Department of Agriculture, the EPA, the local administration, the regulatory agencies (Public Utility or Service Commission – PUC or PSC) and the diverse associations active in the sector, such as the American Water Works Association (AWWA), the Water Environment Federation (WEF), the Association of Metropolitan Water Agencies (AMWA), the National Association of Regulatory Utility Commissioners (NARUC), the National Association of Water Companies (NAWC), the National Rural Water Association (NRWA) and the American Water Works Association Research Foundation (AWWARF). In addition to these entities there are over 100 associations that operate at a state level.

The Department of Agriculture plays an important role in the water sector in the USA, providing subsidies, loans, and loan guarantees in rural areas, i.e. for small communities with a population of fewer than 10,000 inhabitants, making it feasible to provide WWS, which would be hardly sustainable without this support. This department also promotes technical assistance and training in the water sector, especially in rural areas.

The EPA, created in 1970, is an agency of the USA federal government. It is responsible for protecting human health, safeguarding nature and ensuring environmental sustainability. The EPA conducts significant initiatives at the level of science, research, education and assessments of environmental questions in the USA. In this context, the agency develops and reinforces the prevailing environmental regulations, promotes financial assistance to diverse entities and carries out activities in cooperation with national laboratories, with the purpose of solving problems in the environmental sector. The EPA also promotes various educational activities, raising the awareness of the stakeholders about environmental issues and on matters related to the water sector in particular.

The local administration, such as the counties, is responsible for providing WWS of a suitable quality, in conformance with the norms that have been established at a federal level.

The PUCs or PSCs likewise play an important role in the USA water sector, not only concerning the protection of customer interests but also with regard to tariff regulation and the provision of incentives to enhance the sector. When the states lack these kinds of bodies dedicated to the water sector, these responsibilities are entrusted to state governments. At the heart of the PUCs or PSCs, or even operating independently of them, sometimes we can find the so-called Rate Payer Advocate, whose mission is to represent and protect the interests of customers, especially concerning the tariff revisions (rate case) and sometimes hold public hearings. Figure 2.54 provides a simplified overview of the water sector organization in the USA. The main difference between the PUCs and the PSCs is the way in which their board of directors is chosen. They are appointed for the PUCs or elected by means of a democratic vote for the PSCs.

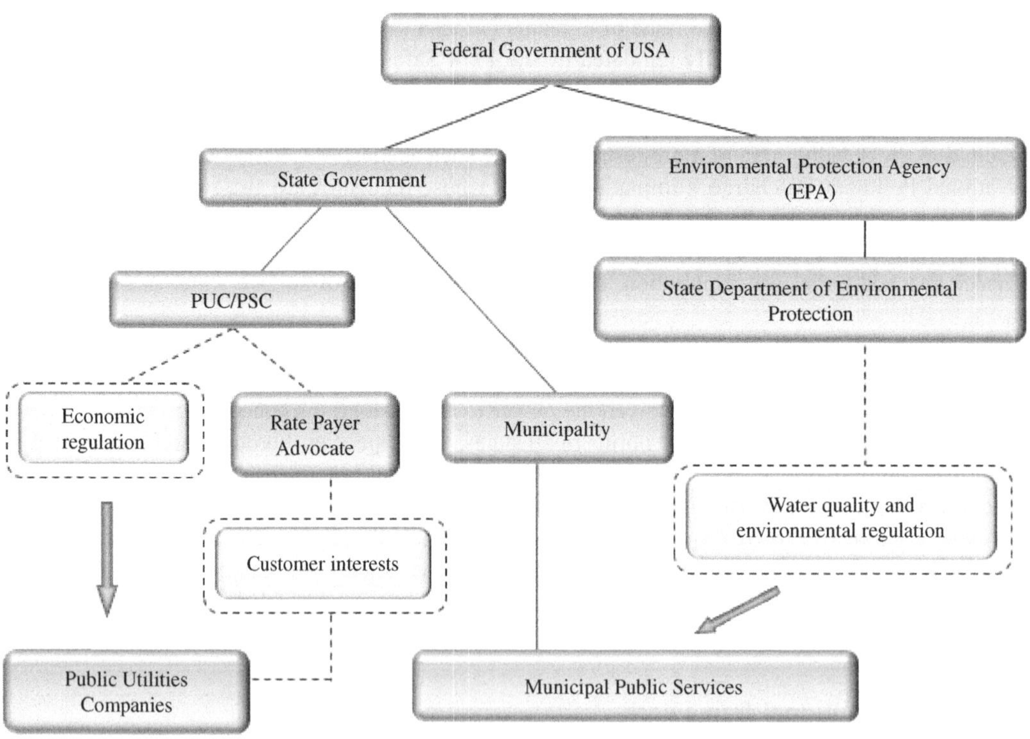

Figure 2.54 Institutional framework of the water sector in the USA

The AWWA is an international professional non-profit organization dedicated to improving water services and the drinking water quality. This association was founded in 1881 and has approximately 60,000 members around the world. AWWA members represent the water community, comprising operators and managers of treatment plants, scientists, environmentalists, manufacturers, academics, regulators and other stakeholders. The AWWA performs its mission by means of communications, conferences, training and local actions, publishing a vast quantity of norms, books, manuals, electronic databases and reports.

The WEF was created in 1928 and is a non-profit association with numerous members both within the USA and internationally. The WEF and its members seek to preserve water resources worldwide. The main mission of the WEF is to preserve and improve the quality of water, promoting informative and educational meetings about the preservation of the quality of water resources among the diverse stakeholders.

The AWWARF is a non-profit organization that promotes research among the WWS, government agencies and other stakeholders. This foundation, created in 1966, aims to achieve scientific progress in areas associated with the water sector.

The NARUC was founded in 1889 and is a non-profit organization that, among its main functions, represents the PUCs/PSCs, which regulate operators providing public services such as energy, telecommunications, water and transport. Its main mission is to defend public interests and promote the quality and effectiveness of regulation in the diverse sectors.

The AMWA is an association of metropolitan operators in the USA, whose members provide WWS to over 127 million Americans, from Alaska to Puerto Rico. The AMWA was created in 1981 and its main function is to defend the interests of its members, working together with the Congress and with federal agencies to monitor the development of the sector.

The NAWC was founded in 1895 and represents the interests of private WWS in the USA. Although the private sector does not represent a very large component given the size of the sector, the NAWC is an extremely important organization in the water sector. Its members encompass all kinds of organizational arrangements as long as they are private operators.

The NRWA is another operators' association, which has more than 26,000 members. The NRWA is a non-profit association whose mission is to defend the interests of its members along with the dissemination of best practices.

2.17.2.2 Regulation

Economic regulation is implemented at a state level by the PUCs or the PSCs. The PUCs/PSCs regulate about 7,000 WWS in 40 states, of which 3,300 are private operators. Normally, the PUCs do not regulate all kinds of WWS. They frequently regulate only private operators and can only act if the operator is above a minimum size.

As the birthplace of economic regulation, the USA is characterized by a very efficient and effective regulatory structure, especially because of its 'almost jurisdictional' and 'almost legislative' functions and due to all the procedures of the regulatory process. The American regulatory tradition has been developed over the course of more than 100 years in diverse sectors and has resulted in a homogeneous set of key regulatory principles. Thus, even for WWS, costs must be prudently incurred and investments must be used and useful. It should be possible to earn a fair rate-of-return and tariffs cannot be unduly discriminatory. These basic principles, among others that are considered in the American context, have been extrapolated to WWS regulation.

A less positive aspect of WWS regulation in the USA is the almost universal use of rate-of-return regulation to set tariffs, which does not provide performance incentives to the WWS. In fact, other methods such as price cap regulation are used only in a handful of cases (e.g. Iowa). This is in line with the specificities of the WWS sector in the USA, which cannot in any way be considered a worldwide benchmark, at least when compared to some European operators. However, the regulatory process in itself works well.

The common functions of regulatory authorities, who frequently only regulate private operators, include *inter alia* the setting of tariffs and service norms, defining areas of jurisdiction, accounting norms, aspects related to safety, the transmission, purchase and merger of assets and regulatory procedures. The interests of customers are frequently represented by an independent body that works with the PUCs/PSCs (Rate Payer Advocate). Table 2.95 presents the regulators of the water sector in each USA state.

However, since most WWS are provided by municipal operators, the role of the state regulators is somewhat limited. Thus, it is a common practice to have municipal commissions or advisory groups whose function is to ensure an appropriate price for these essential services.

Even with the presence of a PUC/PSC, the federal and state governments also regulate the areas of the environment, health and water quality.

Likewise in the regulatory context, associations in the water sector play an important role, not just due to their strong influence within legislative bodies but also because of the self-regulation that they impose on their members, disseminating best practices and encouraging their development.

For example, in the state of Florida, the regulatory agency (FPSC), created in 1887 by the Florida Statutes (chapter 350), is an independent multisector regulatory agency (electricity, gas, telecommunications, water and wastewater) responsible for economic regulation and for safeguarding the quality of service, namely by setting WWS tariffs, supervising competition in the market and monitoring the safety and reliability of the WWS. Despite the independence declared in its statutes, the government still has some powers at the level of economic regulation in the sector, since the FPSC only regulates private operators.

Table 2.95 Regulatory agencies in the water sector in the USA

State	Regulatory agency	State	Regulatory agency
Alabama	Alabama PSC	Montana	Montana PSC
Alaska	Alaska Regulatory Commission	Nebraska	–
Arizona	Arizona Corporation Commission	Nevada	Nevada PUC
Arkansas	Arkansas PSC	New Hampshire	New Hampshire PUC
California	California PUC	New Jersey	New Jersey Board of Public Utilities
Colorado	Colorado PUC	New Mexico	–
Connecticut	Connecticut Department of Public Utility Control	New York	New York State PSC
Delaware	Delaware PSC	North Carolina	North Carolina Utilities Commission
Florida	Florida PSC	North Dakota	–
Georgia	–	Ohio	Ohio PUC
Hawaii	Hawaii PUC	Oklahoma	–
Idaho	Idaho PUC	Oregon	Oregon PUC
Illinois	Illinois Commerce Commission	Pennsylvania	Pennsylvania PUC
Indiana	Indiana Utility Regulatory Commission	Rhode Island	Rhode Island PUC
Iowa	Iowa Utilities Board	South Carolina	South Carolina Office of Regulatory Staff
Kansas	–	South Dakota	–
Kentucky	Kentucky PSC	Tennessee	Tennessee Regulatory Authority
Louisiana	Louisiana PSC	Texas	Utah PSC
Maine	Maine PUC	Utah	Vermont Public Service Board
Maryland	Maryland PSC	Vermont	–
Massachusetts	Department of Public Utilities	Virginia	Virginia State Corporation Commission
Michigan	–	Washington	–
Minnesota	–	West Virginia	West Virginia PSC
Mississippi	Mississippi PSC	Wisconsin	Wisconsin PSC
Missouri	Missouri PSC	Wyoming	Wyoming PSC

The FPSC does not issue legislation but can adopt rules so as to implement legislation. It is financed by the operators by means of a tax of 4.5% on billing (its budget for 2006 was about 3 million euros). The method of economic regulation used to set tariffs is the rate-of-return regulation. The regulatory period is one year.

In California the regulatory agency (CPUC), which was created by the Public Utilities Act of 1911, is an independent multisector regulatory agency that is responsible for economic regulation and the quality of service, such as setting WWS tariffs. The CPUC only regulates private operators in the areas of telecommunications, electricity, natural gas, water, railways and passenger transport companies. Its main mission is to ensure that customers have a safe and reliable public service at reasonable prices, protecting them against frauds and promoting a healthy economy in the Californian WWS. The CPUC is organically independent (staff are selected for their technical skills, with long tenures, and are difficult to dismiss) and is empowered to issue norms and apply penalties and fines in more serious cases of service provision, namely when water quality parameters are constantly disregarded. It intervenes on its own initiative or at the request of any stakeholder. The regulator is accountable to parliament whenever requested and is characterized by transparency and the participation of stakeholders in decision-making processes, by its public visibility and financial accountability. Finally, customers can also appeal to other authorities against decisions. The CPUC is headed by a single individual and is financed by means of a fixed percentage of the tariff that is implemented. The agency publicizes decisions, fines and procedures underway.

The method of economic regulation used to set tariffs is the rate-of-return regulation with a hybrid mechanism of revenue caps. The length of the regulatory period is three years. It is also important to note that in the particular case of California there is also the California Water Association, which represents the interests of about 140 private operators and plays an important role in the water sector.

2.17.3 Market structure

2.17.3.1 Ownership and management of services

The WWS are mainly provided by public operators, within the different institutional arrangements of service provision under public management. In general, private participation includes management but not ownership.

Although the private sector has participated in the American water sector for over a century, WWS have always been viewed as being intrinsically part of the public sector, which has proved to be a constraint in terms of expansion. Thus, only a small percentage of the population is supplied by private operators. There is no reliable data but it has been estimated that nearly 10% of Americans are served by about 3,500 private operators. However, the majority of these private operators are small entities. Some important WWS that were privatized during the 1990s were recently got back, for diverse reasons, by the public sector (e.g. Atlanta). In the meanwhile, many PPPs have been established in the USA, especially in WWTP and WTP.

2.17.3.2 Players in the sector

The water services segment in the USA comprises about 53,000 operators. Research carried out by the EPA has found out that there is still a large number of small systems, i.e. individual systems or collective systems on a limited scale, comprising an estimated total of about 150,000 small operators. Table 2.96 summarizes American water service operators according to the population served (CSS, 2007).

Table 2.96 Characteristics of water services according to size

Population class	Operators (No.)	Population served (10^6)	Operators (%)	American population served per operator (%)
25–500	29,666	1.93	56	2
501–3,300	14,389	20.9	27	7
3,301–10,000	4,748	27.5	9	10
10,001–100,000	3,648	103	7	36
> 100,000	386	126	1	45
Total	**52,837**	**282**	**100**	**100**

There are fewer wastewater services, estimated to be about 50,000 operators. The majority of services are provided along with the water services. It has been estimated that there are approximately 20,000 operators of stormwater services.

The WWS are supplied through municipal systems, which are owned and operated by local governments in different organizational arrangements, or through private systems. The vast majority of the WWS in the USA rely on public systems, which account for about 84% of all water services and about 95% of all WWS.

In general, in urban areas, the municipal systems are directly controlled by municipal governments or by cities (as is the case in New York) or indirectly by companies dedicated to this sector (public utilities), owned by municipal governments or cities. In some cases, the public service companies cover diverse jurisdictions (municipalities or cities). Cooperatives are the main institutional arrangement for the supply of WWS in rural areas.

However, many cities and urban agglomerations select private companies to operate, manage and/or promote investments in the WWS. These contracts, which can range from simple outsourcing (meters, billing, maintenance, etc.) to concession arrangements and the implementation of BOT/BOOT contracts, already represent a relevant segment of operators. Currently, about 11% of the American population, i.e. 33.5 million inhabitants, receive water supplied by private operators. For example, there are 197 private operators in Florida (174 for the water service and 148 for the wastewater service). In California there are 138 private water operators and 12 private wastewater operators, serving 15% of the population of this state.

The largest private operator in the water sector in the USA is the American Water, which is owned by RWE AG (Essen, Germany) and provides WWS to 16.2 million inhabitants in 32 states, representing about 5% of the American population.

Veolia Water North America is the second most important private player in the USA. Presently, the company offers a diverse set of services via PPPs. This company owns the largest existing private water service operator (Indianapolis), the largest BOT contract (WTP in Tampa Bay) and the largest partnership at the level of wastewater services (Milwaukee). Presently, Veolia provides WWS to about 14 million inhabitants in this country.

United Water/Suez also has an important presence in the water sector in the USA, supplying WWS to 6.5 million inhabitants, including PPPs in the states of California, Texas and Florida, among others.

Finally, it is important to note that even though they might not be very expressive concerning population coverage, there is a very significant number of local private players.

In the USA, it is still possible to find some wholesale water operators, even though relatively few in number, especially in the arid southwest region of the country. One of them, the Metropolitan Water District of Southern California (MWD), sells treated water from the Colorado River and from Northern California to its associated companies in Southern California, through California Aqueduct. In this context, 26 operators serving 18 million inhabitants are members of the MWD. Another example is the Central Arizona Water Conservation District, which operates the Central Arizona Project Aqueduct and supplies water from the Colorado River to 80 operators in Central and Southern Arizona.

2.17.3.3 Integration of the sector and services provided

WWS integration tends to be heterogeneous throughout the country. Although it is possible to note a slight superiority of WWS integration as compared to the large percentage of local operators, it is also usual to find the existence of regional operators that ensure integrated services or are responsible for a single component of the productive chain, such as the production of water (wholesale). The latter situation occurs fundamentally with regard to the water supply.

Although the majority of water services are provided along with wastewater services, this is not always the case and separate operators can exist for both services.

2.17.3.4 Numbers for the sector

About 560 billion m^3 of water are annually consumed in the USA. Thermoelectric energy consumes 48%, irrigation 34%, drinking water supply represents 11%, industrial use 5% and the remaining is used for other purposes. Its main source is surface water, accounting for 79% of production, of which 15% is obtained by desalinization. Thus, the volume of drinking water corresponds to 61.6 billion m^3. According to a study carried out by the AWWA, the water used for outdoor spaces (e.g. gardens and swimming pools) represents 58% of domestic consumption.

Although there are no precise data, it has been estimated that there are very high water losses, owing to the poor state of the infrastructures and the low cost of water.

The average per capita consumption of water by USA citizens is 660 litres of water per inhabitant per day, encompassing all types of water usage. There are cities in which the per capita consumption is about 2,500 litres per inhabitant per day. It has also been estimated that indoor consumption is lower, at 260 litres per inhabitant per day.

In the USA, the water service is characterized by the principle of universality, both for the urban population, which represents 80% of the American population, and for the 20% of the population living in rural areas. This is not the case with regard to the wastewater service, due to the limited sustainability in certain areas. Table 2.97 shows the levels of coverage of the two services.

It has been estimated that the water sector turnover in the USA ranges between 65 and 79 billion euros, representing about 1% of the American GDP. Table 2.98 illustrates the origin of these revenues, based on an estimate of about 66 billion euros (Grigg, 2007).

In the American water market the private sector involves a volume of business of approximately 1,120 million euros and has contracts with more than 1,300 local governments.

The water sector in the USA employs about 1 million individuals, although only 445,000 are associated with the water, wastewater and stormwater operators. Figure 2.55 shows the costs that households incur with the different services of general interest.

Table 2.97 Level of WWS coverage in the USA (%)

Service	Nature	Urban population (80%)	Rural population (20%)	Total
Water	Existence of network	100	100	100
	Network connections	100	100	100
Wastewater	Network	100	100	100

Table 2.98 Revenues of WWS in the USA

Service	National total (10^6 €)	Monthly bill (household) – (€)
Water supply	26,339	14.49
Wastewater services	26,339	14.49
Stormwater	13,171	7.24
Total	**65,856**	**36.22**

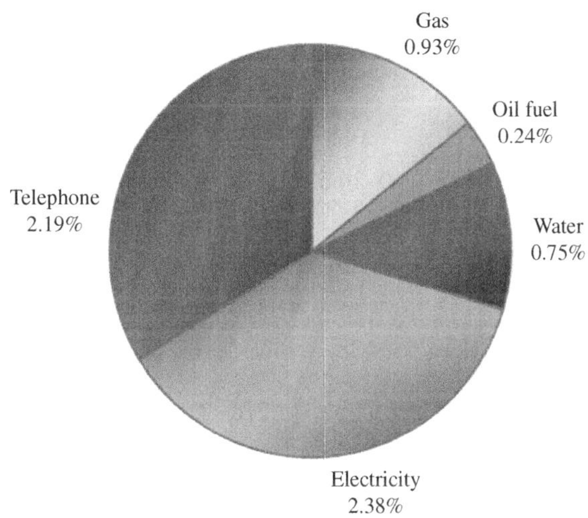

Figure 2.55 Costs of services of general interest per household in the USA

2.17.4 Rules of the game

2.17.4.1 Tariff setting

The price of water in the USA is one of the lowest in developed countries. Despite the successive price increases above the rate of inflation that have been implemented in recent years, owing to a greater awareness of the stakeholders about the excessive consumption of water and the need to transfer all costs to customers, this situation is still a daily reality for Americans. Currently, the average price of water is about 0.33 €/m^3. However, due to a disproportionate level of consumption, which has become deeply rooted in the USA culture, the water bills are more or less identical to those of the majority of OECD countries. According to the EPA, an American family spends, on average, about 312 euros per year on WWS.

As far as tariff systems are concerned, there are great variations among the different states, not only with regard to the average price of the service but also to the different tariff structures implemented by the operators. In this context, a study carried out by the World Bank, in 2000, estimated that approximately one-third of USA states implement tariff systems for the water service based on a fixed component, another one-third of states have instituted progressive tariff systems according to consumption and, finally, the remaining one-third implements decreasing tariffs considering the level of consumption. There are also occasional cases of operators which have opted for tariff systems based on the size of the property. Tariffs that vary with the seasons are also equally common in the USA.

For example, in Florida, the typical tariff structure adopted for domestic customers comprises a fixed component, which encompasses the fixed costs of the service, and a progressive variable component based on consumption, with the purpose of discouraging excessive consumption, despite the low prices.

In California, the tariff structure which is normally implemented for domestic customers is based on a component of fixed prices, defined according to the characteristics of the customers (meter, area, etc.) and on a volumetric component, which is nonetheless fixed. There are special tariffs for low-income customers. Table 2.99 and Table 2.100 highlight some examples of the tariff systems adopted in the states of California and Florida, respectively. Table 2.101 summarizes the tariff system of the city of New York. The figures do not include VAT, which is 17.5%.

Table 2.99 Tariff system in Los Angeles for the period between 2007 and 2008

High season		Low season					High season
Jul–Sep	Oct	Nov	Dec	Jan–Mar	Apr	May	Jun
1st block (€/m^3)							
0.504	0.499	0.499	0.499	0.526	0.537	0.537	0.537
2nd block (€/m^3)							
0.649	0.653	0.539	0.539	0.535	0.539	0.539	0.653

Table 2.100 Tariff systems for Palm Beach and Miami Dade in Florida (2007–2008)

Florida	Water (€/m^3)	Wastewater (€/m^3)
Palm Beach	0–17.6 m^3: 0.108 17.6–44 m^3: 0.231 >44 m^3: 0.549	0–4: 0.144 0.318 (up to 44 m^3)
Miami Dade	0–16.5 m^3: 0.072 16.5–33 m^3: 0.231 33–56 m^3: 0.318 >56 m^3: 0.433	0–16.5 m^3: 0.267 16.5–33 m^3: 0.419 >33 m^3: 0.520

Table 2.101 Tariff system for New York City for the period between 2007 and 2008

Users	Water	Wastewater
With meter (€/m^3)	0.39	159% of the water bill
Without meter (€)	54–188	

The areas supplied outside the city of Los Angeles pay an additional tariff of 0.136 €/m^3. The Los Angeles Department Water and Power Company adopts two types of social tariff for domestic customers, namely the special type and the low-income type. The former includes senior citizens, the disabled and the chronically sick, who receive a subsidy of about 13 euros every two months. The low-income category includes all domestic customers who have limited financial resources, who receive a subsidy of about 6.5 euros every two months, with an additional subsidy of 1.3 euros for each family unit with more than three members.

Not all customers have meters and a significant part is measured by connection. Sometimes there are also minimum values to be paid. Usually tariffs differ, even if slightly, according to the kind of customer, i.e. non-household customers pay more. There are almost always subsidies (to be deducted from the bill) or special tariffs for low-income customers or those with special needs (e.g. senior citizens, the ill and the poor). In addition, the tariff always includes a part to cover the costs of the fire-fighting service and specific tariffs may exist for water used for irrigation, agriculture and industry and for reclaimed water. As mentioned, tariffs frequently vary over the course of the year and the same operator can implement different tariff systems according to the area being supplied.

2.17.4.2 Quality of service

Water consumption in the USA is among the highest in the world and the water tariffs are among the lowest in terms of developed nations. As a result of this situation, along with the population growth and the ageing and poor state of the infrastructure, the WWS sector requires heavy investments.

WWS in the USA are characterized by an acceptable level of quality, including continuity, good pressure and compliance with SDWA norms. However, there is no detailed information about the quality of service provided by the operators. There are, for example, many complaints about billing without a practical impact on the standard of the service provided.

All operators have customer support departments and when disputes are not resolved between customers and operators both parties can apply to the PUC/PSC or the competent government authorities.

Generally minimum quality of service levels have not been established. However, whenever disputes occur and complaints are filed with the PUC/PSC, these entities can stipulate the payment of indemnities to customers or to other stakeholders.

Operators develop benchmarking exercises by their own initiative or guided by decisions of the associations in the sector. In particular, the AWWA and the WEF have played an important role in standardizing the terminology and the definitions of performance indicators, for both water and wastewater services.

For example, the AWWA has annually implemented a programme that seeks to apply a set of 22 performance indicators, known as QualServe, within the scope of its functions to enhance the USA water sector. QualServe is a voluntary programme that assesses performance, designed at various levels: organization development, customer relations, business planning and management, water operations and wastewater operations of wastewater services. By participating, operators can identify their strengths and weaknesses and define objectives to improve their services. The results of the benchmarking exercise are identified and defined in a report, which is later handed out to the participants (QualServe Performance Indicators Report).

Table 2.102 shows the performance indicators used by the AWWA in the context of the QualServe programme.

Table 2.102 Performance indicators adopted by the AWWA

Performance indicators
Organization development
Organizational best practices index
Employee health and safety severity rate
Training hours per employee
Customer accounts per employee, per volume of water delivered and per volume of wastewater treated
Customer relations
Customer service complaints
Disruptions of water service
Residential cost WWS
Customer service cost per account
Billing accuracy
Business planning and management
Debt ratio
System renewal/replacement rate
Return on assets
Water operations
Drinking water compliance rate
Distribution system water loss
Water distribution system integrity
Operation and maintenance cost ratios
Planned maintenance ratio
Wastewater operations
Sewer overflow rate
Collection system integrity
Wastewater treatment effectiveness rate
Operation and maintenance cost ratios
Planned maintenance ratio

2.17.4.3 Public service obligations

There is no federal law establishing public service obligations for the water sector. However, at a state level, with the existence of sector-specific regulation and the ability to issue regulations, priority public service obligations were established, such as universality and accessibility principles in many states (e.g. California and Florida).

The aspects concerning the tariff system have strict procedures and the interests of customers are always protected due to the transparency and participation in decision-making processes. Customers are notified about changes in the tariff systems and can always appeal to the regulatory agency.

2.17.5 Performance

WWS in the USA are provided at a very low cost and with a high level of quality. The majority of the states have independent regulatory agencies, with vast powers and tools, which ensure that the interests of the various stakeholders are protected. However, their scope of action is limited and they generally supervise only private operators.

Normally, concerning economic regulation, the PUCs/PSCs are also based on rate-of-return regulation and thus do not provide incentives for improvement in the performance of the sector. The USA WWS sector requires large investments, since in overall terms the infrastructures present some problems related to their quality. The complex issues of subsidization and the scarce impact of the service cost on customers, along with the enormous per capita consumption, are the main challenges for the sector.

2.18 ZAMBIA

2.18.1 Historical background

Zambia is a democratic republic. The President of Zambia is both the head of state and the head of the government in a plural and multiparty system. While the government is responsible for the executive power, the legislative power is associated not only with the government but also with the Parliament. The country has a population of approximately 11.7 million inhabitants over about 752,600 km^2. It is divided into nine provinces, each of them administered by an appointed governor. In their turn, the provinces are divided into a total of 73 districts.

The water sector in Zambia has faced several challenges that have conditioned its development. The nation's main problem is not the availability of water resources, since Zambia has plenty of water. However, this sector has been managed inadequately over the years, owing to both a lack of financial resources for building and maintaining infrastructure and to insufficient management capabilities and institutional powers.

The first law to be enacted on the water sector was the 1948 Water Act, which later proved to be ineffective and inadequate with regard to the accomplishment of the intended objectives and thus it became imperative to revise it. Currently, the water sector is primarily regulated by the 1991 Local Government Act (Law No. 22), by the 1994 National Water Policy and by the 1997 Water Supply and Sanitation Act (Law No. 28).

Among its various dispositions, Law No. 22 decentralized the control of the WWS sector, attributing this responsibility to the local authorities. Until then the sector had been under the supervision of the central authorities. Apart from being operators, the local authorities were also given normative powers.

The 1994 National Water Policy created the legal framework for implementing the decentralization of responsibility for the WWS sector. This strategic document has attributed equal responsibilities for supplying these services to urban and peri-urban areas. It established the universal service principle with regard to urban and rural populations. Apart from trying to promote the sustainable development of water resources, it tried to facilitate an equitable provision of water with an adequate quantity and quality for all competitive uses, at a fair and acceptable price, ensuring the robustness and reliability of the services. The National Water Policy does not deal directly with peri-urban areas or with informal settlements, although it does not ignore them.

The Water Supply and Sanitation Act delegates responsibility for supplying WWS to the local authorities and provides incentives for commercial organization by creating commercial operators. The introduction of private participation in managing and operating these services was also included amongst the available options for commercial operations in the water sector in Zambia. This normative landmark also created the National Water Supply and Sanitation Council (NWASCO), a national and independent regulator for urban WWS.

In 2004, the government prepared a strategy for rural WWS, implementing decentralization at the level of the districts and based on local planning which was intended to make the WWS more efficient.

In short, the National Water Policy, in compliance with the Water Supply and Sanitation Act, introduced diverse changes in the sector, including separating the management of water resources from WWS management, separating the state as a regulator from the state as an operator at various levels, making local authorities responsible for providing WWS, allowing delegation to private companies, implementing the principle of full cost recovery by means of the tariff systems, professional human resources, using appropriate technologies for local conditions and, finally, placing the water sector at the top of the list of priorities in terms of budgets and policies.

2.18.2 Legal and institutional framework

2.18.2.1 Responsibilities

Responsibility for the water sector is associated with various entities, namely the Ministry of Energy and Water Development (MEWD), the Department of Water Affairs (DWA), the Ministry of Local Government and Housing (MLGH), the Department of Infrastructure and Support Services (DISS), the NWASCO, the Water and Sanitation Association of Zambia (WASAZA), the Environment Council of Zambia (ECZ), the municipalities and NGOs.

The MLGH is the main player in the WWS sector, considering the fact that the responsibility for the WWS provision has been entrusted to local authorities. However, other ministries also have important functions in terms of developing, managing and monitoring the sector, such as the Ministry of Finance and National Development, the Ministry of Environment and the Ministry of Health.

During the process of separating the water resources and the WWS sector, the MEWD was made responsible for establishing WWS policies. The MEWD delegated the management of water resources to the Department of Water Affairs, which it supervises. In this context, policy decision-making processes were separated from services supplied at a central and provincial level, in order to foster more effective regulation and to minimize political interference, thus providing a greater degree of autonomy and independence for the sector.

Through the DISS, the MLGH has responsibilities at the level of planning and mobilizing resources to build infrastructure. In 2003, the DISS established a mission unit for rural WWS systems (RWWSU) and also recently created a unit for WWS in peri-urban areas. In short, according to prevailing legislation, the DISS is responsible for planning the infrastructure, development and management of the WWS sector.

As a sector-specific regulatory agency for the water sector, NWASCO is responsible for regulating WWS supplies in Zambia. Apart from this, it advises the government of the Republic of Zambia about the development of policies, the establishment of norms and guidelines, licensing of WWS operators and, finally, monitoring operator performance. In theory, NWASCO's activities span the entire country but in practice regulated entities only include the WWS in urban areas. NWASCO also oversees the Devolution Trust Fund (DTF), a non-profit-making working structure that manages the donations of international agencies and government subsidies.

WASAZA was created in 1999. Its statutes list its main objectives as organizing a congress where different stakeholders can discuss diverse subjects concerning the water sector and promoting and disseminating best practices amongst stakeholders. The association is open to all professionals from the water sector and related activities, such as activities related to the environment, agriculture and health.

The municipalities, besides being responsible for supplying WWS of suitable quality in the areas under their jurisdiction, also have the power to create additional norms in the context of previous regulation issued by NWASCO. The creation of commercial companies to provide WWS, especially since 2000, does not transfer or diminish the responsibility of the municipalities with regard to the WWS sector.

The ECZ is a statutory body, created in 1990, which aims to protect the environment and control pollution, safeguarding the welfare and health of citizens and the environment. Its functions include regulating and coordinating environmental management, promoting awareness and ensuring environmental protection.

Finally, it is important to mention NGOs such as CARE International or the Japanese International Development Agency, which have contributed significantly towards developing the WWS in peri-urban and rural areas. Figure 2.56 shows the institutional framework of the water sector in Zambia (adapted from Kayaga & Franceys, 2008).

2.18.2.2 Regulation

The water sector in Zambia can benefit from the existence of a sector-specific regulator. NWASCO, an independent agency dedicated exclusively to the water sector, was established by law in 1997, but only effectively started its activities in 2000. Its regulatory scope includes ten commercial companies scattered throughout the country, purely private operators and local authorities. The regulatory tools used by NWASCO with regard to operators include licensing, tariff approval, the supervision of their activities and accounts and the introduction of benchmarking among the regulated operators.

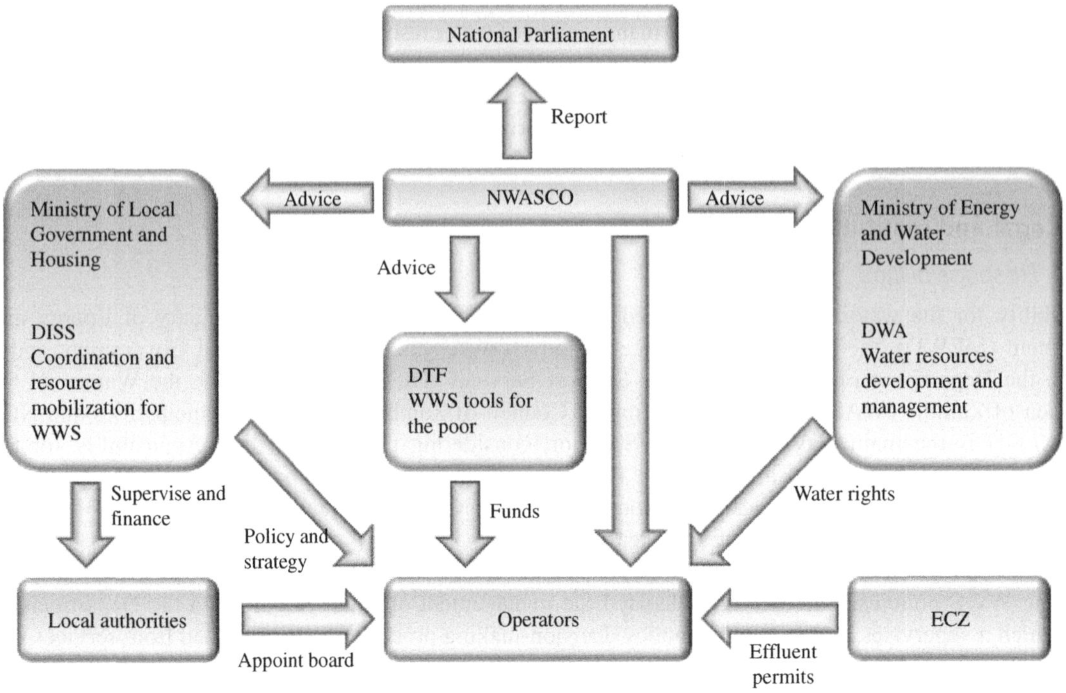

Figure 2.56 Institutional framework of the water sector in Zambia

With regard to customers, NWASCO compares and displays information obtained from operators (sunshine regulation) and the government as well as to resolve disputes between customers and operators. NWASCO presents an annual report of its activities to the MEWD. Figure 2.57 shows the information obtained from the benchmarking exercise for commercial operators (NWASCO, 2008).

	UFW[%]	Trend	Water quality [% compliance]	Trend	Metering ratio [%]	Trend	Water service coverage [%]	Trend	Wastewater service coverage [%]	Trend	Hours of supply (h)	Trend	Staff (no./1000 connections)	Trend	Collection efficiency [%]	Trend	O&M cost coverage by collection [%]	Trend	●	○	●
NWSC	35	↑	70	↑	43	↑	85	↑	69	↑	20	↑	8	↑	80	↑	103	↑	4	5	0
LWSC	51	↑	80	↑	45	↑	64	↑	9	↑	15	↑	13	↑	83	↑	102	↑	1	4	4
KWSC	58	↑	93	↑	11	↑	92	↑	67	↑	15	↑	9	↑	85	↑	114	↑	4	3	2
SWSC	43	↑	94	↑	76	↑	83	↑	23	↑	14	↑	10	↑	102	↑	93	↑	4	3	2
LGWSC	61	...	71	...	1	...	40	...	17	...	15	...	18	...	64	...	62	...	1	1	7
MWSC	56	↑	8	↑	22	↑	91	↑	86	↑	17	↑	8	↑	64	↑	94	↑	4	0	5
WWSC	47	↑	13	↑	15	↑	47	↑	24	↑	8	↑	13	↑	108	↑	86	↑	2	1	6
NWWSC	36	↑	69	↑	100	↑	60	↑	4	↑	20	↑	12	↑	85	↑	90	↑	4	2	3
CHWSC	54	↑	10	↑	3	↑	47	↑	10	↑	9	↑	13	↑	65	↑	69	↑	0	0	9
CWSC	31	↑	58	↑	97	↑	59	↑	33	↑	24	↑	12	↑	121	↑	114	↑	4	3	2
Average	47 (w)	↑	57 (s)	↑	39 (w)	↑	70 (w)	↑	34 (w)	↑	15 (w)	↑	••		84 (w)	↑	102 (w)	↑			

● Worse than the relevant average and benchmark not achieved
○ Better than the relevant average but benchmark not achieved
● At least "acceptable" benchmark achieved

(w) weighted average
(s) simple average

Figure 2.57 Benchmarking sheet presented in NWASCO's annual report

In addition, NWASCO has developed a ranking amongst operators, keeping in mind the performance indicators results weighting according to their importance. Table 2.103 shows the indicators and the weighting adopted (NWASCO, 2008).

The regulator also established target values, considering the scenario and the operational environment. Thus, Table 2.104 and Figure 2.58 show the benchmarking scheme used in the context of water losses (NWASCO, 2008).

Finally, the regulatory agency issues specific recommendations for each commercial operator, trying to improve their performance and, in this sense, to contribute towards the general development of the water sector. Table 2.105 shows the set of performance indicators calculated for the past four years per operator, weighted according to the population.

Table 2.103 Performance indicators used by NWASCO and their weighting

Indicator	Weighting (points)
Water quality	20
Collection efficiency	20
Metering ratio	15
Hours of supply	15
Coverage of operating and maintenance costs	15
Water losses	10
Staff per 1,000 customers	10
Regulators' perception	10
Water service coverage	5
Wastewater service coverage	5

Table 2.104 Target values for the water losses indicator

Benchmark for the water losses indicator (%)	
Good	<20
Acceptable	20–25
Unacceptable	>25

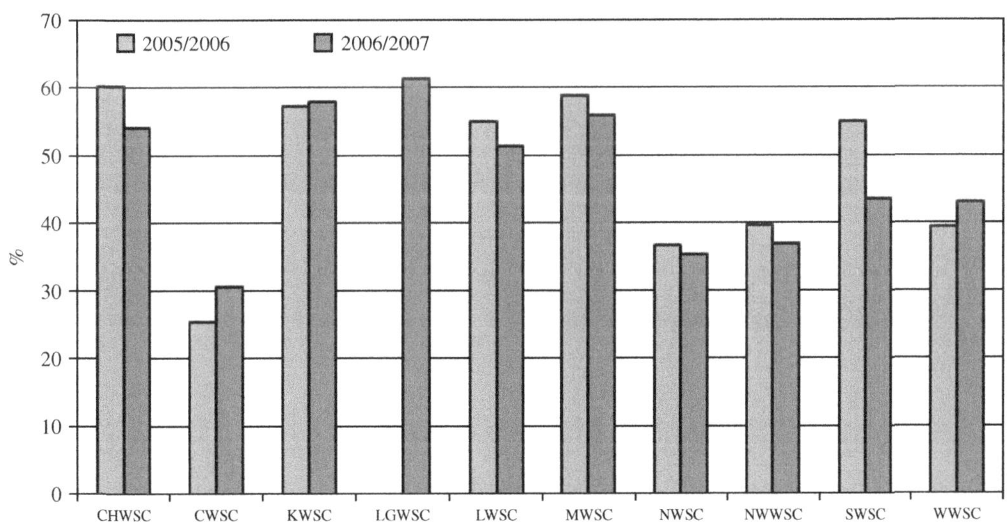

Figure 2.58 Water losses indicator per operator

The performance indicators shown reflect different analyses, especially for 2005/2006 and 2006/2007, owing to the creation of another commercial operator, the Lukanga Water Service Company (LGWSC).

When an operator permanently shows poor performance results, NWASCO places it under Special Regulatory Supervision (SRE). During this period, the operator is compelled to present a monthly report to NWASCO with the indicators referred to and the decisions that have been made, which, in their turn are closely monitored until the moment the regulator is sure that the operators have made real progress.

The budget of NWASCO, including the salaries of its staff (16 employees), is mainly funded by local authorities and operators, by means of licensing taxes, which represent 89% of its total budget. Seven percent of its expenses are funded by the national budget, 1% are obtained from donations and the remaining 3% are obtained from other diverse sources. In this regard, in 2006, the operating costs of NWASCO were about 577,000 euros.

In the scope of NWASCO's primary function, the protection of customer interests, it created the so-called Water Watch Groups (WWGs) in 2001. The WWGs are groups of volunteers responsible for safeguarding customer rights in the water sector and for giving out available information regarding the services provided. NWASCO encourages the WWGs to carry out these functions. WWGs have been established in Lusaka, Kitwe, Chingola and Copper Belt, but a lack of financial resources has hindered national coverage.

Table 2.105 Summary of the indicators used by NWASCO in the last four years

Indicator	2003/2004	2004/2005	2005/2006	2006/2007
Commercial operator				
Water service coverage (%)	54	58	73	71
Wastewater service coverage (%)	32	32	32	34
Household connections (%)	–	86	86	92
Water production (millions of m^3)	–	300.5	303.5	316.3
Water losses (%)	50	49	48	47
Metering (%)	32	32	39	39
Water quality (%)	–	–	77.9	56.6
Hours of supply*	–	–	14.3	16.0
Collection efficiency (%)	68	74	77	84
Staff per 1,000 customers	–	–	9.44	11.54
Average expenditure on staff per month (€)	339.7	420.1	489.5	495.0
Operating costs	–	–	–	–
Cost structure	–	–	–	–
Average tariff	–	–	–	–
Operation/maintenance costs coverage (%)	65	75	77	102
Employee qualifications	–	–	–	–
Local authorities				
Coverage (%)	–	28	24	23
Hours of supply	–	7.0	8.7	5.2
Collection efficiency (%)	–	37	50	33
Private systems				
Coverage*	–	–	99.7	99.8
Hours of supply*	–	–	16.0	17.6
Water quality*	–	–	84.1**	75.4

* Weighted according to the population served; ** One operator did not respond

The WWGs receive complaints and suggestions from customers and compile this information for the respective operator or for NWASCO, in case the commercial operator is not complying with obligations. Apart from this, it also informs customers about the regulations that have been issued, by publicizing them. NWASCO delegated powers to the WWGs to act as the first instance for resolving customer complaints.

In this sense, the WWGs have played a very important and positive role in the water sector developing in Zambia, since the commercial operators, with whom the WWGs intervene, have focused on improving the services they provide to customers. NWASCO also includes their recommendations while assessing the operators' performance. Figure 2.59 shows the procedures of the WWGs while handling complaints (Lwenje, 2007).

In terms of economic regulation, NWASCO has adopted the method of price cap regulation, with a regulatory period of three years, based on diverse objectives such as evaluating the impact of prevailing water tariffs (e.g. in 2007 tariffs were not increased for poorer inhabitants), customer knowledge regarding the cost of supplying water, to encourage customer awareness about the issue of water conservation and verifying the quantity of water consumed by different categories of customers.

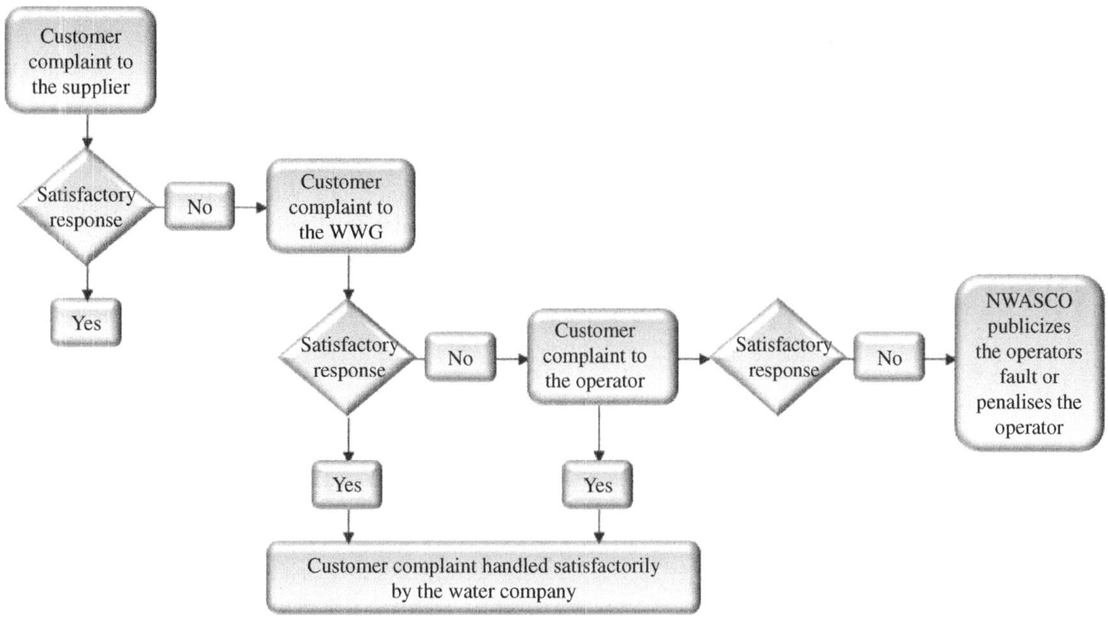

Figure 2.59 WWG regulatory procedure for handling complaints

2.18.3 Market structure

2.18.3.1 Ownership and management of services

The WWS sector in Zambia is characterized by being almost totally provided by the public sector. Despite some private sector participation in the provision of water supplies, this presence is quite limited and only takes place in a few areas.

2.18.3.2 Players in the sector

In Zambia, WWS are available in 87 cities scattered throughout the country. These services are provided by ten commercial operators, who serve approximately 92% of the urban population, and seven private operators, who serve 1% of the urban population. The rest of the population is supplied by local authorities, particularly by 6 licensed operators and 16 operators with provisional licences which supply 7% of the urban population. The remaining customers, namely the poor sections of society, rely mainly on wells and public taps, while the majority of the wealthiest customers have private and individual systems.

Especially since 2000, the government of the Republic of Zambia has encouraged the establishment of commercial operators to provide WWS in urban areas. The institution of these operators is aimed at the corporatization and commercialization of the WWS. All the provinces have opted to create commercial operators, with the exception of the province of Luapula. Commercial operators are established by transferring existing WWS systems to a registered company, owned by the local authorities, and there is no change in the area of jurisdiction. Usually, commercial operators are constituted without additional capital, which severely limits their operational sustainability. In this context, the Lusaka Water Service Company (LWSC) stands out on account of its scale of operations, providing WWS to about 1.6 million inhabitants in the Zambian capital. In areas where there are no commercial operators, WWS remain in the hands of the local authorities. Some peri-urban areas are connected to water networks, but the majority is served by public taps or kiosks.

As referred to above, in large cities water services are mainly provided by commercial operators that have obtained a licence from NWASCO. However, some of these areas are also served by Water Trusts, particularly in the city of Lusaka, as described below. In the Zambian capital, a large percentage of the peri-urban population is supplied by residents' associations, which do not have licences and are not subject to regulation.

In Zambia, the Water Trusts are small network operators created on the initiative of donors (e.g. the Chipata or Kanyama Water Trust), in coordination with local communities, in peri-urban areas. Presently, these entities provide water services without a legal basis, even though they collectively serve almost as many inhabitants as the public operators.

37% of the population of Lusaka is supplied by Water Trusts and 43% of the population is supplied by the LWSC, while the remaining 20% of inhabitants do not have access to the water supply. A recent study regarding social regulation, developed within the scope of the Building Partnerships for Development (BPD) initiative, revealed that the Water Trusts would like to see their service regulated, since they suppose this would 'reinforce transparency and trust'. The solution that is presently being considered by NWASCO is to sign protocols between the Trusts and the LWSC. The preparation of these agreements, supervised by the regulatory agency, is liable to reinforce impartiality and protect each of the concerned parties. This is quite important, since the Water Trusts insist on remaining independent and do not favour proposals to incorporate them into the main supplier or to simply become distributors of the LWSC. Figure 2.60 shows the organizational framework of the Water Trusts (adapted from Kayaga & Kadimba-Mwanamwambwa, 2006).

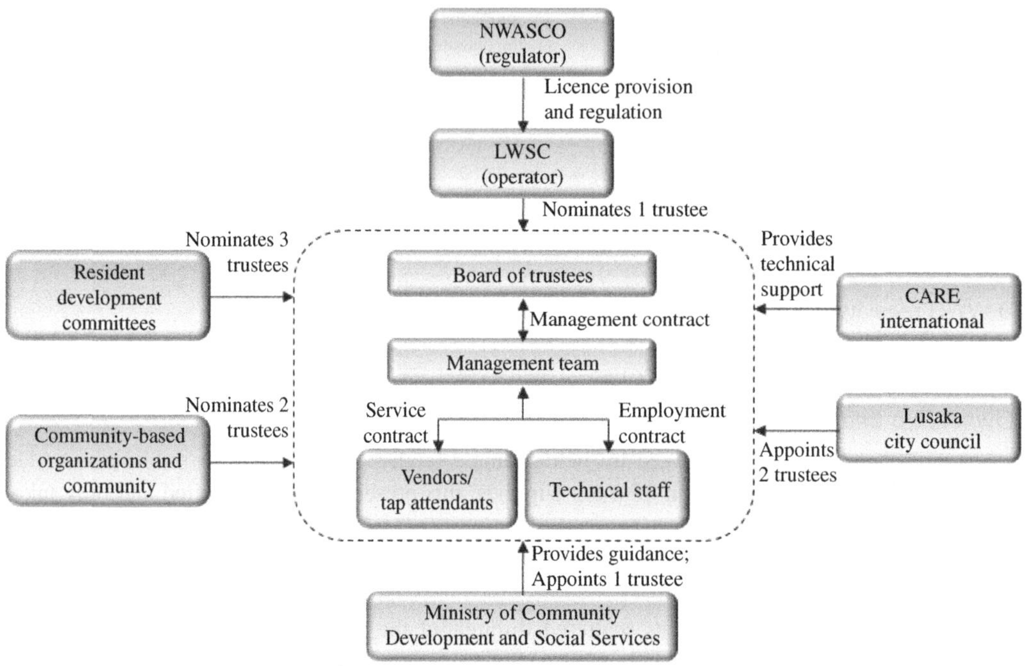

Figure 2.60 Organizational framework of the Water Trust model

The private sector has a fairly limited participation in the provision of the WWS. Zambia has seven private operators, concentrated in isolated areas, namely in industrial zones and other cooperatives. An example of the participation of the private sector is the management contract with SAUR for the mining operations in the city of Copper Belt, likewise providing WWS to the employees of the mining company.

Private operators have systems developed by companies that are in completely different lines of business (e.g. sugar and the production of electricity) and they provide WWS usually free of charge, to their employees who live in the company's units. These systems are licensed by NWASCO according to prevailing legislation.

The development of the wastewater service has lagged far behind the water service. Only a limited percentage of the population has access to this service.

2.18.3.3 Integration of the sector and services provided

The WWS are generally characterized by both horizontal and vertical integration. In this regard, these services are generally supplied by the same operator, which can be commercial operators, local authorities or even private operators.

Commercial operators are specifically dedicated to this area and they do not provide other services. The local authorities provide directly the WWS to customers through their water and wastewater departments.

The private operators consider the WWS merely as a secondary activity, since they are limited to only supplying their own employees and their main activities take place in other sectors.

2.18.3.4 Numbers for the sector

About 316 million m^3 of water are produced every year in Zambia for public consumption. In this context, the per capita consumption for the population that has access to the water service is 110 litres per inhabitant per day, including water losses, which, in Zambia, represent 47% of production.

There are very distinct data regarding WWS coverage, depending on the different sources (especially due to the different criteria adopted), but approximately 68% of the urban population (and only 45% of the peri-urban population) and 40% of the rural population have access to drinking water. The wastewater service coverage is about 34% and all the wastewater collected is treated. Table 2.106 shows the evolution of the water service coverage in urban areas during the last two years.

Table 2.106 Annual evolution of the coverage of the urban water service

	2005/06	2006/07
Urban population	4,903,529	5,046,420
Urban population served	3,270,745	3,428,572
Level of the urban water service coverage	67%	68%

NWASCO and the Central Statistical Office of Zambia estimate that 45% of the total population lives in urban areas. Of the urban population, 80% (3.9 million) live in low-cost and peri-urban areas. It has been estimated that 80% of the Zambian population lives on less than 0.68 euros per day.

The water sector in Zambia obtained annual revenues of about 36.7 million euros in 2006. Although the wastewater segment entails higher costs, it only has revenues corresponding to 20% of that of the water service, i.e. approximately 7.5 million euros. In the same year, only about 552,000 euros were invested in the WWS sector.

2.18.4 Rules of the game

2.18.4.1 Tariff setting

Government policies have been aimed at encouraging a full cost recovery by means of the tariffs charged to customers. Commercial operators have been pursuing this objective, keeping in mind the following principles:

- A social awareness, i.e. tariffs should be 'reasonable' for low-income customers;
- Efficiency of use, where tariff systems must reflect consumption, encouraging the measurement of water (which is still not widespread);
- Conserving resources, whereby tariffs should be higher for levels of consumption that are higher than normal needs;
- Simplicity and transparency, i.e. the tariffs should be established in a transparent and participatory way.

The tariff structures vary considerably according to the different kinds of management and the respective regions. However, the most usual tariff system comprises a fixed component, for the small percentage of inhabitants who have a meter, and a variable component that increases with consumption or remains constant. Owing to the fact that only a small part of the population has meters, the majority of the operators have adopted a fixed tariff, especially in peri-urban areas. Table 2.107 shows the WWS tariff system for Lusaka.

Table 2.107 Tariff system of the LWSC for 2007

Water service	€/m^3
Variable component	
0–6 m^3	0.20
6–30 m^3	0.22
>30 m^3	0.26

NWASCO recently implemented a programme that penalized commercial operators for precarious levels of performance. Under this programme, commercial operators are compelled to reduce the tariffs charged until they achieve an acceptable quality of performance.

Although the process of reforms is currently underway, trying to recover costs and to ensure an efficient service, operators are still not very efficient. Commercial operators in diverse areas have ratios regarding the coverage of operating costs that are less than 1, indicating that WWS revenues continue to be lower than operating costs. It is hoped that the operational indicators will improve drastically by better billing and more efficient charges. Owing to the low level of metering, the volume of water that is not billed can only be estimated and thus it carries the risk of being undervalued.

2.18.4.2 Quality of service

NWASCO establishes directly and indirectly performance objectives for public operators, by introducing virtual competition, calculating performance indicators and using benchmarking as shown in Table 2.107 above.

These performance contracts also include other mechanisms for incentives, in order to promote the development of the services. One of these corresponds to the minimum level of service necessary where, along with the indicators described above for coverage in the area served, all operators are compelled to submit a service charter guaranteeing a specified level of service to NWASCO, within two months of receiving their operating licences. The agreement specifies the actions to be taken within a given time frame (ideally of three years).

Another mechanism is related to the regime of defining tariffs, which is currently being revised. Tariff setting intends to encourage the performance of operators, to increase efficiency and improve the relationship between costs and revenues.

Finally, a third mechanism corresponds to incentives aimed at performance. In this scope, this regime is aimed at increasing the operational performance (mainly the management and development of human resources) of the commercial operators. Examples of some indicators used include reducing expenditure on administrative staff, which can not exceed 25% of total turnover, and the productivity of staff, measured in terms of average revenue obtained per employee.

Apart from the system of performance indicators, NWASCO has introduced various awards aimed at recognizing the operators' efforts, rewarding and encouraging them to maintain a high level of performance. Prevailing legislation also empowers NWASCO to apply penalties to operators who do not uphold the minimum requirements owing to either non-compliance or negligence.

As a result of the recent reforms and the supervision of NWASCO, special attention has been paid to expanding water services to the poor. In this context, NWASCO established the DTF. The sole mission of the DTF is to fund an expansion of the water services in the more underdeveloped areas. Although access to the water supply is considered to be quite high, it has been estimated that after considering the conditions of sustainability and reliability, the level of access corresponds to only 60% of the population.

2.18.4.3 Public service obligations

In Zambia, the creation of a vertical regulatory agency led to the establishment of principles that were not explicitly present in the water sector until then, namely universality, accessibility, the quality of service and equity. Apart from these principles, the legal framework of the National Water Policy also stipulated that peri-urban areas should be treated in the same way as urban areas, thus safeguarding the poorest sections of society against unequal levels of service.

The DTF, created by NWASCO, is an independent financial instrument responsible for funding, which intends to ensure water provision in peri-urban areas. This fund has played a crucial role, enabling subsidization to fund water supply for low-income customers who would not be able to pay for this supply otherwise.

The tariff systems also consider customers' affordability and there are reduced tariffs for low-income customers.

2.18.5 Performance

The reforms that have been implemented in the water sector in Zambia have improved the WWS performance, especially for the poor. The regulator (NWASCO) has played an extremely important role in this process, leading the development of the water sector. Another very positive aspect is the existence of the WWGs, which, even though they only cover a limited part of the water sector, have contributed towards improving the quality of service and the compliance with public service obligations, allowing participation and increasing the accountability of all the players in the water sector. The method of economic regulation based on price caps, with incentives for performance, is another significant aspect.

However, the WWS sector faces various challenges. These include the need to improve the level of coverage of the water service and, mainly, its quality, allowing a 24-hour service with a suitable pressure and quality. To this end, massive funds for investments are required, which the sector does not have. The situation is even more serious with regard to the wastewater service. It is necessary to increase and improve the tariff systems, making them more suitable for the country's situation. A greater efficiency in charges, simultaneously along with a reduction in operating and maintenance costs, by means of greater productive efficiency, would achieve greater revenues for the sector. For example, the volume of water that is not billed is very high in Zambia. The gradual and sustained introduction of the private sector in the country could also attract more resources to the sector. A better institutional and operational management of water resources, especially concerning groundwater and environmental sustainability, could lead to significant savings.

Part 3
Worldwide water and wastewater services regulation overview

3.1 INTRODUCTION

This investigation of the worldwide regulation of the water sector adopted the following methodology. First, a research was carried out to find the existence of WWS regulators all over the world, the entities responsible for the regulatory activities (prices and tariffs, quality of service and public service obligations), the main stakeholders in the water sector (operators, associations, etc.), and all the relevant information on the entities and the regulatory processes at work. The contact information for all the regulators and the main bodies with regulatory functions are shown in Appendix 3.

Secondly, a survey was created worldwide and sent to all the regulators and main entities with WWS regulatory functions. This survey was composed of six chapters in the main body, namely the general WWS aspects, the regulatory environment, the general aspects of the regulatory agency, the organizational aspects of the regulatory agency, the economic regulation and public service obligations, and the quality of service regulation. Contextual information and the opportunity to make close comments on the strong and weak points of the regulatory system were included with the survey as well. Each of the 46 questions was multiple choice, and often included the possibility of an open-ended response. The survey was sent to 279 entities, with only 63 responses received. One of the problems affecting the number of responses was the fact that the survey was only sent out in English. Certain regulators in Africa (francophone origin), in Europe (above all in Eastern Europe and in the Eurasian region) and America (Hispanic origin) have encountered language difficulties. The response time was also much longer than originally imagined. The two months that were first allotted quickly became nearly six months. Appendix 1 contains the survey sent to each country's regulators and different WWS stakeholders. Appendix 2 presents the results and a survey summary in table format.

It should also be pointed out that other countries, as shown in the previous section, were included in the detailed analysis, although they did not respond to the survey. Because they are interesting case studies or representative of a certain region, technology or other specific aspect.

In the research which was carried with respect to January 2008, there were 136 water regulatory agencies detected worldwide. These regulatory agencies are spread between 57 countries, of which 12 are located in Africa, 5 in Asia, 16 in Europe (2 in Eurasia), 2 in Oceania and 22 in America. In four countries in the Americas, WWS regulation did not cover the entire territory, namely in Canada, the USA, Brazil and Argentina. In fact, little more than 22% of countries in the world have WWS regulators. Considering the population, of those covered by regulated WWS, the percentage falls to 20%. If China and India are excluded, the regulated population climbs to 37% of the global population. Of the 136 regulatory agencies, 106 cover multiple sectors, involving other SGEI beyond drinking water and wastewater, and 30 operate solely within the sector. When there are regulators covering multiple sectors, they normally include the energy sectors (electricity and gas), although they can also include transport services, telecommunications, postal services, urban waste and others.

Concerning the surveys, which will be dealt with in great detail in the following subsections, there are, as mentioned, 63 responses corresponding to 39 countries. Of the responses obtained, 53 were from regulators corresponding to 33 different countries. The remaining responses were given by operator associations or governmental bodies working in the water sector. Of the 53 regulators, 31 were multisector regulatory agencies and 22 were sector-specific regulators. Besides drinking water and wastewater, other activities included in the regulatory scope were electricity, gas, solid waste, telecommunications, heating, lighting, trains, ports, highways, oil and gaming.

ISBN: 9781843393412. Published by IWA Publishing, London, UK.

Of the 53 regulators, 28 are national/federal, 18 are state/regional and 7 are municipal. Of these 53 regulators, only 8 are not independent. In 39 of the cases, there is a board of directors in the regulatory agencies (collegial body). In the remaining cases, regulatory command is held by one person. 32% of the regulators are financed by the government, while the remaining are financed by the operators. In these latter cases, only 30% are indexed to the tariff. 44 regulators have economic regulatory functions, while 9 only regulate the quality of service because there is either contractual regulation (for example, concession contracts) or because the price and tariff controls are handled by the government. 79% of the regulators perform functions related to quality of service supervision and 72% apply benchmarking to control the quality of service or to the establishment of tariffs and prices. 72% of regulators perform sanctioning functions, although only 40% can produce legislation. In 49% of the cases studied, there are mechanisms for appealing decisions of the regulatory bodies. Regulatory accountability is guaranteed in many ways, with the publication of decisions as the most important means, occurring 66% of the time, followed by transparency and participation, with 62%, followed by a Parliamentary response, with 51%. Of the regulatory agencies with economic regulatory functions, 22 (50%) have instituted methods based on performance incentive regulation, almost always price cap regulation. In more than half the cases, the regulatory period is 5 years, with the remaining being lower.

The following illustrations, Figure 3.1, Figure 3.2, Figure 3.3, Figure 3.4, Figure 3.5 and Figure 3.6 show countries with regulators in blue and those that responded to the survey in red, organized by continent.

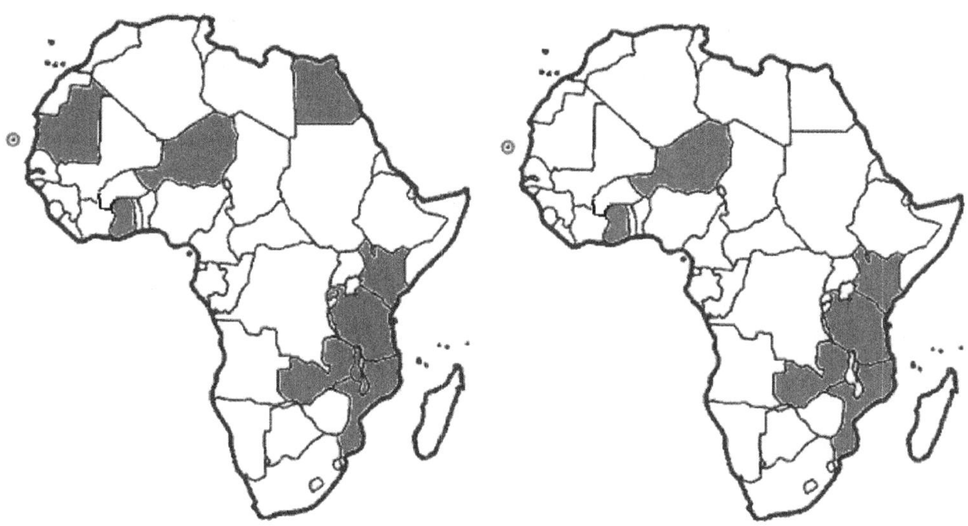

Figure 3.1 Countries with regulators and survey responses obtained from Africa

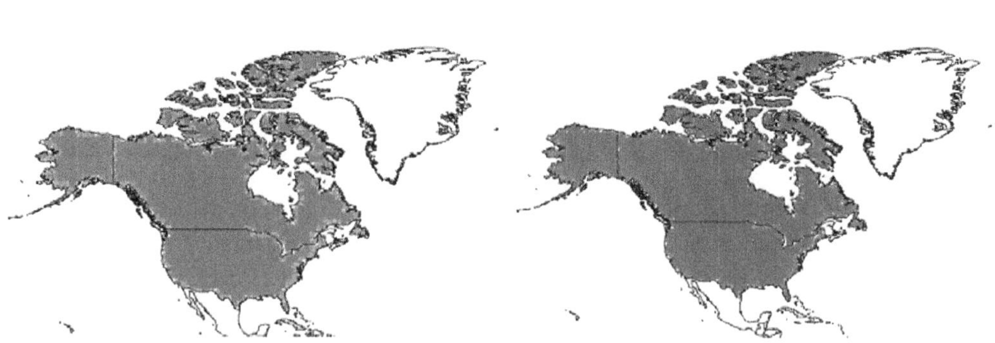

Figure 3.2 Countries with regulators and survey responses obtained from North America

3.2 PUBLIC SERVICE OBLIGATIONS

3.2.1 General considerations

Public service obligations are the backbone of the WWS. In many countries around the world, it is only through the establishment of universal service that such an important step towards assuring the survival and quality of life of the populations can be taken. The achievement of the World Health Organization's (WHO) Millennium Development Goals will be difficult. Even though the poorest populations in low-income countries pay for water, often more than what is paid by richer populations, they continue to be deprived of this essential service. Wastewater service is an even more serious situation. The negative externalities caused by its absence are very important. There is a problem of bad governance in the water sector, particularly due to deficiencies in, or a complete lack of, regulation. State failures, along with WWS market failures and other characteristics, jeopardize the provision of a service with the appropriate quality. In the case studies that are presented, especially in Zambia and Mozambique, even though they are developing countries with countless problems, significant progress has been made in the quality of WWS, thanks to the instituted regulation, which is supported by their governments. In the poorest countries, the circumstances where the infrastructures (networks) do not exist and where the customers are unable to pay for the service, including the connections, must be separately considered.

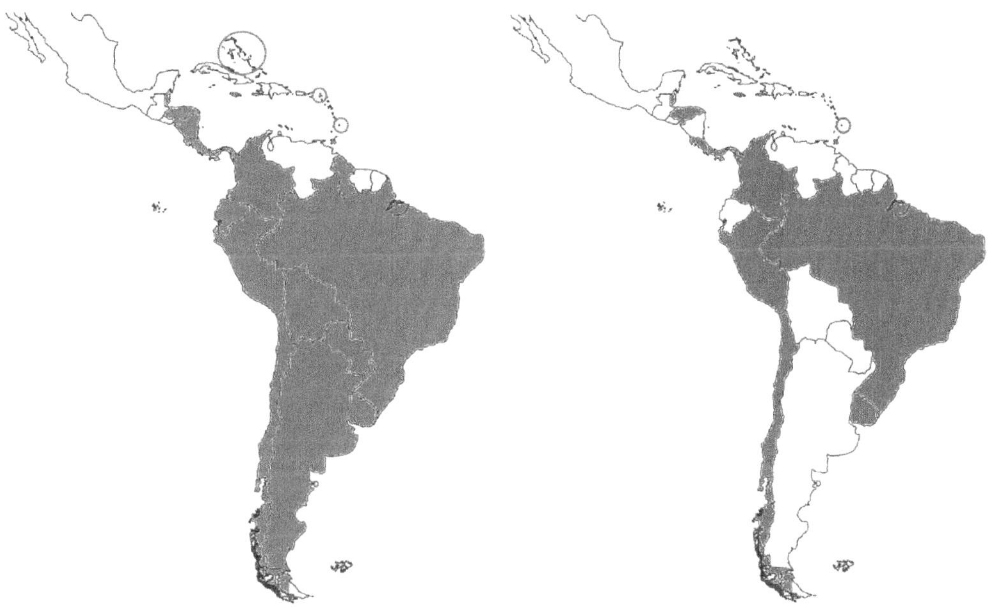

Figure 3.3 Countries with regulators and survey responses obtained from South America

Figure 3.4 Countries with regulators and survey responses obtained from Asia

In developing countries (and even in developed ones), the problems related to public service obligations are of the highest importance. Strengthening the economic and social cohesion of the populations justifies the definition by law of the public service obligations and the supervision of their compliance by regulators or other competent bodies. In these countries, there is also a part of the population that, even though the infrastructure gets to their door, cannot pay the costs of the WWS. In addition, the populations and customers demand other requirements, which include uninterrupted service, high quality, alternative tariff schemes and the protection of their interests, even at the individual level.

In the investigation that was carried out, and particularly in the case studies analysed, a group of disparate situations was found, depending on the degree/level of the countries' development and their respective individual situations.

The following points show some of the strategies and practices used, above all the most important with respect to the different public service obligations. There are also summary tables and figures which lay out the public service obligations of the countries investigated or adopted as case studies in this document.

Figure 3.5 Countries with regulators and survey responses obtained from Oceania

Figure 3.6 Countries with regulators and survey responses obtained from Europe

3.2.2 Universal service

Each country offers universal service in different ways. Even though all the countries have included this principle into their legislation in one way or another, its implementation varies greatly. Several countries in Europe, such as Italy, France and the Netherlands, assume that it is compulsory to provide service to all customers and, therefore, the principle is not enacted by law. Others, such as Greece and Portugal, despite having the obligation explicitly defined by law, do not have a definition as to how it should be obtained, thereby giving the operators the freedom to make it happen, namely through cross-subsidization among services (water supply and wastewater treatment) and customers, either relative to their type (domestic or non-domestic) or relative to the consumption highlighted (customers with higher consume subsidize the customers with lower consume or vice versa). The institution of social tariffs is usual where there is a block (normally priced towards lower consumption) where the corresponding tariff is greatly reduced. In Table 3.1 and Table 3.2, this situation is shown for the WWS in Athens, Greece (EYDAP), and for the WWS in Rio de Janeiro, Brazil (CEDAE), in the year 2007, respectively.

It should be pointed out that, in the poorest countries, the cross-subsidization hypothesis is not sufficient, as there is a need for subsidies of the government or donors in order to establish universal service.

In other countries or states/regions, there is a social fund for financing the most disadvantaged customers and operators to which all the customers contribute. This practice is carried out in a variety of countries, such as Belgium (in two of the three regions) for the customers, and in France for the operators. This fund, which can be centrally managed or handled by the operator itself, allows for the customers in need to be integrated into the system or for a reduction in the tariff for certain operators. The merit of this system is closely linked to the quality of the social safety net, which is to say that the poor people or operators are the ones who benefit from these special social privileges.

Table 3.1 Cross-subsidization in the WWS in Athens, Greece (EYDAP)

Water	Value (€/m³)
Domestic	
0–5 m³	0.40
5–20 m³	0.63
20–27 m³	1.80
27–35 m³	2.52
Greater than 35 m³	3.14
Industrial	
Up to 1000 m³	0.81
Greater than 1000 m³	0.96
Wastewater	60% of the water bill

Table 3.2 Cross-subsidization in the WWS in Rio de Janeiro, Brazil (CEDAE)

	Water (€/m³)	Water and wastewater (€/m³)
Domestic		
0–15 m³	0.62	0.64
16–30 m³	1.36	1.41
31–45 m³	1.85	1.93
46–60 m³	3.70	3.86
Greater than 60 m³	4.95	5.15
Commercial		
0–20 m³	2.10	2.19
21–30 m³	3.70	3.85
Greater than 30 m³	3.95	4.12
Industrial		
0–20 m³	3.21	3.34
21–30 m³	3.38	3.51
Greater than 30 m³	3.95	4.11

In the region of Wallonia, Belgium, for example, the water bill includes 0.0125 €/m³ to be used in constructing a social fund for the region, called '*Fonds Social Wallon de l'Eau de la Région Wallonne de Belgique*'. Its main objective is the creation of WWS access conditions for the customers characterized by difficulties in paying the WWS bill and by the impossibility of assuming the increasing cost of drinking water. This WWS payment difficulty or impossibility in extreme cases is evaluated by the municipal (*communes*) social service workers. Roughly 0.5% of the population is guaranteed by this Social Fund.

In France, the 2004 Financial Rectification Law No. 1485 removed the *Fonds National de Développement des Adductions d'Eau* (FNDAE) tax, which corresponded to a payment of 21.34 € for every 1000 m³ of water consumed.

The Water Agencies are currently responsible for financing the rural WWS areas (which were previously assured by the State through revenue obtained by the FNDAE). The suppression of the FNDAE, in order to benefit from an increase in tariffs, had as its objective the transfer of revenues charged by the State for the Water Agencies, without an increase in the cost of service for the customer. In fact, the tariff increase due to the Agencies is compensated by the elimination of the FNDAE tax. It should be noted that this item disappeared from the water bill. There is a minimum limit from which the operators are exempted from this payment (for example, in the Rhône-Méditerranée et Corse basins, the limit was 30,000 m^3 per year, in 2005). Every year, the operators are informed by the Water Agency of the anticipated amount of tax for the following year. The rural areas (according to the current legislation) that satisfy the economic criteria defined by the Agency can benefit from financial help for WWS rehabilitation investments. These funds are established within the limits stipulated in an annual agreement set by the Agency, and can reach 30% subsidization, with a maximum of 50%, considering all the assistance.

Another universal service possibility that takes place in certain countries, such as Chile, Zambia, South Africa and Belgium's Flanders region, consists of a direct subsidy to the customers in the volume of water. Below is an overview of the situation in Chile.

Among the countless successes that characterize Chile's water sector, the universality of the WWS is especially relevant. This accomplishment, considering the country's specific geographic composition, is even more noteworthy given the state of affairs in the surrounding countries (and the entire Latin American region). Among all the factors that sustain WWS universality, subsidization of the water sector, especially in the rural areas, stands out as a dominant factor, being considered a benchmark for other countries with similar conditions. This subsidy arrangement, operational since the 1990s, was brought forth to deal with the adverse social effects of the substantial increase in tariffs (characterized by privatization and by the need for large investments). The subsidization program is managed by the Social Planning Ministry, in partnership with local administration, which determines the amount of subsidies to give to each region. Afterwards, the handling of these subsidies is under the responsibility of the local administration.

The subsidy arrangement gives access to WWS at reduced prices to domestic customers, within certain limits, which are legally set between 20 and 85% of the bill. The governmental body sets a consumption cap that the subsidy will cover (currently 15 m^3 per month in all regions) and the percentage of the discount for this first step on the scale (which varies between regions). The number of subsidies and corresponding amounts are determined annually and the associated expenses are included in the State budget. The subsidy is entirely financed by the taxpayers. The SISS (regulator), responsible for setting the tariffs, does not have any responsibility in this area. To be eligible for this subsidy, the customers should request it to the municipality. This appliance is presented considering the socio-economic needs of the customers according to the criteria established by the Communal Social Assistance Committees (CAS), which is known as the CAS score. This determination is based on information obtained during an interview at the customers' household. The interview consists of 50 questions, divided into 9 groups (general information, environmental conditions, size of the household, health conditions, family make-up, occupations and earnings, monetary subsidies, education and wealth). The resulting determination holds for two years. Once the subsidy is assigned to the customer, the operator is notified and, in the following month's bill, the customer will see their subsidy's discount. Afterwards, once per month, the operator will charge the municipality for the subsidies provided.

In Zambia and South Africa, the first 6 m^3 are given free to all the families. In the Flanders region in Belgium, the first 15 m^3 of water are made available to domestic customers without any cost. In extreme cases, such as in Ireland, water is given free of charge to all domestic customers.

The subsidies can also be assigned directly to the customers through, for example, fixed amounts per month which indirectly allow for the customers to fulfil their obligations with the WWS (for example, vouchers). This is not normally the most appropriate policy because the customers can use the obtained funds for other ends, thereby failing to benefit from this service and the positive externalities that it brings.

Another way of guaranteeing universal service is direct subsidization of the operators, allowing them to offer a below-cost tariff system. The capital costs are rarely recovered. This is the most common form of WWS subsidization in the world. Almost all countries, in one way or another, have one kind of WWS subsidization. In the analysed case studies, only Australia, the UK, the Netherlands and Sweden do not receive subsidies directly from the State.

Table 3.3 presents WWS universality approaches in the analysed cases and countries surveyed.

3.2.3 Accessibility and equality

Table 3.4 shows the coverage level of WWS for the countries surveyed and case studies between 2005 and 2007. For sewage, the value in parentheses is the coverage level of residential wastewater treatment.

In accordance with that shown in the previous chapter, there is a serious concern among the different countries to provide their populations with access to WWS in appropriate conditions. The number of countries adopting cross-subsidization policies is rather high. This fact, which shapes the WWS equality, is shown in particular by the fact that consumption at the lowest scale is provided without cost or with a highly reduced tariff in comparison with the others, which is also known as a social tariff. In fact, nearly 60% of the countries analysed have social tariffs. There is also a growing concern with equity in developed countries, which seek to specifically avoid the occurrence of cross-subsidization. For example, Table 3.5 shows the tariff system for two Australian operators in 2006. Here it can be seen that water supply does not subsidize wastewater service, industrial customers do not subsidize domestic customers and the higher customers do not subsidize those who consume less. The goal for this situation, which is still scarce, will become dominant in the future, mainly in countries where it could most easily occur (for example, Europe and the USA). Otherwise, a wrong signal will continue to be sent to the market, instigating consumption and wastefulness.

Table 3.3 Universal service for the case studies and countries surveyed

Country	Universal Service	Country	Universal Service
Asia		Panama	Cross-subsidization
Armenia	Cross-subsidization and direct subsidy to the customers	Peru	Cross-subsidization
Philippines	Cross-subsidization	Trinidad and Tobago	Customer subsidies
Indonesia	Cross-subsidization and operator subsidies	Uruguay	Cross-subsidization
Singapore	Light cross-subsidization between only two blocks	**Europe**	
Africa		Belgium	Social fund, and monetary and in volume subsidies to customers
Cape Verde		Scotland	Discounts for special customers
Ghana	Cross-subsidization and operator subsidies	France	Cross-subsidization and social fund
Mozambique		Greece	Cross-subsidization
Niger	Cross-subsidization	Netherlands	Cross-subsidization among customers
Kenya	Cross-subsidization	E&W	Discounts for special customers and large families
Tanzania	Cross-subsidization	Ireland	Cross-subsidization (absence of tariffs for domestic customers)
Zambia		Northern Ireland	Does not exist
America		Italy	Cross-subsidization
Barbados	Cross-subsidization and operator subsidies	Kosovo	Customer subsidies
Belize	Operator subsidies	Lithuania	Operator subsidies
Brazil	Largely cross-subsidized	Portugal	Cross-subsidization and operator subsidies
Canada	Operator subsidies	Czech Republic	Cross-subsidization
Chile	Customer volume subsidization	Slovak Republic	Operator subsidies
Colombia	Cross-subsidization and operator subsidies	Romania	Cross-subsidization
Costa Rica	Cross-subsidization	Sweden	Does not exist
USA	Cross-subsidization	**Oceania**	
Honduras	Cross-subsidization and operator subsidies	Australia	Subsidization among customers
Jamaica	Cross-subsidization	Papua New Guinea	Cross-subsidization
Mexico	Cross-subsidization		

Table 3.4 Coverage level for case studies and countries surveyed (%)

Country	Water	Wastewater	Country	Water	Wastewater
Asia			Panama	86	51 (39)
Armenia	90	70 (0)	Peru	83.8	76 (28.1)
Philippines	85	72 (10)	Trinidad and Tobago	92	21 (21)
Indonesia	18	3 (2.3)	Uruguay	92	48 (38)
Singapore	100	100 (100)	**Europe**		
Africa			Belgium	99	82 (50)
Cape Verde	85	16 (15)	Scotland	100	100 (100)
Ghana	57	7 (0)	France	99	80 (80)
Mozambique	26	4.6 (3)	Greece	99	96 (85)
Niger	63	20 (10)	Netherlands	100	98 (97)
Kenya	55	25 (20)	E&W	99	96 (93)
Tanzania	60	3 (3)	Ireland	99.9	93 (92)
Zambia	68	34 (34)	Northern Ireland	99	83 (83)
America			Italy	96	84 (75)
Barbados	99	90 (60)	Kosovo	74	55 (0)
Belize	90	40 (40)	Lithuania	66	60 (58)
Brazil	81	42 (30)	Portugal	91.1	76 (67)
Canada	85	85 (80)	Czech Republic	92.4	80 (75)
Chile	99	95 (82)	Slovak Republic	86.3	58 (55)
Colombia	88	74 (25)	Romania	70	50 (30)
Costa Rica	98	21 (4.5)	Sweden	99	99 (95)
USA	99	90 (90)	**Oceania**		
Honduras	75	69 (36)	Australia	99	90 (90)
Jamaica	73	30 (17)	Papua New Guinea	30	22 (22)
Mexico	89	86 (36)			

Table 3.5 Tariff system equality in Australia

	Gosford City Council		Power and Water Corporation	
Tariff system	Water	Sewage	Water	Sewage
Fixed part (€)	46.1	207.7	58.7	183.5
Variable part (€/m^3)	0.53	–	0.39	–

3.2.4 Continuity

WWS continuity makes up another of the fundamental public service obligations. This principle, which is related to the principle of universal service, should be analysed from two different perspectives. On one hand, the WWS should be available 24 hours a day, 365 days a year. On the other hand, this circumstance, which is usual in developed countries, is still a rather ambitious goal for developing countries. Of the countries analysed and surveyed, nearly 50% provide intermittent water service, even in urban areas. Table 3.6 shows the WWS continuity in the case studies and countries surveyed.

Intermittent water supply, in addition to the obvious quality of service problems that it causes, has public health consequences and makes the system hydraulically inefficient and ineffective, generating high losses (e.g. water losses).

Conversely, the situation is different in developed countries. The supervision of this public service obligation is concentrated on the issue of reliability and control of the interruptions in the water supply due to failures in the network. In various countries, when non-planned interruptions occur, there are penalties aimed at directly or indirectly compensating the customers. Table 3.7 shows examples of this situation in three different countries. Of the 60 countries included in this study, only three of them encompass operators that commit themselves to compensate their customers for an interruption in the water supply service, while eight refer to this right but do not specify an amount or a means of carrying this out. Some countries, such as the UK and Australia, also have special non-interruptible plans for customers in which this aspect takes on heightened importance (e.g. hospitals and schools). These situations can be predefined or can occur by agreement (e.g. industries whose water consumption is essential), with or without a corresponding special tariff.

Table 3.6 Continuity for case studies and countries surveyed

Country	Continuity	Country	Continuity
Asia		Panama	A majority of the population
Armenia	No, 8 hours in 2003	Peru	No, 17 hours in urban areas
Philippines	No, 19 hours in Manila	Trinidad and Tobago	No
Indonesia	No (Jakarta is next)	Uruguay	Yes, except for rural areas
Singapore	Yes	**Europe**	
Africa		Belgium	Yes
Cape Verde	No	Scotland	Yes
Ghana	No in 75% of Accra. 30%, 12 hours, 5 days	France	Yes
Mozambique	No	Greece	Yes
Niger	No	Netherlands	Yes
Kenya	Variable (e.g. Nyeri is continuous and Mombasa 19 hours)	E&W	Yes
Tanzania	No	Ireland	Yes
Zambia	15 hours on average, 11 hours in Lusaka	Northern Ireland	Yes
America		Italy	Yes
Barbados	No	Kosovo	Yes
Belize	Yes	Lithuania	Yes
Brazil	Variable	Portugal	Yes
Canada	Yes	Czech Republic	Yes
Chile	Yes	Slovak Republic	Yes
Colombia	Yes, except for rural areas	Romania	Yes
Costa Rica	No	Sweden	Yes
USA	Yes	**Oceania**	
Honduras	No	Australia	Yes
Jamaica	No	Papua New Guinea	No
Mexico	No, <50% of the population		

3.2.5 Special customer protection

In addition to the social tariffs, which mostly include customers who consume little or are faced with economic difficulties, several countries have special tariffs for certain groups of customers, such as large families. This happens in the UK and in Portugal, for example. Other countries have social protection plans for customers with special needs. Australia and the UK are good examples that will be shown below.

The Australian state of Victoria uses a subsidization plan known as Utility Relief Grant Scheme, whose objective is to help the neediest customers. Customers who hold a current concession card or who are registered with the utility company's hardship program can sign up for this program. To be eligible, the household income must be lower than a certain percentage indexed to the value of the retirement pension. A significant increase in water consumption, a reduction in the family earnings and an unexpected increase in essential spending are criteria for making customers eligible for this system, provided they will occur for a maximum of 12 months. Those whose rent is greater than 30% of their income or whose WWS charges are greater than 10% of their earnings can also apply for this customer subsidization plan. This scheme is aimed at the complete payment of the WWS bill over a period of one year; if these conditions continue, it can be renewed. Afterwards, the Victoria Department of Human Services pays the operators the charge amounts related to these customers.

Table 3.7 Water service interruption commitments and compensations

City	Operator	Country	Commitment	Compensation
Lincoln	Anglian Water	England	Service interruption without prior notice	24.75 € + 12.38 €/day
Sydney	Sydney Water	Australia	Service not restored within 5 hours or without prior notice	10% discount on the bill (minimum 5.7 €)
Valongo	Veolia	Portugal	Intervention in the water network within 2 hours and in the wastewater network within 4 hours	10,000 litres of water

In 1999, the Water Industry Act was introduced in the UK. This Act is a protection plan for so-called 'vulnerable groups', which limits the WWS bill of an eligible customer to the average of the regional customers' bills. In order to benefit from this plan, the customer must belong to one of two categories. The first one relates to the customer or a member of their household receiving social benefits, which could include exemption from municipal taxes, subsidized housing credit, minimum wage, unemployment payments, child support, pensions or any other social service, as well as those with three or more children all under the age of 19. The second one is related to the existence of health problems, in which the customer or a member of his household has been diagnosed with serious health problems such as chronic kidney failure, incontinence, etc. In these circumstances, the operators are reimbursed for the costs relative to these customers by the Department for Work and Pensions. All customers who show signs of unnecessary use, such as pools or automatic watering systems, will not be covered by this system. At present, close to 0.4% of the population in the UK (nearly 440,000 inhabitants) benefit from this system. In addition, the cited legislation also establishes the impossibility of disconnecting domestic customers in the UK. As such, in the event that customers do not pay the WWS, the operators will have to go to court in order to force the respective payment or proceed with the disconnection.

3.2.6 Protecting customer interests

Protecting customer interests is carried out in different ways depending on the countries. In some cases, this task belongs to the regulators, where they might or might not have economic regulation or quality of service regulation functions. Other countries, such as Australia and Sweden, use an Ombudsman whose chief mission is to defend customers in the face of operators and other public administration bodies. In other cases, there are customer groups who carry out this function, such as in Niger and the Netherlands. It should be pointed out that in the regulatory processes of certain countries the customers are represented in the Advisory Councils that allow for the issue of their opinion regarding ongoing regulatory procedures. This is the case, for example, in Portugal.

In certain countries, for instance Australia and Sweden (pioneers in the field), the defence of customer interests is guaranteed and safeguarded by an independent body known as Ombudsman. For example, in Australia's different states and territories there is generally a specific Ombudsman for WWS and for energy services (electricity and gas). Their services are triggered through a citizen or customer complaint relative to a WWS, but the complaint must first be filed with the Ombudsman. The Ombudsman will investigate the merits of this complaint and take actions in order to fix the problem. This service is offered for free, without the need for intermediary parties. In broad terms, the Ombudsman is generally responsible for helping and assisting the organizations about whom the complaint was issued, especially when they are repeat incidents. The majority of the complaints are resolved within a period of 2 to 4 weeks, although a more complex situation may need more time. The Ombudsman's decision is binding for the operator, but not for the complainer.

In E&W, there is a body exclusively dedicated to protecting WWS customers, known as the Consumer Council for Water (CCW). It is an independent and decentralized agency, with delegations in several regions. This policy of proximity allows for a more effective relationship with customers. The CCW's main activities are raising awareness of the customers regarding their rights and strengthening its role as an informative organization which is influential and effective in the defence of customers with respect to WWS and the other entities associated with the water sector. The CCW, established in 2005 and replacing WaterVoice, carries out countless activities in the water sector, wholly dedicated to the preservation and amplification of the customers' collective voice. The organization of national debates on water issues, investigation work to be published, the handling of customer complaints in the event of a failed resolution with the corresponding operator, involvement in the price review process, safeguarding value for money and the promotion of customer awareness about the correct modes of consumption with the aim of assuring high quality and reliable WWS for all customers are, among others, some of the most important CCW attributions.

Particularly in developing countries, questions related to customer defence take on greater proportions given the associated difficulty of access and, sometimes, excessive or improper costs. For example, in Niger, the *Wadata – Association de Défense des Droits des Consommateurs* plays a very important role. This association, which has few resources, seeks to protect the environment and the management of natural resources, providing legal assistance to the customers, carrying out impact studies in all areas and working against the spread of infectious diseases, such as AIDS and malaria. In the water sector, Wadata is leading various investigations into the sector's price and quality of service, as well as the impact of their respective reductions. Beyond this, they also carry out drinking water quality and water resource analyses, occurring mainly in the urban community of Niamey. Wadata seeks to safeguard and guarantee an appropriate management of the WWS by the *Société des Patrimoines des Eaux du Niger* (SPEN) and also the promotion of workshops that deal with customer rights and the quality of service provided, among other topics.

3.2.7 Participation and transparency

Participation in the WWS regulatory and operational processes takes on an important role for their respective credibility and acceptability. Lack of participation takes accountability away from the customers with respect to the decisions that are made and puts the sustainable development of the WWS at risk. Customers should be able to participate in the setting of the tariff systems. The forms of customer participation in the regulatory processes vary from country to country. Participation is also fundamental for process transparency. All WWS and stakeholder activities should be clear, public and easily accessible. Although all the inquired entities affirmed that the water sector was governed by transparency and stakeholder participation in their activities, when a more detailed analysis was carried out, it was noticed that in only 45% of the cases were the customers able to offer input into the tariff system. In 30% of the examined cases, the regulators did not have a website or did not provide much information about the WWS there, such as information related to the tariff system. It should also be noted that only 40% of the regulators provided information on their websites about the quality of service provided. A variety of relevant cases are summarized below.

The regulatory processes should be completely open to participation from the public, since the decisions that are made directly affect their lives. One example is the case of Porto Alegre, Brazil, whose municipal administration model includes public participation in decision-making processes, known as 'procedural democracy', with successful results referred to in the relevant literature as best practices. This type of operation and innovation has allowed the city to demonstrate very high WWS quality levels, even though respective treatment is lower, with 99.5% coverage in drinking water services at a relatively reduced price and 84% in wastewater collection (Viero & Cordeiro, 2003). This process, known as a Participatory Budget Process (PBP), was implemented in 1989. In order to carry it out, the city was divided into 16 regions, whose limits were drawn based on the presence of resident groups and the layout of the social movements. Later in 1994, as a means of promoting a discussion of the social questions, the government substantially rewrote the PBP constitution, thereby allowing for the participation of union groups, governmental and non-governmental organizations, political parties, businessmen, students and cultural and community groups, thus forming Plenary Sessions.

The PBP includes three stages. First, there is the 'discussion in the regional meetings and the plenary meetings regarding certain topics', with the former carried out in each region in April and the latter performed in the month of May. Afterwards another round of regional meetings is carried out. The second step consists of a 'budget compatibility process' where the budget proposal is developed through a group of proposals and priorities presented by the community, incorporated in order to make up the minutes of the budgetary matrix to be discussed by the Participatory Budget Council (COP). The final proposal is delivered to the City Council in September. Lastly, a 'budgetary specification' is included where, during the last months of the year, various regional investments or general investments proposed during the specific plenary sessions are detailed.

In this process, after a few years of experience, activity criteria came to be defined, namely those regarding the necessities of the urban regions services and infrastructures; the neediest populations in the area of service; and the regional infrastructures, overall population and priorities. A concern for objectivity in the distribution of resources brought about the use of a value scale based on weights and points, in order to guarantee a fair distribution based on the allotments available between the regions (Larangeira, 1996). As such, each criterion is given a classification between

1 and 4 (for example, if a region has paving needs, it receives a 4 classification in terms of infrastructure needs). Besides this, a weight (between 1 and 3) is attributed to each criterion, representing the respective relative importance according to the PBP evaluation (the criterion 'urban services or infrastructure needs of the region' has received the strongest weight). Finally, the classifications and weights are added and compared with each region's priorities, thereby forming the base for calculating the distribution of investments (Orçamento Participativo, 1995). However, any criterion can be reviewed upon the establishment of a new Council.

In Jakarta, Indonesia, customer participation has mainly come through the Jakarta Water Supply Regulatory Body (JWSRB), created in November 2001. In January 2002, the JWSRB carried out the first meeting between the various water sector stakeholders with the objective of creating synergies in order to increase the water service reliability in Jakarta. This meeting led to the creation of the Consumer and Community Communication Forum (CCCF). From that point onward, the CCCF became a valuable platform for the sharing of know-how among the various stakeholders, including the JWSRB, the customer and community representatives, NGOs, governmental authorities, PAM Jaya (the local administration) and the water service operator. Simultaneously, the JWSRB developed an internal system called the Community Opinion Investigation Committee (COIC) in order to safeguard forum operations and administer customer satisfaction levels. Moreover, in March 2003 the JWSRB promoted the establishment of the Water Supply Customer Advisory Committees (WCC) in five municipalities in the province of Jakarta. The WCC are non-governmental organizations dedicated exclusively to the defence of customer interests. Among their responsibilities are the monitoring of service levels, the handling of complaints related to service quality, communication between customers and operators and the promotion of access to water services for the lower income customers. The JWSRB maintains the responsibility for disseminating the information obtained from the WCCs and CCCF through the media. Figure 3.7 below shows the organization of the Jakarta participation scheme.

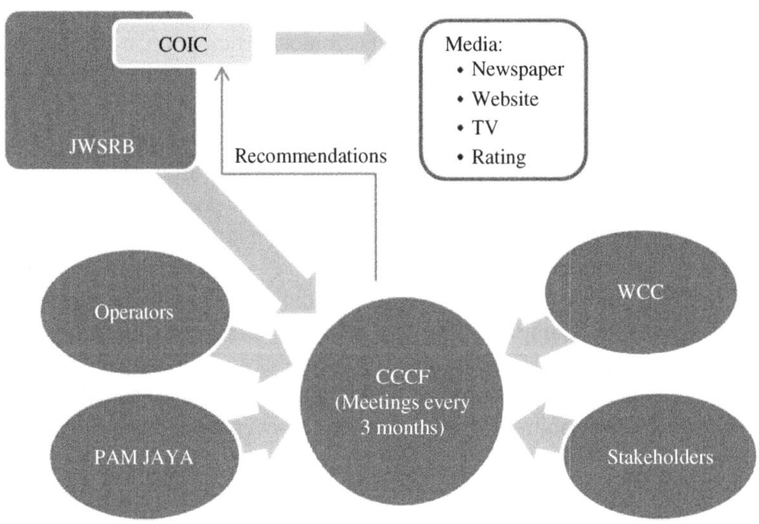

Figure 3.7 Customer participation plan for water service in Jakarta

In E&W, the price review regulatory process is defined by the participation of all stakeholders and by the existence of various appeal hypotheses. Figure 3.8 shows the regulatory process calendar adopted by OFWAT for the last tariff review.

Figure 3.9 highlights the example of IPART in Australia for the Sydney Water tariff definition regulatory process.

3.3 QUALITY OF SERVICE REGULATION

Regulation of WWS quality of service is unquestionably an important activity. Although not all countries and/or regulators deal with this issue appropriately and frequently, mainly due to the existence of other priorities (access to service), all of them recognize its fundamental nature and assure that it will be instituted or improved in the future. Of the analysed case studies, 79% monitor and supervise the quality of service. The objectives are clearly different based on the country development and the WWS. However, the methods are not very different. Performance indicators are predominantly used in 95% of the cases in which there is quality of service regulation. The strategy is almost always the application of sunshine regulation, which relies on comparing, disclosing and publicly discussing the operator's performance in order to put pressure on those that show poorer performance and to recognize the merits of the best performers. Frequently the comparison through benchmarking is related to performance during the previous years. In about 57% of the cases of study, there is a comparison with other operators, in 15% with operators from other countries, and in roughly 20% with reference benchmarks. In some countries (roughly 25%) penalties are imposed, including consequences for the tariff system when the performance or levels of service are not the most appropriate. In some cases, there is a direct reimbursement to the customers, while in others it goes to the government (roughly 12%). In other countries, rankings are issued relative to operator performance (8% of the cases). Table 3.8 provides a breakdown of the results obtained relative to service quality regulation for the case studies and countries surveyed. Next, a more detailed description of the quality of service regulation in Scotland, Portugal, Jamaica and Ghana is provided.

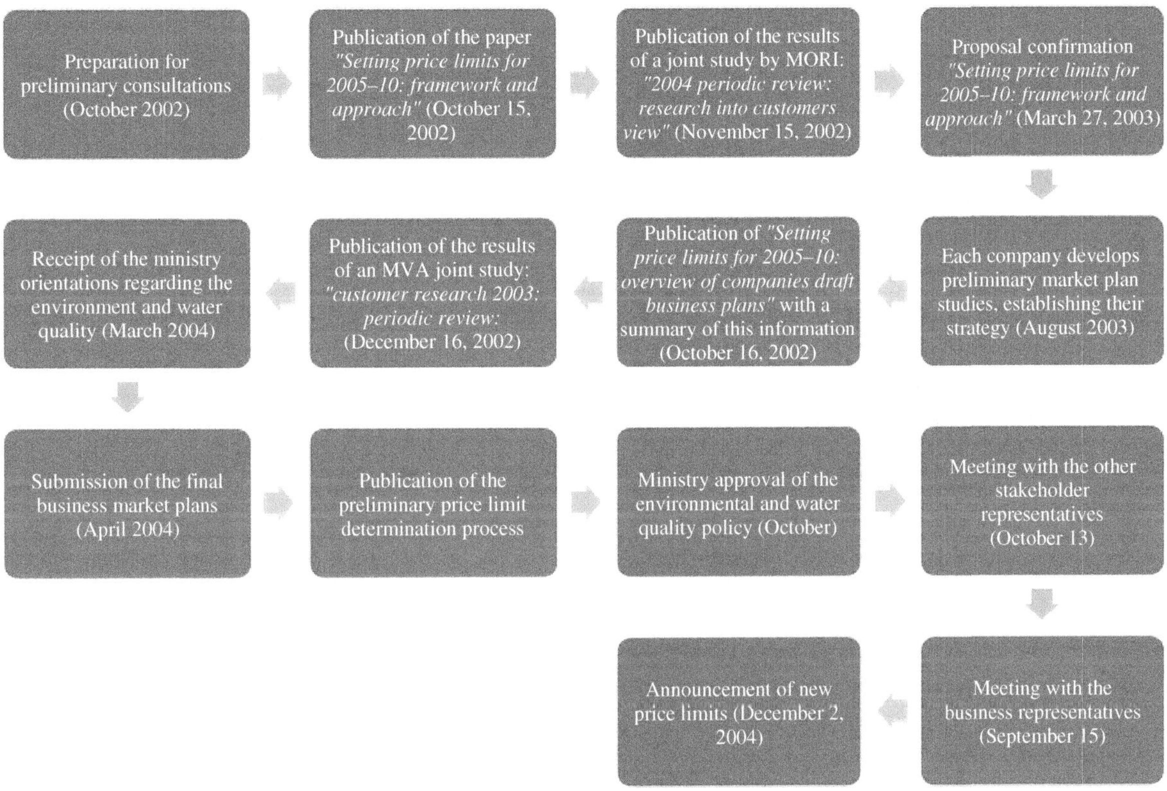

Figure 3.8 Tariff review for the 2005–2010 regulatory period in England/Wales

a) Scotland

In Scotland, the introduction of a sector-specific regulatory agency (WICS) brought countless improvements, from the reduction of operating costs (by roughly 40%), to a significant increase in the quality of service, resulting in the reduction, year after year, of the cost of service and a general increase in customer satisfaction. Figure 3.10 shows the change in the average bill in Scotland and in E&W. Given the existence of a single WWS for the entire country, WICS used information from the WWS in E&W for benchmarking purposes.

WICS adopted a group of performance indicators which allowed them to evaluate the levels of service provided. In addition, the evolution of the quality of service is evaluated by the WICS using an overall performance assessment (OPA). As stated, WWS have evolved significantly, mainly in terms of the quality of service, which is reflected in an OPA increase of 40% between 2005–2006 and 2006–2007.

The OPA is an aggregated index which considers the performance measurement that customers feel to be the most important, such as the speed with which water service is re-established, the response to billing contacts, the quality of drinking water supply and the compliance with environmental regulations. Whenever possible, the WICS uses the same measurements that are adopted by OFWAT to evaluate the performance of the companies in E&W, in order to have benchmarks.

The OPA uses the customer information about the service offered, which is given on a quarterly basis. Although the precision and reliability of this information has substantially improved, there is still room for some progress in this area. As a means of increasing the incentive for a positive OPA evolution, the regulator instituted goals for this indicator, which have been exceeded. Figure 3.11 shows the (positive) evolution of the OPA between 2002–2003 and 2006–2007, as well as the goals set for Scottish Water.

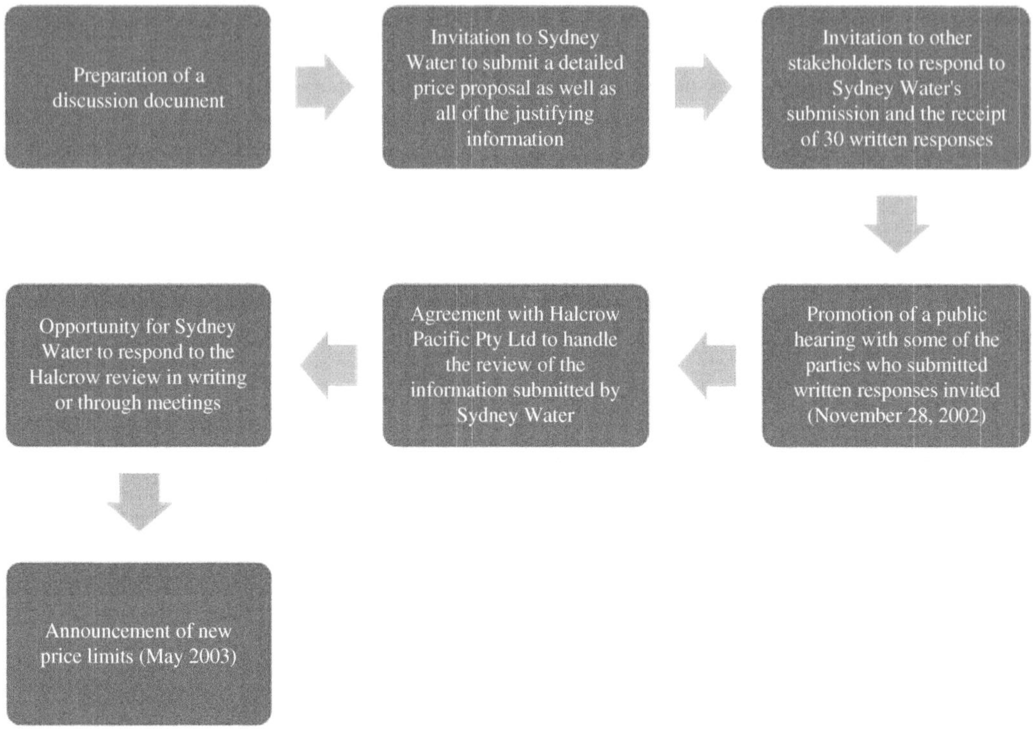

Figure 3.9 Tariff review for the 2003–2005 regulatory period in Sydney Water

The OPA can be broken down into four areas, namely water service (pressure, interruptions, non-domestic use restrictions and water quality), wastewater services (number of floods and risk of floods), environmental impacts (compliance with the wastewater treatment regulations, water losses and pollution) and, finally, customer relations (complaints, billing questions and telephone contact information). Figure 3.12 shows the OPA evolution between 2005–2006 and 2006–2007 by individual performance measurement.

b) Jamaica

In Jamaica, the existence of a regulator with responsibilities in the water sector has resulted in important developments. Despite the fact that the WWS show numerous problems (e.g. universality), the performance of the Office of Utilities Regulation (OUR) has clearly been positive concerning the regulation of quality of service, not only through its supervision, but also through the calculation of certain performance indicators (e.g. the volume of unbilled water) and the institution of GSS. However, the drinking water quality aspects remain under the auspices of the Health Ministry.

In terms of fostering improvement in the quality of service, the OUR also establishes targets for the state-owned operator, the National Water Commission (NWC). Figure 3.13 presents the metering and unbilled water volume targets established for the NWC by the OUR.

As mentioned, besides the development of performance indicators, the OUR established minimum quality of service indicators for the water sector as a means of promoting the quality of service enhancement and compensating the customers when the service is not the most appropriate. Table 3.9 shows the quality of service levels imposed by the OUR. Non-compliance of commitments results in a 10€ compensation for the customers.

c) Ghana

In Ghana, a country characterized by severe socio-economic problems, the role of the WWS regulatory agency, the Public Utilities Regulatory Commission (PURC), also has notable importance. Besides the valuable economic regulation and all the care in terms of costs (which have been a subject of concern), the quality of service is also an important aspect receiving the regulators attention. As such, the PURC applies a set (restricted, given the limitations) of performance indicators, which seek to promote the development of the service provided to the customers (e.g. the volume of unbilled water). These indicators are later published in a report that analyses the previous five years, thereby demonstrating their evolution. Table 3.10 shows some of the performance indicators determined by the PURC.

Table 3.8 Quality of service regulation for the case studies surveyed

Country	Quality of Service	Country	Quality of Service
Asia		Panama	No
Armenia	Performance indicators	Peru	Performance indicators
Philippines	No	Trinidad and Tobago	Performance indicators
Indonesia	Performance indicators	Uruguay	No
Singapore	–	**Europe**	
Africa		Belgium	Indicators in Wallonia
Cape Verde	No	Scotland	Several techniques
Ghana	Performance indicators	France	–
Mozambique	Performance indicators	Greece	–
Niger	Performance indicators	Netherlands	–
Kenya	Performance indicators	E&W	Several techniques
Tanzania	Performance indicators	Ireland	–
Zambia	Performance indicators	Northern Ireland	No
America		Italy	Performance indicators
Barbados	No	Kosovo	Performance indicators
Belize	Total factor productivity	Lithuania	Total factor productivity
Brazil	Partial indicators	Portugal	Performance indicators
Canada	No	Czech Republic	–
Chile	Performance indicators	Slovak Republic	–
Colombia	Several techniques	Romania	
Costa Rica	No	Sweden	–
USA	No	**Oceania**	
Honduras	Performance indicators	Australia	Performance indicators
Jamaica	Yes	Papua New Guinea	Performance indicators
Mexico	–		

Some of the most outstanding results, obtained through work in cooperation with the Ghana Water Company, Ltd. (GWCL), are the reduction in water losses and the elimination of illegal connections by 10% (today roughly 40%) and the increase of about 20% in the billing collection capacity (presently at 95%). Figure 3.14 shows the evolution of GWCL water production and water losses (PURC, 2005).

Additionally, the PURC publishes the performance results regarding complaints, breaking down the information into complaints placed and resolved. Additionally, the comparison with results from other countries such as Senegal, Malawi, Chile, Poland, France, the UK and Germany with respect to water losses adds value as a form of promoting better performance within the regulated agency.

d) Portugal

The presence of a sector-specific regulatory agency in Portugal for the WWS (which also includes urban waste), known as the IRAR, brought meaningful benefits for the water sector. The IRAR has had a particularly active role in the promotion of the quality of service and the satisfaction of public service obligations.

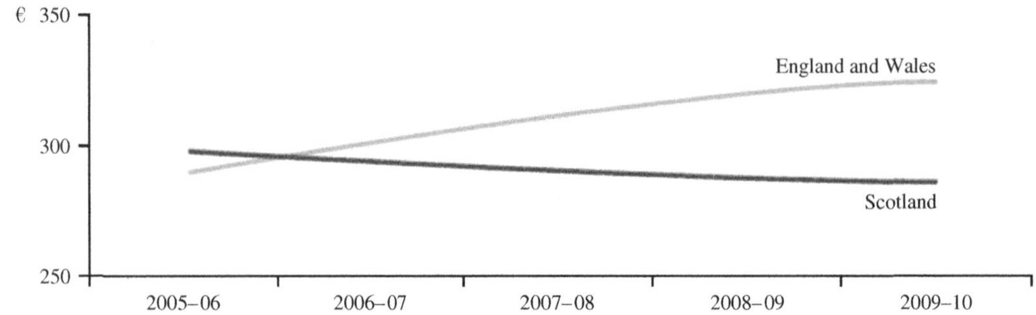

Figure 3.10 Average bill change (predicted) in Scotland and in E&W

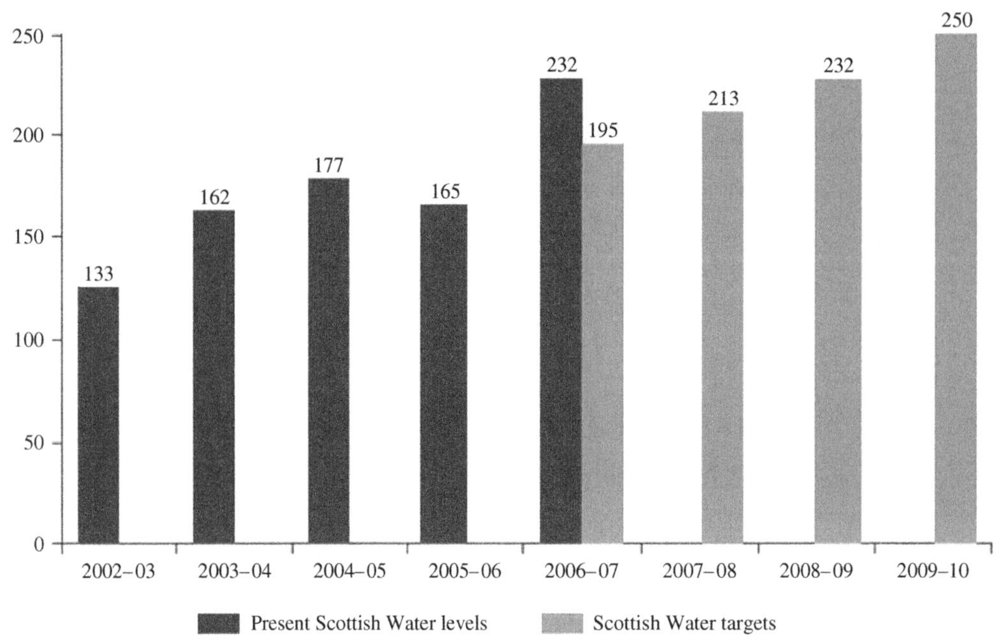

Figure 3.11 Evolution of the OPA indicator and the goals established for Scottish Water

Among the important contributions, namely the regulation of quality of service and drinking water quality, sunshine regulation has come out as one of the main tools of the IRAR. In this way, in order to promote the quality of service, the IRAR relies on a set of performance indicators that are published in its annual report. These indicators measure the effectiveness and efficiency of the operators, relative to specific aspects of the activity being carried out or the behaviour of WWS.

In this regard, the IRAR implemented a set of 20 performance indicators, divided into three groups and specific to each service (broken down into water and wastewater, and, in turn, into wholesale and retail segments), namely those relative to the defence of customer interests, sustainability of the operator and environmental sustainability.

Finally, the IRAR, in addition to these roles, evaluates and issues certain comments related to the result of each indicator by operator, through the annual report referred to. This evaluation is carried out according to a classification system that compares the value obtained with the corresponding reference value. These reference values are optimum values, or close to them, which the regulator considers 'reachable' by the regulated operators. However, as the name itself indicates, these reference values cannot be analysed by themselves nor can they be taken out of context. The IRAR publishes the results, ranking them based on the performance indicators. Besides this, it justifies, by operator, the results obtained through the performance indicators calculation, quantifying and classifying the quality of service as unsatisfactory, average or good and attributing, respectively, a red, yellow or green ball.

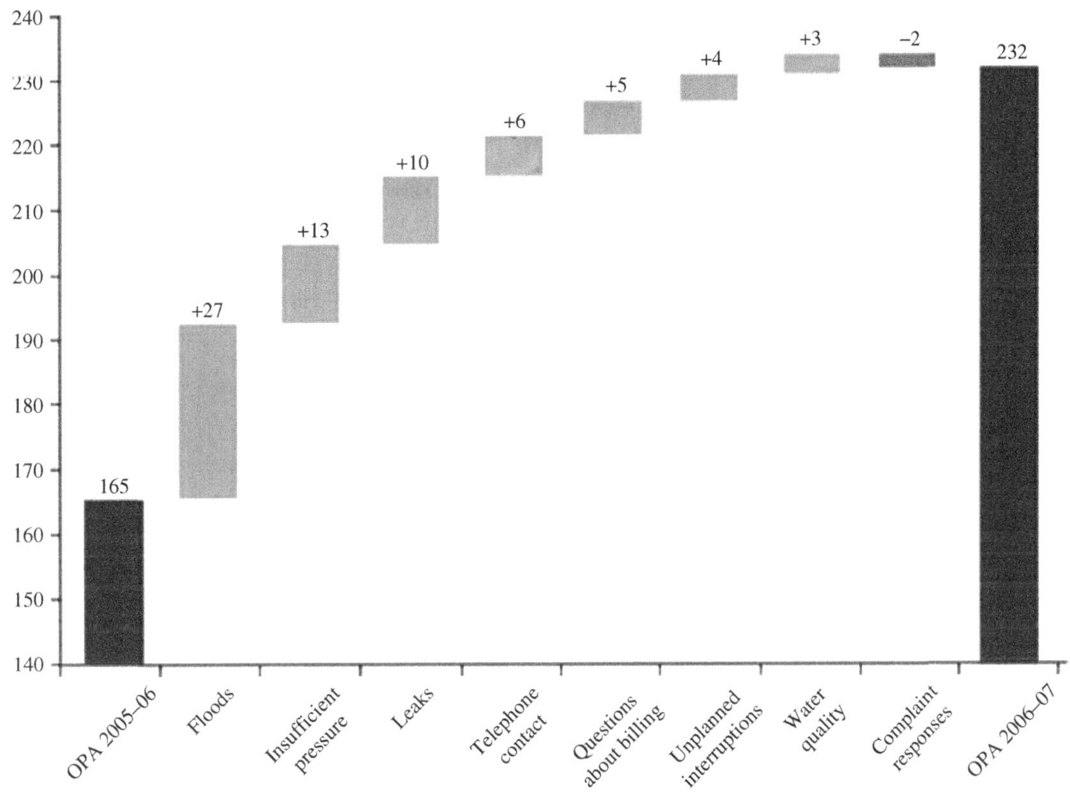

Figure 3.12 OPA indicator development between 2005–2006 and 2006–2007 in scotland

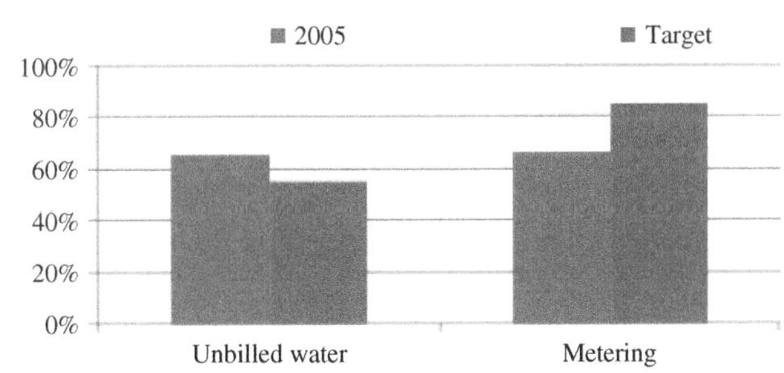

Figure 3.13 Scores and targets for the unbilled water and metering indicators in Jamaica

Figure 3.15 and Figure 3.16 are examples of evaluation tables published by the IRAR, such as an evaluation score table and the benchmarking of an indicator used in quality of service evaluation, respectively (IRAR, 2007).

3.4 ECONOMIC REGULATION

The economic evaluation methods applied in the different case studies and countries surveyed oscillate between rate-of-return regulation and price cap regulation. Although the former option predominates, it has been shown that in recent years there has been a growing preference for economic regulatory methods based on performance incentives. In the analysed case studies and inquiries, 35 countries (58%) had, in some way, an explicit regulation implemented and 60% adopted an economic regulatory method based on rate-of-return regulation. As for the countries that adopt performance-based incentive regulation, almost all of them adopt price cap regulation and 41% of these used benchmarking techniques to determine the X factor, therefore based on yardstick competition. Two countries adopted the efficient company model and four others employed, completely or partially, the revenue cap regulatory method. Table 3.11 lays out the regulatory methodologies for the case studies and countries surveyed. In the following section, a detailed analysis of the economic regulatory method adopted in four of the case studies, namely Costa Rica, Australia (New South Wales), Peru and Colombia, is provided. The first case corresponds to a rate-of-return regulation and the remaining cases show performance incentive regulation. Among them there is revenue cap regulation (Australia) without the use of benchmarking and yardstick competition (efficient company model and price cap regulation, using benchmarking methods to determine the X factor).

Table 3.9 Quality of service levels established by the NWC

Description	Commitment
Access	
Service connection	New connection in 5 business days
Basic service connection (existing meter)	Connection in 4 business days
Emergency response	Response to emergency calls within 6 hours
Billing	First bill sent within 45 business days of the connection
Complaints	
Recognition	Recognition of written question in 5 business days
Investigation	Complete investigation within 30 business days
Third-party investigation	Complete investigation within 60 business days
Reconnection	
After payment of debt (urban area)	1 day
After payment of debt (rural area)	2 days
Meter replacement (requested by customer)	20 business days
Compensation (monetary)	45 days

Table 3.10 Variation of PURC performance indicators between 1996 and 2003

Description	1996	1997	1998	1999	2000	2001	2002	2003
Water produced*	100.0	97.4	96.1	101.9	96.8	98.6	107.3	107.4
Water charged*	100.0	106.1	111.6	107.2	109.7	109.7	103.7	107.5
Water billed*	100.0	108.9	116.1	105.2	113.3	111.3	96.6	100.0
Revenue*	100.0	76.7	130.2	145.5	146.4	193.6	221.6	264.1
Collection*	100.0	67.8	100.7	119.6	131.1	155.1	167.4	203.4
Average tariff*	100.0	72.3	116.6	135.7	133.4	176.4	213.7	245.7
Operating costs*	100.0	86.7	88.4	127.9	175.6	188.6	177.4	189.0
Unit operating costs*	100.0	95.8	125.0	148.9	141.4	185.6	177.6	197.3
Unit energy costs*	100.0	102.0	154.5	221.7	204.0	259.6	309.1	353.0
Unit operating costs (excluding energy)*	100.0	94.1	115.9	126.4	121.9	162.8	137.3	149.3
Unit costs*	100.0	89.0	92.0	125.5	181.4	191.3	165.3	175.9
Water losses	57.1	53.2	50.2	54.8	51.4	52.2	58.5	57.1
Real revenue	42.1	40.1	37.8	36.4	42.7	37.5	30.7	32.4
CPI*		100.0	118.2	133.2	162.6	219.4	249.0	
Inflation*	100.0	59.9	39.1	27.3	47.4	74.9	29.0	21.5

*Determined as a % with respect to 1996

a) Costa Rica

In Costa Rica, WWS economic regulation is based on rate-of-return regulation. This methodology determines the tariff system using the following formula:

$$CSP = CO + CA + D + R$$

where *CSP* corresponds to the cost of service provided, *CO* to the operating and maintenance costs, *CA* to the administrative costs, *D* to the depreciations and *R* with the amount relative to the reimbursement for the service provided.

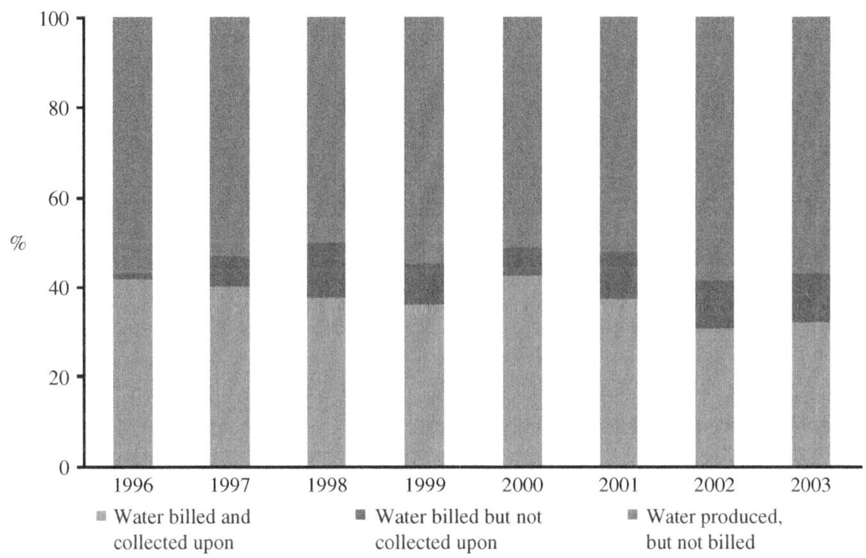

Figure 3.14 Evolution of the GWCL water production and water losses indicators

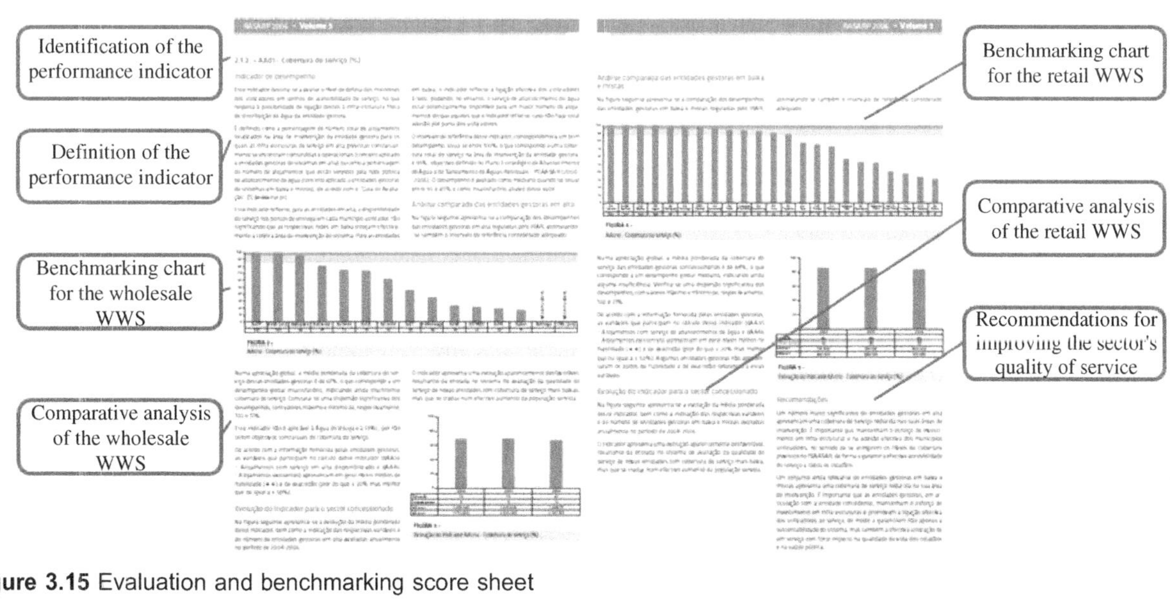

Figure 3.15 Evaluation and benchmarking score sheet

The agreed amount for covering all the items referred to sets the so-called tariff level (total value of the revenues that are necessary to cover the costs). These revenues are distributed in the form of tariffs, according to the tariff structure in effect (which is the attribution of different prices among services and blocks of consumption identified from within the universe of customers) and the results of the market analysis (demand). The application of these criteria, together with the guide for defining the service conditions and prices, is known as the connection.

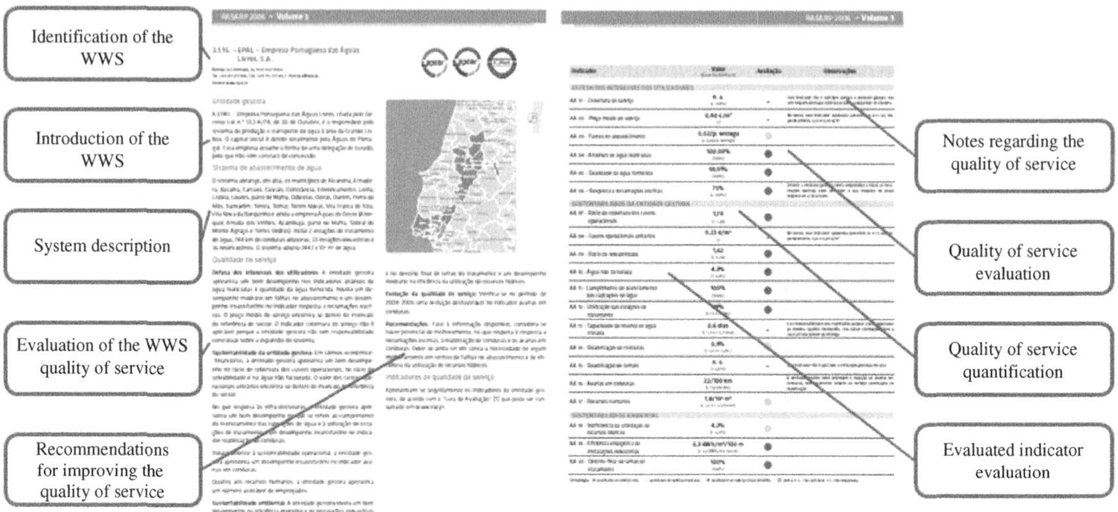

Figure 3.16 Annual quality of service evaluation score sheet

In order to determine the components of the formula the following elements are used (present and forecasted situation for at least five years, although usually applied for two years), namely the revenues and costs, the tariff base and the origin and application of the tariff funds. It should also be noted that even in the case of public companies, obtaining an excess (R) is allowed as a compensation for the delivery of services.

b) New South Wales, Australia

In regulating the metropolitan WWS in New South Wales, Australia, which included Sydney Water and Hunter Water, IPART adopted a building block strategy in order to determine the revenue level required by the WWS. This revenue level is defined by the following formula:

$$Revenues = Operating\ Costs + Depreciation + Asset\ Return$$

Asset Return is obtained through the following formula:

$$Asset\ Return = Rate\text{-}of\text{-}Return \times Asset\ to\ Reimburse$$

In particular, the operating costs are determined considering the past costs and their change over the years due to uncontrollable factors and assuming some efficiency gains. Generally benchmarking methodologies are used for determining these OPEX (efficient). The depreciation is calculated considering the life cycle of the assets, which, in the case of water supply and wastewater infrastructures in Australia, corresponds to 70 years. Asset return is determined based on the weighted average cost of capital (WACC).

From these values, the regulator (IPART) uses a CPI-X formula to review the revenues allowed during the regulatory period. The X factor is determined by the regulator, considering the future expectations and past results, as well as the social and environmental aspects, without using any special methodology. In this way, during the regulatory period (three years), the operator has incentives to become more efficient, retaining the profits reached and, as a result, increasing their respective return.

c) Peru

The economic regulatory process in Peru, based on the efficient company model, is structured in three phases: the Preparatory Stage, the Improvement Stage and the Definitive Stage. In the Preparatory Stage, the emphasis is on economic and financial viability, in order to guarantee that the tariffs collected by the company providing sanitation services (EPS) will at least cover the costs. During this phase, the regulator evaluates the operator's tariff structure and adopts corrective measures. The Improvement Stage seeks to improve the economic efficiency of the EPS, proposing alterations at the operator's organic and institutional levels, which will be reflected in setting the Management Objectives that must be complied with. For each EPS, tariff formulas are defined for five years, so that the annual tariffs allow for the recovery of average annual costs (AAC). The AAC cover the costs included in its numerator relative to the operation and maintenance of the systems, the capital costs, which include those associated with the return on the company's equity and the debt and the charges from taxes, divided by the volume of billed water, presented in its denominator.

Table 3.11 Economic regulation methods for the case studies surveyed

Country	Economic regulation	Country	Economic regulation
Asia		Panama	Price cap (5 years)
Armenia	Price cap	Peru	'Model company' (5 years)
Philippines	ROR	Trinidad and Tobago	Revenue cap (5 years)
Indonesia	ROR	Uruguay	ROR
Singapore	–	**Europe**	
Africa		Belgium	ROR (Wallonia)
Cape Verde	Price cap (5 years)	Scotland	Revenue cap (4 years)
Ghana	ROR/Price cap (4 years)	France	–
Mozambique	ROR	Greece	–
Niger	ROR	Netherlands	–
Kenya	ROR	E&W	Price cap (5 years)
Tanzania	ROR/Price cap (2 years)	Ireland	–
Zambia	Price cap (3 years)	Northern Ireland	Price cap (5 years)
America		Italy	Price cap (5 years)
Barbados	ROR	Kosovo	ROR
Belize	ROR	Lithuania	ROR
Brazil	ROR/Price cap/Revenue cap	Portugal	ROR
Canada	ROR	Czech Republic	–
Chile	'Model company' (5 years)	Slovak Republic	ROR
Colombia	ROR/Price cap (5 years)	Romania	ROR
Costa Rica	ROR	Sweden	–
USA	ROR	**Oceania**	
Honduras	ROR	Australia	Revenue cap and Price cap
Jamaica	Price cap (3 years)	Papua New Guinea	Price cap (5 years)
Mexico	–		

ROR – Rate-of-return regulation

The AAC is calculated using the EPS Business Plan, which will be established based on the Management Objectives and the corresponding Strategic Plan (SP). The Business Plan will be a five-year term. The SP should contain medium- and long-term schedules of investments for the systems rehabilitation, renovation and expansion, as well as an operation and institutional service improvement plan, where the quality and efficiency targets for WWS are set for the medium term (efficient company model).

The tariff formula mentioned in the Improvement Stage, which refers to each year's average tariff, is defined by the following formula:

$$T_t = T_{t-1}(1 + k_t)(1 + Ø)$$

where T_t corresponds to the average tariff for year t, T_{t-1} represents the average tariff of year $t-1$, k_t is associated with an annual adjustment factor for expressing the real tariff increase and Ø refers to the CPI.

During the Improvement Stage, a reduction in cross-subsidization is sought, above all among the different customer categories. The adjustment process should be considered and included in the corresponding Tariff Study.

In the Final Stage, the orienting principles of the tariff regulation laid out by the General Law will be met. As such, concerning the principle of economic efficiency, the tariff charges, depending on the volume consumed, will correspond to the long-term marginal cost, calculated according to the assumptions previously referred to. The average long-run incremental cost (LRIC) will be used as estimate. The LRIC is defined with the following formula:

$$LRIC = \frac{\sum_{t=1}^{j} \frac{I_t}{(1+r)^t} - \frac{VR}{(1+r)^n} + \sum_{t=1}^{n} \frac{(C_t - C_0)}{(1+r)^t}}{\sum_{t=1}^{n} \frac{(Q_t - Q_0)}{(1+r)^t}}$$

where I_t is the investment in year t written in the SP, VR corresponds to the residual value of the investments therein defined for the year n, C_t is equal to the operating costs without depreciations in the year t, C_0 refers to the operation costs without depreciation in the year 0, Q_t is associated with consumption in the year t in m^3 and Q_0 is equal to consumption in year 0 in m^3, r is the adjustment rate (defined by the SUNASS, the peruvian regulator), j is the number of years in the investment program and n is the number of years on the planning horizon.

The AAC is defined with the following formula:

$$AAC = \left[\sum_{t=1}^{5} \frac{AACa_t}{(1+r)^t}\right] \times \left[\frac{r(1+r)^5}{(1+r)^5 - 1}\right]$$

in which $AACa_t$ corresponds to the average annual cost of year t and r is the adjustment rate.

If the tariff studies conclude that the tariff charges by volume, calculated based on the LRIC, are less than the AAC, fixed tariffs will be applied to meet the costs referred to. As such, the tariff for each service will be made up of two parts:

(a) Volumetric tariff by m^3, equal to the LRIC;
(b) Connection tariff, by which the AAC is met.

If the tariff studies show that the volumetric tariffs, determined in accordance with the LRIC, are greater than the AAC, there will be a differentiation based on the block of consumption, where the value charged for the last block will correspond with the LRIC and the average value of all the blocks will be equal to the AAC.

The parameter Ø will be calculated through the application of a polynomial readjustment formula, whose parameters will reflect the cost structure of each service, obtained from the corresponding Tariff Study. The tariff will be automatically adjusted, provided that the Ø parameter varies more than 3%.

The present regulatory period (2006–2011) is in effect for five years. Before the end of the regulatory period, the SP and the Financial Plan will be adjusted, new tariff studies will be carried out and new tariff structures will be determined for the following period.

d) Colombia

Colombia's regulatory method is based on the performance incentive regulatory approach, using yardstick competition. In the price cap system adopted, the X factor is estimated using benchmarking techniques, particularly in this case data envelopment analysis (DEA). The criteria and methodologies used for determining the WWS tariffs were established in 2004 by the CRA Resolution no. 287. The adopted regulatory process is based on a system of maximum prices for a five-year period. The maximum prices are determined separately, considering the water and wastewater services and the size of the company (up to 25,000 customers and greater than 25,000 customers). These prices are limited in a three-pronged approach, which specifically considers the fixed administrative costs and variable costs with respect to operations and maintenance (OPEX), capital costs (CAPEX) and taxes.

The administrative costs (AC) for the water service (WS) and for the wastewater service (SS) are the following:

$$CA_{WS} = \frac{TAC^e \times (1 - S_{SS})}{N_{WS}} \quad CA_{SS} = \frac{TAC^e \times S_{SS}}{N_{SS}}$$

where TAC^e is the total administrative cost value of the WWS, N_{WS} is the number of water service customers, N_{SS} is the number of wastewater service customers and S_{SS} is the percentage (%) of the TAC^e relative to the delivery of the SS.

The TAC^e will be obtained through $TAC^e = TAC_{DEA} + T$, where T are the costs related to monthly taxes and other contributions included in the administrative costs (excluding environmental taxes). $TAC_{DEA} = AC \times E$ is made up of AC, which represents the wastewater service administrative costs (without taxes) that are used in the DEA technical analysis, and by E, which represents the administrative costs, including the wastewater service value included in the tariff. The E value is given by $E = \min(P_{DEA} \times (1 + 0.046);\ 1.03)$, where P_{DEA} is the efficiency obtained by the DEA. The 0.046 value seeks to cover the costs and the known variables that are not considered in the DEA model. The 1.03 value is the maximum value adopted for the comparison of a 100% cost value and a maximum return of 3% on the capital expenditures. This procedure controls the outliers through a statistical rule and the existence of minimum prices.

The variable cost item is obtained through the sum of the three previously mentioned parts. In particular, the OPEX is obtained through the following formula:

$$OPEX_{SS} = OPEX_{SS}^{i} + OPEX_{SS}^{e} \quad OPEX_{WS} = OPEX_{WS}^{i} + OPEX_{WS}^{e}$$

where $OPEX_{SS}$ corresponds to the wastewater service operation and maintenance costs, $OPEX_{SS}^{i}$ are the wastewater service operation and maintenance costs specific to each company, $OPEX_{SS}^{e}$ is the wastewater service operation and maintenance cost value, $OPEX_{WS}$ corresponds to the water service operation and maintenance costs, $OPEX_{WS}^{i}$ is associated with the water service operation and maintenance costs specific to each company and $OPEX_{WS}^{e}$ is the efficient water service operation and maintenance value.

The wastewater service OPEX is obtained using the following formula (there is a similar formula for water service):

$$OPEX_{SS}^{i} = \frac{(EC_{SS} + RC_{SS})}{VW_{SS} \times (1 - L)} + \frac{T}{VBW_{SS} + [(VBW_{SAR}/1 - UW) \times 0.57 \times (UW - L)]}$$

where VW is the annual volume of water within the system, VBW is associated with the annual volume of billed water, L corresponds to a value of 30% relative to the maximum water losses limit, EC are the energy costs, RC are the reagent costs, T are the taxes associated with the OPEX and UW is the percentage of unbilled water. In the event that the company purchased water from third parties, an additional term is added to the formula.

The OPEX obtained through benchmarking are calculated using the following formulas:

$$OPEX_{SS}^{e} = \frac{TOPEX_{DEA} \times S_{SS}}{VW_{SS} \times (1 - L)} \quad OPEX_{WS}^{e} = \frac{TOPEX_{DEA} \times (1 - S_{SS})}{VW_{WS} \times (1 - L)}$$

$TOPEX_{DEA} = OPEX \times E$, being, $E = \min(S_{DEA} \times (1 + 0.088);\ 1.03)$ where S_{DEA} is the obtained DEA efficiency. Guidelines are also defined for the identification of outliers.

It should be noted that the variable item, besides the OPEX, includes investment costs based on development plans approved by the regulator as well as the associated environmental charges and taxes.

3.5 BEST PRACTICES

In conclusion, several suggestions and recommendations are listed regarding the best practices that were found during this investigation, which should be considered by the governments, regulators and operators of different countries and, if possible, should be followed. The ten principles and conclusions presented, while naturally reflecting the opinion of the author of the study, also lay out the results of the investigation. Hence, the following observations are brought together regarding good governance of the WWS, considering the different aspects analysed for each country, namely the WWS market structure and organization, as well as the respective regulation, the regulatory environment and characteristics of the regulatory agency, the regulatory principles and procedures adopted, the public service obligations, the quality of service regulation and the economic regulation.

a) The need for sector/multisector regulation

Due to all that has been pointed out, the institution of a WWS regulator is essential. ***Regulation not only allows for the mitigation of market failures, but can also improve WWS governance through an analysis of their operations in a more technical, professional and apolitical manner.*** Regulation benefits can be seen early on in the existing available information about WWS. When there are regulators, the performance and main statistical and operational elements of the WWS are known. In the absence of regulation, as in many European countries, although the WWS might work without major problems for a variety of reasons (water quality, environmental quality and reduced tariffs), their performance is not transparent, accountability is not placed, the customers do not play an active role and there are, without a doubt, abuses, wastefulness and inefficiency. Regulation seeks to function as a market substitute, stimulating competition in the non-existent market through yardstick competition and thereby incentivizing an improvement in performance and a reduction in their monopoly power. It should be noted that the existence of a regulator does not have any relation to the privatization of WWS. In Europe, this idea is often promoted, mainly by an industry that fears regulation, and bears no relation to reality. The most elucidating example is Australia, where the WWS are public (in terms of ownership and management) and, by law, the implementation of independent regulation in monopolies is compulsory.

b) The regulatory environment should include aspects relative to the public service obligations, quality of service and the establishment of prices and tariffs

The sector/multisector regulator ***should be present in the three sides of the regulation triangle, which consists of public service obligations, the quality of service provided and the establishment of prices and tariffs***, even though there might be other entities with functions in these areas. The complementary nature and relationship between these attributes are evident, allowing the regulator to take more effective and efficient decisions. The quality of service provided cannot be disassociated from the economic regulation, just as the fulfilment of public service obligations also has relevant consequences. As such, when there are several entities with responsibilities in these three aspects of regulation, besides the difficulty of analysing the problems as a whole and the problem of having a partial view of the regulation (based on the aspect under its responsibility), their individual decisions have consequences that might put at risk the other's decisions and success.

c) The regulator should be endowed with independence and the tools necessary to carry out its role

The regulator should be independent of all involved interests, including the operators, governments and customers. As such, it should ***have organic, functional and financial independence***. The organic independence is conferred upon the regulator through the permanence of its members; they should be appointed for a long, fixed period, without chance of dismissal except in special cases anticipated by law. Functional independence means that the regulator is only subject to the law and the control of the courts, and not submitted to the government supervision. Finally, financial independence means that the budget does not depend on the actions of outside parties. Regulatory independence has several advantages (Marques, 2005), namely the separation from politics and the economy and, as a result, the making of decisions without direct government involvement, which makes a more professional, impartial, apolitical and technical regulation possible; greater agility, de-bureaucratization and flexibility in the regulatory processes, given that these entities do not belong to the traditional State administration; safety, stability and credibility in the regulatory sphere; extrajudicial conflict resolution among various parties; and a greater accountability concerning the financial balance.

In order that they fulfil their missions, regulators should have ***regulatory powers*** (issuing regulatory documents), ***executive powers*** (implementing regulations) and ***supervision and sanctioning powers*** (supervising and punishing possible infractions of the regulatory elements), as well as ***conflict resolution powers*** between the regulated parties or between the regulated parties and the customers or other stakeholders. The effectiveness of the regulation is often endangered by the lack of some of these powers, such as the lack of ability to issue regulatory documents or the power to apply sanctions.

d) Regulator independence does not mean impunity and there should always be appeal mechanisms in place

While the regulators are independent, they should be accountable to the various stakeholders. Assuming accountability allows for the verification and questioning of the regulatory acts considered to be unfair or incompetent, thereby democratizing, in some way, the regulatory process. In this way, ***regulatory processes should be endowed with mechanisms that promote the regulator's accountability***, namely the clear definition of the scope of its functions, the way in which the board of directors is appointed and the possibility of dismissal, the jurisdictional control of the acts practised (court cases) and the submission of entities to the rule of law, as well as the financial and budgetary control of the Audit Court and other independent entities, with specific abilities for the control of political power (for example, a Magistrate, access to administrative documents, protection of personal data, etc.) Other instruments, such as the asking for opinions, the analysis and accompanying of the regulatory activity by the Parliament, contraditory proceedings and appeal, provided that they are written into law, and credible and independent external entities composed of respectable and independent people taking part in the re-analysis of conflicting decisions and intervention in the regulatory activity can also give credibility to the regulatory processes (e.g. advising councils). The implementation of clear regulatory procedures, which include mediation, information and the participation of interested parties, as well as transparency, legitimize the regulatory process.

e) The regulatory processes should be transparent, participatory and easy to understand

Transparency, participation and intelligibility are extremely important for a successful regulatory process. In the cases analysed, there is, on the whole, a strong correlation between effectiveness, the development and success shown by the WWS, and the existence of transparent and participatory regulatory processes. Regulatory processes should be ***clear, justifiable, easily verified*** (by the courts, if necessary) and ***public***. All the relevant regulatory information should be divulged from the beginning on the regulator's (and operator's) website. ***Transparency*** encourages participation and grounds the legitimacy and responsibility of the regulation. The regulatory process should allow for the ***participation*** of all the stakeholders. Participation allows for the improvement of the regulatory process through the received criticism, comments and suggestions; it helps and supports the decision-making process, increases the transparency and responsibility of those involved, therefore contributing to a reduction in the respective contention, fostering a greater understanding of the sector and the interests involved. Finally, the regulatory processes should be ***easy to understand*** and presented in an accessible language that is known to all the stakeholders. Even though the regulatory processes might be complex and rely on sophisticated techniques, the regulators should always present the non-technical results and summaries that allow for the greatest level of understanding.

f) The public service obligations, particularly universality, should be met, thereby sending the appropriate message for the WWS market

Public service obligations are an unequivocal part of regulation. It is not possible for WWS to experience sustainable development if the ***public service obligations are not properly cared for***. It is unthinkable, particularly in the private sector, that there is a future for the WWS if part of the population does not have access to the service or if the quality of service offered by the same operator varies according to service area or customer. Past failures prove this fact (e.g. Buenos Aires). As such, regulation has a fundamental role to play here as well. For example, when seeking WWS universality, cross-subsidization arrangements often send the wrong message to the market, causing wastefulness and leading the main customers to search for new sources of water. The subsidy could be granted in different ways, for example in volume, through social tariffs or operator subsidies, through the constitution of a social fund or the government itself. The most effective solutions prove this point (e.g. Chile). In any event, it must be acknowledged that countries require solutions that fit their realities. Often advanced, high-quality solutions are sought in poor socio-economic conditions, which make them impossible to carry out and translate into a waste of resources. The public service obligations should be regulated by law in each country, with sanctions for non-compliance.

g) The service levels should be defined by law and periodically analysed and published

Through legislative means or not, there should be an ***issuance of service charters*** from the operators, which identify the levels of service that will be reached by the WWS. These levels of service, which represent targets to be met and exceeded, should be associated with compensations for the customers in the case of non-compliance or other types of sanctions (GSS). These types of tools, while more often seen in developed countries, should also be promoted in less wealthy countries, with a variation in their degrees. The ***levels of service*** are warranties that the customers should receive in return for paying the WWS. Only with WWS providing high-quality service levels will the value for money be assured and the customers show a willingness to pay for the real cost of the service.

h) The regulator should promote the use of benchmarking

The use of benchmarking has several advantages that are reinforced with WWS due to their specific characteristics. Within WWS, the use of benchmarking by the regulators is not just one of the few instruments available for creating some type of competitive environment in this market but, rather, its potential grows with the existence of a significant group of players. The reality mitigates one of the main benchmarking problems, which is the comparability between operators. Actually, the application of benchmarking creates ***strong incentives for the operators to be efficient and innovative, mitigating their operating costs and capital costs***; it promotes the efficiency in terms of capital expenses, guarantees a 'fair' recuperation of the costs and a 'fair' remuneration of the capital invested and an increase in the sharing and transparency of the information, minimizing the traditional asymmetric information present among stakeholders. The fact is that this tool has much potential, both in the supervision of levels of service and quality, and in the establishment of prices and tariffs (in determining the efficiency and productivity of the operators).

i) Providing performance incentives should be one of the main regulatory objectives

There is limited competition in the WWS market. Much like any other monopoly entity operating in the public sector, WWS are susceptible to the quiet life, which consists in the lack of pressure to offer good results and to the inefficiency-X (e.g. waste), which is to say that they are indifferent to costs. As a result, WWS should be oriented to manage according to objectives, in terms of both the quality of service and costs (economic efficiency).

Quality performance must be rewarded and performance that is lacking should be penalized. The allocation of objectives to the different parties involved and the respective meritocracy should always exist, independent of the development of the WWS. As such, ***performance-based incentives are recommended, particularly with respect to economic regulation***. Remuneration that is independent of the results obtained by the operator should not be acceptable, even if they are under control. This circumstance is particularly worsened in the case of WWS due to the fact that they require significant investments (capital intensive industry) and operate close to the construction industry, which can encourage the A–J effect.

Carrot-and-stick policies in the regulation of quality of service, where incentives are provided for excellence (or penalizations for poor performance), guarantee always positive results. ***Performance incentives are the best tools that governments and regulators have for promoting WWS efficiency and innovation*** and therefore to provide value for money.

j) Regulation role and objectives should be clearly defined

Regulation roles and objectives should be ***clear and clearly defined***. The regulatory functions should be ruled by stability (maintaining the rules of the game), which provides a foreseeable nature to the regulation. Given that the WWS infrastructures have a high life cycle, short-term strategic orientations (and policies) should be avoidable. The roles should be clearly defined for those operating in the political and circumstantial arena, and those in the more technical and apolitical arena. There is often a ***lack of definition, overlap or even incompatibility*** between the government functions and those of the various regulators. These circumstances occur between the government (and possibly the Parliament) and the regulator, above all with respect to structural regulation, although other issues can also arise (e.g. the definition of prices and tariffs, the guarantee of certain public service obligations). When there is more than one WWS regulator (e.g. one competition authority and another for the defence of customer rights), it is possible that conflicts will reduce the effectiveness of the processes and harm the image of the different parties involved, and, consequently, the regulation itself.

REFERENCES

ABAR (2006). *Regulação – Indicadores para a Prestação de Serviços de Água e Esgoto* (eds Alceu Junior and Alexandre Silva). Associação Brasileira de Agências de Regulação (ABAR), Fortaleza, Brazil.

ADB (2001). Regulatory System and Networking Water Utilities and Regulatory Bodies. In *Proceedings of the Regional Forum*, 26–28 March, Asian Development Bank, Manila, Philippines.

Alpaos, C. & Valbonesi, P. (2008). *Una Valutazione delle Ipotesi di Revisione del Metodo Tariffario Normalizzato per il Servizio Idrico Integrato*. Dipartimento di Scienze Economiche, Università degli Studi di Brescia, Italy.

Alvarinho, A. (2002). *Water Regulation in Mozambique – Thoughts and Challenges*. Conselho de Regulação do Abastecimento de Água, May, Washington, DC, USA.

Aqualibrium (2003). *European Water Management between Regulation and Competition* (eds Shahrooz Mohajeri, Bettina Knothe, David-Nicolas Lamothe and Jean-Antoine Faby), The European Commission Community Research, Brussels, Belgium.

Barraqué, B. & Bris, C. (2007). Water Sector Regulation in France. *CESifo DICE Report*, **5**(2), 4–12.

BELGAQUA (2005). Data from the site http://www.belgaqua.be/homeEng.htm.

CCE (2006). *Étude Relative` l'Élaboration d'Indicateurs de Performance des Services de Distribuition d'Eau – Rapport Final*. Comité de Contrôle de l'Eau, Liege, Belgium.

Chavez, C. & Quiroga, M. (2002). Regulatory Schemes for Water Provision in Theory and Practice. In *Second Expert Workshop on the Tariff Reform in the Municipal Water Supply in the NIS*, 2–3 June, Moscow, Russia.

Cheng, H. (2007). *Singapore: Water Treatment and Wastewater Recycling Systems*. The US Commercial Service, Tanglin, Singapura, 9 pp.

CNE (2006). *Comptes de l'Eau – Compterendu de Reunion du 5 Janvier 2006, Ministere de l'Ecologie et du Developpement Durable, Direction de l'Eau*, Comite National de l'Eau, Paris, France.

Cour des Comptes (2003). *La Gestion des Services Publics d'Eau et d'Essainissement*. Report presented to the President of the Republic, Cour des Comptes, Paris, France.

CRA (2006). *Desafios da Regulação da Água em Moçambique*. Conselho de Regulação do Abastecimento de Água, Maputo, Mozambique.

Crase, L., Dollery, B. & Byrnes, J. (2008). An intersectorial comparison of Australian water reforms. *Water Pol.*, **10**(1), 43–56.

CSS (2007). *US Water Supply and Distribution*. Center for Sustainable Systems, Michigan, USA.

Danesi, L., Passarelli, M. & Peruzzi, P. (2007). Water services reform in Italy: its impacts on regulation, investment and affordability. *Water Pol.*, **9**(1), 33–54.

DNA (2007). *Recursos Hídricos/Abastecimento de Água – Políticas e Estratégias*. Direcção Nacional de Água, Maputo, Mozambique.

Econometria (2004). Estudio de Regulation Tarifa de Aseo. *Resumen Ejecutivo*, Bogotá, Colombia.

ESC (2008). *Water Performance Report – Performance of Urban Water Sewerage Business 2006–07*. Essential Services Commission, Melbourne, Australia.

Euromarket (2003). *Analysis of the European Union explicit and implicit policies and approaches in the larger water sector*. Work Package IV. Water Liberalization Scenarios. Energy, Environment and Sustainable Development. The European Commission Community Research, Brussels, Belgium.

FP2E (2008). *Les Services Collectifs d'Eau et d'Assainissement en France: Données Économiques, Socials et Environnementales*. Fédération Professionnelle des Entreprises de l'Eau (FP2E), Paris, France.

Fraquelli, G. & Moiso, V. (2005). *Cost Efficiency and Economies of Scale in the Italian Water Industry*. In *Pavia University Presentation*, 15–16 September, Pavia, Italy.

Gouveia, R. (2001). *Os Serviços de Interesse Geral em Portugal*. Coimbra (Portugal): Centro de Estudos de Direito Público e Regulação (CEDIPRE), Coimbra Editorial, Coimbra, Portugal, 177 pp.

Grigg, N. (2007). Water sector structure, size, and demographics. *J. Water Resour. Plann. Manag.*, **133**(1), 60–66.

Guasch, L. (2004). *Granting and Renegotiating Infrastructure Concessions: Doing it Right*. The International Bank for Reconstruction and Development – the World Bank, Washington, DC, USA.

HERA (2005). The Italian Water Sector Between Market Opening and Societal Concerns: Hera's experience. In *Euromarket Conference*, 14 December, Lausanne, France.

IFEN (2005). *La Gestion de l'Eau Potable en France en 2001, Serie etudes et travaux, no. 44*. Institut Français de l'Environnement, Paris, France.

IFEN (2007). *Les Services Publics de l'Eau en 2004 – Volet Eau Potable*. Institut Français de l'Environnement, Paris, France.

IRAR (2003). *As Linhas Estratégicas do Modelo de Regulação a Implementar pelo Instituto Regulador de Águas e Resíduos*. Instituto Regulador de Águas e Resíduos, Lisbon, Portugal.

ISBN: 9781843393412. Published by IWA Publishing, London, UK.

IRAR (2007). *Relatório Anual dos Serviços de Água e de Resíduos em Portugal 2006*. Instituto Regulador de Águas e Resíduos (IRAR), Lisbon, Portugal.

IWA (2006). *International Statistics for Water Services*. International Water Association, Beijing, China.

Kayaga, S. & Franceys, R. (2008). Water Service Regulation for the Urban Poor: Zambia. *Water Manag.*, **161**(2), 65–71.

Kayaga, S. & Kadimba-Mwanamwambwa, C. (2006). Bridging Zambia's Water Service Gap: NGO/Community Partnerships. *Water Manag.*, **159**(3), 155–160.

Larangeira, S. (1996). *Gestão Pública e Participação: a Experiência do Orçamento Participativo em Porto Alegre/Public Management and Participation: Participative Budget Experience of Porto Alegre, Brazil. São Paulo perspect*, **10**(3), 129–137.

Littlechild (1983). *Regulation of British Telecom's Profitability*. London (Inglaterra): Her Majesty's Stationery Office (HMSO), Secretary of State, Department of Industry, 40 pp.

Lwenje, L. (2007). *Lições dos Water Watch Groups*. In *Conferência sobre Regulação*, 11–14 September, Maputo, Mozambique.

Marques, R. (2005). *Regulação de Serviços Públicos*, Edições Sílabo, Lda, Lisbon, Portugal.

Marques, R. (2006a). Performance benchmarking in water and sewerage services regulation: Principles and experiences. *Water Utility Management International*, **1**(2), 6–21.

Marques, R. (2006b). A yardstick competition model for Portuguese water and sewerage services regulation. *Utilities Policy*, **14**(3), 175–184.

Marques, R. & Contreras, F. (2007). Performance-based potable water and sewer service regulation: the regulatory model. *Cuadernos de Administración*, **20**(34), 283–298.

Marques, R. & De Witte, K. (2010). Towards a benchmarking paradigm in the European drinking water sector. *Public Money and Management*, **30**(1), 42–48.

Marques, R. & Monteiro, A. (2004). A privatização dos serviços de água e de águas residuais. *Águas & Residuos*, série II, no. 4, pp. 43–49.

Moran, D. & Dann, S. (2008). The economic value of water use: implications for implementing the water framework directive in Scotland. *J. Environ. Manag.*, **87**(3), 484–496.

Moreira, V. (1997). Auto-regulação Profissional e Administração Pública. Livraria Almedina, Coimbra, Portugal, 427 pp.

MPO-SEPURB (1995). *Regulação da Prestação de Serviços de Saneamento – Análise comparada da Legislação Internacional*. NH Consultoria e Planejamento – Ministério do Planejamento e Orçamento – Secretaria de Política Urbana – Instituto de Pesquisa Econômica Aplicada, Brasília, Brazil.

NWASCO (2008). *Urban and Peri-urban Water Supply and Sanitation – Sector Report 2006/2007*. National Water Supply and Sanitation Council, Lusaka, Zambia.

NWC (2005). Australian Water Resources. National Water Commission, Australia Government, Australia.

OFWAT (2006). *Levels of Service for the Water Industry in E&W – 2005–06 Report*. Water Services Regulation Authority (OFWAT), Birmingham, England.

OFWAT (2007a). *Facts and Figures – May 2007*. Water Services Regulation Authority, Birmingham, England.

OFWAT (2007b). *Regulatory Director Letter 2007*. Water Services Regulation Authority, Birmingham, England.

Oliveira, A. (2008). Private Provision of Water Service in Brazil: Impacts on Access and Affordability. *MPRA Paper* No. 11149.

Orçamento Participativo (1995). *Você é Quem faz uma Cidade de Verdade*. Prefeitura Municipal de Porto Alegre, Porto Alegre, Brazil.

PURC (2005). *Review of the Performance of GWCL 1998–2003*. Public Utilities Regulatory Commission (PURC), Accra, Ghana.

Reynaud, A. (2007). *Social Policies and Private Sector Participation in Water Supply – The Case of France*. Developed for Organização das Nações Unidas para a Educação, a Ciência e a Cultura (UNRISD) project in 'Social Policy, Regulation and Private Sector Involvement in Water Supply', Genoa, Switzerland.

RIONED (2006). *Urban Drainage Statistics 2005–2006 – Figures on urban drainage in the Netherlands*. Fundação RIONED, Tilburg, Holanda.

Rovizzi, L. & Thompson, D. (1995). The regulation of product quality in the public utilities. In *The Regulatory Challenge* (eds M. Bishop, J. Kay & C. Mayer), Oxford University Press, Oxford, UK.

SISS (2007a). *Informe de Gestión del Sector Sanitario Año 2006*. Superintendencia de Servicios Sanitarios, Santiago of Chile, Chile.

SISS (2007b). *Superintendencia de Servicios Sanitarios (SISS) Entrego Evaluación de la Calidad de Servicio Año 2006*. Superintendencia de Servicios Sanitarios, Santiago of Chile, Chile.

SNIS (2006). *Diagnóstico dos Serviços de Água e Esgotos*. Sistema Nacional de Informações sobre Saneamento, Brasília, Brazil.

Speck, S. (2006). *Financial Aspects of Water Supply and Sanitation in Transboundary Waters of South-Eastern Europe*. Relatório para o German Federal Ministry for the Environment, Nature Conservation and Nuclear Safety, Berlin, Germany.

SSPD (2002). Superintendencia de Servicios Publicos Domiciliarios. *Revista SuperCifras en m^3*, no. 5, Bogotá, Colombia.

Stern, J. & Holder, S. (1999). Regulatory governance: criteria for assessing the performance of regulatory systems. An application to infrastructure industries in the developing countries of Asia. *Util. Pol.*, **8**(1), 33–50.

SUI (2008). Sistema Unico de Informacion de Servicios Publicos: www.sui.gov.co.

Swami, K. (2004). *Evolution of Legislation and Regulations in the Dutch Water Supply and Watewater Sector*. Master of Science Thesis. UNESCO-IHE Institute for Water Education, Delft, The Netherlands.

UNDP (2004). *Assessment and Development of Municipal Water and Wastewater Tarifs and Effluent Charges in the Danube River Basin. National Profile of Municipal Water and Wastewater Management in Romania*. Project for UNDP 'Danube Regional Project', Vienna, Austria.

VEOLIA (2005). *The French Water Services: Main Present Challenges*. May, Veolia Water.

VEWIN (2005). *Water Supply Statistics 2005*. Association of Dutch Water Companies, Rijswijk, The Netherlands.

VEWIN (2006). *Reflections on Performance 2006 – Benchmarking in the Dutch Drinking Water Industry*. Association of Dutch Water Companies, Rijswijk, The Netherlands.

Viero, O. & Cordeiro, A. (2003). *The Case for Public Provisioning in Porto Alegre*. WaterAid/Tearfund, London, England.
Water Aid (2005). *WaterAid – National Water Sector Assessment*. WaterAid – water for life, ONG Internacional, London, England.
WaterTime (2004). *Standard Format for Case Studies and City in Time Reports*. Coordinated by Public Services International Research Unit (PSIRU), University of Greenwich, Greenwich, England.
WICS (2002). *Commissioner's Corporate Plan*. Water Industry commissioner for Scotland, UK.
Williamson, B. (2001). Incentives for service quality: getting the framework right. *The Electricity Journal,* **14**(5), 62–70.
World Bank (1999). *Project Appraisal Document on a Proposed Credit in The Amount of SDR 55.4 Million (US$ 75 Million Equivalent) to the Republic of Mozambique for a National Water Development II*. Document of The World Bank Report no. 19131 MOZ, Washington DC, USA.
World Bank (2005). World Bank Indonesia: www.publicfinanceindonesia.org.

ANNEXES – EXPLORATORY STUDY

ANNEX I – QUESTIONNAIRE ABOUT REGULATION

QUESTIONNAIRE 'WATER SERVICES REGULATION'

Country or Region Profile

Notes:

- Please fill in the questions below, referring to the guidelines at the end of the questionnaire if applicable;
- If your country (state/region) does not have any regulatory agency for the water sector, please ignore sections 3 and 4;
- If you have any queries please contact Pedro Simões by phone +351 218418316 or email: psimoes@civil.ist.utl.pt. We are glad to answer any questions or comments you might have regarding the questionnaire;
- We intend to use the information you provide within the framework of a survey. Once data are provided we assume that we are allowed to use the data for this specific purpose;
- We will be happy to provide you with the results of the survey after completion.

Elements of the questionnaire:

Reference time for the survey: .... 5 December 2007
Latest date for submission of the questionnaire: .... 19 December 2007
Estimated time for response: .... 45 minutes

The questionnaire was filled in by:

Name: ______________________________

Position: ______________________________

Contacts: ______________________ (Phone) ______________________ (Email)

Please send the completed questionnaire by email or by post to:

Address:

Prof. Rui Cunha Marques
Centre for Urban and Regional Systems
Instituto Superior Técnico
Technical University of Lisbon
Av. Rovisco Pais
1049-001 Lisbon
Portugal

Additional contacts:

Phone: +351 218418319
Fax: +351 218409884
Mobile: +351 969076504
Email: rcmar@civil.ist.utl.pt

Thank you very much for your collaboration!

ISBN: 9781843393412. Published by IWA Publishing, London, UK.

Country or Region Profile

a) Information about the water services industry:

Geographical scope of the data provided (name of country or region): ______________________

Coverage level of drinking water supply service (%): ______________________

Coverage level of wastewater collection service (%): ______________________

Coverage level of wastewater treatment service (%): ______________________

Annual drinking water volume produced (million m^3): ______________________

Non-revenue water volume (%): ______________________

Annual collected wastewater volume (million m^3): ______________________

Annual treated wastewater volume (million m^3): ______________________

Number of drinking water supply service providers (no.): ______________________

Number of wastewater service providers (no.): ______________________

Total revenue from public water supply service (million US$): ______________________

Total revenue from wastewater collection and treatment services (million US$): ______________________

Investments on the water supply infrastructure/service (million US$/last 3 years): ______________________

Investments on the wastewater collection and treatment infrastructure/service (million US$/last 3 years): ________

b) Information about the regulator:

Creation of the regulator (legal document and date of beginning of activity): ______________________

Scope of the regulation (industries, activities and territory): ______________________

Total expenditures for the year 2006 (US$): ______________________

Sources of the regulator revenues (% by source): ______________________

Staff (no.): ______________________

1. General aspects of the water services

A) Who are the responsibles for the provision of water services?

- ☐ Government
  - ☐ Local
  - ☐ Regional
  - ☐ National
  - ☐ Other: ____________________
- ☐ Other: ____________________

B) What are the roles of the private sector in these industries?

- ☐ Ownership
- ☐ Management (concession, lease or other type of contract)
- ☐ Outsourcing. Please specify the scope of activities: ____________________
- ☐ Other: ____________________

C) How is the relationship between the water assets owner and the service provider established?

- ☐ Existence of a contract
- ☐ Existence of a licence
- ☐ Other: ____________________
- ☐ None. They are the same entity

D) What kind of regulatory agency is responsible for quality of service and economic regulation of water services?

- ☐ Independent agency
- ☐ Non-independent agency
- ☐ Transversal agency (e.g. competition authority). Please specify: ____________________
- ☐ Self-regulation
- ☐ Other: ____________________
- ☐ None

E) Who are the main actors of the water sector?

- ☐ Minister responsible for the water issues
- ☐ Regulator
- ☐ Service provider
- ☐ Service providers association
- ☐ Customers association
- ☐ Competition authority
- ☐ Ombudsman
- ☐ Other: ____________________

2. Scope of the regulatory activity

A) Who establishes or approves the water services tariffs?
☐ Government
☐ Regulator
☐ Service provider
☐ Contract
☐ Other: ____________________
B) Is there any procedure of appeal against decisions taken regarding the tariff setting?
☐ No
☐ Yes, to the Government
☐ Yes, to the regulator
☐ Yes, to the competition authority (or similar agency)
☐ Yes, to the Courts
☐ Other: ____________________
C) How is the quality of service supervised?
☐ By the application of sanctions and awards
☐ By means of the tariff system
☐ By publicizing the best and worst practices
☐ By threatening to revoke the licence or to cancel the contract
☐ Other: ____________________
☐ There are no explicit procedures established
D) How do customers claim their rights?
☐ Directly to the sector regulator
☐ Directly to the Government or to a Government agency
☐ Directly to the customer's association
☐ Directly to the ombudsman or similar entity
☐ Other: ____________________
E) What is the role of benchmarking in the regulatory activity?
☐ To compute the tariffs
☐ To supervise the quality of service
☐ Other: ____________________
F) Are there any public service obligations defined in the law?
☐ Yes
☐ No

3. General aspects of the regulatory agency

A) Are the role, tasks and competences of the regulatory agency clearly defined and published?
☐ Yes
☐ No
B) How is independence from political influence ensured?
☐ Decisions are taken without being subject to Government assessment
☐ Staff are chosen without political criteria, with long mandates and are hard to dismiss
☐ Its budget does not depend on the Government
C) How does the regulatory agency carry out investigations?
☐ Periodically, imposed by its statutes
☐ By request of a stakeholder
☐ Other: ______
D) Are the decisions of the regulatory agency legally binding?
☐ Yes, immediately
☐ Yes, but with suspensory effect if decision is appealed
☐ No
E) Can the regulatory agency impose fines or other sanctions?
☐ Yes
☐ No
F) Can the regulatory agency enact legislation?
☐ Yes
☐ No
G) How is the regulators' accountability guaranteed?
☐ Answering to Parliament whenever required
☐ Transparency and participation in the decisions by the stakeholders
☐ Publicizing the decisions and the regulator financial statements
☐ Possibility of appeal to other authorities
☐ Other: ______
H) How high is the legal certainty of *ex ante* decisions taken by the regulatory agency?
☐ Legal certainty ensured (*ex ante* decisions are not revised)
☐ Legal certainty not ensured (*ex ante* decisions are revised)
☐ No prior experience of these cases

4. Organizational aspects of the regulatory agency

A) Does the agency cover various sectors or is it specific to water services?

- ☐ Various sectors
- ☐ Specific to water services
- ☐ Other: ____________________

B) How many employees are responsible for the regulation of the water sector in your agency?

Number: ____________________

C) Who is the head of the regulatory agency?

- ☐ Singular person
- ☐ Collegiate board
- ☐ Other: ____________________

D) Are there any scientific/stakeholders' advisory committees that assist the agency?

- ☐ Yes. Please specify: ____________________
- ☐ No

E) How is the regulator financed?

- ☐ By the Government
- ☐ By the service provider
  - ☐ Depending on the volume
  - ☐ Depending on the number of customers
  - ☐ Fixed percentage of the tariff
  - ☐ Other: ____________________
- ☐ Other ____________________

F) Are the decisions and other important activities of the agency published?

- ☐ No
- ☐ Yes, namely:
  - ☐ Decisions
  - ☐ Exemptions
  - ☐ Powers
  - ☐ Fines
  - ☐ Ongoing procedures
  - ☐ Other administrative matters, for example: ____________________

G) Are the criteria and methods of the regulation transparent and understandable?

- ☐ Yes. Please specify: ____________________
- ☐ No

5. Economic regulation

A) What is the economic regulation method used for tariff setting?	
☐ Rate-of-return regulation (cost-based)	
☐ Price cap regulation	
☐ Revenue cap regulation	
☐ Other: ____________________	
B) How long does the regulatory period take?	
☐ 1 years	☐ 2 years
☐ 3 years	☐ 4 years
☐ 5 years	☐ Other: __________
C) What kind of incentives does the regulatory method have?	
☐ The service provider retains all the profits	
☐ Sharing mechanism	
☐ Other: ____________________	
☐ None	
D) How is the quality of service encompassed in the tariff setting procedure?	
☐ Change in tariffs for up and down	
☐ Rewards for out-performance	
☐ Penalties for under-performance	
☐ None	
☐ Other: ____________________	
E) How are the investment plans defined?	
☐ Presentation of master plans (short-, medium- and long-term)	
☐ Based on the network quality of service	
☐ Based on minimum investment levels	
☐ Other: ____________________	
F) Are the investments approved by the regulator?	
☐ Yes	
☐ No	
G) Who determines the cost of capital?	
☐ The regulator. Please specify the procedure (CAPM, DGM, ...): __________	
☐ The country's central bank	
☐ The Government	
☐ The service provider	
☐ Other: ____________________	

H) How are unpredictable situations included in the regulatory process?
☐ By means of extraordinary tariff revisions
☐ By means of the contract extension
☐ Service providers responsibility
☐ Compensation by the Government
☐ Other: ________________
I) If benchmarking is applied, which methodologies are adopted?
☐ Total factor productivity
☐ Non-parametric techniques (Data envelopment analysis – DEA)
☐ Stochastic parametric techniques (Stochastic frontier analysis – SFA)
☐ Deterministic frontier analysis (Regression, ...)
☐ Performance indicators
☐ Other: ________________
J) How is the quality of information supervised?
☐ By auditors
☐ By periodical inspections
☐ Other: ________________
K) What kind of tariff structure is typically adopted for residential customers?
☐ Fixed charge equalized for each customer
☐ Fixed charge depending on customer characteristics (meter, area, ...)
☐ Volumetric charge:
☐ Fixed
☐ Increasing block charge
☐ Decreasing block charge
☐ Other: ________________

6. Public service obligations and quality of service regulation

A) Is the universal service requirement established in the law?
☐ Yes
☐ No
B) How is the universal service intended to be guaranteed?
☐ Cross-subsidization
☐ Compensation fund
☐ Government subsidizes service providers
☐ Government subsidizes some customers directly
☐ Other: ______
C) Do special tariffs exist for special customers? If yes, please tick the correct answer.
☐ Numerous families
☐ Poor
☐ Customers with special needs (disabilities and health problems, elderly, ...)
☐ Other: ______
D) What is the procedure adopted for customers' debts?
☐ Interrupting the service
☐ Negotiating, for all customers
☐ Negotiating, for special customers
☐ Other: ______
E) Are there information obligations of the service provider defined in the law?
☐ Yes. Please specify: ______
☐ Only about water quality data
☐ No
F) Are there automatic compensations when the supply is interrupted without notice and due to the service provider fault?
☐ Yes
☐ No
G) In which circumstances are there other kind of compensations?
☐ Water quality problems
☐ Water pressure problems
☐ Complaints not answered
☐ Other: ______

H) Which levels of service are defined in law?
☐ Water pressure
☐ Interruptions
☐ Restrictions in water use
☐ Flooding from sewers
☐ Billing contacts
☐ Written complaints
☐ Ease of telephone contact
☐ Other: ____________________
I) What are the procedures adopted when there is a poor quality of service?
☐ Automatic compensations
☐ Compensations when required by the particular stakeholders
☐ Penalties in tariff systems
☐ Other: ____________________
☐ None

Final Comment

How would you assess the performance of the regulatory model applied to water service provision?

Key strengths
■
■
■
■
■
■
Areas for improvement
■
■
■
■
■

Comments on 'Water services regulation'

Please cross-reference your comments by question number. If you need extra space, please make a copy of this page and number each copy in the space provided.

Question number	Comment

Copy number

GUIDELINES FOR COMPLETING THE QUESTIONNAIRE

- This questionnaire asks you to provide details about various aspects of water services regulation in your country. It is important that you complete it fully and accurately.
- Each box should be completed where possible. Any boxes that do not apply should be marked as not applying (n.a.) in your case.
- Sections 3 and 4 only apply if your country (state/region) has a regulatory agency for the water services sector.
- If the space provided for your answer is insufficient, write 'see comments' and provide the extended answer in the comments question at the end of the questionnaire.

Comments for answering the questions according to their sections:

1. General aspects of the water service:

(a) The entity responsible for the provision of water services can be either public or private owned. The public entity may correspond to several kinds of government, such as local, regional, provincial, national or federal, including other bodies (e.g. river basin authorities). Please separate the responsibilities for the water supply and wastewater services if they are not carried out by the same entity.
(b) Discriminate the type of the management contract. Even if only a particular activity is subject to the intervention of the private sector, please refer it (e.g. BOT of wastewater treatment plant).
(c) Sometimes it is the same entity, other times there are different entities but no contract established and other times the rights and obligations of the service provider and owner of the assets are established.
(d) The regulatory functions can belong to a transversal agency, that is, an agency that encompasses several sectors with only a particular responsibility (e.g. customers' protection or competition rules). In the case of self-regulation, please indicate if it corresponds to the service provider itself or to the association of the sector.
(e) Please write the names. When more than one entity exists or if they are different for water and wastewater please specify their names.

2. Scope of the regulatory activity

(a) Specify the government level if applicable.
(b) The ombudsman or similar entity assumes itself as independent from the Government.
(c) Please point out on which terms the regulator uses benchmarking to set tariffs and/or to supervise the quality of service.
(d) Please specify what are the public obligations established in the law, namely if universal service, continuity, quality of service, price affordability, users and customers' protection are defined in the legislation.

3. General aspects of the regulatory agency

(a) Refer if the role, tasks and competences do not overlap those of other agencies.
(b) Please specify how long is the mandate of the board of directors (or director-general) and how it is appointed (sector ministry, Cabinet, Parliament ...).
(c) Identify the situations in which investigations can be developed.
(d) Please explain in which situations the fines or other sanctions can be applied.
(e) Please specify what kind of legislation (and its subject) the regulator can enact.
(f) Identify the forms of appeal of the regulated firms if they exist and if there are any kinds of advisory board that give advice before the decisions are taken. Explain the participation of stakeholders in the regulatory process.

4. Organizational aspects of the regulatory agency

(a) Specify under which situations and on what topics the advisory committees can be heard and how binding are their recommendations.
(b) Point out how transparency is provided.

5. Economic regulation

(a) Specify how the rate-of-return is calculated, if it adopts rate-of-return regulation and how factor X in price or revenue formulas is defined.
(b) Indicate how the quality of service is evaluated, if by performance indicators, satisfaction surveys or other mechanisms.
(c) Describe the tariff system features.

6. Public services obligations and quality of service regulation

(a) Detail the procedure adopted.
(b) Point out the consequences of infringing the levels of service.

Glossary

- Customers – Clients, end-users;
- Responsible body for service provision – Entity with legal obligation to provide the service;
- Regulatory agency – Regulatory commission, body, authority;
- Service provider – Operator, water/wastewater company;
- Water assets owner – Owner of the water services infrastructure;
- Water services – Drinking water supply and wastewater collection and treatment services.

In case of questions or suggestions please do not hesitate to contact the following CESUR team:

Prof. Rui Cunha Marques
Project Team Leader
Centre for Urban and Regional Systems
Technical University of Lisbon
Phone: +351 218418319
Fax: +351 218409884
Mobile: +351 969076504
Email: rcmar@civil.ist.utl.pt

Eng. Pedro Simões
Assistant Researcher
Centre for Urban and Regional Systems
Technical University of Lisbon
Phone: +351 218418316
Fax: +351 218409884
Mobile: +351 963628321
Email: psimoes@civil.ist.utl.pt

CESUR, DECivil
Instituto Superior Técnico
Technical University of Lisbon
Av. Rovisco Pais
1049-001 Lisbon
Portugal

We are aware that it is not possible to capture all the specificities and richness of detail through a standardized survey. Therefore prior to data treatment and processing, our team will contact you (phone interview/email) in order to validate our understanding of your answers and add any additional commentary and clarification.

Thank you very much for your collaboration!

ANNEX II – ANSWERS TO THE QUESTIONNAIRE

ARMENIA

Regulatory Model

In Armenia, there is a multisector regulatory agency (Public Service Regulatory Commission – PSRC) with competencies in economic regulation, quality of service and public service obligations for the WWS. It also has regulatory functions in the areas of electricity, gas and telecommunication services. Economic regulation, performed by PSRC, is based on rate-of-return regulation for a regulatory period of one year. In carrying out its functions, PSRC is an independent authority and has vast operational powers including the capacity of applying sanctions and enacting legislation. The regulator plays a relevant role in the quality of service supervision, acting according to the principles of transparency and participation of various stakeholders in the decision-making process. PSRC uses its website as an essential tool for interacting with stakeholders. The main legal documents of the water sector in Armenia are the Water Code, the Domestic Water Policy and the National Water Program.

Market Structure

The WWS are fundamentally provided by five public regional companies of commercial law. Additionally, there is a significant group of small public operators, local in nature, that act in rural zones with less representation in terms of population. From these five companies, the two main ones, which serve nearly 85% of the population, have management contracts in the private sector, specifically with Veolia. The WWS find themselves vertically integrated in just one operator and include only activities related to the water and wastewater activities.

Tariff Setting

The tariff system adopted by the operator is based on a variable charge, established according to the volume of water consumed. The tariffs are greatly reduced and insufficient. The collection system is inadequate and ineffective. The access for monitoring meters is limited. The government subsidizes low-income customers. The operators set distinct tariffs depending on the type of customer.

Quality of Service

The infrastructures are highly deteriorated with a great need for investments. The wastewater treatment coverage is null. Service is available for a limited number of hours per day and the levels of water losses are extremely high. The regulator has been making efforts to improve quality of service by publicizing and making performance information available.

The Regulator

Characteristics
- Multisector regulator
- Independent regulatory agency

Legal basis
Decree issued by the President of the Republic of Armenia in 1997

Number of employees
94 employees

Budget
5,414,000 euros (2006)

Contacts
Public Service Regulatory Commission
22 Saryan St, Yerevan, 375002
Armenia
Tel: +374 1 522 522
Fax: +374 1 525 563
http://www.psrc.am

Relevant Entities

State Committee for Economy
- http://www.mineeconomy.am

Ministry for the Protection of Nature
- http://www.mnpiac.am

Water Resource Management Agency
- http://www.wrma.am/

Ministry for the Spatial Planning
- http://www.mta.gov.am/

The Sector

Country
- Area – 298,000 km^2
- Population – 3.2 million
- Population density – 101 inhab/km^2

Service Coverage
- Water – 90%
- Wastewater service – 70%
- Wastewater treatment – 0%

Number of operators
- Water – 5
- Wastewater – 5

Ownership and management of services

	Ownership		Management	
	Pub.	Priv.	Pub.	Priv.
Water	100%	0%	15%	185%
Wastewater	100%	0%	15%	85%

Volume of activity
- Abstracted water volume – 588 million m^3
- Drinking water – 88 million m^3
- Average consumption – 84 litres per person per day
- Main source – 96% groundwater
- Water losses – 85%

Economic and financial data
- Turnover – 16.6 million euros
- Average price water service – 0.36 €/m^3
- Average annual invoice per customer –
- Employment –

AUSTRALIA (VICTORIA)

Regulatory Model

In Victoria, there is a multisector regulatory agency (Essential Services Commission – ESC) responsible for the regulation of public services in the state, such as electricity, gas, ports, rail and road transportation, insurance, renewable energy and WWS. However, the regulatory framework is very distinct and dependent upon the sector regulated. The regulatory agency was created with the primary purpose of protecting customer interests in the state of Victoria, in terms of pricing, quality and reliability of public services. To meet this end, the ESC has to provide an efficient, viable (economically), competitive and consistent service, promoting the economic and social cohesion in the territorial range. Up until January 2004, the ESC competencies for the water sector include the quality of service regulation and the fulfilment of public service obligations, but only for metropolitan entities. The activity carried out by the ESC over nearly ten years was based on sunshine regulation. Starting from this date, its powers were expanded to include economic regulation and all the operators of the state of Victoria. In compliance with tariff regulation, ESC established the maximum prices (price-cap regulation) for a period of five years for all WWS services. The Essential Services Commission Act of 2001 and the Water Act of 1989 are the major legal documents related to the WWS.

Market Structure

Melbourne Water, the main operator in Victoria, was restructured in 1994, and was divided into four companies (with public management and ownership), one with the responsibility of providing wholesale water services and the main WWTP and three others with end-user WWS competencies, respectively City West Water, South East Water and Yarra Valley Water. Currently, Melbourne Water, besides supplying these three metropolitan companies, also provides wholesale water to Western Water, Gippsland Water and Southern Rural Water operators. In addition to the metropolitan operators, there are also a total of 13 regional and rural operators that provide WWS in the State of Victoria, comprising approximately 400 cities. In opposition to the metropolitan operator, which was subject to unbundling, the regional operators are generally vertically integrated.

Tariff Setting

The tariff system adopted by the operators varies between a single fixed part, charged uniformly to all domestic customers, and a system consisting of a fixed part and a proportional part priced per increasing block rate set according to the volume of water consumed. The operators, especially the urban, set distinct tariffs depending on the customer.

Quality of Service

The WWS are characterized by providing an outstanding service. The high level of coverage and continuity of service prove the great quality of the WWS, which is a feature of the state. The GSS and the levels of quality of service established by each operator as well as the sunshine regulation model applied by the ESC guarantee the excellent service provided.

The Regulator

Characteristics
- Multisector regulator
- Independent regulatory agency

Legal basis
Essential Services Commission Law, 2001

Number of employees
65 employees (13 water sector)

Budget
8.4 million euros (2007)

Contacts
Essential Services Commission
Level 2, 35 Spring Street, Melbourne Victoria 3000, Australia
Tel: +61 3 9651 0222
Fax: +61 3 9651 0222
http://www.esc.vic.gov.au

Relevant Entities

Ombudsman
- http://www.ombudsman.vic.gov.au

Environment Protection Authority
- http://www.epa.vic.gov.au

Australian Water Association
- http://www.awa.asn.au

National Water Commission
- http://www.nwc.gov.au

The Sector

State
- Area – 237,000 km^2
- Population – 5.2 million
- Population density – 22.9 inhab/km^2

Service Coverage
- Water – 94%
- Wastewater service – 86.7%
- Wastewater treatment – 86.7%

Number of operators
- Water – 17
- Wastewater – 17

Ownership and management of services

	Ownership		Management	
	Pub.	Priv.	Pub.	Priv.
Water	100%	0%	100%	0%
Wastewater	100%	0%	100%	0%

Volume of activity
- Abstracted water volume – 671 million m^3
- Drinking water – 607 million m^3
- Average consumption – 180 litres per person per day
- Main source – 94% superficial
- Water losses – 9.5%

Economic and financial data
- Turnover – 1,070 million euros
- Water service average price – 1.24 €/m^3
- Average annual invoice per customer – 444 euros
- Employment – 1,700 employees

AUSTRALIA (WESTERN AUSTRALIA)

Regulatory Model

In Western Australia, there is a multisector regulatory agency (Economic Regulation Authority – ERA) which has the functions of regulating WWS, electricity, gas and railways. In the scope of WWS regulation, its authority is centred in the quality of service regulation and the licensing of operators. The ERA also has a relevant role in defending customer rights as well as in safeguarding the sustainability and viability of the systems by making recommendations to the government. The ERA performs its activity independently. Although it is endowed with sanctioning powers, it does not have legislative powers. The regulator, being a collegiate body, self governs based on the principles of responsibility, independence, integrity, professionalism, transparency, participation and excellence. The quality of service regulation is based, above all, on the comparison and publication of the performance results of regulated operators (sunshine regulation). Pricing and tariff setting are proposed by the operators in conformity with their respective licence, and are always subject to the opinion of customers and other stakeholders, as well as supervised by governmental entities. The main legislation of the water sector in the state of Western Australia is derived from the Water Services Licensing Act of 1995 and the Economic Regulation Authority Act of 2003.

Market Structure

The WWS are provided by public operators in Western Australia. Assets ownership is always public, even though it may belong to the state government or local administration. The services have a high level of horizontal integration and usually include all activities of the water cycle, in the water service (including supplying non-drinking water), the wastewater service or the stormwater service. The main WWS operator in Australia is the Water Corporation, a public state company that supplies 90% of the total population. The second operator is Aqwest, which covers around 2% of the population. The remaining operators are of limited size.

Tariff Setting

The WWS tariffs are set by the operator based on the licence granted through the contract agreed. The tariff system adopted by the operators consists of a fixed part according to the characteristics of the property. The tariffs are fairly reasonable, taking into account the cost recovery. State subsidization is considerably low. The operators set distinct tariffs depending on the type of customer (domestic, industry, etc.).

Quality of Service

The WWS are recognized for having an excellent quality of service. Beyond the high level of coverage and service continuity and reliability that characterize the WWS, service charters and quality of service levels are established for each operator. The sunshine regulation model carried out by ERA also contributes for the noteworthy quality provided.

The Regulator

Characteristics
- Multisector regulator
- Independent regulatory agency

Legal basis
Economical Regulation Authority Law of 2003

Number of employees
43 employees (4 water sector)

Budget
4.5 million euros (2007)

Contacts
Economic Regulation Authority
PO Box 8469, Perth Business Centre 6849
Western Australia
Australia
Tel: +61 8 9213 1900
Fax: +61 8 9213 1999
http://www.era.wa.gov.au

Relevant Entities

Customers' Association
- http://www.customers.asn.au

Ombudsman
- http://www.ombudsman.wa.gov.au

Department of Water
- http://www.water.wa.gov.au

National Water Commission
- http://www.nwc.gov.au

The Sector

State
- Area – 2,645,000 km^2
- Population – 2.1 million
- Population density – 0.83 inhab/km^2

Service Coverage
- Water – 98%
- Wastewater service – 94%
- Wastewater treatment – 94%

Number of operators
- Water – 4
- Wastewater – 28

Ownership and management of services

	Ownership		Management	
	Pub.	Priv.	Pub.	Priv.
Water	100%	0%	100%	0%
Wastewater	100%	0%	100%	0%

Volume of activity
- Abstracted water volume – 311 million m^3
- Drinking water – 280 million m^3
- Average consumption – 373 litres per person per day
- Main source – 50% groundwater
- Water losses – 10%

Economic and financial data
- Turnover – 1,680 million euros
- Water service average price – 1.25 €/m^3
- Average annual invoice per customer – 452 euros
- Employment – 2,406 employees (Water Corporation)

BARBADOS

Regulatory Model

In Barbados, there is no a regulatory agency dedicated to WWS economic regulation, quality of service and public service obligations. In fact, the Fair Trading Commission – FTC – which is dependent upon the state budget, takes on an even larger role, having relevant competencies of transversal regulation of the different sectors. This authority, intervening in all aspects related to competition in the water sector, predominantly performs the role of its supervision, especially in aspects of economic nature. The FTC defines the criteria for setting tariffs, which includes the application of price caps, monitoring the compliance with these tariffs, promoting the efficiency of regulated operators, determining the quality of service levels and their fulfilment and, above all, protecting customer interests and guaranteeing the service provision with quality and at a fair price. However, only the sectors of electricity and telephones are liable to the action of FTC. In the future it is foreseen that water and gas will be also included. Currently, the regulator does not control the quality of service of the state-owned WWS company. The service is primarily regulated by the Barbados Water Authority 1980 Act that created the operator, and by the FTC Act of 2001.

Market Structure

The Barbados Water Authority is the sole WWS operator in the country. It is a public entity. Within this framework, the services are characterized by vertical integration, in both water and wastewater services. Within this scope, the operator performs all the services related to water supply, from abstraction to distribution, and the collection and treatment concerning the wastewater services.

Tariff Setting

The tariff system adopted by the operator consists of a proportional tariff that can be either constant or priced per increasing block rate, both according to the volume of water consumed. The domestic customers with consumption higher than 34 m^3 are considered commercial customers. The most low-income customers are subsidized by the state. The government also subsidizes the operator directly.

Quality of Service

Traditionally, the Barbados Water Authority has provided a reasonable quality of service, with emphasis on the provision of universal service. However, the country shows some weaknesses on water resources protection, reliability of service and the non-revenue water volume is substantial. Wastewater service has a significantly low level of coverage and requires huge investments.

The Regulator

Characteristics
- Multisector regulator
- Independent regulatory agency

Legal basis
FTC Law of 2001

Number of employees
30 employees (1 water sector)

Budget
1.4 million euros (2007)

Contacts
Fair Trading Commission
Good Hope, Green Hill, St. Michael
Barbados
Tel: + 246 424 0260
Fax: + 246 424 0300
http://www.ftc.gov.bb

Relevant Entities

Ministry for Trade and Industry
- http://www.commerce.gov.bb

Ministry for the Energy and the Environment
- http://www.barbados.gov.bb

Caribbean Water and Wastewater Association
- http://www.cwwa.net

Caribbean Water Management Programme Inc.
- http://www.cbwmp.org

The Sector

Country
- Area – 431 km^2
- Population – 279,000
- Population density – 647 inhab/km^2

Service Coverage
- Water – 99%
- Wastewater service –
- Wastewater treatment –

Number of operators
- Water – 1
- Wastewater – 1

Ownership and management of services

	Ownership		Management	
	Pub.	Priv.	Pub.	Priv.
Water	100%	0%	100%	0%
Wastewater	100%	0%	100%	0%

Volume of activity
- Abstracted water volume – 42.7 million m^3
- Drinking water – 21.4 million m^3
- Average consumption – 210 litres per person per day
- Main source – 87% groundwater
- Water losses – 50%

Economic and financial data
- Turnover – 46 million euros (projected)
- Water service average price – 0.55 €/m^3
- Average annual invoice per customer –
- Employment –

BELIZE

Regulatory Model

Belize has a multisector regulatory agency (Public Utilities Commission – PUC) with the responsibility of regulating WWS, electricity and telecommunications. The regulator has organic and functional independence. It is responsible for ensuring the provision of public services with quality, reliability and fair tariffs to the customers and, similarly, guaranteeing the sustainability of the operators. The regulator basically has the role of safeguarding the public interest by ensuring that there are correct, viable and clear methods for handling questions, complaints and technical assistance when requested. The multipurpose role of the PUC creates a dynamic but well-sustained regulation. To this end, the PUC adopted, as its main method of quality of service regulation, a system of benchmarking that determines the total factor productivity as a way of evaluating and encouraging the quality of service development. Concerning the economic regulation, the PUC adopted the method of rate-of-return regulation for a period of five years. The Public Services Commission Act of 2001 is the main regulatory tool in the sector.

Market Structure

The government privatized the Belize Water and Sewerage Authority in 2000, and today the WWS are mostly managed by the private sector, of which Belize Water Services is emphasized. The water sector is characterized by a horizontal and vertical integration of its respective services into a single operator, providing all services for the whole water cycle, from abstraction to water distribution and from collection to treatment of wastewater. In areas where wastewater service is available, the operator simultaneously manages both services.

Tariff Setting

The tariff system adopted by the operator is priced per increasing block rate, according to the volume of water consumed. The rate values are very low and insufficient. The government subsidizes the operators. Special tariffs exist for low-income customers. The tariffs are set differently depending on the area and type of customer.

Quality of Service

The WWS provides a reasonable quality of service in spite of the constraints that characterize the country. The water service coverage is good enough but there is a great need for improvement in the wastewater service. Some performance indicators, such as water losses, require enhancement in the future.

The Regulator

Characteristics
- Multisector regulator
- Independent regulatory agency

Legal basis
Commission of Public Services Law of 2001

Number of employees
20 employees (7 water sector)

Budget

Contacts
Public Utilities Commission
41 Gabourel Lane, PO Box 300, Belize City
Belize
Tel: +501 223 4938
Fax: +501 223 6818
http://www.puc.bz

Relevant Entities

Ministry for Natural Resources and the Environment
- http://www.mnrei.gov.bz

Ministry of Public Works
- http://www.governmentofbelize.gov.bz

Ministry of Public Services
- http://www.governmentofbelize.gov.bz

Caribbean Water and Wastewater Association
http://www.cwwa.net

The Sector

Country
- Area – 23,000 km^2
- Population – 314,300
- Population density – 13 inhab/km^2

Service Coverage
- Water – 90%
- Wastewater service – 40%
- Wastewater treatment – 40%

Number of operators
- Water – 2
- Wastewater – 1

Ownership and management of services

	Ownership		Management	
	Pub.	Priv.	Pub.	Priv.
Water	17.3%	82.7%	0%	100%
Wastewater	17.3%	82.7%	0%	100%

Volume of activity
- Abstracted water volume – 65 million m^3
- Drinking water – 25 million m^3
- Average consumption – 240 litres per person per day
- Main source – 87% superficial
- Water losses – 38%

Economic and financial data
- Turnover – 8.9 million euros (water)
- Water service average price – 0.15 €/m^3
- Average annual invoice per customer –
- Employment – 550 employees

BRAZIL (ALAGOAS)

Regulatory Model

In the state of Alagoas, there is a regulatory agency (the Public Services Regulatory Agency of the State of Alagoas – ARSAL), which came about as part of the decentralization of the federal government and the process of State government modernization. The main mission of the regulator is to protect the rights and interests of customers by overseeing concessionaires and to ensure the quality of public services, particularly in the sectors of electricity, natural gas, inter-city transportation and WWS. The ARSAL also has the task of ensuring the economic and financial balance of the concessionaires, providing mediation between customers and operators of public services. The role of ARSAL is to promote and ensure the economic and technical efficiency of public services and protect the interests of customers against abuse of power by concessionaires. Currently, the regulator does not yet cover (entirely) the WWS, as its capacity does not extend to regulating quality of service. As far as economic regulation is concerned, the regulator is in the process of implementing a rate-of-return regulation method. The main legal documents of the water sector are the Law No. 11,445 of 2007, and the Program of Modernisation of the Water and Wastewater Sector.

Market Structure

The WWS are mainly managed by private operators of which CASAL is the main one. The participation of public operators is limited, providing WWS to only about 12% of the state population. Regardless of the type of service management, the WWS are vertically integrated, and the operators perform all services of the production chain, from abstraction to the water distribution. The situation is similar for wastewater services with regard to the operations of collection and treatment, although the wastewater treatment is almost non-existent.

Tariff Setting

The tariff system usually adopted by the operators is priced per increasing block rate, established according to the volume of water consumed. Universal service is ensured by cross-subsidization. There are also social tariffs. The tariffs for domestic, commercial, industrial and public customers are distinct.

Quality of Service

The regulated water services, in general, provide a reasonable quality of service. However, the quality gradually decreases as we move away from large urban centres. The WWS have a high number of service interruptions and present a significant non-revenue water volume. There is a great need for investment, especially with regard to wastewater, in order to improve service coverage.

The Regulator

Characteristics

- Multisector regulator
- Independent regulatory agency

Legal basis

State Law No. 6267 of 2001

Number of employees

112 employees (2 water sector)

Budget

1.5 million euros (2006)

Contacts

Agência Reguladora dos Serviços Públicos
Rua Cincinato Pinto, 226, 2.° Andar Edf. Ipaseal – Centro – Macció – AL. CEP: 57.020-050,
Brazil
Tel: +55 82 3315 2515
Fax: +55 82 3315 2501
http://www.arsal.al.gov.br

Relevant Entities

Brazilian Association of Regulatory Agencies

- http://www.abar.org.br

Ministry for the Environment

- http://www.mma.gov.br

National Information System on Sanitation

- http://www.snis.gov.br

Brazilian Association of Water Resources

- http://www.abrh.org.br

The Sector

State

- Area – 27,800 km^2
- Population – 3 million
- Population density – 101.8 inhab/km^2

Service Coverage

- Water – 86% (77 of the 102 Municipal Areas)
- Wastewater service – 13%
- Wastewater treatment – 0%

Number of operators

- Water – 17
- Wastewater – 17

Ownership and management of services

	Ownership		Management	
	Pub.	Priv.	Pub.	Priv.
Water	100%	0%	12%	88%
Wastewater	100%	0%	12%	88%

Volume of activity

- Abstracted water volume – 107.8 million m^3
- Drinking water – 51.7 million m^3
- Average consumption – 55 litres per person per day
- Main source –
- Water losses – 52%

Economic and financial data

- Turnover – 67.5 million euros (2007)
- Water service average price – 0.73 €/m^3
- Average annual invoice per customer –
- Employment – 2,887 employees

BRAZIL (BRASILIA)

Regulatory Model

In Brasilia, there is a sector-specific regulatory agency (Regulatory Agency for Water and Wastewater of the Federal District – ADASA) that has, as its basic goals, the regulation, control and monitoring of the quality and quantity of water resources in the district, or those that are delegated by the federal government and the states, as well as the WWS of the Federal District. The Federal District has both state and city powers, and this feature makes the ADASA the only one of its type in Brazil, a regulatory agency that simultaneously acts in both the regulation of water resources and the WWS, monitoring and regulating the complete water cycle. The ADASA is a regulatory agency endowed with organic and functional independence. It aims to ensure the sustainable use of the water resources of the Federal District and the provision of high-quality WWS. It is in this context that the ADASA attempts to disseminate, by means of effective action, the concept of water as an economic good, subject to conflicts among its potential customers, and, considering its important social role, one that needs to be protected and managed with efficiency and effectiveness. In order to promote quality of service, the regulator has developed a system of benchmarking based on a set of performance indicators. Concerning the economic regulation, the regulator has adopted the method of revenue-cap regulation for a period of three years. The main legal documents of the water sector are the Law No. 11,445 of 2007 and the Program of Modernisation of the Water and Wastewater Sector.

Market Structure

In the Federal District there is a single public operator, the CAESB. Therefore, the wholesale and end-user water services of the WWS are integrated, which is to say the same operator performs all services, from abstraction to the water distribution concerning the water service, and from collection to the wastewater treatment regarding the wastewater service.

Tariff Setting

The tariff system adopted by the operator consists of a fixed part, based on metering, and a proportional part, priced per increasing block rate according to the volume of water consumed. Universal service is ensured by cross-subsidization and the establishment of direct subsidies to customers from the state. The operator sets distinct tariffs depending on whether the customer is domestic, commercial, industrial, public or irrigation.

Quality of Service

The CAESB is characterized by a high quality of service, especially with regard to service coverage. However, there are some problems concerning continuity and reliability of the water networks. The quality of service is safeguarded by the role played by the benchmarking system carried out by the regulator, which applies of a set of performance indicators.

The Regulator

Characteristics
- Sector-specific regulator
- Independent regulatory agency

Legal basis
Law No. 3565 of 2004

Number of employees
70 employees

Budget
2.65 million euros (2006)

Contacts
Agência Reguladora de Águas e Saneamento
SCN 04 Bloco 'B' Sala 1302 Ed. Varig Brasília/DF
CEP: 70.710-5000, Brazil
Tel: + 55 61 3961 4956
Fax: + 55 61 3961 5050
http://www.adasa.df.gov.br

Relevant Entities

Brazilian Association of Regulatory Agencies
- http://www.abar.org.br

Ministry for the Environment
- http://www.mma.gov.br

National Information System on Sanitation
- http://www.snis.gov.br

Brazilian Association of Water Resources
- http://www.abrh.org.br

The Sector

State
- Area – 5,800 km^2
- Population – 2.4 million
- Population density – 423.3 inhab/km^2

Service Coverage
- Water – 98%
- Wastewater service – 91%
- Wastewater treatment – 91%

Number of operators
- Water – 1
- Wastewater – 1

Ownership and management of services

	Ownership		Management	
	Pub.	Priv.	Pub.	Priv.
Water	100%	0%	100%	0%
Wastewater	100%	0%	100%	0%

Volume of activity
- Abstracted water volume – 195 million m^3
- Drinking water – 136.5 million m^3
- Average consumption – 159 litres per person per day
- Main source – Superficial
- Water losses – 30%

Economic and financial data
- Turnover – 117 million euros
- Water service average price – 0.34 €/m^2
- Average annual invoice per customer –
- Employment – 3,550 employees

BRAZIL (CACHOEIRO DE ITAPEMIRIM)

Regulatory Model

In Cachoeiro de Itapemirim, there is an independent regulatory agency (Regulatory Agency of Water and Wastewater Services – AGERSA) exclusively dedicated to WWS, which is autonomous in carrying out its duties. The main objective of this independent regulator is ensuring the goals of the Water and Wastewater Municipal Policy, which include the universal access to an adequately reliable WWS, the quality of service provided and the fairness of tariffs set by operators. In this scope, the regulator fulfils the roles of regulation, supervision and control of the concessionaire (a single operator of WWS). However, the AGERSA is currently developing a regulatory process for regulating the water sector, which includes a system of benchmarking aimed at improving the quality of service provided by the concessionaire. Concerning the economic regulation, a rate-of-return regulation method is being implemented for a regulatory period of one year. The main legal documents of the water sector are the Law No. 11,445 of 2007, and the Program of Modernisation of the Water and Wastewater Sector.

Market Structure

In accordance with jurisdiction, the WWS are solely managed by the private sector, with only one operator, Citágua. The services are vertically integrated, either in water or in wastewater, and include the whole production chain from abstraction to distribution in the water service and from collection to treatment, in the wastewater service.

Tariff Setting

The tariff system adopted by the operator is priced per increasing block rate, depending on the volume of water consumed. Customers with limited financial resources benefit from social tariffs. There are distinct values for the tariffs of domestic, commercial, industrial and public customers.

Quality of Service

Citágua has generally provided a high quality of service, especially at the level of service coverage. However, there is evidence of frequent interruptions in the water service and blockages in the wastewater service.

The Regulator

Characteristics

- Sector-specific regulator
- Independent regulatory agency

Legal basis

Law No. 4789 of 1999

Number of employees

9 employees

Budget

194,000 euros (2006)

Contacts

Agência de Regulação dos Serviços de Saneamento
R. Prof. Quintiliano de Azevedo, 31 – Ed. Guandu Center – 6° andar – Guandu – Cachoeiro de Itapemirim – ES – CEP 29300-803
Brazil
Tel/Fax: + 55 28 3511 7077
http://www.agersa.com.br

Relevant Entities

Brazilian Association of Regulatory Agencies

- http://www.abar.org.br

Ministry for the Environment

- http://www.mma.gov.br

National Information System on Sanitation

- http://www.snis.gov.br

Brazilian Association of Water Resources

- http://www.abrh.org.br

The Sector

Municipality

- Area – 877 km^2
- Population – 195,300
- Population density – 222.7 inhab/km^2

Service Coverage

- Water – 99.7%
- Wastewater service – 96.4%
- Wastewater treatment – 85%

Number of operators

- Water – 1
- Wastewater – 1

Ownership and management of services

	Ownership		Management	
	Pub.	Priv.	Pub.	Priv.
Water	100%	0%	0%	100%
Wastewater	100%	0%	0%	100%

Volume of activity

- Abstracted water volume – 14 million m^3
- Drinking water – 10.4 million m^3
- Average consumption – 142 litres per person per day
- Main source –
- Water losses – 25.5%

Economic and financial data

- Turnover – 14.43 million euros
- Water service average price – 0.9 €/m^3
- Average annual invoice per customer –
- Employment – 275 employees

BRAZIL (GOIÁS)

Regulatory Model

In the State of Goiás, there is a multisector regulatory agency (Goiana Regulatory Agency – AGR) that controls and monitors the public services delegated by the State of Goiás and/or the arrangements between the federal government and municipalities. The AGR objectives include the safeguarding of public service obligations (gas, electricity, transportation and WWS) adequate to the needs of customers, the guaranteeing of customers' rights, the assurance of quality of service at fair prices, the mediation of conflicts between public service operators and customers, the granting of concessions, permits and authorizations for the execution of public services by approval of the State Council of Privatization and, finally, the promotion of universal access and modernization of public services. In addition to the aforementioned duties, the AGR is responsible for leading the Forum for Protection of Public Service Customers. Concerning the economic regulation, the regulator adopted the method of price-cap regulation for a period of four years. The main normative documents for the water sector are the Law No. 11,445 of 2007, and the Program of Modernisation of the Water and Wastewater Sector. At this time, the regulator does not apply any methodology of benchmarking as a means of supervising the quality of service.

Market Structure

The WWS are all publicly managed, the most important of which, the state company, Saneago, provides water services to approximately 4.5 million inhabitants in the state. With few exceptions, the WWS are horizontally integrated, which is to say there is a single operator for the same jurisdiction that provides both services. The same occurs at the level of vertical integration of these services, whereby the same operator has the capacity of managing wholesale and end-user water services, as well as in the scope of the wastewater service, where the operator is responsible for services from collection to wastewater treatment. However, in some municipalities, the WWS provide only end-user services.

Tariff Setting

The tariff system adopted by the operator consists of a variable charge priced per increasing block rate, according to the volume of water consumed. The government subsidizes the operators and low-income customers directly. There are distinct tariffs for domestic, commercial, industrial and public customers.

Quality of Service

In general the WWS provide a reasonable quality of service. However, there is a need for large investments, especially for the wastewater services. Poor reliability, associated with large water losses, is the main concern of the water service. In terms of wastewater service, the priority is increasing the levels of service coverage.

The Regulator

Characteristics

- Multisector regulator
- Independent regulatory agency

Legal basis

Law No. 13,569 of 1999

Number of employees

500 employees (7 water sector)

Budget

8.6 million euros (2006)

Contacts

Agência Goiana de Regulação
Av.Goiás no. 305, Edifício Visconde de Mauá – Centro-CEP: 74005-010- Goiânia – Goiás, Brazil
Tel : +55 62 226 6400
Fax: +55 62 226 6523
http://www.agr.go.gov.br

Relevant Entities

Brazilian Association of Regulatory Agencies

- http://www.abar.org.br

Ministry for the Environment

- http://www.mma.gov.br

National Information System on Sanitation

- http://www.snis.gov.br

Brazilian Association of Water Resources

- http://www.abrh.org.br

The Sector

State

- Area – 340,100 km^2
- Population – 5.6 million
- Population density – 16.5 inhab/km^2

Service Coverage

- Water – 80.8%
- Wastewater service – 33.6%
- Wastewater treatment – 27.3%

Number of operators

- Water – 10
- Wastewater – 6

Ownership and management of services

	Ownership		Management	
	Pub.	Priv.	Pub.	Priv.
Water	100%	0%	100%	0%
Wastewater	100%	0%	100%	0%

Volume of activity

- Abstracted water volume – 309.9 million m^3
- Drinking water – 189 million m^3
- Average consumption – 142 litres per person per day
- Main source – Superficial
- Water losses – 39%

Economic and financial data

- Turnover – 250 million euros
- Water service average price – 0.9 €/m^3
- Average annual invoice per customer –
- Employment – 5,900 employees

BRAZIL (JOINVILLE)

Regulatory Model

In the city of Joinville, there is an independent regulatory agency (Regulatory Agency of Joinville – AMAE) responsible for the regulation, monitoring and supervision of the WWS, which are provided by a single concessionaire (Águas de Joinville) from Joinville. The activity of the AMAE is focused on defence of public interest in the provision of WWS. It acts with impartiality, autonomy, transparency and independence. Within its authority is the ability to create rules for the regulation of services, which are approved by the Joinville City Council, to monitor the implementation of services, and to oversee the quality of the WWS. Furthermore, the regulator deals with complaints, and issues binding opinions on conflicts between customers and the concessionaire, which are heard by the Council and subject to the adequate provision of services in accordance with the laws, norms and established conditions. It also carries out activities for regulating the WWS based on the principles of universal access, continuity and quality of service. As a way of encouraging the concessionaire to improve the quality of service, the AMAE uses a system of benchmarking based on a set of performance indicators. As far as economic regulation is concerned, the regulator adopted the method of rate-of-return regulation for a period of one year. The main legal documents of the water sector are the Law No. 11,445 of 2007, and the Program of Modernisation of the Water and Wastewater Sector.

Market Structure

With only one operator, the WWS are entirely managed by the private sector, specifically by Aguas de Joinville. This company is vertically integrated, encompassing wholesale and end-user systems for either water or wastewater. Both of them are provided by this operator that performs services from abstraction to the water distribution and from collection to the wastewater treatment.

Tariff Setting

The tariff system adopted by the operator is based on a fixed part and a proportional part priced per increasing block rate, both established according to the volume of water consumed. The government subsidizes low-income customers. The tariffs for domestic customers are distinct from those of commercial, industrial and public customers.

Quality of Service

Wastewater services need huge investments in order to increase the level of service coverage. The water service presents an acceptable quality. The main problems of water services are the interruptions in supply and the significant water losses level presented.

The Regulator

Characteristics
- Sector-specific regulator
- Regulatory agency without independence

Legal basis
Law No. 4341 of 2001

Number of employees
16 employees

Budget
429,000 euros (2006)

Contacts
Agência Reguladora de Joinville
Rua Paraná, 420 Cep: 89202-160 – Joinville – SC
Brazil
Tel/Fax: + 55 47 3433 1158
http://www.amae.sc.gov.br

Relevant Entities

Brazilian Association of Regulatory Agencies
- http://www.abar.org.br

Ministry for the Environment
- http://www.mma.gov.br

National Information System on Sanitation
- http://www.snis.gov.br

Brazilian Association of Water Resources
- http://www.abrh.org.br

The Sector

Municipality
- Area – 1,100 km^2
- Population – 487,000
- Population density – 430.6 inhab/km^2

Service Coverage
- Water – 99%
- Wastewater service – 16%
- Wastewater treatment – 16%

Number of operators
- Water – 1
- Wastewater – 1

Ownership and management of services

	Ownership		Management	
	Pub.	Priv.	Pub.	Priv.
Water	100%	0%	0%	100%
Wastewater	100%	0%	0%	100%

Volume of activity
- Abstracted water volume – 57 million m^3
- Drinking water – 26 million m^3
- Average consumption – 146 litres per person per day
- Main source –
- Water losses – 54%

Economic and financial data
- Turnover – 35.2 million euros
- Water service average price – 1.2 €/m^3
- Average annual invoice per customer –
- Employment – 850 employees

BRAZIL (MANAUS)

Regulatory Model

In Manaus, there is a multisector regulator, the Regulatory Agency of Delegated Public Services (ARSAM), which follows the policies and guidelines established by the state. It may also act at a municipal or federal level in cases where, by arrangement, it has been delegated to cooperate technically with entities or bodies related to essential public services. Within this context, the regulator has, as its main objective, the assurance of the quality of public services through their monitoring and supervision, and the evaluation and fulfilment of the goals and standards set forth in legislation and contracts. As a multisector regulator, the ARSAM has responsibilities, within the scope of its powers, in the sectors of electricity, inter-city road transportation and WWS. The regulatory agency adopted a model of price-cap regulation. However, the regulatory process is currently under review, specifically the calculation of the rate-of-return on capital. The main legal documents of the water sector are the Law No. 11,445 of 2007, and the Program of Modernisation of the Water and Wastewater Sector. Regarding the quality of service and promotion of incentives, the regulator, until now, has not applied any benchmarking methodology.

Market Structure

The WWS are entirely managed by the private sector, particularly Águas do Amazonas. In this framework, services are characterized by vertical integration, both in water and wastewater activities. Here, the operator performs all services related to water production, from abstraction to the distribution of water, as well as collection and treatment with regard to the wastewater service.

Tariff Setting

The tariff system adopted by the operator is based on a variable charge priced per increasing block rate, according to the volume of water consumed. There are no social tariffs for WWS. The service is guaranteed by cross-subsidization. The operator sets separate tariffs depending on whether the customer is domestic, commercial, industrial or public.

Quality of Service

Águas do Amazonas provides a reasonable quality of service, where the water service coverage stands out. However, there are some problems with regard to the number of interruptions in the water service and to the blockages in the wastewater service. The levels of water losses are still very high. The wastewater services require important investments to improve the level of service coverage both in collection and treatment.

The Regulator

Characteristics

- Multisector regulator
- Regulatory agency without independence

Legal basis

Law No. 2568 of 1999

Number of employees

69 employees (11 water sector)

Budget

2 million euros (2006)

Contacts

Agência Reguladora de Serviços Públicos Concedidos
Rua Prof. Ernani Simão, 1314 – Cachoeirinha,
CEP: 69020-050, Manaus – AM
Brazil
Tel/Fax: + 55 92 3234 8787
http://www.arsam.am.gov.br

Relevant Entities

Brazilian Association of Regulatory Agencies

- http://www.abar.org.br

Ministry for the Environment

- http://www.mma.gov.br

National Information System on Sanitation

- http://www.snis.gov.br

Brazilian Association of Water Resources

- http://www.abrh.org.br

The Sector

State

- Area – 11,400 km^2
- Population – 1.6 million
- Population density – 144 inhab/km^2

Service Coverage

- Water – 80.2%
- Wastewater service – 8.4%
- Wastewater treatment – 6%

Number of operators

- Water – 1
- Wastewater – 1

Ownership and management of services

	Ownership		Management	
	Pub.	Priv.	Pub.	Priv.
Water	100%	0%	0%	100%
Wastewater	100%	0%	0%	100%

Volume of activity

- Abstracted water volume – 194 million m^3
- Drinking water – 66 million m^3
- Average consumption – 138 litres per person per day
- Main source – 77.4% superficial
- Water losses – 65.8%

Economic and financial data

- Turnover – 72.6 million euros (2007)
- Water service average price – 0.9 €/m^3
- Average annual invoice per customer –
- Employment – 765 employees

BRAZIL (MAUÁ)

Regulatory Model

In the city of Mauá, there is a regulatory agency, the Regulatory Agency for Water and Wastewater Services (ARSAE), dedicated to WWS that performs functions as a technical body and coordinates the Local Utility of Regulation of Public Water Service (SAMA Authority) and the Wastewater Company of the City of Mauá (Concessionaire ECOSAMA). The main objective of the regulator is to monitor services, promote the implementation of the Master Plan for the Management of Service (PDGS) and ensure the provision of an adequate service, that is, a service that meets the conditions of regularity, equity, continuity, efficiency and fair tariffs. Furthermore, the regulator works to ensure a consensus between the interests of customers and WWS providers in the municipality of Mauá. It monitors the WWS, which includes using the powers it has been assigned to in the concession contract or the program agreement and ensures that the economic and financial balance is consistent with both the program agreement and the WWS concession contract in the municipality of Mauá. It exercises normative power over the grant, monitors the fulfilment of WWS services, and has the ability to impose penalties on operators in cases where there is a breach of the standards imposed by the service regulations. Concerning the economic regulation, the regulator has adopted the method of rate-of-return regulation for a period of one year. The main legal documents of the water sector are the Law No. 11,445 of 2007, and the Program of Modernisation of the Water and Wastewater Sector. At this time, the regulator does not apply any methodology of benchmarking as a way of enhancing the quality of service and fostering the productive efficiency.

Market Structure

Atypically, taking into account the general services of the country, the WWS are provided separately by both the public and private sectors, depending on whether it is water service (SAMA) or wastewater service (ECOSAMA), respectively. In this scope the services are vertically integrated, that is, the operator (the local authority) provides all services from the production chain, from abstraction to distribution in the water activity. The situation is similar for wastewater where there is integration of services related to the wastewater collection and treatment.

Tariff Setting

The tariff system adopted by the operator consists of a proportional variable priced per increasing block rate based on the volume of water consumed. The government subsidizes some low-income customers directly. The tariffs differ depending on the type of customer: domestic, commercial, industrial or public.

Quality of Service

The water service has, in general, provided a high quality of service. However, the wastewater service, especially at the level of wastewater treatment needs large investments. Levels of water losses remain a problem that requires attention.

The Regulator

Characteristics
- Sector-specific regulator
- Independent regulatory agency

Legal basis
Municipal Law No. 3263 of 2000

Number of employees
2 water sector

Budget

Contacts
Agência Reguladora dos Serviços de Água e Esgoto
Av. Queiroz Pedroso, 235 – Jardim Pedroso – Mauá – SP Brazil
Tel: +55 11 4513 5972
Fax: +55 11 4513 5983
http://www.arsae.maua.sp.gov.br

Relevant Entities

Brazilian Association of Regulatory Agencies
- http://www.abar.org.br

Ministry for the Environment
- http://www.mma.gov.br

National Information System on Sanitation
- http://www.snis.gov.br

Brazilian Association of Water Resources
- http://www.abrh.org.br

The Sector

Country
- Area – 62.3 km^2
- Population – 402,600
- Population density – 6,463.7 inhab/km^2

Service Coverage
- Water – 92%
- Wastewater service – 75%
- Wastewater treatment – 2%

Number of operators
- Water – 1
- Wastewater – 1

Ownership and management of services

	Ownership		Management	
	Pub.	Priv.	Pub.	Priv.
Water	100%	0%	100%	0%
Wastewater	100%	0%	0%	100%

Volume of activity
- Abstracted water volume – 32 million m^3
- Drinking water – 17 million euros
- Average consumption – 127 litres per person per day
- Main source –
- Water losses – 47%

Economic and financial data
- Turnover – 20.2 million euros
- Water service average price – 0.78 €/m^3
- Average annual invoice per customer –
- Employment – 666 employees

BRAZIL (NATAL)

Regulatory Model

In the city of Natal, there is a sector-specific regulatory agency, the Water and Wastewater Services Regulatory Agency of Natal (ARSBAN), whose purpose is to regulate the provision of WWS. The regulator is linked to the Municipal – Secretary for Planning and Strategic Management and is endowed with financial independence. The ARSBAN exercises its powers according to the policies and guidelines established by the City Council for Water and Wastewater. In this scope, the regulator's main objective is to monitor services to ensure the provision of an adequate service, that is, the one which meets the conditions of regularity, equity, continuity, efficiency and fair tariffs. Its objectives also include ensuring conformity between the interests of customers and public service providers of WWS in the city of Natal, the WWS monitoring and guaranteeing the economic-financial balance of the operator. Concerning the economic regulation, the regulator has adopted the method of rate-of-return regulation. The main legal documents of the water sector are the Law No. 11,445 of 2007, and the Program of Modernisation of the Water and Wastewater Sector. The ARSBAN is in the process of developing a benchmarking study, which provides for the use of deterministic techniques in the multivariate analysis of data, associated with performance indicators.

Market Structure

With only one operator, the WWS are entirely managed by the public sector, particularly by state company CAERN. The WWS are vertically integrated encompassing the wholesale and end-user systems. Irrespective of being the water service or the wastewater service, they are provided by the same operator. The operator is responsible for services from abstraction to the water distribution for the water service and from wastewater collection to treatment for the wastewater service.

Tariff Setting

The tariff system adopted by the operator consists of a variable charge priced per increasing block rate based on the consumption of water. There is cross-subsidization between customers. There are social tariffs for customers with scarce financial resources. Domestic consumption (social, temporary, regular) is priced similarly with the exception of minimum tariff. Moreover, there are different tariffs for commercial, industrial and public customers.

Quality of Service

The regulated water services have, in general, provided an acceptable quality of service. However, the quality in rural areas decreases sharply, compared to urban centres. The WWS have a significant number of service interruptions and the non-revenue water volume is relevant. There is a need for large investments, especially in the area of wastewater service.

The Regulator

Characteristics
- Sector-specific regulator
- Independent regulatory agency

Legal basis
Law No. 5346 of 2001

Number of employees

Budget

Contacts
Agência Reguladora dos Serviços de Água e Esgoto
Rua Doutor Poty Nóbrega, no. 344, Lagoa Nova – CEP 59056-180 – Natal/RN
Brazil
Tel: +55 84 3232 9316
Fax: +55 84 3232 9311
http://www.natal.rn.gov.br/arsban

Relevant Entities

Brazilian Association of Regulatory Agencies
- http://www.abar.org.br

Ministry for the Environment
- http://www.mma.gov.br

National Information System on Sanitation
- http://www.snis.gov.br

Brazilian Association of Water Resources
- http://www.abrh.org.br

The Sector

Municipality
- Area – 170.3 km^2
- Population – 774,200
- Population density – 4,546.3 inhab/km^2

Service Coverage
- Water – 99%
- Wastewater service – 32.8%
- Wastewater treatment – 15.6%

Number of operators
- Water – 1
- Wastewater – 1

Ownership and management of services

	Ownership		Management	
	Pub.	Priv.	Pub.	Priv.
Water	100%	0%	100%	0%
Wastewater	100%	0%	100%	0%

Volume of activity
- Abstracted water volume – 81 million m^3
- Drinking water – 41.7 million m^3
- Average consumption – 147 litres per person per day
- Main source –
- Water losses – 49%

Economic and financial data
- Turnover – 48.5 million euros
- Water service average price – 0.50 €/m^3
- Average annual invoice per customer –
- Employment – 974 employees

BRAZIL (PARAÍBA)

Regulatory Model

In the State of Paraíba, there is an independent regulator, the Regulatory Agency of the State of Paraíba (ARPB), that aims to regulate, control and monitor the public services of electricity, piped gas distribution, WWS and other public services that fall under the jurisdiction of the state of Paraíba, but whose regulation, control and monitoring are assigned to the ARPB by executive power or by delegation. It is the ARPB's responsibility to ensure compliance with legislation and concessionary agreements for the public services it monitors. The regulator aims for a desirable balance between the concession-granting authority, the concessionaire and the customers and in this scope it supervises the services, prevents or settles conflicts and guides concessionaires and customers. Based on a system of benchmarking (performance indicators) that addresses the quality of service, the regulator has the ability to impose fines on an operator that consistently provides poor scores. The ARPB, concerning the economic regulation, has adopted the rate-of-return regulation. The main legal documents of the water sector are the Law No. 11,445 of 2007, and the Program of Modernisation of the Water and Wastewater Sector.

Market Structure

The WWS are provided by public operators. The most important is CAGEPA, the state company which provides water to 2.8 million people. The WWS are generally characterized by their respective integration, either horizontal or vertical. In this context, such services are usually provided by the same operator, both in the case of the state company and in the case of local authorities.

Tariff Setting

The tariff system adopted by the operators consists of a variable charge priced per increasing block rate according to the volume of water consumed. The government subsidizes some customers directly. There are social tariffs for low-income customers. Cross-subsidization is also a common practice. The tariffs, including a minimum charge, are distinct for domestic, commercial, industrial and public customers.

Quality of Service

The WWS are characterized by a reasonable quality of service, where water service coverage stands out. However, the services present some problems not only due to the high number of interruptions of water service, but also to the degree of blockages at the level of wastewater service. Water losses issue is a problem that requires further attention. Wastewater service needs of huge investments.

The Regulator

Characteristics
- Multisector regulator
- Independent regulatory agency

Legal basis
Law No. 7032 of 2001

Number of employees
41 employees (3 water sector)

Budget
819,000 euros (2006)

Contacts
Agência de Regulação do Estado da Paraíba
Av. Min. José Américo de Almeida, s/n Ed. DER – 1° andar,
Cep: 58040-300 – João Pessoa – PB,
Brazil
Tel: + 55 83 3218 6681
Fax: + 55 83 3218 6644
http://www.arpb.pb.gov.br

Relevant Entities

Brazilian Association of Regulatory Agencies
- http://www.abar.org.br

Ministry for the Environment
- http://www.mma.gov.br

National Information System on Sanitation
- http://www.snis.gov.br

Brazilian Association of Water Resources
- http://www.abrh.org.br

The Sector

State
- Area – 56,400 km^2
- Population – 3.7 million
- Population density – 63.7 inhab/km^2

Service Coverage
- Water – 90.9%
- Wastewater service – 29%
- Wastewater treatment – 26%

Number of operators
- Water – 6
- Wastewater – 6

Ownership and management of services

	Ownership		Management	
	Pub.	Priv.	Pub.	Priv.
Water	100%	0%	100%	0%
Wastewater	100%	0%	100%	0%

Volume of activity
- Abstracted water volume – 192 million m^3
- Drinking water – 182.4 million m^3
- Average consumption – 196 litres per person per day
- Main source – Superficial
- Water losses – 42.6%

Economic and financial data
- Turnover – 133.1 million euros
- Water service average price – 0.56 €/m^3
- Average annual invoice per customer –
- Employment – 2,850 employees

BRAZIL (PERNAMBUCO)

Regulatory Model

In the State of Pernambuco, there is a regulatory agency (Agency for the Regulation of Delegated Public Utilities – ARPE) that has the function of regulating all public utilities delegated by the State of Pernambuco, including gas, public transportation, energy and WWS. The regulator was created in order to establish a balance in the relations between customers, government and the concessionaires of public services. The regulatory intervention of the ARPE in the water sector includes the economic-financial areas, management of complaints and technical-operational aspects. With regard to the technical and operational aspects, the ARPE supervises the WWS, monitoring the efficiency of wastewater treatment, metering indicators and water losses. In each review, the regulator makes a substantiated report that is later sent to COMPESA along with the required questionnaire, so that necessary corrective actions can be implemented. Besides the evaluation of quality of service, another important function of the regulator is monitoring the water quality through a system of benchmarking based on performance indicators. Concerning the economic regulation, the regulator has adopted the rate-of-return regulation method for a period of one year. The main legal documents of the water sector are the Law No. 11,445 of 2007, and the Program of Modernisation of the Water and Wastewater Sector.

Market Structure

The WWS are mostly run by private operators, mainly by COMPESA, which services more than 8 million inhabitants, and also by public operators, who provide service for about 5% of the state population. Regardless of the type of service management, the operators are vertically integrated and perform all services on the production chain, from abstraction to the water distribution and similarly, concerning the wastewater service, from collection to the wastewater treatment.

Tariff Setting

The tariff system usually applied by the operators consists of a fixed part and a proportional part priced per increasing block rate established according to the volume of water consumed. The operator also defines social tariffs for customers with scarce financial resources. Domestic, commercial, industrial and public customers are charged distinct tariffs, including minimum charges.

Quality of Service

The regulated water services generally provide a reasonable quality of service. However, the quality decreases gradually as we move away from urban centres. The WWS showed a significant number of service interruptions. There is a need for large investments, especially in wastewater services. Water losses contribute to the inefficiency of the water service.

The Regulator

Characteristics
- Multisector regulator
- Independent regulatory agency

Legal basis
Law No. 11,742 of 2000

Number of employees
87 employees (12 water sector)

Budget
2.1 million euros (2006)

Contacts
Agência de Regulação dos Serviços Públicos
Delegados do Estado de Pernambuco (ARPE)
Av. Conselheiro Rosa e Silva, 975 – Aflitos –
CEP 52.050-020 – Recife – PE,
Brazil
Tel: +55 81 3382 9700
Fax: +55 81 3303 4725
http://www.arpe.pe.gov.br

Relevant Entities

Brazilian Association of Regulatory Agencies
- http://www.abar.org.br

Ministry for the Environment
- http://www.mma.gov.br

National Information System on Sanitation
- http://www.snis.gov.br

Brazilian Association of Water Resources
- http://www.abrh.org.br

The Sector

Country
- Area – 98,900 km^2
- Population – 7.9 million
- Population density – 80.37 inhab/km^2

Service Coverage
- Water – 94% (173 of the 184 Municipal Areas)
- Wastewater service – 30%
- Wastewater treatment – 17%

Number of operators
- Water – 10
- Wastewater – 2

Ownership and management of services

	Ownership		Management	
	Pub.	Priv.	Pub.	Priv.
Water	100%	0%	5%	95%
Wastewater	100%	0%	5%	95%

Volume of activity
- Abstracted water volume – 331.1 million m^3
- Drinking water – 212.4 million m^3
- Average consumption – 80 litres per person per day
- Main source – Superficial
- Water losses – 60%

Economic and financial data
- Turnover – 220.3 million euros
- Water service average price – 0.8 €/m^3
- Average annual invoice per customer –
- Employment – 5,500 employees

BRAZIL (RIO DE JANEIRO)

Regulatory Model

In Rio de Janeiro, there is an independent regulatory agency (Regulatory Agency for Energy and Water and Wastewater – AGENERSA) that has the function of monitoring, controlling and overseeing the existing concessionary contracts of the public utilities of gas and WWS in the region of Lagos. In this area, the regulator aims to stimulate the enhancement, universality and modernization of services, monitor the compliance with the standards, deadlines, duties and rights established in the contracts, ensure the economic and financial balance of concessionary contracts, correct failures or problems in services, mediate differences between customers and concessionaires, defend the rights of customers against the interests of concessionaires and preserve the fairness of tariffs. The regulatory process is reviewed and arbitrated by the Council Director in public regulatory meetings that are attended by representatives of the Government, the concessionaire, the public and civil society organizations. The Council may determine or recommend changes in procedures, notice and even apply sanctions to the concessionaires. Concerning the economic regulation, the regulator has adopted the rate-of-return regulation method. The main legal documents of the water sector are the Law No. 11,445 of 2007, and the Program of Modernisation of the Water and Wastewater Sector. Apart from these aspects, the regulator applies a system of benchmarking, based on a set of performance indicators, as a way of enhancing the quality of service.

Market Structure

The WWS are managed mostly by the public sector and especially by the state company, CEDAE, which provides services to about 13 million customers. Private sector participation is found in only eight municipalities, through Águas de Juturnaíba (1% of the state) and Prolagos (2% of the state). The WWS are vertically integrated. As such, the operator performs all services related to the water cycle, from abstraction to the distribution of water and the services of collection and treatment of wastewater.

Tariff Setting

The tariff system usually applied by operators consists of an increasing block rate defined according to the volume of water consumed. There are no social tariffs. There is cross-subsidization between customers. The tariffs, including minimum charges, are distinct between domestic, commercial, industrial and public customers.

Quality of Service

In general, the water services are characterized by a high quality of service provided. Universal access, continuity of service and moderation of water losses are the main features that sustain the good quality of service. However, rural areas still need special attention and require a good deal of investment.

The Regulator

Characteristics
- Multisector regulator
- Independent regulatory agency

Legal basis
Law No. 2686 of 1997

Number of employees
60 employees (4 water sector)

Budget
3.9 million euros (2006)

Contacts
Agência Reguladora de Energia e Saneamento Básico
Rua 13 de Maio, 23 – 23° Andar – Centro
CEP: 20.031-902 – Rio de Janeiro – RJ, Brazil
Tel: +55 21 2299 4895
Fax: +55 21 2299 4899
http://www.agenersa.rj.gov.br

Relevant Entities

Brazilian Association of Regulatory Agencies
- http://www.abar.org.br

Ministry for the Environment
- http://www.mma.gov.br

National Information System on Sanitation
- http://www.snis.gov.br

Brazilian Association of Water Resources
- http://www.abrh.org.br

The Sector

State
- Area – 43,700 km^2
- Population – 15.4 million
- Population density – 352.6 inhab/km^2

Service Coverage
- Water – 85% (8 municipal areas of the state)
- Wastewater service – 70%
- Wastewater treatment – 70%

Number of operators
- Water – 17
- Wastewater – 17

Ownership and management of services

	Ownership		Management	
	Pub.	Priv.	Pub.	Priv.
Water	100%	0%	97%	3%
Wastewater	100%	0%	97%	3%

Volume of activity
- Abstracted water volume – 48 million m^3
- Drinking water – 33.6 million m^3
- Average consumption – 216 litres per person per day
- Main source – Superficial
- Water losses – 30%

Economic and financial data
- Turnover – 36.3 million euros
- Water service average price – 0.65 €/m^3
- Average annual invoice per customer –
- Employment – 14,260 employees (total)

BRAZIL (RIO GRANDE DO SUL)

Regulatory Model

In the state of Rio Grande do Sul, there is a regulator (Regulatory Agency of Delegated Public Services – AGERGS) endowed with organic independence and multisector responsibility that includes WWS, road and maritime passenger transportation, roadways, irrigation and electricity. The objective of the regulator is to ensure the quality of public services provided to customers by private sector concessionaires and to ensure the balance of economic and financial power in contracts between the concession granting authority (government) and the concessionaires. It also has as its objectives the provision of adequate services, particularly those that meet the conditions of regularity, continuity, efficiency, safety, equity and fair tariffs, the conformity between the interests of concessionaires and customers of public services and the economic and financial balance of delegated public services. Based on Law No. 11,445 of 2007, and Law No. 11,107 of 2005, the regulator makes agreements with municipalities in order to regulate the water sector. Currently, despite the limited number of municipalities involved, the regulator is implementing the regulatory process of the water sector. In this context of economic regulation, the WWS tariffs are set by rate-of-return regulation for a period of one year. The Program of Modernization of the Water and Wastewater Sector is an important normative document of the water sector. The AGERGS is developing a system of benchmarking, based on a non-parametric technique (data envelopment analysis – DEA) in order to promote better quality of service and productive efficiency.

Market Structure

The WWS are managed by public operators, the most important of which is the state company CORSAN, which serves about 70% of the state population. The services are vertically integrated in both water service and wastewater service, and include the whole production chain, from abstraction to the distribution of water, as well as collection and wastewater treatment. However, there are some particular situations where this does not happen.

Tariff Setting

The tariff system normally applied by operators in the state of Rio Grande do Sul is based on an increasing block rate set according to the volume of water consumed and/or the characteristics of the property. There are social tariffs for the poor and cross-subsidization is common between customers. The operators of WWS set distinct tariffs (including minimum tariffs), depending on whether the customers are domestic, commercial, industrial or public.

Quality of Service

The water services, namely in the urban areas, provide a reasonable quality of service. However, this is not the case with wastewater service where more investments are needed to improve the service coverage. The WWS, particularly in rural areas, show a significant number of service interruptions and non-revenue water volume.

The Regulator

Characteristics

- Multisector regulator
- Independent regulatory agency

Legal basis

Law No. 10,930/97

Number of employees

85

Budget

2,125,500 euros (2006)

Contacts

Agência Estadual de Regulação dos Serviços Públicos Delegados do Rio Grande do Sul
Av. Borges de Medeiros, 659 – 13° andar
90020-023 – Porto Alegre – RS, Brazil
Tel. +51 3288 8880
Fax: + 512 3288 8881
http://www.agergs.rs.gov.br/

Relevant Entities

Brazilian Association of Regulatory Agencies

- http://www.abar.org.br

Ministry for the Environment

- http://www.mma.gov.br

National Information System on Sanitation

- http://www.snis.gov.br

Brazilian Association of Water Resources

- http://www.abrh.org.br

The Sector

State

- Area – 281,748 km^2
- Population – 10,582,000
- Population density 37.4 inhab/km^2

Service Coverage

- Water – 78%
- Wastewater service – 23%
- Wastewater treatment – 9.5%

Number of operators

- Water – 8 (327 municipal areas)
- Wastewater – 8 (327 municipal areas)

Ownership and management of services

	Ownership		Management	
	Pub.	Priv.	Pub.	Priv.
Water	100%	0%	100%	0%
Wastewater	100%	0%	100%	0%

Volume of activity

- Abstracted water volume – 811.6 million m^3
- Drinking water – 538.3 million m^3/month
- Average consumption – 170 litres per person per day
- Main source –
- Water losses – 33.7%

Economic and financial data

- Turnover – 826.7 million €
- Water service average price – 1.18 €/m^3
- Average annual invoice per customer –
- Employment – 10,150 employees

BRAZIL (STATE OF CEARÁ)

Regulatory Model

In the state of Ceará, there is a multisector regulatory agency (Regulatory Agency of Delegated Public Services – ARCE) endowed with organic, functional and financial independence, and whose main objective is to promote and ensure the economic and technical efficiency of public services, which include natural gas, inter-city road transportation, electricity and WWS. It intends to provide customers with adequate conditions of regularity, continuity and universality. The ARCE performs the regulatory tasks of standardizing and monitoring the public services provided by the concessionaires. The ARCE also acts in the mediation of possible conflicts between service operators and customers, seeking a balance between the parties. In this scope, the goal of the regulator is to endow the sector with principles such as transparency, to protect the interests of customers, governments and operators of delegated public services and, finally, to ensure the WWS excellence. Quality of service is evaluated through a system of benchmarking based on a set of performance indicators. Regarding the economic regulation, the ARCE adopted the method of rate-of-return regulation for a period of one year. The main legal documents of the water sector are the Law No. 11,445 of 2007, and the Program of Modernisation of the Water and Wastewater Sector.

Market Structure

The WWS are managed solely by the public sector, mainly by the state company CAGECE, which provides drinking water supply services to about 4.7 million inhabitants of the state. The water sector is characterized by a vertical integration of its services into a single operator that provides all the services for the entire water cycle from abstraction to water distribution and from wastewater collection to its treatment. Only a small percentage of the population is connected to wastewater services. Therefore, in areas where this service is available the water operator simultaneously manages the wastewater service.

Tariff Setting

The tariff system adopted by the operator is based on an increasing block rate set by the volume of water consumed. There are special tariffs for low-income customers. Domestic, commercial, industrial and public customers are charged differently.

Quality of Service

The water services as a whole provide an acceptable quality of service. However, this scenario is present mainly in urban centres. In general, the WWS have reasonable performance indicators (e.g. continuity) concerning the water supply but low levels regarding the wastewater service where meaningful investments are required.

The Regulator

Characteristics
- Multisector regulator
- Independent regulatory agency

Legal basis
Law No. 12,786 of 1997

Number of employees
40 employees (8 water sector)

Budget
2.6 million euros (2006)

Contacts
Agência Reguladora de Serviços Públicos Delegados
Av. Santos Dumont 1789, 14° andar, Cep 60.150.160 Fortaleza – CE Brazil
Tel: + 55 85 3101 1027
Fax: + 55 85 3101 1000
http://www.arce.ce.gov.br

Relevant Entities

Brazilian Association of Regulatory Agencies
- http://www.abar.org.br

Ministry for the Environment
- http://www.mma.gov.br

National Information System on Sanitation
- http://www.snis.gov.br

Brazilian Association of Water Resources
- http://www.abrh.org.br

The Sector

State
- Area – 148,800 km^2
- Population – 8.2 million
- Population density – 54 inhab/km^2

Service Coverage
- Water – 95.7% (149 of the 184 municipal areas)
- Wastewater service – 36.5%
- Wastewater treatment – 36.5%

Number of operators
- Water – 22
- Wastewater – 22

Ownership and management of services

	Ownership		Management	
	Pub.	Priv.	Pub.	Priv.
Water	100%	0%	100%	0%
Wastewater	100%	0%	100%	0%

Volume of activity
- Abstracted water volume – 370 million m^3 (regulated)
- Drinking water – 244 million m^3
- Average consumption – 85 litres per person per day
- Main source – Superficial
- Water losses – 33.7%

Economic and financial data
- Turnover – 133.9 million euros
- Water service average price – 0.50 €/m^3
- Average annual invoice per customer –
- Employment – 6,100 employees

CANADA

Regulatory Model

In Canada, the water sector is generally characterized by self-regulation. As such, the associations of the sector play a considerably important role, having the ability to disseminate best practices among their members and thereby encouraging the progress of the sector. However, there are specific issues requiring action by the provincial or territorial regulatory agencies, namely the multisector agencies (WWS, electricity, gas, oil and transportation), such as the drinking water quality supplied by operators, the discharge of effluents into the environment and price setting. The criteria and regulation methods are transparent and understandable and, in most provinces, the adopted regulation is published in provincial newspapers as official news. The method of economic regulation used in the setting of tariffs between operators is almost always based on rate-of-return regulation. However, due to the lack of regulatory agencies, the decisions regarding tariff setting are carried out exclusively by the elected municipal councils. This regulatory method is extended for a period of two years. The legal framework for the sector is provided in the Water Act of 1970, and in the Environmental Protection Law of 1988.

Market Structure

Public operators primarily provide the WWS. The presence of the private sector is insignificant. Even when present, it is only in the management and operation of services, and never in their ownership. The WWS are provided mostly by multisector operators (WWS, urban waste, energy, etc.). In general, the services are vertically integrated, either in water or in wastewater services, and include the whole production chain from abstraction to the water distribution as well as the collection and wastewater treatment. However, there are some situations where this does not apply (e.g. in Vancouver).

Tariff Setting

Canada has not introduced a national pricing structure for the WWS. Tariff systems implemented among WWS operators vary between the implementation of a fixed amount regardless of consumption, a fixed part according to consumption, an increasing/decreasing block rate depending on consumption, or a combination of the above. In general, customers have different tariffs depending on their characteristics.

Quality of Service

The WWS have been characterized by the universal service provided and by their general high quality. The quality of service is ensured by the threat of revoking licence or contract cancellation. In the scope of quality of service supervision the national agencies have developed benchmarking studies, but not in a systematic way. In general, they are guided by high transparency and participation of stakeholders.

The Regulator

Characteristics

Legal basis

Number of employees

Budget

Contacts

Relevant Entities

Sector Association
- http://www.cwwa.ca

Water Resources Institute
- http://www.nwri.ca

Ministry for the Environment
- http://www.ec.gc.ca

Ministry of Health
- http://www.hc-sc.gc.ca

The Sector

Country
- Area – 9,984,700 km^2
- Population – 33.3 million
- Population density – 3.2 inhab/km^2

Service Coverage
- Water – 85%
- Wastewater service – 85%
- Wastewater treatment – 80%

Number of operators
- Water – 9,000
- Wastewater – 9,000

Ownership and management of services

	Ownership		Management	
	Pub.	Priv.	Pub.	Priv.
Water	100%	0%	97%	3%
Wastewater	100%	0%	97%	3%

Volume of activity
- Abstracted water volume – 5,400 million m^3
- Drinking water – 4,200 million m^3
- Average consumption – 343 litres per person per day
- Main source – 60% superficial
- Water losses – 22.5%

Economic and financial data
- Turnover – 3,000 million (distribution)
- Water service average price – 0.72 €/m^3
- Average annual invoice per customer – 235 euros
- Employment – 300,000 employees

CANADA (PRINCE EDWARD ISLAND)

Regulatory Model

In the province of Prince Edward Island (PEI), as an atypical situation for Canada, there is a multisector regulatory agency (PEI Regulatory and Appeals Commission – IRAC), which acts as an independent tribunal on matters relating to tariffs of land use, property and revenue obtained from public services. At the level of economic regulation, the commission, which regulates only a limited number of operators, has adopted the method of rate-of-return regulation for a period of one year. Currently, the regulator does not have any authority over the control and enhancement of the quality of service. However, the IRAC does make an annual report, which publishes (sunshine regulation) the prices charged by the WSS operators. The Environmental Department, complementing the duties of the commission, has the ability to regulate the quality of drinking water that is supplied to domestic customers. As a form of control, the regulatory agency is compelled to report all the activities performed to the Assembly of PEI through the Ministry of Education. The Water Act of 1970 and the IRAC Act of 1991 provide the main legal framework governing the WWS.

Market Structure

The WWS are organized in different ways between the various areas, but always under public ownership and management. Within the area of the province, service management ranges from the integration of the full water cycle, with operators providing services from water abstraction to treatment of wastewater (e.g. in North Rustico), to the specific provision of a determined service, whether it is the water (e.g. in Alberton) or wastewater service (e.g. in Seawood Water). In horizontal terms, the WWS provision is equally divided between the number of operators that provide both services and the operators that provide only one service.

Tariff Setting

The tariff system usually adopted by operators is based on a fixed rate, according to the characteristics of the property. The government subsidizes the operators in order to promote universal service. The tariffs are set according to the customer (e.g. an individual customer pays a different rate from a family or an industrial customer).

Quality of Service

Despite the limited information, the WWS have a high quality of service. The lack of authority of IRAC is balanced by the threat of revoking the licence or contract cancellation. However, the sector is characterized by transparency and participation in terms of both financial and environmental quality aspects.

The Regulator

Characteristics

- Multisector regulator
- Independent regulatory agency

Legal basis

IRAC Law of 1991

Number of employees

23 employees (3 water sector)

Budget

1.3 million euros (2007)

Contacts

PEI Regulatory and Appeals Commission
PO Box 577, Charlottetown, PE C1A 7L1
Canada
Tel: +902 892 3501
Fax: +902 566 4076
http://www.irac.pe.ca

Relevant Entities

Sector Association

- http://www.cwwa.ca

Water Resources Institute

- http://www.nwri.ca

Water Resources Association

- http://www.cwra.org

Ministry for the Environment

- http://www.gov.pe.ca/enveng

The Sector

Province

- Area – 5,700 km^2
- Population – 139,000
- Population density – 23.9 inhab/km^2

Service Coverage

- Water –
- Wastewater service –
- Wastewater treatment –

Number of operators

- Water – 14
- Wastewater – 25

Ownership and management of services

	Ownership		Management	
	Pub.	Priv.	Pub.	Priv.
Water	100%	0%	100%	0%
Wastewater	100%	0%	100%	0%

Volume of activity

- Abstracted water volume –
- Drinking water –
- Average consumption –
- Main source –
- Water losses –

Economic and financial data

- Turnover – 1.7 million euros
- Water service average price –
- Average annual invoice per customer – 123 euros
- Employment –

CAPE VERDE

Regulatory Model

In Cape Verde there is a regulatory agency (Economic Regulation Agency – ARE) with administrative, financial and patrimonial independence, which is endowed with regulatory functions, including duties with regard to the regulation, supervision and sanctioning of violations. ARE aims to promote economic efficiency and financial stability of the regulated sectors (energy, WWS, telecommunications and transportation) to ensure the provision of services of general interest. Furthermore, ARE intends to fulfil public service obligations, protect the rights and interests of customers, and guarantee the objectivity of the rules of regulation and the transparency of commercial relations between operators and customers. It also tries to guarantee the implementation and monitoring of compliance with applicable laws and norms of the regulated sectors. ARE also works to coordinate with the national authority in the application of competition law in regulated sectors, to contribute to the gradual improvement of economic conditions in the regulated sectors and to encourage information and transparency to customers in coordination with the appropriate authorities. Concerning the economic regulation, the regulator has adopted the price cap regulation method for a period of five years. The Water Code and the Decree No. 8 of 2004 provide the main legal framework for the water sector.

Market Structure

In the water service there is a relatively equal distribution in the percentage of participation by both private and public sectors. Wastewater treatment service is provided by the private sector (ELECTRA), with a WWTP in the city of Praia, and by the public sector (directly by the municipality) in the city of Mindelo. Anyway, where it exists, it is vertically integrated and includes all stages of the production chain, which are, however, horizontally separated (several municipal companies). The exception occurs in the city of Praia, where the company ELECTRA vertically and horizontally integrates the WWS, managing all the water cycle services, when they are available, from abstraction and desalination to wastewater treatment. ELECTRA provides the electricity service as well.

Tariff Setting

The tariff system adopted by the operator corresponds to the application of an increasing block rate according to the consumption of water. The prices are high for the majority of the population, given the cost of the desalination. The government subsidizes new connections. The operator sets social tariffs as well as tariffs for domestic customers, industry and tourism, commercial customers and cistern trucks.

Quality of Service

The WWS quality of service provided in Cape Verde is generally poor. There are interruptions in situations of increased demand. Issues related to the quality of service, which is limited and poorly supervised, are mainly economic and financial in nature. The priority has been to increase the service coverage and the quantity of water available.

The Regulator

Characteristics
- Multisector regulator
- Independent regulatory agency

Legal basis
Decree No. 26 of 2003

Number of employees
19 employees (4 water sector)

Budget

Contacts
Agência de Regulação Económica
Av. Cidade de Lisboa, no. 785,
Cape Verde
Tel: +238 260 0430
Fax: +238 261 1440
http://www.are.cv

Relevant Entities

National Institute for the Management of Water Resources
- http://www.ingrh.cv

Ministry for the Environment and Agriculture
- http://www.maap.cv

Ministry of Health
- http://www.minsaude.gov.cv

African Development Bank
- http://www.afdb.org

The Sector

Country
- Area – 4,000 km^2
- Population – 500,000
- Population density – 118 inhab/km^2

Service Coverage
- Water – 85% (includes trucks and fountains)
- Wastewater service – 16%
- Wastewater treatment –

Number of operators
- Water – 15
- Wastewater – 3

Ownership and management of services

	Ownership		Management	
	Pub.	Priv.	Pub.	Priv.
Water	100%	0%	50%	50%
Wastewater	100%	0%	40%	60%

Volume of activity
- Abstracted water volume – 6.7 million m^3
- Drinking water – 4.6 million m^3
- Average consumption – 50 litres per person per day
- Main source – 85% superficial
- Water losses – 31% (ELECTRA)

Economic and financial data
- Turnover – 8.4 million euros
- Water service average price – 2.72 €/m^3
- Average annual invoice per customer – 74 euros
- Employment – 679 employees (ELECTRA)

CHILE

Regulatory Model

In Chile, the WWS operators (urban) are under the control of SISS (Superintendence of Water and Sanitation Services), an autonomous, decentralized government agency, to which they report all their activities. The SISS main responsibilities are the analysis, development and supervision of compliance with technical standards for the design, construction and operation of WWS. The SISS also has the task of guaranteeing compliance with the existing legal framework and concessionary contracts, regulating the process of setting tariffs for a period of five years, resolving any cases of disputes between operators and customers, applying fines and sanctions on companies in accordance with the law and, finally, controlling discharges of WWS operators and discharges of industrial wastewater. The system of economic regulation, based on the efficient company model, was adopted in all sectors of public utility services in Chile for a period of five years, based on price caps. As a standard of comparison, this approach uses an efficient virtual business model (a purely theoretical company), created by the regulator to meet the projected demand in the best possible way. The Water Code of 1981 and the General Sewerage Services Act of 1988 constitute the main legal framework for the water sector.

Market Structure

The WWS are, in the majority, provided by the private sector, despite the presence of a public partner (CORFO) in the main operators. Nevertheless, there are two main models of privatization. First, until 2001, there was mainly a material privatization of operators through the sale of assets. Beginning in 2001, the model they followed was that of concession (privatization of management). The first option includes the four major Chilean companies (ESVAL, Aguas Andinas, ESSAL and ESSBIO). Private participation always has a majority shareholding. For operators in rural areas the situation is mixed, with as many private as public operators. The WWS are usually merged in the same operator and are vertically integrated, which is to say the operators manage all activities of the production chain, from abstraction to the water distribution, as well as wastewater collection and treatment. The degree of horizontal integration is also high.

Tariff Setting

The tariff system normally applied by operators consists of a fixed part and an increasing block rate set according to the volume of water consumed. The government directly subsidizes low-income customers, providing a given volume of water free of charge. The tariffs are set differently by area.

Quality of Service

The WWS in Chile represent one of the benchmarks in Latin America, particularly at the level of service coverage. To encourage the quality of service, SISS established a benchmarking system based on a set of performance indicators. In short, the WWS generally provide a high quality of service.

The Regulator

Characteristics
- Sector-specific regulator
- Regulatory agency without independence

Legal basis
Law No. 18,902 of 1990

Number of employees
157 employees

Budget
8.5 million euros (2006)

Contacts
Superintendencia de Servicios Sanitarios
Oficina Regiốn Metropolitana, Moneda N°673,
piso 1, Santiago, Chile
Tel: +56 2 382 4000
Fax: +56 2 382 4003
http://www.siss.cl

Relevant Entities

National Commission for the Environment
- http://www.conama.cl

Ministry of Health
- http://www.minsal.cl

Ministry of Public Works
- http://www.mop.cl

General Water Board
- http://www.dga.cl

The Sector

Country
- Area – 757,000 km^2
- Population – 16.6 million
- Population density – 21.3 inhab/km^2

Service Coverage
- Water – 99.8%
- Wastewater service – 95.2%
- Wastewater treatment – 81.9%

Number of operators
- Water – 52
- Wastewater – 50

Ownership and management of services

	Ownership		Management	
	Pub.	Priv.	Pub.	Priv.
Water	35%	65%	0%	100%
Wastewater	35%	65%	0%	100%

Volume of activity
- Abstracted water volume – 1,451 million m^3
- Drinking water – 905 million m^3
- Average consumption – 150 litres per person per day
- Main source – 51% superficial
- Water losses – 37.6%

Economic and financial data
- Turnover – 910 million euros
- Water service average price – 0.59 €/m^3
- Average annual invoice per customer – 225 euros
- Employment – 9,566 employees (13.5 million inhab.)

COLOMBIA

Regulatory Model

In Colombia, there is an independent regulatory agency (Commission for Regulation of Water, Wastewater and Waste – CRA) dedicated to issues of economic regulation, quality of service and the public service provision of the WWS, which also has regulatory functions over urban waste services. The CRA is a special administrative unit, under the authority of the Ministry for the Environment, with technical, administrative and patrimonial independence, but without legal nature. The economic regulation developed by the CRA is based on regulation by comparison (yardstick competition) *stricto sensu*, i.e. the price systems of the operators are determined based on the performance of operators with the best practices in the sector. The system of price caps, adopted for a regulatory period of five years, consists of three parts: one related to administrative costs, another related to operational and maintenance costs, and another related to capital costs, which also includes tariffs and environmental taxes. The first two parts are based on cost efficiency (best practice), while the third relies on the approval of the operators' development plans. This calculation is carried out separately for the water service and for the wastewater service. The determination of best practices is based on the benchmarking non-parametric frontier technique of data envelopment analysis (DEA). The Law of Domestic Public Services of 1994 and the Departmental Planning of WWS are the main legal documents of the water sector.

Market Structure

The WWS are publicly owned and are mainly characterized by public sector management. Despite the low representation of the private sector in the national provision of WWS, the sector has always been defined by great stability and sustainability, and, in the context of Latin America, is considered a success story. The WWS are usually characterized by vertical integration and horizontal deverticalization, with both services being provided by the same operator. However, some public operators have some stages of the production chain (the water cycle) contracted externally, such as EAAB. With few exceptions (e.g. EPM), the WWS operators in Colombia do not provide other activities.

Tariff Setting

The tariff system adopted by operators usually consists of a fixed part and an increasing block rate based on the volume of water consumed. The government subsidizes the operators. There are social tariffs for low-income customers. The operator sets distinct tariffs according to whether the customer is domestic, commercial, industrial or public.

Quality of Service

Despite numerous advances the water sector has made, notably at the level of service coverage and the reliability of services, there are still some problems in terms of water quality, especially in rural areas. The sector needs a system of quality of service analysis and monitoring. The water losses are a particularly serious problem. However, taking into account its geography, the country demonstrates a high level of quality of service.

The Regulator

Characteristics
- Sector-specific regulator
- Independent regulatory agency

Legal basis
Law No. 142 of 1994

Number of employees
105 employees

Budget
3.1 million euros (2006)

Contacts
Comisión de Regulación de Agua Potable y Saneamiento Básico, Edificio Palma Real Cra.13 No. 28-01 Pisos 5 y 6, Bogotá, Colombia
Tel: +571 327 2800
Fax: +571 350 9393
http://www.cra.gov.co

Relevant Entities

Environment Authority
- http://www.car.gov.co

Development Board
- http://www.findeter.gov.co

Public Services Overseeing
- http://www.superservicios.gov.co

Ministry for the Environment, Housing and Regional Development
- http://www.minambiente.gov.co

The Sector

Country
- Area – 1,139,000 km^2
- Population – 44.4 million
- Population density – 39 inhab/km^2

Service Coverage
- Water – 88.3%
- Wastewater service – 74.1%
- Wastewater treatment – 25%

Number of operators
- Water – 2,886
- Wastewater – 1,071

Ownership and management of services

	Ownership		Management	
	Pub.	Priv.	Pub.	Priv.
Water	100%	0%	88.5%	11.5%
Wastewater	100%	0%	88.5%	11.5%

Volume of activity
- Abstracted water volume – 2,329 million m^3
- Drinking water – 1,188 million m^3
- Average consumption – 59 litres per person per day
- Main source – Superficial
- Water losses – 49%

Economic and financial data
- Turnover – 712.4 million euros
- Water service average price – 0.36 €/m^3
- Average annual invoice per customer – 92.3 euros
- Employment – 18,633 (50% of the population)

COSTA RICA

Regulatory Model

In Costa Rica there is an independent multisector regulatory agency (Regulatory Authority of Public Services – ARESEP) comprising the sectors of telecommunications, electricity, WWS, oil, irrigation and transportation. It has the duty of an economic regulator, which includes the tasks of setting tariffs for operators, monitoring and controlling the efficiency and quality of service, and protecting the interests of customers. The pricing methodology applied to the water sector is based on rate-of-return regulation. The regulatory period corresponds to one year. The process of setting tariffs involves public participation, where customers can express their views on price changes. Concerning the customer protection, they can submit their complaints to the regulator, directly to the government or they may use the services of a customer protection organization such as the Office of the Ombudsman. The regulator also intends to mediate and resolve conflicts. The water sector is essentially regulated by the Drinking Water Supply Act of 1953, and the General Health Act of 1973.

Market Structure

The water sector is mainly managed by public entities, specifically by AyA, which plays an important role in the direct and indirect management of WWS both inside and outside the respective areas of jurisdiction. The participation of some private operators, which provide the water service to individual systems or on a small scale, is limited. The water sector is characterized by a horizontal and vertical integration of their respective services into a single operator that provides all services for the whole water cycle from abstraction to water distribution and from wastewater collection to its treatment. However, the wastewater service is still only available to a small percentage of the population. In areas where this service is available, it is the same operator that simultaneously provides both water and wastewater services.

Tariff Setting

The tariff system normally applied by operators is based on an increasing block rate according to the volume of water consumed. There is no policy of social tariffs although in some areas customers are provided with water free of charge. The rate values are very low and insufficient. Domestic, commercial, industrial and public customers are charged differently.

Quality of Service

Despite a considerable increase in prices, the WWS have a poor quality of service, especially with regard to the limited wastewater service coverage and to the non-revenue water volume. Since its implementation, ARESEP has collected data on the activities of the regulated operators. However, until now, this information has not been used in the regulatory process.

The Regulator

Characteristics
- Multisector regulator
- Independent regulatory agency

Legal basis
Law No. 7593 of 1996

Number of employees
175 employees (16 water sector)

Budget
4.1 million euros (2006)

Contacts
Autoridad Reguladora de los Servicios Públicos
Apartado 936-1000, San José,
Costa Rica
Tel: +506 222 0102
Fax: +506 222 0374
http://www.aresep.go.cr

Relevant Entities

Ministry for the Environment and Energy
- http://www.minae.go.cr

Ministry of Health
- http://www.ministeriodesalud.go.cr

Water Resources Institute
- http://www.aya.go.cr

Customer Protection
- http://www.dhr.go.cr

The Sector

Country
- Area – 51,100 km^2
- Population – 4.1 million
- Population density – 85 inhab/km^2

Service Coverage
- Water – 98%
- Wastewater service – 21%
- Wastewater treatment – 4.5%

Number of operators
- Water – 1,800
- Wastewater – 5

Ownership and management of services

	Ownership		Management	
	Pub.	Priv.	Pub.	Priv.
Water	100%	0%	96%	4%
Wastewater	100%	0%	100%	0%

Volume of activity
- Abstracted water volume – 350 million m^3
- Drinking water – 175 million m^3
- Average consumption – 117 litres per person per day
- Main source – 98% groundwater
- Water losses – 50%

Economic and financial data
- Turnover – 66.2 million euros (50% of the population)
- Water service average price – 0.26 €/m^3
- Average annual invoice per customer –
- Employment – 3,000 employees

CZECH REPUBLIC

Regulatory Model

In the Czech Republic, as in most developed countries, there is no sector-specific regulation for the water sector. However, the association of operators (SOVAK), while not having explicit regulatory power, has considerable influence in the political sector, particularly in the drafting of legislation, through its cooperation with the national and regional authorities and institutions, as well as with businesses and other important organizations in the water sector. As such, it plays a major role in the progress and modernization of the sector. SOVAK has provided incentives for the use of benchmarking among the WWS operators in the Czech Republic, producing an annual report that outlines the performance of the participants. On the economic front, the Ministry of Finance is officially responsible for the approval of WWS tariffs, including the persistent dichotomy between the consideration of 'justified cost' charges and ones that do not lead to 'inappropriate profits'. However, due to the lack of transparency, this competency is very limited. The main legal documents governing the water sector are the Law of Privatisation of 1991 and the Water Act of 2001.

Market Structure

The WWS are mostly run by the private sector (serving 70% of Czech population). The remaining population is served by municipal companies or public companies. The ownership of the infrastructures remains under the control of the public sector, except in Ostrava, where they are held by Aqualia, a Spannish company. The management of WWS is essentially found in three organizational arrangements: municipal companies, public companies and commercial companies. Currently there are around 20 municipal companies, representing 3% of the water market. Commercial companies may or may not have mixed capital. The remainder of the customers is provided by municipal services or by individual entities that service about 10% of the total population. In the Czech Republic the WWS are characterized by their vertical and horizontal integration, i.e. the operators, in general, provide wholesale and end-user services at the same time, together with WWS.

Tariff Setting

The tariff system usually implemented by operators for the WWS consists of a variable charge applied to both services. Cross-subsidization is used as a means of ensuring universal access to the WWS. The operators set different tariffs for domestic and industrial customers.

Quality of Service

The WWS performance in general is acceptable, with service coverage and reliability in particular standing out. Poor performance is prevented through contracts established with private operators. Currently a new benchmarking system (WaBe) based on 24 performance indicators is being developed.

The Regulator

Characteristics

Legal basis

Number of employees

Budget

Contacts

Relevant Entities

Operators' Association
- http://www.sovak.cz

Association for Consumer Protection
- http://www.customers.cz

Ministry for the Environment
- http://www.env.cz

Ministry of Health
- http://www.mzcr.cz

The Sector

Country
- Area – 78,900 km^2
- Population – 10.3 million
- Population density – 132 inhab/km^2

Service Coverage
- Water – 92.4%
- Wastewater service – 80%
- Wastewater treatment – 75%

Number of operators
- Water – 1,211
- Wastewater – 1,211

Ownership and management of services

	Ownership		Management	
	Pub.	Priv.	Pub.	Priv.
Water	94%	6%	30%	70%
Wastewater	94%	6%	30%	70%

Volume of activity
- Abstracted water volume – 698.7 million m^3
- Drinking water – 532.4 million m^3
- Average consumption – 100 litres per person per day
- Main source – 75% superficial
- Water losses – 23.8%

Economic and financial data
- Turnover – 973 million euros
- Water service average price – 1.8 €/m^3
- Average annual invoice per customer – 66 euros
- Employment – 15,000 employees

ENGLAND AND WALES

Regulatory Model

In E&W, the regulatory model is very effective, proving why it is an international benchmark. It is based on three independent regulators (economic regulator, environmental regulator and drinking water quality regulator). These three regulators are complemented by the existence of other bodies with regulatory duties such as DEFRA, the Competition Authority and the Consumer Council. The economic regulator, OFWAT, has the responsibility of safeguarding the activities that are an integral part of businesses, protecting the interests of customers, and ensuring the financial viability, economic efficiency and competitiveness of businesses. The Environment Agency (EA) has the functions of an environmental regulator, which includes the licensing of abstractions and discharges, and, finally, the Drinking Water Inspectorate (DWI) has the duties of regulating drinking water quality. Concerning the economic regulation, the methodology adopted by OFWAT is based on regulation by price caps for a period of five years. The formula applied in establishing the price caps is RPI $\pm$ K, where K represents the difference between Q (new investments) and X (productivity), allowing best practices to retain some of the gains obtained and penalizing the worst practices (the carrot-and-stick approach). In addition to these tools, OFWAT encourages inset appointments where, for a given geographical area, the incumbent operator is replaced by another, thereby encouraging the competition, the third-party access to the network (common carriage), the competition for commercialization and the use of private and unregulated systems. The Water Act of 1983 and the Water Industry Act of 1991 provide the main legal framework for the water sector.

Market Structure

The WWS in E&W have private ownership. The water market structure in these countries although always private since 1989 has undergone significant changes over the last five years. Currently, in E&W there are 12 companies that provide WWS, 10 of which are large companies of water and wastewater services (WaSCO), and 13 companies that offer only the service of water, 12 of which are known as water only companies (WoC). The main players referred (in large areas) vertically and horizontally integrate the WWS. However, the water sector includes approximately a dozen companies that only provide water services relative to the water production chain.

Tariff Setting

The tariff system is composed of a fixed part and a proportional part and, even though only 25% of domestic customers have a meter, this is based on the level of consumption and the characteristics of the property. There are social tariffs for large families and for customers with special needs. Domestic and industrial customers are charged differently.

Quality of Service

The WWS demonstrate an excellent quality of service (e.g. service coverage and continuity). The quality is assured by service charters made by each operator and the sunshine regulation model of OFWAT, which quantifies the quality of service levels. Minimum levels of service lead to a compensation payment (GSS).

The Regulator

Characteristics
- Sector-specific regulator
- Independent regulatory agency

Legal basis
Water Law of 1989

Number of employees
190 employees

Budget
14.5 million euros (2007)

Contacts
Water Services Regulation Authority (OFWAT)
Centre City Tower, 7 Hill Street,
Birmingham B5 4UA
England
Tel: +44 121 625 1300
Fax: +44 121 625 1400
http://www.ofwat.gov.uk

Relevant Entities

Department of the Environment
- http://www.defra.gov.uk

Consumer Council Association
- http://www.ccwater.org.uk

Operators' Association
- http://www.water.org.uk

Water Quality Inspectorate
- http://www.dwi.gov.uk

The Sector

Country
- Area – 156,200 km^2
- Population – 53.7 million
- Population density – 343.8 inhab/km^2

Service Coverage
- Water – 99%
- Wastewater service – 96%
- Wastewater treatment – 93%

Number of operators
- Water – 25
- Wastewater – 10

Ownership and management of services

	Ownership		Management	
	Pub.	Priv.	Pub.	Priv.
Water	0%	100%	0%	100%
Wastewater	0%	100%	0%	100%

Volume of activity
- Abstracted water volume – 5,469 million m^3 (2007)
- Drinking water – 4,267 million m^3
- Average consumption – 147 litres per person per day
- Main source – 70% superficial
- Water losses – 22%

Economic and financial data
- Turnover – 10,987.5 million euros
- Water service average price – 1.5 €/m^3
- Average annual invoice per customer – 390.3 euros
- Employment – 35,000 employees

GHANA

Regulatory Model

In Ghana, there is an independent regulatory agency (Public Utilities Regulatory Commission – PURC) that has the competencies of regulating the urban WWS, which includes economic regulation, quality of service and public service obligations. In this area, the regulator has no authority over the water systems managed by local communities and regulates only the state company (GWCL). Besides providing guidelines for establishing, analysing and approving tariffs, the PURC simultaneously protects the interests of both customers and operators, promotes the principles of competition, and defines and monitors service levels. In order to encourage efficiency and quality of GWCL services, the regulator issues an annual report, which publishes (sunshine regulation) a set of performance indicators (benchmarking). As far as economic regulation is concerned, the regulator uses the method of price cap regulation for a period of four years. As a complement to the actions of the PURC, the government established the Water Resources Commission, which has the authority to grant licences for abstraction to operators of WWS. The PURC Act of 1997 and the Water Resources Act of 1996 provide the main legal framework for the water sector.

Market Structure

Currently, following the establishment of a management contract in 2005, the WWS are primarily managed by the private sector. The WWS are vertically integrated, that is, the operator, particularly the state company GWCL, performs all services related to the production chain, from abstraction to the distribution of water in towns with populations greater than 5,000 inhabitants. Presently, GWCL, through a five-year contract, is being managed by Vitens, which created the operator Aqua Vitens Rand Limited. In addition, the company is responsible for establishing, operating and controlling wastewater systems. In rural areas the Community Water and Sanitation Agency (CWSA) is responsible for coordinating (but not operating) and implementing WWS programs, which are directly provided by communities, and providing technical assistance to all WWS operating in these areas (e.g. NGOs).

Tariff Setting

The tariff system normally applied by operators consists of an increasing block rate set according to the volume of water consumed. Some low-income customers are subsidized directly by the government. The operator sets different tariffs depending on the type of customer.

Quality of Service

The WWS require major developments, not only due to the limited level of service coverage and the poor reliability, even in the capital city of Accra (25% of service continuity), but also due to the low efficiency caused by high water losses. The wastewater services are almost non-existent.

The Regulator

Characteristics
- Multisector regulator
- Independent regulatory agency

Legal basis
Energy Commission Law of 1997

Number of employees
48 employees (5 water sector)

Budget
712,000 euros

Contacts
Public Utilities Regulatory Commission
51 Liberation Road, African Liberation Circle,
PO Box CT 3095, Cantonments, Accra,
Ghana
Tel: +233 21 244 181
Fax: +233 21 244 188
http://www.purc.com.gh

Relevant Entities

Sewerage Agency
- http://www.cwsagh.or

Environmental Agency
- http://www.epa.gov.g

Water Resources Commission
- http://www.wrc-gh.org

Water Sector Reconstruction Department
- http://www.waterforghana.org

The Sector

Country
- Area – 238,500 km^2
- Population – 23 million
- Population density – 93 inhab/km^2

Service Coverage
- Water – 59%
- Wastewater service – 7%
- Wastewater treatment –

Number of operators
- Water – 2 + rural operators
- Wastewater – 1

Ownership and management of services

	Ownership		Management	
	Pub.	Priv.	Pub.	Priv.
Water	100%	0%	0%	100%
Wastewater	100%	0%	0%	100%

Volume of activity
- Abstracted water volume – 205 million m^3
- Drinking water – 103 million m^3
- Average consumption – 20 litres per person per day
- Main source – 95% superficial (urban areas)
- Water losses – 50%

Economic and financial data
- Turnover – 46 million euros
- Water service average price – 0.43 €/m^3
- Average annual invoice per customer –
- Employment – 3,200 employees

GREECE

Regulatory Model

In Greece, there is no regulatory agency with explicit functions of economic regulation and/or quality of service and/or supervision of public service obligations in the water sector. However, self-regulation does exist which contributes to the provision of WWS with quality. In terms of economic regulation, operators of WWS (EYDAP S.A. and EYATH S.A., DEYA and municipalities directly) set their own respective prices, which are subsequently approved by the national authorities. The state companies set their prices with the approval of both the state itself and the joint approval of the Ministers for the Economy and Finance and the Ministry of Public Works. In the scope of municipal companies, prices are set by their own corporations. However, their adoption requires approval by the municipality and by the regional representative of the state. Specifically, the DEYA set their prices for WWS, and require approval by the Municipal Assembly. The association of Greek operators, EDAYA, performs functions of self-regulation, encouraging the dissemination of best practices among its members and its respective representation abroad. The main normative documents are the Water Resources Management Act of 1987, and the Environmental Protection Act of 1986.

Market Structure

With few exceptions, the WWS, particularly in big cities such as Athens and Thessalonica, are integrated (water and wastewater). Thus, there is a single manager for the same jurisdiction that provides all services. The same occurs as regards vertical integration of these services, where the same operator manages both wholesale and end-user water services, as well as the services of wastewater, where it is responsible for the operations from collection to wastewater treatment. However, in some municipalities, only end-user WWS services are provided, while the wholesale services are provided by the state-run companies of the country (EYDAP S.A. in Athens, and EYATH S.A. in Thessalonica). These companies have some of their capital (< 40%) quoted on the stock exchange.

Tariff Setting

Water service charges are usually based on progressive volumetric tariffs and priced per cubic metre, that is, they increase directly with the consumption level. There are social tariffs for customers with special needs. Domestic, industrial, commercial customers and public institutions are charged differently.

Quality of Service

In Greece, there is limited information on the quality of service provided. There are no details published on the quality of the sector, except for the aspects related to drinking water quality and the wastewater discharges quality, which have shown a very positive development in recent years. Anyway, customers do not complain about the quality of service.

The Regulator

Characteristics

Legal basis

Number of employees

Budget

Contacts

Relevant Entities

Municipal Enterprises Association
- http://www.edeya.gr

Customer Protection Association
- http://www.kepka.org

Ministry for the Economy and Finance
- http://www.mnec.gr

Ministry for the Environment and Public Works
- http://www.minenv.gr

The Sector

Country
- Area – 132,000 km^2
- Population – 11 million
- Population density – 84 inhab/km^2

Service Coverage
- Water –
- Wastewater service –
- Wastewater treatment –

Number of operators
- Water – > 1,000
- Wastewater – > 1,000

Ownership and management of services

	Ownership		Management	
	Pub.	Priv.	Pub.	Priv.
Water	100%	0%		
Wastewater	100%	0%		

Volume of activity
- Abstracted water volume – 800 million m^3
- Drinking water – 600 million m^3
- Average consumption – 200 litres per person per day
- Main source – 80% superficial
- Water losses – 25%

Economic and financial data
- Turnover –
- Water service average price –
- Average annual invoice per customer –
- Employment –

HONDURAS

Regulatory Model

In Honduras, there is an independent regulatory agency (Regulatory Agency of Water and Wastewater Services – ERSAPS) dedicated solely to the regulation of WWS. As an independent authority (organic and functional), it has the task of regulating the WWS in the national territory regarding economic regulation, quality of service and public service obligations as imposed by the Law of WWS in 2003. Its competencies include also the compliance and the enforcement of compliance with the law, the promotion of efficiency in service provision, the establishment of standards, efficiency criteria, indicators and representative models for evaluating the technical, environmental, financial and administrative management of services, the maintenance and promotion of transparency in the sector and, mainly, the defence of customers' interests. As far as economic regulation is concerned, the regulator has adopted the method of revenue-cap regulation for a period of five years. The Water Framework Law of 2003 and the General Law of the Environment of 1993 constitute the legal framework governing the water sector.

Market Structure

The WWS are primarily provided by operators in the public sector, with the only situations of privatization taking place in the cities of San Pedro Sula and Puerto Cortés. However, the organization of the sector is under a changing process as a consequence of the enactment of the new WWS law in 2003, which requires the delegation of services provided by SANNA (a state-run company) to the municipalities. Generally, the WWS are integrated, that is, the operator, particularly the state company in urban areas, provides all services both in the water service (from abstraction to the water distribution) and in the wastewater service (wastewater collection and treatment). In rural areas, WSS are usually run by community organizations (Water Management Boards – JAA).

Tariff Setting

The tariff system adopted by most operators is based on a fixed part charged to all the customers, or by the characteristics of the property. The rate values are very low and insufficient. Most customers are charged a fixed rate, given the limited metering. Domestic, commercial, industrial customers and public institutions are charged differently.

Quality of Service

The WWS have been characterized by a significant increase in service coverage. However, the sector remains unreliable and inefficient, with a very low quality of service especially in rural areas. Water losses are a serious issue. The regulator has developed a benchmarking system based on a set of performance indicators with the main purpose of enhancement the quality of service of the water sector.

The Regulator

Characteristics
- Sector-specific regulator
- Independent regulatory agency

Legal basis
Decree No. 118 of 2003

Number of employees
15 employees

Budget

Contacts
Ente Regulador de los Servicios de Agua Potable y Saneamiento
Colonia Palmira, edificio Roma No. 1601, Tegucigalpa,
Honduras
Tel: +504 238 8151
Fax: +504 238 2598
http://www.ersaps.gob.hn

Relevant Entities

National Water and Sanitation Board
- http://www.sanaa.hn/conasa/conasa.htm

Association of Water Councils
- http://www.rashon.org.hn/web_ahjasa/index.html

Ministry of Health
- http://www.salud.gob.hn

Water Sector Forum
- http://www.rashon.org.hn

The Sector

Country
- Area – 112,500 km^2
- Population – 7.5 million
- Population density – 64 inhab/km^2

Service Coverage
- Water – 75%
- Wastewater service – 36%
- Wastewater treatment –

Number of operators
- Water – >2,000
- Wastewater –

Ownership and management of services

	Ownership		Management	
	Pub.	Priv.	Pub.	Priv.
Water	100%	0%	87%	13%
Wastewater	100%	0%	87%	13%

Volume of activity
- Abstracted water volume –
- Drinking water –
- Average consumption –
- Main source – 96% superficial
- Water losses – 50%

Economic and financial data
- Turnover –
- Water service average price – 0.10 €/m^3
- Average annual invoice per customer – 25 euros
- Employment –

INDONESIA

Regulatory Model

In Indonesia, there is a sector-specific regulator for the water service (Jakarta Water Supply Regulatory Board – JWSRB). The JWSRB is local in nature, with assignments only in the city of Jakarta. One of the JWSRB functions is to ensure the balance of interests between customers and water operators, with a focus on protecting the customers' right to obtaining good drinking water quality at a reasonable price. The JWSRB also plays an important role in the sector credibility and in the promotion of private sector participation in these services. In order to carry out the respective tasks, committees were established within the JWSRB, namely the Tariff Committee and the Committee of Public Opinion, which, among other functions, carry out a survey on the quality of service, and the Expert Committee. Concerning the economic regulation, there are two types of payment: the lease and water tariffs. The lease, relative to the payment to concessionaires, is adjusted in each semester using an indexing formula, which includes several factors such as exchange rates (e.g. CPI). The lease is reformulated every five years. On the other hand, the tariffs on water are directly related to the services provided by the operator. The legal framework for the water service is provided by the Water Resources Law of 2004, and the Environmental Management Act of 1997.

Market Structure

The WWS are mainly provided by the public sector. The main exception occurs in Jakarta where water services were privatized by two companies, namely PALJAYA (West) and TPJ (East), dividing the city into two parts. Additionally, there is an operator dedicated to wastewater service, PALJAYA. The majority of WWS are provided by public multisector operators (PDAM) that perform the services of water, wastewater and urban waste. Although the WWS are generally vertically integrated, there are some exceptions where the wholesale water service can be purchased from other operators such as the Perum Jasa Tirta II (PJT II), in the area of Jakarta.

Tariff Setting

The tariff system adopted by PDAM consists of volumetric rates, with or without blocks (increasing). In Jakarta, the tariff system adopted by the two water service concessionaires comprises an increasing block rate. PALJAYA adopted a system based on a fixed part depending on the property.

Quality of Service

Besides the low level of service coverage, the water sector presents diverse weaknesses (e.g. low water pressure and intermittent service and high water losses). In order to monitor the performance of the operator and the satisfaction of customers, JWSRB applies and monitors a set of performance indicators. WWS need huge investments.

The Regulator

Characteristics

- Sector-specific regulator
- Regulatory agency without independence

Legal basis

Decree No. 95 of 2001

Number of employees

30 employees

Budget

257,000 euros (2006)

Contacts

Jakarta Water Supply Regulatory Board
Jl Pejompongan Raya No. 57, Jakarta Pusat, Indonesia
Tel: +62 21 570 9732
Fax: +62 21 570 9723
http://jakartawater.org

Relevant Entities

National Planning Development Agency

- http://www.bappenas.go.id

Operators' Association

- http://www.perpamsi.org

Ministry for the Environment

- http://www.menlh.go.id

Asian Development Bank

- http://www.adb.org

The Sector

Country

- Area – 1,900,000 km^2
- Population – 234.7 million
- Population density – 134 inhab/km^2

Service Coverage

- Water – 18%
- Wastewater service – 3%
- Wastewater treatment – 2.25%

Number of operators

- Water – >320
- Wastewater – >320

Ownership and management of services

	Ownership		Management	
	Pub.	Priv.	Pub.	Priv.
Water	100%	0%	75%	25%
Wastewater	100%	0%	75%	25%

Volume of activity

- Abstracted water volume – 6,600 million m^3 (2006)
- Drinking water – 3,300 million m^3
- Average consumption – 146 litres per person per day
- Main source – 85% superficial
- Water losses – 50%

Economic and financial data

- Turnover – 101.6 million euros
- Water service average price – 0.49 €/m^3
- Average annual invoice per customer – 25 euros
- Employment – 3,162 employees (Jakarta)

ITALY

Regulatory Model

In Italy, there is a regulatory agency, the Committee for the Supervision of the Water Resources Use (COVIRI), whose functions include monitoring the principles of the water sector reform law. This Committee regulates the setting and adjustment of tariffs and the protection of customer interests. The most problematic question the Committee has been facing is the formula for setting tariffs based on price caps. A key concern of COVIRI is to ensure the efficiency, effectiveness and self-sustainability of the WWS operators. The Committee may propose actions against the operators, or even against their board of directors, if they are found to be in violation of some of the principles outlined in the Galli Law. The water sector regulatory process is composed of different levels, namely the Ministry for the Environment through the proposals of COVIRI, especially with regard to the national rate system, the regional areas, which provide a draft contract and, finally, the AATOs, which have the power to guarantee, monitor and set tariffs within each respective jurisdiction and control and monitor the quality of service and contractual commitments. The water sector has the Galli Law of 1994, and the Merli Law of 1976, as its main normative documents.

Market Structure

In the past, the WWS in Italy were mainly provided by public operators. With the implementation of the Galli Law and the constitution of 91 ATOs, there was a proliferation and a growing tendency for the presence of the private sector in providing these services, in particular large multinational companies (e.g., Veolia, Suez and FCC). However, despite the established contracts with private operators, infrastructures and facilities remain under public ownership. From the 91 ATOs, 57 were expected to be in operation in 2006, 13 of which were public (in-house), 11 private and 33 of mixed capital with a majority of public capital. In the past, the WWS were mostly performed by the public sector, particularly by municipalities. With the restructuring of the water sector, operators began providing the service in an integrated way, whether vertically or horizontally in the area of jurisdiction of the ATO, where a single operator provides the whole water cycle (water and wastewater).

Tariff Setting

The tariff system practised by operators is a two-part rate structure based on a fixed part, depending on the size of the meter or property, and a volumetric rate, depending on consumption and divided into blocks (increasing) or units. The operator sets distinct tariffs according to the type of customer: domestic, industrial, commercial or public.

Quality of Service

The WWS has primarily been known for the good quality of service provided. Continuity and service coverage are two aspects to highlight but the level of water losses is still not acceptable. In order to enhance the performance of the sector and the protection of customers, COVIRI started to use a system of benchmarking based on 55 performance indicators.

The Regulator

Characteristics
- Sector-specific regulator
- Regulatory agency without independence

Legal basis
Law No. 36 of 1994 (Galli Law)

Number of employees
25 employees (7 water sector)

Budget
640,000 euros (2006)

Contacts
Comitato per la Vigilanza sull'Uso delle Risorse Idriche, Via Cristoforo Colombo, 44 – 00147, Rome
Italia
Tel: +39 06 5722 5200
Fax: +39 06 5722 5290
http://www.coviri.it

Relevant Entities

Operators' Association
- http://www.federutility.it

National Environmental Protection Agency
- http://www.apat.gov.it

ATO National Association
- http://www.associazioneanea.it

Ministry for the Environment and Land Use
- http://www.minambiente.it

The Sector

Country
- Area – 301,300 km^2
- Population – 59.4 million
- Population density – 197 inhab/km^2

Service Coverage
- Water – 96%
- Wastewater service – 83.6%
- Wastewater treatment – 74.8%

Number of operators
- Water – 91
- Wastewater – 91

Ownership and management of services

	Ownership		Management	
	Pub.	Priv.	Pub.	Priv.
Water	100%	0%	75%	25%
Wastewater	100%	0%	75%	25%

Volume of activity
- Abstracted water volume – 7,600 million m^3
- Drinking water – 4,500 million m^3
- Average consumption – 230 litres per person per day
- Main source – 50% superficial/groundwater
- Water losses – 40%

Economic and financial data
- Turnover – 4,231 million euros
- Water service average price – 1.23 €/m^3
- Average annual invoice per customer – 250 euros
- Employment – 64,000 employees

JAMAICA

Regulatory Model

In Jamaica, there is a regulatory agency (Office of Utilities Regulation – OUR), which, according to the law, approves tariffs, sets targets for efficiency improvements, processes all licence applications for utility services and makes recommendations to the Minister. In this context, the regulator supervises the compliance and the enforcement of compliance with general legislation, environmental regulations and competition. It promotes transparency in the sector, specifies licensing and financial sanctions and, mainly, defends the interests of customers. The regulatory agency also establishes targets for the main WWS operators, thereby encouraging the enhancement of quality of service. Concerning the economic regulation, the regulator has adopted the method of regulation by price cap for a period of three years. The main normative documents of the water sector are the Water Resources Act of 1995 and the OUR Act, also of 1995.

Market Structure

The WWS are primarily managed by public operators. The management of these services by the private sector is carried out only by small operators. In this context, the WWS are, in most cases, vertically integrated, i.e. operators, especially the state company (NWC), have all services on the production chain, from abstraction to the distribution of water. The situation is similar for the wastewater service where there is integration of wastewater collection and treatment. Private operators specifically provide the WWS. In rural areas, services are usually shared between NWC and local authorities.

Tariff Setting

The tariff system adopted by operators is usually composed of a fixed part based on the size of the meter (33% of the population) and an increasing block rate according to the volume of water consumed. There are no special tariffs. The tariffs adopted by operators vary according to whether the customer is domestic or industrial/commercial.

Quality of Service

Despite numerous efforts, WWS still require major developments, not only to increase the level of service coverage, but also for the many areas with reduced pressure. The water service presents significant water losses, thus requiring large investments in the replacement and rehabilitation of the infrastructures. The wastewater service also needs large investments to increase its coverage.

The Regulator

Characteristics
- Multisector regulator
- Independent regulatory agency

Legal basis
Law No. 13 of 1995

Number of employees
49 employees (4 water sector)

Budget
2.3 million euros (2006)

Contacts
Office of Utilities Regulation
3rd Floor, PCJ Building,
36 Trafalgar Road,
Kingston 10, Jamaica
Tel: +876 968 6053
Fax: +876 968 6057
http://www.our.org.jm

Relevant Entities

Ministry of Water and Housing
- www.jis.gov.jm/water_housing

Ministry of Health and Environment
- www.moh.gov.jm

Environmental Agency
- www.nepa.gov.jm

Caribbean Water and Wastewater Association
- http://www.cwwa.net

The Sector

Country
- Area – 11,000 km^2
- Population – 2.7 million
- Population density – 252 inhab/km^2

Service Coverage
- Water – 73%
- Wastewater service – 30%
- Wastewater treatment –

Number of operators
- Water – 8
- Wastewater – 4

Ownership and management of services

	Ownership		Management	
	Pub.	Priv.	Pub.	Priv.
Water	100%	0%		
Wastewater	100%	0%		

Volume of activity
- Abstracted water volume – 287 million m^3 (2005)
- Drinking water – 95 million m^3
- Average consumption – 132 litres per person per day
- Main source – 92% groundwater
- Water losses – 67%

Economic and financial data
- Turnover – 92.3 million euros
- Water service average price – 0.75 €/m^3
- Average annual invoice per customer –
- Employment –

KENYA

Regulatory Model

In Kenya, there is an independent regulatory authority (Water Services Regulatory Body – WRSB) that oversees the regular functions of the WWS, with the overall aim of protecting customer rights, and safeguarding the sustainability and economic viability of systems. This council, acting as the national authority, also has the responsibility to monitor and control the drinking water quality in all the municipalities in the country. In order to regulate the quality of service, the WRSB developed a system of benchmarking based on a set of performance indicators. Additionally, the council establishes guidelines for Minimum Service Levels, ensuring the provision of a reliable service. The regulator also has the task of licensing the Water Service Boards (WSB) and ensuring that the Water Service Providers (WSP) follow the conditions established and the development of guidelines for the service provision agreements between the WSB and the WSP. Regarding the economic regulation, the regulator adopted the method of rate-of-return regulation for a period of one year. The National Water Policy, created in 1999, and the Water Act of 2002 represent the main guidelines for the water sector.

Market Structure

The WWS are mostly provided by public operators. Despite the existence of some private sector participation (Community Based Organizations) in the water service provision, it is very limited and only found in isolated areas (urban and rural). Recently, the WWS have undergone major changes, particularly due to the decentralization of powers over the sector. Currently, the ownership of infrastructure, which previously belonged to local governments, is under the WSB which then appoints a WSP to provide services.

Tariff Setting

The tariff system normally applied by the operators is based on an increasing block rate according to the volume of water consumed. Low-income customers are aided by special tariffs. Cross-subsidization is also used by operators between the different type of customers.

Quality of Service

In general, the WWS require major developments, with particular emphasis in rural areas (80% of the population). Given the recent reforms, the information about the service is very limited. The service is not very reliable and infrastructure degradation is a great problem which leads to the existence of high water losses. Wastewater service requires significant investments to enhance its coverage.

The Regulator

Characteristics
- Sector-specific regulator
- Independent regulatory agency

Legal basis
Presidential Circular No. 1/2003

Number of employees
23 employees

Budget
780,000 euros

Contacts
Water Services Regulatory Body
9th Floor, NHIF Building,
PO Box 41621 00100-GPO,
Nairobi, Kenya
Tel: +254 0 20 273 3559
Fax: +254 0 20 273 3558
http://www.wasreb.or.ke

Relevant Entities

Ministry for Water and Irrigation
- http://www.water.go.ke

African Development Bank
- http://www.afdb.org

Environmental Protection Agency
- http://www.nema.go.ke

The Sector

Country
- Area – 538,400 km^2
- Population – 34.7 million
- Population density – 59 inhab/km^2

Service Coverage
- Water – 55%
- Wastewater service – 25%
- Wastewater treatment – 20%

Number of operators
- Water – 100
- Wastewater – 50

Ownership and management of services

	Ownership		Management	
	Pub.	Priv.	Pub.	Priv.
Water	100%	0%		
Wastewater	100%	0%		

Volume of activity
- Abstracted water volume –
- Drinking water –
- Average consumption –
- Main source – 95% superficial
- Water losses – 60%

Economic and financial data
- Turnover –
- Water service average price –
- Average annual invoice per customer –
- Employment –

KOSOVO

Regulatory Model

In Kosovo, there is a regulatory agency (Water and Waste Regulatory Office – WWRO) that is responsible for the regulation of WWS and municipal waste, intending, above all, to protect customer rights. It also safeguards the sustainability and economic viability of the systems. The main method adopted by WWRO to regulate the quality of service is the sunshine regulation model, based on published results of the operators' performance. For this purpose, the WWRO produces an annual report that analyses the quality of service of the operators based on a set of performance indicators. The preparation of this document, rooted in benchmarking, includes a joint evaluation of performance in which comparisons are made between the operators and the individual assessment of the quantitative evolution of each operator's performance. The regulator has the ability to impose penalties on operators with lower performance in terms of quality of service. In the field of economic regulation, the regulator adopts the method of regulation by price caps for a period of one year. The legal framework of the water sector consists of two main laws, namely the Water Act of 2004 and the Regulation No. 49 of 2004.

Market Structure

The WWS are primarily provided by seven regional companies. There is no private sector participation in the water sector in Kosovo. The end-user and wholesale services of WWS are vertically integrated, i.e. the same operator performs all services, from abstraction to the water distribution for the water services, and from collection to the wastewater treatment for the wastewater services.

Tariff Setting

The tariff system normally applied by operators is based on a fixed part and a proportional part according to the volume of water consumed. The government subsidizes some customers directly. There are no social tariffs. However, industrial/business customers are subject to higher tariffs. Despite a recent 22% increase, tariff rates remain insufficient.

Quality of Service

The WWS are characterized by reasonable quality of service, which underlines the need for large investments in operational efficiency (large water losses) and wastewater treatment. As the main tool to encourage an improved quality of service in the sector, the regulator established a system of benchmarking based on a set of performance indicators.

The Regulator

Characteristics
- Sector-specific regulator
- Independent regulatory agency

Legal basis
Regulation no. 49 of 2004

Number of employees
17 employees (8 water sector)

Budget
153,000 euros (2007)

Contacts
Water and Waste Regulatory Office
Ferat Dragaj str. No. 68, Sunny Hill, Prishtina, Kosovo
Tel: + 381 38 249 165
Fax: + 381 38 249 168
http://www.wwro-ks.org

Relevant Entities

Operators' Association
- www.shukos.org

Competition Authority
- http://kta-kosovo.org/html

Ministry for the Environment
- http://www.ks-gov.net/mmph

United Nations Mission in Kosovo
- http://www.unmikonline.org

The Sector

Country
- Area – 10,900 km^2
- Population – 2.1 million
- Population density – 220 inhab/km^2

Service Coverage
- Water – 74%
- Wastewater service – 55%
- Wastewater treatment – 0%

Number of operators
- Water – 7
- Wastewater – 7

Ownership and management of services

	Ownership		Management	
	Pub.	Priv.	Pub.	Priv.
Water	100%	0%	100%	0%
Wastewater	100%	0%	100%	0%

Volume of activity
- Abstracted water volume – 138.8 million m^3
- Drinking water – 57 million m^3
- Average consumption – 180 litres per person per day
- Main source – 70% groundwater
- Water losses – 59%

Economic and financial data
- Turnover – 19.2 million euros
- Water service average price – 0.14 €/m^3
- Average annual invoice per customer – 108 euros
- Employment – 1,500 employees

LITHUANIA

Regulatory Model

In Lithuania, the multisector regulatory agency (National Control Commission for Prices and Energy – NCCPE) is responsible for the regulation of the 47 largest WWS operators, as well as the services of heating, electricity and transportation. It aims to guarantee the rights of customers and safeguard the sustainability and economic viability of operators. Other competencies include the compliance and enforcement of compliance with general legislation and environmental and competition regulations and to promote efficiency in service provision. Moreover, as its regulatory method, the regulator has adopted the sunshine regulation model, based on published prices for the regulated operators. To this end, the NCCPE publishes through its annual report the prices charged by operators and a set of performance indicators, which characterize the general evolution of the water sector. Regarding the economic regulation, the regulator has adopted the method of rate-of-return regulation for a period of one year. However, this methodology is undergoing a study with the aim of changing this period to three years. The Water Act of 1997 represents the main normative document regulating the water sector.

Market Structure

The WWS are only managed by the public sector. Although municipalities are legally responsible for these services, they are primarily provided by 47 regional operators. With few exceptions, the WWS are horizontally integrated, i.e., there is a single manager for the same jurisdiction that provides both services. The same occurs at the level of vertical integration of these services, where the same operator has the powers of managing both wholesale and end-user water services, and also in the scope of wastewater, where it is simultaneously responsible for services from wastewater collection to its treatment. However, in some municipalities, the WWS provide only end-user services.

Tariff Setting

The tariff system adopted by the operators comprises a fixed part, fairly divided among customers and based on the characteristics of the property, and a proportional part depending on the volume of water consumed. Municipalities subsidize some low-income customers. Domestic and industrial customers are charged differently.

Quality of Service

The WWS are generally characterized by a reasonable quality of service in urban areas. However, this does not happen in rural areas. In these areas, the level of service coverage and the quality of service show a clear deficit compared to urban areas. As a way of encouraging the development of the sector, the regulator publishes the progress of some performance indicators.

The Regulator

Characteristics
- Multisector regulator
- Regulatory agency without independence

Legal basis
Decree issued by President of Lithuania in 1997

Number of employees
46 employees (7 water sector)

Budget

Contacts
National Control Commission for Prices and Energy
Ochil Algirdo g. 31, LT-03219, Vilnius,
Lithuania
Tel: +370 5 213 5166
Fax: +370 5 213 5270
http://www.regula.is.lt

Relevant Entities

Operators' Association
- http://www.lvta.lt

Ministry for the Environment
- http://www.am.lt

Association of Local Authorities
- http://www.lsa.lt

Competition Authority
- http://www.konkuren.lt

The Sector

Country
- Area – 65,200 km^2
- Population – 3.4 million
- Population density – 52 inhab/km^2

Service Coverage
- Water –77%
- Wastewater service –66%
- Wastewater treatment –60%

Number of operators
- Water – 300
- Wastewater – 300

Ownership and management of services

	Ownership		Management	
	Pub.	Priv.	Pub.	Priv.
Water	100%	0%	100%	0%
Wastewater	100%	0%	100%	0%

Volume of activity
- Abstracted water volume – 140 million m^3
- Drinking water – 98 million m^3
- Average consumption –63 litres per person per day
- Main source – groundwater
- Water losses –27%

Economic and financial data
- Turnover –119 million euros
- Water service average price –1.12 €/m^3
- Average annual invoice per customer – 40 euros
- Employment – 6,431employees

MEXICO

Regulatory Model

In Mexico, there is no explicit regulation of WWS. However, the National Association of Water and Wastewater Companies (ANEAS) plays a very positive role in WWS self-regulation and leads the sector by encouraging the performance of operators concerning both the quality of service and the main economic and financial indicators. The ANEAS, acting in a proactive way and using benchmarking as its main tool, has filled the absence of a sector-specific regulator. Furthermore, the association is active in the dissemination of best practices. Currently, the ANEAS is developing a new federal law for water and wastewater that will include the creation of a regulator. In regard to economic regulation, the local authorities set tariffs. The main legal documents of the water sector are the National Water Act of 1992, and the National Water Program, established in 2007.

Market Structure

The WWS are mainly provided by public sector operators. Private sector participation is limited compared to public management and ownership. In some areas, the state operators directly provide WWS, which are integrated both horizontally and vertically for the whole water cycle. In this context, the services from abstraction to the water distribution (water production) and the services of collection and treatment (wastewater service) are provided by the same operator. In rural areas, the water boards (*Juntas*) are responsible for the provision of water services.

Tariff Setting

The tariff system adopted by operators usually consists of a variable value charged according to the volume of water consumed. There is no national policy for social tariffs. More than two-thirds of the population has a meter. Cross-subsidization is adopted between the different type of customers.

Quality of Service

While the main operators are among the best in Latin America, the WWS are characterized, in general, by poor technical and commercial efficiency (high water losses), inadequate quality of service, especially in wastewater treatment, and a lack of service coverage in poor rural areas. Wastewater treatment requires meaningful investments.

The Regulator

Characteristics

Legal basis

Number of employees

Budget

Contacts

Asociacíon Nacional de Empresas de Água e Saneamiento
Palenque 287, Col. Navarte, C.P. 03020, México, D.F., Mexico
Tel: +52 55 5543 6000
Fax: +52 55 5543 6605
http://www.aneas.com.mx

Relevant Entities

National Water Commission
- http://www.conagua.gob.mx

Mexican Water Association
- http://www.amh.org.mx

Environmental and Natural Resources Department
- http://www.semarnat.gob.mx

Competition Authority
- http://www.cfc.gob.mx

The Sector

Country
- Area – 1,953,000 km^2
- Population – 108.7 million
- Population density – 252 inhab/km^2

Service Coverage
- Water – 89.2%
- Wastewater service – 85.6%
- Wastewater treatment – 36%

Number of operators
- Water – 10,500
- Wastewater – 250

Ownership and management of services

	Ownership		Management	
	Pub.	Priv.	Pub.	Priv.
Water	100%	0%	98.5%	1.5%
Wastewater	100%	0%	98.5%	1.5%

Volume of activity
- Abstracted water volume – 10,300 million m^3
- Drinking water – 5,100 million m^3
- Average consumption – 144 litres per person per day
- Main source – 63% superficial
- Water losses – 50%

Economic and financial data
- Turnover – 1,900 million euros
- Water service average price – 0.15 €/m^3
- Average annual invoice per customer – 115 euros
- Employment – 96,803 employees

MOZAMBIQUE

Regulatory Model

In Mozambique, there is a regulatory authority (Water Supply Regulatory Council – CRA) dedicated to the water sector. The CRA is 'a body which, by its independence and technical expertise, ensures a balance of interests involved in the provision of public service'. Additionally, the CRA has responsibilities in the protection of customer rights, economic regulation (only the delegated systems), the conciliation of interests between the concession-granting authority and the operators, the monitoring and support for the process of delegated management, the identification of the needs of current and future customers and, finally, the provision and improvement of the Delegated Management Framework operation. Thus, the scope of CRA's action is located between the government, on the one hand, and the municipalities, on the other hand, which are the local authorities acting in the defence of customer interests. Concerning the economic regulation, the method adopted is the rate-of-return regulation for a period of one year. Besides the regulator, the Water Supply Investment and Patrimony Fund (FIPAG) holds important responsibilities over the WWS, implementing investment programs and ensuring the sustainability of services in terms of the coverage of operating and maintenance costs. In 1998, the government established the Framework for Delegated Management (QGD), which created new entities in the sector. The institutionalization of the QGD aimed to ensure the efficient management of WWS. Furthermore, the Water Act of 1991 is the main normative document of the water sector.

Market Structure

The water sector is largely dominated by the private sector, by the Águas de Moçambique (Águas de Portugal) and by Vitens, providing, together, the water service to 2.5 million inhabitants. The remaining population is served by the public sector. The water sector, in particular the water service, is vertically integrated within the same operator (e.g. Águas de Moçambique) which provides the whole production chain, ranging from abstraction to the distribution of water, including its treatment and storage. In Mozambican cities where wastewater services exist, they are managed by the same operator responsible for the water service.

Tariff Setting

The tariff system adopted globally by operators is composed of an increasing block rate according to the volume of water consumed. There are special tariffs for customers with scarce financial resources. The operator sets different tariffs (minimum) depending on whether the customers are domestic, commercial, industrial or public taps/standpipes.

Quality of Service

The water services, and especially the wastewater, generally lack large investments. The level of coverage and the reliability and efficiency of the services are very low. The regulator has developed a system of benchmarking, through a set of performance indicators with the goal of encouraging performance.

The Regulator

Characteristics

- Sector-specific regulator
- Independent regulatory agency

Legal basis

Decree No. 74 of 1998

Number of employees

26 employees

Budget

1.7 million euros (2006)

Contacts

Conselho de Regulação do Abastecimento de Água
Av. Amilcar Cabral, 757, C.P. 235, Maputo
Mozambique
Tel: +258 21 312 825
Fax: +258 21 312 826
http://www.cra.org.mz

Relevant Entities

Investment Fund and Water Supply Directorate

- http://www.fipag.co.mz

Ministry for Public Works and Housing

- http://www.moph.gov.mz

National Water Board

- http://www.dnaguas.gov.mz

Ministry of Health

- http://www.misau.gov.mz

The Sector

Country

- Area – 801,600 km^2
- Population – 21.4 million
- Population density – 25 inhab/km^2

Service Coverage

- Water – 26%
- Wastewater service – 4.6%
- Wastewater treatment – 3% (estimated)

Number of operators

- Water – 220
- Wastewater – 20 (19 municipal authority septic tanks)

Ownership and management of services

	Ownership		Management	
	Pub.	Priv.	Pub.	Priv.
Water	100%	0%	50%	50%
Wastewater	100%	0%	%	%

Volume of activity

- Abstracted water volume – 85 million m^3 (delegated)
- Drinking water – 42 million m^3
- Average consumption – 21 litres per person per day
- Main source – 75% superficial
- Water losses – 51%

Economic and financial data

- Turnover –
- Water service average price –
- Average annual invoice per customer –
- Employment –

NIGER

Regulatory Model

In Niger, there is an independent regulatory agency (Multisector Regulatory Authority – ARM) that has the duty of regulating the quality of service and public service obligations of WWS, telecommunications, energy and national transportation. The regulator's competencies are to comply with and enforce compliance with legislation and to promote the efficiency of service provision. It also works to establish standards, efficiency criteria, indicators and representative models for the evaluation of the environmental, financial and administrative management of service provision. Furthermore, the ARM fosters transparency and participation in the sector, and above all defends customer interests. Concerning the economic regulation, the regulator has no responsibilities regarding the setting of tariffs, as this is the responsibility of the government. However, the ARM has the task of implementing and monitoring the guidelines for setting tariffs. The main method of quality of service regulation adopted by the ARM is the sunshine regulation model, based on published results of the operators' performance. For this purpose, the regulator publishes an annual report with the performance scores, obtained from a set of performance indicators. Law No. 12 of 2000 and Regulation No. 44 of 1999 constitute the legal framework governing the water sector.

Market Structure

The WWS are only managed by the private sector, particularly by the Association for the Development of the Water of Niger. In this context, the services are vertically integrated, i.e. the operator performs all services related to the production chain, from abstraction to the water distribution. Moreover, there is another private operator that manages investments and assets of the water sector, the Association for the Heritage of the Water of Niger. The wastewater services are practically non-existent.

Tariff Setting

The tariff system adopted by operators includes the implementation of an increasing block rate according to the volume of water consumed. There are special tariffs for low-income customers. The tariffs vary with the type of customer: domestic, industrial and public taps/standpipes.

Quality of Service

The WWS are inadequately provided and need large developments. Despite major reforms in the sector (e.g. privatization), it still shows poor levels of service coverage. There is no wastewater service. The WWS requires huge investments. There is little information about the water sector.

The Regulator

Characteristics
- Multisector regulator
- Independent regulatory agency

Legal basis
Decree No. 44 of 1999

Number of employees
25 employees (3 water sector)

Budget
2 million euros (2006)

Contacts
Autorité de Régulation Multisectorielle
64, Rue des Bâtisseurs, BP 13179 Niamey
Niger
Tel: +227 739 008/11
Fax: +227 738 591
http://www.arm-niger.org

Relevant Entities

Ministry of Water and the Environment
- http://www

African Development Bank
- http://www.afdb.org

The World Bank
- http://web.worldbank.org

French Development Agency
- http://www.afd.fr

The Sector

Country
- Area – 1,267,000 km^2
- Population – 14 million
- Population density – 11 inhab/km^2

Service Coverage
- Water – 63.2%
- Wastewater service –
- Wastewater treatment –

Number of operators
- Water – 2
- Wastewater – 0

Ownership and management of services

	Ownership		Management	
	Pub.	Priv.	Pub.	Priv.
Water	100%	0%	0%	100%
Wastewater	0%	0%	0%	0%

Volume of activity
- Abstracted water volume – 43 million m^3
- Drinking water – 37 million m^3
- Average consumption – 12 litres per person per day
- Main source –
- Water losses –

Economic and financial data
- Turnover – 14 million euros
- Water service average price – 0.33 €/m^3
- Average annual invoice per customer –
- Employment –

NORTHERN IRELAND

Regulatory Model

In Northern Ireland, the regulatory model is still being implemented, although it is based on the same principles of the other countries that compose the United Kingdom, namely the establishment of three independent regulators (tripartite regulation): the economic regulator, the environmental regulator and the drinking water quality regulator. These three regulators are complemented by the existence of other bodies with important regulatory functions, such as the EHS, the Competition Authority and the Consumer Council. The Northern Ireland Authority for Utility Regulation (NIAUR), holding functions as economic regulator, has the responsibility for safeguarding the activities that are an integral part of businesses, protecting the interests of customers and ensuring financial viability, economic efficiency and competitiveness of businesses. The EHS has the functions of an environmental regulator, including the licensing of abstractions and discharges, and the DWI regulates drinking water quality. Concerning the economic regulation, NIAUR has implemented the methodology of calculation based on price-cap regulation for a period of five years. The legal framework of the water sector is mainly provided by the Water Order of 1999 and the Water Act of 1972.

Market Structure

In Northern Ireland there is only one operator run by the public sector: Northern Ireland Water. Therefore, the end-user and wholesale services of WWS are vertically integrated, that is the same operator performs all services, from abstraction to the water distribution within the scope of the water service, and the services of collection and wastewater treatment within the scope of the wastewater service.

Tariff Setting

The tariff system adopted by the operator is based on a fixed part and on a proportional part, according to the volume consumed and the characteristics of the property regardless of the existence of a meter. There are specific tariffs for low-income customers and those with special needs.

Quality of Service

The WWS are generally characterized by a high quality of service. Contributing to this is the important role of the international benchmarking system, carried out by the regulator in E&W, as the regulatory method in Northern Ireland is still undergoing consolidation.

The Regulator

Characteristics
- Multisector regulator
- Independent regulatory agency

Legal basis
Article 3 Part II of the Energy Order 2003

Number of employees
65 employees (12 water sector)

Budget
1.5 million euros (2007)

Contacts
Northern Ireland Authority for Utility Regulation
Queens House, 14 Queen Street, Belfast, BT1 6ER,
Northern Ireland
Tel: +44 289 031 1575
Fax: +44 289 031 1740
http://ofreg.nics.gov.uk

Relevant Entities

Environmental Protection Agency
- http://www.ehsni.gov.uk

Consumer Council Association
- http://www.consumercouncil.org.uk

Operators' Association
- http://www.water.org.uk

Water Quality Inspectorate
- http://www.ehsni.gov.uk/environment/drinkWater/drinkWater.shtml

The Sector

Country
- Area – 13,800 km^2
- Population – 1.7 million
- Population density – 122 inhab/km^2

Service Coverage
- Water – 99%
- Wastewater service – 83%
- Wastewater treatment – 83%

Number of operators
- Water – 1
- Wastewater – 1

Ownership and management of services

	Ownership		Management	
	Pub.	Priv.	Pub.	Priv.
Water	100%	0%	100%	0%
Wastewater	100%	0%	100%	0%

Volume of activity
- Abstracted water volume – 226 million m^3
- Drinking water – 165 million m^3
- Average consumption – 266 litres per person per day
- Main source – 93% superficial
- Water losses – 27%

Economic and financial data
- Turnover – 354.4 million euros
- Water service average price – 1.45 €/m^3
- Average annual invoice per customer – 235 euros
- Employment – 1,800 employees

PANAMA

Regulatory Model

In Panama, there is an independent multisector regulator (National Authority of Public Services – ASEP), which has the responsibility of regulating the WWS, electricity, telecommunications, radio and television. This authority is responsible for guaranteeing that customers are provided with high quality of service and reliability at fair prices, while ensuring the sustainability of the operators. The regulator also has the task of safeguarding the public interest by ensuring that their needs and concerns are accounted for, and assuring adequate procedures for the routines and problems, dealing with complaints and providing technical assistance. The ASEP publishes on its website a set of financial and operational performance indicators in order to increase the transparency of the sector and encourage operators to enhance their services. With regard to economic regulation, the regulator adopted the method of price-cap regulation for a period of five years. The legal framework of the water sector consists of the Water Act of 1997, Law No. 77 of 2001 and Law No. 41 of 1998.

Market Structure

The country has a national public manager of the WWS, the IDAAN, with responsibilities in urban areas with populations over 1,500 inhabitants. The only exception is the municipality of Boquete, which directly provides these services in its jurisdiction. Besides the above, there are over 3,000 water systems in rural areas where more than half of them are managed by the Rural Water Boards (Juntas). The remaining is managed by Health Committees. The largest operator in the country, IDAAN, is characterized by a horizontal and vertical integration of the whole water cycle where both services co-exist. Private sector participation is merely figurative, providing specific services in areas outside the jurisdiction of IDAAN.

Tariff Setting

The tariff system adopted by operators consists of a proportional tariff that can be either constant or priced per increasing block rate, both according to the volume of water consumed. The government finances both operators and customers. There are social tariffs for the poor. The operators set different tariffs for domestic, commercial and public customers.

Quality of Service

The WWS are characterized by high levels of access and the quality of service is reasonable. However, they need major developments, especially in rural areas. Despite the little information available, the drinking water service seems to have good quality in comparison with other Latin American countries. The continuity and reliability of services are also adequate for the majority of the urban population.

The Regulator

Characteristics
- Multisector regulator
- Independent regulatory agency

Legal basis
Law No. 26 of 1996

Number of employees
5 employees for the water sector

Budget
3.65 million euros (2006)

Contacts
Autoridad Nacional de los Servicios Públicos
Vía España, Edificio Office Park
Panama
Tel: +507 508 4500
Fax: +507 508 4600
http://www.asep.gob.pa

Relevant Entities

National Environmental Authority
- http://www.anam.gob.pa

National Water and Wastewater Institute
- http://www.idaan.gob.pa

Ministry of Health
- http://www.minsa.gob.pa

Panama Canal Authority
- http://www.acp.gob.pa

The Sector

Country
- Area – 75,500 km^2
- Population – 3.3 million
- Population density – 43 inhab/km^2

Service Coverage
- Water – 86%
- Wastewater service – 51%
- Wastewater treatment – 39%

Number of operators
- Water – 11 + private operators
- Wastewater – 2

Ownership and management of services

	Ownership		Management	
	Pub.	Priv.	Pub.	Priv.
Water	100%	0%	99%	<1%
Wastewater	100%	0%	100%	0%

Volume of activity
- Abstracted water volume – 497.9 million m^3
- Drinking water – 290 million m^3
- Average consumption – 280 litres per person per day
- Main source – 96% superficial
- Water losses – 41.7%

Economic and financial data
- Turnover – 64.2 million euros
- Water service average price – 0.22 €/m^3
- Average annual invoice per customer – 71 euros
- Employment – 2,500 employees

PAPUA NEW GUINEA

Regulatory Model

In Papua New Guinea, atypical in international terms, there is a state agency, the PNG Waterboard, with functions of operation and regulation. To regulate private operators, the agency has altered its statutes with the purpose of registering all water systems not operated by it, in order to ensure consistency in the use of technology in the water sector and enforce environmental regulations. To date, under the regulatory regime, the PNG Waterboard regulates and monitors the private systems of WWS (systems not operated by PNG Waterboard) to ensure compliance with the standards. As a way of encouraging the development of the sector, the council established a benchmarking system, composed of a performance indicators set, focused mostly on the quality of service. Concerning the economic regulation, the agency has adopted the method of price cap regulation for a period of five years. The main normative document of the water sector is the Environment Act 2000.

Market Structure

The WWS are mostly provided by private (local) operators. However, the largest operator is a state-owned company, the PNG Waterboard. With few exceptions, the WWS are integrated, i.e. a single operator for the same jurisdiction provides both services. The same occurs at the level of vertical integration of these services, where the same operator has the responsibility for managing both wholesale and end-user water services and is simultaneously responsible for the wastewater services, performing operations from collection to treatment. However, in some municipalities, the WWS provide the only end-user services, while the wholesale services are provided by the state-owned company.

Tariff Setting

The tariff system adopted by operators usually consists of a variable charge priced per increasing blocks, according to the volume of water consumed and the characteristics of the property. There are no social tariffs. The tariff system is set similarly for domestic and non-domestic customers.

Quality of Service

The WWS requires great development. The simple fact of limited service coverage demonstrates some of these weaknesses. The quality of service presented is in general poor. In this context, the benchmarking system, developed by the regulator based on a set of performance indicators will be essential to encourage development and improvement of the quality of service.

The Regulator

Characteristics
- Sector-specific regulator
- Regulatory agency without independence

Legal basis
National Water and Wastewater Law of 1986

Number of employees
30 employees

Budget
202,000 euros (2006)

Contacts
Papua New Guinea Waterboard
Level 1, Garden City Building, Angau Drive,
PO Box 2779 Boroko, National Capital District,
Papua New Guinea
Tel: +675 323 5700
Fax: +675 325 6298
http://www.pngwater.com.pg

Relevant Entities

Competition Authority
- http://www.iccc.gov.pg

Department of Health
- http://www.health.gov.pg

Pacific Water Association
- http://www.pacificwaterassociation.org

Department of the Environment and Preservation
- http://www.pngonline.gov.pg/government

The Sector

Country
- Area – 462,800 km^2
- Population – 6.3 million
- Population density – 13 inhab/km^2

Service Coverage
- Water – 30%
- Wastewater service – 22%
- Wastewater treatment – 22%

Number of operators
- Water – 141
- Wastewater – 3 (urban areas)

Ownership and management of services

	Ownership		Management	
	Pub.	Priv.	Pub.	Priv.
Water	100%	0%	20%	80%
Wastewater	100%	0%	30%	70%

Volume of activity
- Abstracted water volume – 21.2 million m^3
- Drinking water – 14.2 million m^3
- Average consumption – 21 litres per person per day
- Main source – groundwater
- Water losses – 33%

Economic and financial data
- Turnover – 14.5 million euros
- Water service average price – 0.86 €/m^3
- Average annual invoice per customer – 120 euros
- Employment – 376

PERU

Regulatory Model

In Peru, the urban WWS operators are under the control of a sector-specific regulator (National Superintendence of Water and Wastewater Services – SUNASS), which regulates the services over 53.3% of the population. The main responsibilities of SUNASS, a body with organizational, financial and functional independence, include the supervision of compliance with technical standards for the design, construction and operation of WWS. SUNASS also has the power to ensure compliance with the current legal framework, regulate procedures for setting tariffs, resolve disputes between operators and customers, and finally impose legal fines and penalties on companies. The system of economic regulation is based on an efficient company model. This approach uses, as a standard of comparison, a virtual efficient company (benchmark company), created by the regulator, to meet the projected demand in the optimal way. Apart from its activities of supervision, SUNASS established a system of benchmarking in order to monitor the performance of operators. The General Law of WWS of 1994 is the main normative document of the water sector.

Market Structure

The WWS are almost entirely provided by the public sector. There is only one private operator (Tumbes). In urban areas, the WWS are merged into the same operator and are vertically integrated, i.e. operators manage all activities of the production chain, from abstraction to the distribution of water and in terms of the wastewater service, from collection to treatment of wastewater, as in the case of the state-owned company, SADAPAL. In rural areas the services are usually promoted by small Water Boards (*Juntas*) or directly by small municipalities.

Tariff Setting

The tariff system adopted by the operators consists of a fixed part and a proportional part priced per increasing block rate based on the volume of water consumed. There are social tariffs. Domestic, commercial/industrial and public customers are charged differently. Cross-subsidization is adopted by SUNASS.

Quality of Service

The WWS have shown significant progress over the past years, especially at the level of service coverage and continuity. This enhancement has been sustained by a system of benchmarking promoted by SUNASS, which is based on a set of performance indicators. There are, however, some operational problems (high water losses) in the water service which require significant investments. The wastewater service also needs investments to increase its coverage.

The Regulator

Characteristics
- Sector-specific regulator
- Independent regulatory agency

Legal basis
Decree No. 25,965 of 1992

Number of employees
139 employees

Budget
2.8 million euros (2006)

Contacts
Superintendencia Nacional de Servicios de Saneamiento
Av. Bernardo Monteagudo 210 – 216, Magdalena del Mar, Peru
Tel: +511 264 1440
Fax: +511 264 4020
http://www.sunass.gob.pe

Relevant Entities

Operators' Association
- http://www.anepssa.tripod.com

Ministry for Housing, Construction and Sanitation
- http://www.vivienda.gob.pe

Ministry of Health
- http://www.minsa.gob.pe

The Sector

Country
- Area – 1,285,000 km^2
- Population – 28.7 million
- Population density – 22 inhab/km^2

Service Coverage
- Water – 83.8%
- Wastewater service – 75.5%
- Wastewater treatment – 28.1%

Number of operators
- Water – 50
- Wastewater – 50

Ownership and management of services

	Ownership		Management	
	Pub.	Priv.	Pub.	Priv.
Water	100%	0%	99.5%	0.5%
Wastewater	100%	0%	99.5%	0.5%

Volume of activity
- Abstracted water volume – 1,264 million m^3
- Drinking water – 717 million m^3
- Average consumption – 96.8 litres per person per day
- Main source – 65.7 % superficial
- Water losses – 42.3 %

Economic and financial data
- Turnover – 240.8 million euros
- Water service average price – 0.41 €/m^3
- Average annual invoice per customer – 60 euros
- Employment – 8,000 employees

PHILIPPINES

Regulatory Model

In the Philippines, there is an agency (National Water Resources Board – NWRB) with regulatory powers over water resources. In 2002, the responsibilities of the Local Water Utilities Administration (LWUA) were added to the jurisdiction of the NWRB, thus ending the conflict of interest that had existed until then, due to the LWUA's previous simultaneous functions of both financial supporter and regulator. The NWRB sets tariffs based on the price-cap regulatory method for a period of five years. In addition to the NWRB, there is another operator (Metropolitan Waterworks and Sewerage System Regulatory Office – MWSSRO) that was created when the Manila WWS were privatized. It is a sector-specific regulator and has regulatory functions relating to pricing, quality of service and public service obligations. The MWSSRO adopted the sunshine regulation model as a way of encouraging compliance with the targets established in the concessionary contracts. The WWS are primarily regulated by the National Water Code of 1976 and the Clean Water Act of 2004.

Market Structure

The WWS have historically been provided by operators associated with the public sector. However, due to the crisis in the country that occurred in 1995, the government initiated the development of legal conditions for private sector participation in WWS. As a result, there was the privatization of Metro Manila which was split into two private operators, the Manila Water Company, Inc. (MWCI), in the eastern zone, and Maynilad Water Services, Inc. (MWSI), in the western zone, currently providing the water service. However, assets ownership remains in the public sector. In the Philippines, the WWS are usually merged into a single operator, with the various activities of the production chain vertically integrated. Horizontally, the WWS are poorly integrated, with the exception of the region of Manila. It is noteworthy that in Manila the service of wastewater treatment is provided only by MWCI, which therefore also receives the wastewater collected by MWSI.

Tariff Setting

The tariff system usually implemented by operators in the Philippines includes, for domestic customers, two fixed parts one for the water service and the other for wastewater service, and a proportional part for the water service priced per increasing block of volume of water consumed. The operator imposes different tariffs depending on the characteristics of the customers.

Quality of Service

The WWS need large investments to improve the limited level of wastewater treatment as well as the degradation of water infrastructures and facilities. The creation of MWSSRO, and especially the benchmarking system based on a series of performance indicators, constitutes an important tool for progress in the sector.

The Regulator

Characteristics
- Multisector regulator
- Independent regulatory agency

Legal basis
Concession Agreement of 1997

Number of employees
72 employees

Budget
1.5 million euros (2006)

Contacts
Metropolitan Waterworks and Sewerage System
Regulatory Office, 3rd Floor Engineering Bldg.
MWWS Complex, Katipunan Road,
Balara Quezon City 1105
Philippines
Tel: +632 435 8900
Fax: +632 922 2568
http://www.mWWSro.org.ph/

Relevant Entities

National Water Resources Board
- http://www.nwrb.gov.ph

Local Water Utilities Administration
- http://www.lwua.gov.ph

Department of Environment and Natural Resources
- http://www.denr.gov.ph/

Asian Development Bank
- http://www.adb.org

The Sector

Country
- Area – 300,000 km^2
- Population – 90.5 million
- Population density – 295 inhab/km^2

Service Coverage
- Water – 85.3%
- Wastewater service – 72%
- Wastewater treatment – 10% (Manila)

Number of operators
- Water – > 1,600
- Wastewater – > 1,600

Ownership and management of services

	Ownership		Management	
	Pub.	Priv.	Pub.	Priv.
Water	100%	0%	75%	25%
Wastewater	100%	0%	75%	25%

Volume of activity
- Abstracted water volume – 4,800 million m^3
- Drinking water – 2,400 million m^3
- Average consumption – 133 litres per person per day
- Main source – 80% superficial
- Water losses – 49.9% (Manila)

Economic and financial data
- Turnover – 143.4 million euros (Manila)
- Water service average price – 0.22 €/m^3 (Manila)
- Average annual invoice per customer –
- Employment – 4,000 employees (Manila)

PORTUGAL

Regulatory Model

In Portugal, there is a regulatory agency, the Institute for the Regulation of Water and Solid Waste (IRAR), which, according to its statutes, exercises important regulatory functions over regional systems and municipal concessions of WWS and urban waste. Its primary duty is the protection of customer rights and the safeguarding of sustainability and economic viability of the systems. As a national authority, IRAR also has the duty of monitoring and controlling the drinking water quality for all the municipalities in the country. IRAR adopted the sunshine regulation model as the main method of quality of service regulation, which is based on the published results of the performance of regulated operators. To this end, it publishes in an annual report the performance results of operators, obtained from a set of performance indicators. The preparation of this document, based on benchmarking, includes a joint assessment of performance in which comparisons are made between the operators and an individual quantitative and qualitative evaluation of each operator's performance. Law No. 379 of 1993, Law No. 147 of 1995 and Law No. 58 of 1998 provide the main legal framework for WWS regulation in Portugal.

Market Structure

The ownership and management of the WWS is mainly in the public sector. The private sector provides WWS to around 20% of the population. The WWS are mostly deverticalized, i.e. the wholesale and end-user systems of the WWS are provided by separate entities. In horizontal terms, the operators have a low degree of integration, and even without including the local councils, there are 300 operators for almost 11 million inhabitants. Regarding available services, the operators almost always provide WWS together, frequently including the services of urban waste and, less frequently, other activities. There is only a limited number of operators that individually provide the service of water or wastewater.

Tariff Setting

The tariff system adopted by operators usually consists of a fixed part and a proportional part based on the volume of water consumed and the characteristics of the property. There are special tariffs for large families (in a small number of cases), for social service institutions and low-income customers. Domestic and industrial customers are charged distinctly and there is cross-subsidization between them.

Quality of Service

The WWS as a rule provide a high quality of service. To encourage quality of service between the operators under concession contracts, IRAR performs an annual benchmarking exercise, which determines, compares and publishes the performance of regulated operators. Some operators require large investments to upgrade their infrastructures.

The Regulator

Characteristics
- Multisector regulator
- Regulatory agency without independence

Legal basis
Decree No. 230 of 1997

Number of employees
62 employees

Budget
5 million euros (2007)

Contacts
Instituto Regulador de Águas e Resíduos
Centro Empresarial Torres de Lisboa
Rua Tomás da Fonseca, Torre G–8°,
1600-209 Lisboa,
Portugal
Tel: + 351 210 052 200
Fax: + 351 210 052 259
http://www.irar.pt

Relevant Entities

Water Resources National Institute
- http://www.inag.pt

Operators' Association
- http://www.apda.pt

Portuguese Water Company
- http://www.adp.pt

Ministry for the Environment, Land Use Planning and Rural Development
- http://www.maotdr.gov.pt

The Sector

Country
- Area – 92,300 km^2
- Population – 10.3 million
- Population density – 111 inhab/km^2

Service Coverage
- Water – 91.1%
- Wastewater service – 75.1%
- Wastewater treatment – 66.3%

Number of operators
- Water – 523
- Wastewater – 314

Ownership and management of services

	Ownership		Management	
	Pub.	Priv.	Pub.	Priv.
Water	100%	0%	80%	20%
Wastewater	100%	0%	80%	20%

Volume of activity
- Abstracted water volume – 862 million m^3
- Drinking water – 560 million m^3
- Average consumption – 153 litres per person per day
- Main source – 65% superficial
- Water losses – 35%

Economic and financial data
- Turnover – 1,650 million euros
- Water service average price – 0.33 €/m^3
- Average annual invoice per customer – 130 euros
- Employment – 17,500 employees

REPUBLIC OF IRELAND

Regulatory Model

In the Republic of Ireland, as in most European countries, there is no sector regulator for the WWS. The Environmental Protection Agency (EPA) was established with the responsibilities of licensing, enforcing, monitoring and evaluating activities related to environmental protection. Although acting as the environmental regulator, the EPA also oversees some aspects related to the quality of service. There are also other bodies with regulatory functions that extend across the whole economy, such as those responsible for the defence and protection of customers (Citizens' Information Board) and for leveraging competition (Competition Authority). The most important regulations of WWS are found in the normative documents relating to Sewerage Services and Public Health (1878 and 1964), Water Pollution (1977 and 1990) and in regulations produced since 1992 by the EPA regarding urban wastewater treatment.

Market Structure

The current legislation specifies two types of operators for the water sector: public and private. However, the sector shows that, in reality, there are four distinct categories of operators in the water service sector in Ireland: Public Water Supplies (PWS), the 'Public' Group Water Schemes (PuGWS), the 'Private' Group Water Schemes (PrGWS) and the Small Private Supplies (SPS). According to legislation, the last three operators are classified as 'private'. The major operators in Ireland such as the Dublin City Council's Water Services Division and Cork County Council's Divisional Services have integrated water services, producing and distributing water to all customers, including domestic and non-domestic customers and also providing the wastewater service. In the Irish water sector publicly managed operators integrate the respective services of WWS, while privately managed operators, acting mainly in rural areas, normally provide only one of the services.

Tariff Setting

Due to legislation, operators only charge for services provided to non-domestic customers and only for the service of water. Consequently, the volume of water consumed by domestic customers is 100% funded by the government as are the wastewater services.

Quality of Service

The WWS are characterized by their high quality. However, there are not minimum quality of service requirements nor penalties for non-compliance. Quality of service is supervised, albeit weakly, by EPA, mainly regarding the water quality.

The Regulator

Characteristics

Legal basis

Number of employees

Budget

Contacts

Relevant Entities

Environmental Protection Agency
- http://www.epa.ie

National Federation of Group Water Schemes
- http://www.nfgws.ie

Citizens Information Board
- http://www.citizensinformationboard.ie

Environment, Heritage and Local Government Department
- http://www.environ.ie

The Sector

Country
- Area – 70,300 km^2
- Population – 4.3 million
- Population density – 60.3 inhab/km^2

Service Coverage
- Water – 99.9%
- Wastewater service – 93%
- Wastewater treatment – 92%

Number of operators
- Water – 34 + small private operators
- Wastewater – 34 + small operators

Ownership and management of services

	Ownership		Management	
	Pub.	Priv.	Pub.	Priv.
Water	100%	0%	90%	10%
Wastewater	100%	0%	65%	35%

Volume of activity
- Abstracted water volume – 1,200 million m^3
- Drinking water – 780 million m^3
- Average consumption – 130 litres per person per day
- Main source – 83% superficial
- Water losses – 35%

Economic and financial data
- Turnover –
- Water service average price – 0 €/m^3 (domestic customers)
- Average annual invoice per customer – 0 euros
- Employment –

SCOTLAND

Regulatory Model

In Scotland, the regulatory model currently adopted based on three independent regulators (tripartite regulation), the economic regulator, the environmental regulator and the regulator of water quality, seems to be considerably effective. These three regulators are complemented by the existence of other bodies with important regulatory functions, such as the Scottish Environment Protection Agency (SEPA), the Competition Authority and the Consumer Association. The economic regulator, the Water Industry Commission for Scotland (WICS), has the responsibility over WWS, safeguarding their appropriate integral activities, protecting their interests, ensuring their financial viability, promoting their economic efficiency and leveraging competition. SEPA has the duties of an environmental regulator, including the licensing of abstraction and discharges. Finally, the Drinking Water Quality Regulator has the role of regulating water quality. In the scope of economic regulation, WICS adopted the price-cap regulation for a period of four years. The Water Act of 1980 and the Water Services Act of 2005 are the main legal documents in the water sector.

Market Structure

With only one operator, WWS are managed solely by the public sector, in particular, by Scottish Water. In this context, the services are vertically integrated, i.e. the operator performs all services on the production chain, from abstraction to the distribution of water. The situation is similar for the wastewater service, specifically the integration of services related to wastewater collection and treatment. In the past, there were three operators that merged to form Scottish Water in 2004. Private sector participation is significant, mainly in management contracts, especially as far as treatment plants are concerned.

Tariff Setting

The tariff system adopted by the operators is divided between a fixed part and a proportional part based on the volume of water consumed or the characteristics of the property. Some low-income customers benefit from a 25% discount on their water bill. Cross-subsidization is used to guarantee universal access to the service.

Quality of Service

The WWS have generally provided an excellent quality of service, particularly regarding the high levels of universal service, continuity and reliability. The important role played by the benchmarking system contributes to this situation through the use of a set of performance indicators carried out by the regulator to control the quality of service. However, some operational problems still remain (e.g. water losses).

The Regulator

Characteristics
- Sector-specific regulator
- Independent regulatory agency

Legal basis
Water Services Law of 2005

Number of employees
18 employees

Budget
6.3 million euros (2007)

Contacts
Water Industry Commission for Scotland
Ochil House, Sprinkerse Business Park,
Stirling FK7 7XE,
Scotland
Tel: +44 1786 430200
Fax: +44 1786 462018
http://www.watercommissioner.co.uk

Relevant Entities

Environmental Department
- http://www.sepa.org.uk

Association for Consumer Protection
- http://www.waterwatchscotland.org

Operators' Association
- http://www.water.org.uk

Water Quality Regulator
- http://www.dwqr.org.uk

The Sector

Country
- Area – 78,800 km^2
- Population – 5.1 million
- Population density – 65 inhab/km^2

Service Coverage
- Water – 100%
- Wastewater service – 100%
- Wastewater treatment – 100%

Number of operators
- Water – 1
- Wastewater – 1

Ownership and management of services

	Ownership		Management	
	Pub.	Priv.	Pub.	Priv.
Water	100%	0%	100%	0%
Wastewater	100%	0%	100%	0%

Volume of activity
- Abstracted water volume – 838 million m^3
- Drinking water – 523 million m^3
- Average consumption – 160 litres per person per day
- Main source – Superficial
- Water losses – 37%

Economic and financial data
- Turnover – 1,231 million euros
- Water service average price – 1 €/m^3
- Average annual invoice per customer – 380 euros
- Employment – 3,700 employees

SLOVAK REPUBLIC

Regulatory Model

In the Slovak Republic, there is a multisector regulatory agency (Regulatory Office for Network Industries – URSO), which is responsible for the regulation of WWS, gas, electricity and heating, and which also has functions in the areas of economic regulation, public service obligations and quality of service (still in progress). Furthermore, the regulator aims, on the one hand, to ensure the rights of customers and, on the other hand, to safeguard the sustainability and economic viability of systems, while maintaining quality and reliable service. The economic regulation in this country primarily focuses on price control and is based on the method of price cap for a period of five years. As a complement to these two functions, URSO, with the goal of controlling the quality of service and providing incentives, is at present developing a system of benchmarking as a way to expand its performance. Currently, the regulator only publishes the prices adopted by regulated operators. The water sector is mainly regulated by Law No. 185 of 2006 and Law No. 684 of 2006.

Market Structure

The WWS are largely managed by the public sector, particularly by public companies under private law, and under the ownership of the local administration (e.g. The Bratislava Water Company). Through this framework, services are characterized by vertical integration of both water and wastewater services, whereby the operator performs all services related to the water services from abstraction to the water distribution and the collection and wastewater treatment with regard to the wastewater service.

Tariff Setting

The WWS prices are primarily based on a variable charge according to the volume of water consumed. There are no social tariffs. In 2004, a uniform pricing structure for different customers (domestic and non-domestic) was implemented.

Quality of Service

The WWS are generally known for providing a reasonable quality of service. In the current situation of the country, the WWS present an acceptable level of service coverage and reliability. The industry requires a system of benchmarking as a means of promoting quality of service and efficiency of utilities.

The Regulator

Characteristics
- Multisector regulator
- Regulatory agency without independence

Legal basis
Law No. 276 of 2001

Number of employees
90 employees (5 water sector)

Budget
2.6 million euros (2006)

Contacts
Regulatory Office for Network Industries
Bajkalská 27, PO Box 12, 820 07 Bratislava 27,
Slovak Republic
Tel: +2 581 004 11
Fax: +2 581 004 79
http://www.urso.gov.sk

Relevant Entities

Operators' Association
- http://www.avssr.sk

Customers' Association
- http://www.zss.sk/

Ministry for the Environment
- http://www.enviro.gov.sk

Competition Authority
- http://www.antimon.gov.sk

The Sector

Country
- Area – 49,000 km^2
- Population – 5.45 million
- Population density – 111 inhab/km^2

Service Coverage
- Water – 86.3%
- Wastewater service – 57.5%
- Wastewater treatment – 55%

Number of operators
- Water – 14 (95% of the population)
- Wastewater – 14 (95% of the population)

Ownership and management of services

	Ownership		Management	
	Pub.	Priv.	Pub.	Priv.
Water	100%	0%	100%	0%
Wastewater	100%	0%	100%	0%

Volume of activity
- Abstracted water volume – 324.8 million m^3
- Drinking water – 225 million m^3
- Average consumption – 121 litres per person per day
- Main source – 81% groundwater
- Water losses – 32%

Economic and financial data
- Turnover – 443.3 million euros
- Water service average price – 0.65 €/m^3
- Average annual invoice per customer –
- Employment – 8,736 employees

TANZANIA

Regulatory Model

In Tanzania, there is an autonomous multisector regulatory agency (Energy and Water Utilities Regulatory Authority – EWURA) that is responsible for the technical and economic regulation of the services of electricity, oil, natural gas and WWS. The EWURA adopted the sunshine regulation method, which is based on the published results of the performance of regulated operators. For this purpose, it compiles an annual report of performance results, obtained from a set of performance indicators of WWS operators. The preparation of this document, based on benchmarking, includes a joint assessment of performance that makes comparisons between the operators. The economic regulation is based on price-cap regulation for a period of two years. In addition to the above, the authority is responsible for resolving disputes between customers and operators. The main legal documents of the sector are the Water Act of 2006 and the National Water Policy of 2002.

Market Structure

The WWS are provided almost entirety by the public sector. Despite the existence of some private sector involvement in the provision of water services, it is very limited and only in isolated areas (urban and rural). The services are vertically integrated, either in water or in wastewater, and include the whole production chain, from abstraction to distribution in the water service and from collection to wastewater treatment in the wastewater service. However, there are some situations where this does not happen.

Tariff Setting

The tariff system adopted by WWS operators consists of a proportional tariff that can be either constant or priced per increasing block rate, both according to the volume of water consumed. Special tariffs are available for low-income customers or those with special needs.

Quality of Service

The WWS need large investments, especially in rural areas (more than 80% of the population). The services are characterized by a reduced level of service coverage, reliability and other operational problems (e.g. water losses). The quality of service is only moderate in urban centres.

The Regulator

Characteristics
- Multisector regulator
- Independent regulatory agency

Legal basis
Law of 2001, Chap. 414

Number of employees
53 employees (14 water sector)

Budget
2.3 million euros (2006)

Contacts
Energy and Water Utilities Regulatory Authority
PO Box 72175, 6th Floor, Harbour View Towers,
Samora Avenue, Dar es Salaam,
Tanzania
Tel: +255 22 2123850/3/4/6
Fax: +255 22 2123180
http://www.ewura.com

Relevant Entities

National Environment Council
- http://www.nemctan.org

Ministry of Water and Irrigation
- http://www.maji.go.tz

African Development Bank
- http://www.afdb.org

The World Bank
- http://web.worldbank.org

The Sector

Country
- Area – 945,100 km^2
- Population – 37.8 million
- Population density – 41 inhab/km^2

Service Coverage
- Water – 60%
- Wastewater service – 3% (only urban)
- Wastewater treatment – 3% (only urban)

Number of operators
- Water – 20
- Wastewater – 10

Ownership and management of services

	Ownership		Management	
	Pub.	Priv.	Pub.	Priv.
Water	100%	0%		
Wastewater	100%	0%		

Volume of activity
- Abstracted water volume – 185 million m^3
- Drinking water – 111 million m^3
- Average consumption – 14 litres per person per day
- Main source – groundwater
- Water losses – 40%

Economic and financial data
- Turnover – 34.3 million euros (urban)
- Water service average price – 0.28 €/m^3
- Average annual invoice per customer –
- Employment –

THE NETHERLANDS

Regulatory Model

In the Netherlands there is no explicit regulation of the WWS. The postponement of the creation of a regulatory agency is made up for by the very positive role played by VEWIN, the Association of Dutch Water Companies, in the self-regulation of the sector. The VEWIN behaves, *stricto sensu*, as a regulator, encouraging the performance improvement concerning the quality of service and the main financial and economic indicators. The VEWIN, playing an important and proactive role, and using benchmarking as a primary tool, has maintained the water companies at a top level of excellence on a worldwide scale. In the wastewater area, the Water Council, RIONED and VNG also have self-regulation duties, encouraging the use of benchmarking and performance improvement in wastewater services. Their results are, however, less evident. Additionally, the Consumer Association also has relevant regulatory functions, specifically the protection of customer interests in everything related to WWS, and particularly with regard to quality of service issues. The water sector is mainly regulated by the Water Supply Act of 1957 and the Water Services Act of 1992.

Market Structure

The WWS in the Netherlands are carried out in the public sector. In the area of wastewater treatment, the scenario is different because in 2002 the Delfland Water Board established a PPP to build a WWTP in The Hague. The water services in the Netherlands are vertically integrated. PLC provides all stages of the production chain, from abstraction to commercialization. The level of horizontal integration is very high, thus 10 companies exist for 16.5 million inhabitants. However, PLC only provides the service of water, with the exception of Waternet, which also provides the wastewater service and other activities related to water resources. The service of wastewater is characterized by deverticalization. In this sense, the end-user services (wastewater collection) are provided by the municipalities, and are separate from the wholesale services (wastewater treatment), which are provided by the Water Council. The level of horizontal integration is also smaller, especially with regard to the collection of wastewater.

Tariff Setting

The tariff system usually adopted by operators is a two-part system for the water service, based on a fixed part (differing depending on the area supplied) according to the meter or the property, and a proportional part charged by levels of consumption. Concerning the wastewater service the situation is alike, that is there is a two-part system as well.

Quality of Service

The quality of the WWS stands out for its excellence at all levels (service, reliability, etc.). With regard to water losses, it is one of the most efficient countries in the world. In the quality of service scope, VEWIN has a key role in developing a system of voluntary benchmarking for its members (PLC).

The Regulator

Characteristics

Legal basis

Number of employees

Budget

Contacts

Relevant Entities

Association of Water Companies
- http://www.vewin.nl

RIONED Foundation
- http://www.rioned.org

Association of Netherlands Municipalities
- http://www.vng.nl

Association for Consumer Protection
- http://www.consumentenbond.nl

The Sector

Country
- Area – 41,500 km^2
- Population – 16.5 million
- Population density – 395 inhab/km^2

Service Coverage
- Water – 100%
- Wastewater service – 98%
- Wastewater treatment – 97%

Number of operators
- Water – 10
- Wastewater – 25 (treatment); 443 (collection)

Ownership and management of services

	Ownership		Management	
	Pub.	Priv.	Pub.	Priv.
Water	100%	0%	100%	0%
Wastewater	100%	0%	100%	0%

Volume of activity
- Abstracted water volume – 1,209 million m^3 (2006)
- Drinking water – 1,099 million m^3
- Average consumption – 124 litres per person per day
- Main source – 60% groundwater
- Water losses – 10%

Economic and financial data
- Turnover – 1,500 million euros (water)
- Water service average price – 1.34 €/m^3
- Average annual invoice per customer – 177 euros
- Employment – 4,900 employees (water)

TRINIDAD AND TOBAGO

Regulatory Model

In Trinidad and Tobago, there is an independent multisector regulatory agency (Regulated Industries Commission – RIC), which is responsible for safeguarding the provision of high-quality services at fair prices to customers, and which also works to ensure the self-sustainability of the operators. The regulator is responsible for promoting quality of service and for resolving disputes between customers and operators. In order to develop the quality of service, RIC adopted the sunshine regulation model, which is based on publishing a set of performance indicators of the regulated operator. To this end, it composes an annual report of the operator's performance results. The economic regulation is based on the revenue-cap regulation for a period of five years. The WWS Law of 1980 and the Law of Environmental Management of 2000 provide the present regulatory framework for the water sector.

Market Structure

With only one operator, the WWS are managed solely by the public sector, namely by WASA. The water sector is characterized by a horizontal and vertical integration of the services into a single operator that provides all the water cycle services, from abstraction to the water distribution and from collection to the wastewater treatment. However, it should be noted that the wastewater service is still only available to a small percentage of the population. In areas where this service is available, it is the same operator that simultaneously provides the services of water and wastewater.

Tariff Setting

The tariff system adopted by the operator consists of a fixed part charged quarterly and an increasing block rate, based on the volume of water consumed. The government directly subsidizes low-income customers and, additionally, there are also social tariffs. The operator sets different tariffs for domestic, commercial/industrial and irrigation customers and for charity institutions.

Quality of Service

WASA provides, in general, an acceptable quality of service. The high level of service coverage (particularly in the water service) contrasts with the low level of reliability, where only 50% of the population has water service for 24 hours and with high levels of water losses. The regulator uses benchmarking as an important tool for promoting development in the sector.

The Regulator

Characteristics
- Multisector regulator
- Independent regulatory agency

Legal basis
Decree No. 26 of 1998

Number of employees
30 employees (water and electricity)

Budget
1.2 million euros (2006)

Contacts
Regulated Industries Commission
1st and 3rd Floors Furness House
Corner Wrightson Road and Independence Square
Port of Spain
Republic of Trinidad and Tobago
Tel: +868 625 5384
Fax: +868 624 2027
http://www.ric.org.tt

Relevant Entities

Water Resources Institute
- http://www.wasa.gov.tt/Witt_main.htm

Caribbean Water and Wastewater Association
- http://www.cwwa.net

Ministry for Public Services
- http://www.gov.tt/directory/govministries.aspx

Ministry for the Environment
- http://www.gov.tt/directory/govministries.aspx

The Sector

Country
- Area – 5,100 km^2
- Population – 1.3 million
- Population density – 208 inhab/km^2

Service Coverage
- Water – 92%
- Wastewater service – 21%
- Wastewater treatment – 21%

Number of operators
- Water – 1
- Wastewater – 1

Ownership and management of services

	Ownership		Management	
	Pub.	Priv.	Pub.	Priv.
Water	100%	0%	100%	0%
Wastewater	100%	0%	100%	0%

Volume of activity
- Abstracted water volume – 365 million m^3
- Drinking water – 160 million m^3
- Average consumption – 360 litres per person per day
- Main source – 100% superficial
- Water losses – 55%

Economic and financial data
- Turnover – 47.3 million euros
- Water service average price – 0.12 €/m^3
- Average annual invoice per customer – 48 euros
- Employment – 2,500 employees

URUGUAY

Regulatory Model

In Uruguay, there is a multisector regulatory agency (Regulatory Unit of Energy and Water Services – URSEA) that has responsibility over the regulation of WWS and energy (electricity, gas and fuel). Among its competencies, the main objectives are to protect the rights of customers, monitor compliance with the standards and ensure that the services are provided with quality and reliability at a fair price. The regulator is endowed with organic and functional independence. URSEA has the responsibility to foster competition in the industry and has the legal authority to regulate monopolies, establish the minimum quality levels and set prices based on principles of efficiency and sustainability. Acting as the national authority, URSEA also has the duties of supervising and controlling drinking water quality in the whole Uruguayan territory, and also has the ability to impose sanctions on operators that fail to comply with regulations. Concerning the economic regulation, the regulator has adopted the rate-of-return regulation. The regulation of the water sector is fundamentally governed by the Water Code of 1992 and the General Law of Environmental Protection of 2000.

Market Structure

The WWS are provided almost entirely by the public sector. In this country the private sector presence is still quite weak. There is a state-run company, the OSE, which provides WWS throughout the country with the exception of the wastewater service in Montevideo, where it is provided directly by the municipality. Another exception is the existence of a mixed capital company, the *Aguas de la Costa*, which provides WWS in the Department of Maldonado. The WWS are usually provided by the same operator and are vertically integrated, i.e. operators manage all activities of the production chain, from abstraction to water distribution and from collection to wastewater treatment in the scope of the wastewater services. Montevideo is the only exception. The degree of horizontal integration is also high, with a limited number of operators for the entire urban population of Uruguay.

Tariff Setting

The tariff system normally applied by the operators is a variable charge priced per increasing block rate based on the volume of water consumed. Operators also provide social tariffs. The sector includes a high percentage of metering. Domestic, commercial, industrial and public customers are charged differently.

Quality of Service

The water service is proudly provided (almost) universally to the population. The quality of service is considered high, and most areas are provided with a reliable service. However, the country needs to enhance the level of service coverage in the wastewater service as well as to upgrade the water infrastructures and facilities.

The Regulator

Characteristics
- Multisector regulator
- Independent regulatory agency

Legal basis
Decree No. 17,598 of 2002

Number of employees
30 employees (5 water sector)

Budget
360,000 euros (2006)

Contacts
Unidad Reguladora de Servicios de Energía y Agua
Yaguarón 1407 – Piso 8, 11100 Montevideo,
Uruguay
Tel: +59 82 908 2221
Fax: +59 82 900 8795
http://www.ursea.gub.uy

Relevant Entities

Caribbean Water and Wastewater Association
- http://www.cwwa.net

Ministry for Housing, Land Use Planning and the Environment
- http://www.mvotma.gub.uy

Ministry of Health
- http://www.msp.gub.uy

WWS Advisory Commission (COASAS)
- http://www.mvotma.gub.uy/dinasa/

The Sector

Country
- Area – 176,200 km^2
- Population – 3.5 million
- Population density – 19 inhab/km^2

Service Coverage
- Water – 92%
- Wastewater service – 48%
- Wastewater treatment –

Number of operators
- Water – 2
- Wastewater – 3

Ownership and management of services

	Ownership		Management	
	Pub.	Priv.	Pub.	Priv.
Water	100%	0%	97%	3%
Wastewater	100%	0%	97%	3%

Volume of activity
- Abstracted water volume – 320 million m^3
- Drinking water – 147 million m^3
- Average consumption – 125 litres per person per day
- Main source – 85% superficial
- Water losses – 54%

Economic and financial data
- Turnover – 140 million euros
- Water service average price – 0.85 €/m^3
- Average annual invoice per customer – 171 euros
- Employment – 4,362 employees (OSE)

USA (ARKANSAS)

Regulatory Model

In Arkansas, there is a regulatory agency (Arkansas Public Service Commission – APSC) that is proactive and focuses on the customer interests. Its regulatory duties include the electricity, telephone, gas and water services. Currently, it does not regulate any operator that provides wastewater services. The focus of the APSC is on ensuring safe and reliable public services at a reasonable price and guaranteeing the sustainability of the operators, as well as increasing public awareness of the various problems facing the water sector. The regulator has functional independence and the capacity to set standards. The APSC regulates inter-state tariffs and public services in Arkansas. Concerning economic regulation, the APSC has adopted the rate-of-return regulation for a period of one year. At this time, the monitoring of quality of service (through a system of benchmarking) is beyond the scope of its powers. The Safe Drinking Water Act of 1974 and the Clean Water Act of 1977 are the main regulatory documents governing the water sector.

Market Structure

The WWS are mostly provided by public operators within the different institutional arrangements of their respective services and are publicly managed. As a rule, private sector participation includes only management and not ownership. WWS are heterogeneously integrated across the state. Although the integration of the WWS demonstrates a slight superiority, compared to the large percentage of local operators, there also exist regional operators, such as United Water Arkansas, which provide all the services related to water production.

Tariff Setting

The tariff system adopted by operators mainly consists of a fixed part and a proportional part priced per decreasing block rate based on the volume of water consumed. The government subsidizes operators. The operators set different tariffs depending on whether the customer is domestic, commercial or industrial.

Quality of Service

The WWS are characterized as a whole, by an acceptable quality, including continuity, good pressure conditions and compliance with current standards. However, no detailed information exists on the quality of service and efficiency provided by the operators.

The Regulator

Characteristics
- Multisector regulator
- Independent regulatory agency

Legal basis
Law of 1899

Number of employees
99 employees (3 water sector)

Budget
6.4 million euros (2006)

Contacts
Arkansas Public Service Commission
PO Box 400, Little Rock, AR 72203-0400,
USA
Tel: +1 501 682 2051
Fax: +1 501 682 5731
http://www.arkansas.gov/psc

Relevant Entities

Water Association
- http://www.awwwea.org

Rural Water Association
- http://www.arkansasruralwater.org

National Association of Regulators
- http://www.naruc.org

American Water Works Association
- http://www.awwa.org

The Sector

State
- Area – 137,000 km^2
- Population – 2.7 million
- Population density – 19.8 inhab/km^2

Service Coverage
- Water – ≈100%
- Wastewater service – ≈100%
- Wastewater treatment – ≈100%

Number of operators (private)
- Water –
- Wastewater –

Ownership and management of services

	Ownership		Management	
	Pub.	Priv.	Pub.	Priv.
Water	100%	0%		
Wastewater	100%	0%		

Volume of activity
- Abstracted water volume –
- Drinking water –
- Average consumption –
- Main source –
- Water losses –

Economic and financial data
- Turnover –
- Water service average price –
- Average annual invoice per customer –
- Employment –

USA (CALIFORNIA)

Regulatory Model

In California, the independent multisector regulator (California Public Utilities Commission – CPUC) has authority at the level of economic and quality of service regulation. The CPUC regulates private operators of telecommunication services, electricity, natural gas, water, railways and passenger airline companies. Its primary mission is to ensure that customers have access to safe, reliable and affordable services, protecting them against fraud and promoting the development of Californian economy. The CPUC has functional and organic independence and is able to set standards and impose penalties and fines in cases of serious service violations, especially in cases of consistent non-compliance with water quality parameters. It intervenes on its own initiative or upon request of any stakeholder. The regulator reports to the congress whenever requested and is guided by transparency and participation of stakeholders in decisions and by information dissemination and financial responsibility. The information it publishes includes decisions, fines and ongoing procedures. The method of economic regulation used to set tariffs is the rate-of-return regulation with a hybrid mechanism of revenue caps. The regulatory period is three years. The Safe Drinking Water Act of 1974 and the Clean Water Act of 1977 are the main legal documents for the water sector.

Market Structure

Public operators, within a variety of institutional arrangements primarily provide WWS. In general, private sector participation includes only management and not ownership. The integration of WWS is divergent throughout the state. Although the integration of the WWS demonstrates a slight superiority, compared to the large percentage of local operators, there are also private regional operators that provide the whole WWS, such as the California Water Service Company.

Tariff Setting

The tariff system adopted by operators usually consists of a fixed part according to the meter, and a proportional rate based on the volume of water consumed. The operators provide special tariffs for low-income customers, and domestic, industrial and irrigation customers are charged differently.

Quality of Service

The WWS levels of quality of service are considerably high. As a way of encouraging quality of service among the WWS operators, CPUC conducts an annual benchmarking exercise that determines, compares and publishes the performance of regulated operators.

The Regulator

Characteristics
- Multisector regulator
- Independent regulatory agency

Legal basis
Public Services Law of 1911

Number of employees
1,000 employees (52 water sector)

Budget
58.3 million euros (2006)

Contacts
California Public Utilities Commission
505 Van Ness Avenue San Francisco, CA 94102, USA
Tel: +1 415 703 2782
Fax: + 1 415 703 1758
http://www.cpuc.ca.gov

Relevant Entities

California Water Association
- http://www.calwaterassn.com

Water Environmental Association
- http://www.cwea.org

California Rural Water Association
- http://www.calruralwater.org

National Association of Regulators
- http://www.naruc.org

The Sector

State
- Area – 424,000 km^2
- Population – 36.5 million
- Population density – 90.3 inhab/km^2

Service Coverage
- Water – ≈100%
- Wastewater service – ≈100%
- Wastewater treatment – ≈100%

Number of operators (private)
- Water – 138
- Wastewater – 12

Ownership and management of services

	Ownership		Management	
	Pub.	Priv.	Pub.	Priv.
Water	100%	0%		
Wastewater	100%	0%		

Volume of activity
- Abstracted water volume – 59,000 million m^3 (20% of the pop.)
- Drinking water –
- Average consumption –
- Main source –
- Water losses –

Economic and financial data
- Turnover – 780 million euros (private)
- Water service average price –
- Average annual invoice per customer –
- Employment –

USA (FLORIDA)

Regulatory Model

In the state of Florida, there is an independent multisector regulator (Florida Public Service Commission – FPSC), a regulatory agency responsible for economic regulation and for safeguarding the quality of service of electricity, gas, telecommunication services and WWS. The duties of the agency include the setting of WWS tariffs, the supervision of competition in the market and the monitoring of the safety and reliability of service. The FPSC regulates only the private operators. The FPSC is unable to issue legislation. Tariffs are set by the rate-of-return regulation. The regulatory period is one year. The Safe Drinking Water Act of 1974 and the Clean Water Act of 1977 are the main regulators of the water sector. Currently, the regulator does not apply any methodology of benchmarking as a means of assessing quality of service.

Market Structure

The WWS are mostly provided by public operators within the different institutional arrangements of service provision under public management. However, there is already a considerable proliferation and participation of the private sector, which, as a rule, includes only management but not ownership. The WWS integration is dissimilar across the state, with a slight superiority of the integrated services. Although local public providers are the more common institutional arrangement, in Florida there are also private regional operators, such as Aqua Utilities Florida, Inc., which provide the whole WWS and also private operators that provide services in a single area, such as the Crooked Lake Park Sewerage Company, which provides services related to wastewater.

Tariff Setting

The tariff system adopted by the operators consists of a fixed part and a proportional part priced per increasing block rate based on the volume of water consumed or the characteristics of the property. The operators set distinct tariffs according to whether the customer is domestic, commercial, industrial or public.

Quality of Service

The water consumption in Florida is one of the highest in the world which, combined with population growth and the ageing infrastructure, constitutes a major factor for its degradation. The WWS require substantial investments. However, overall the WWS present a high quality of service.

The Regulator

Characteristics

- Multisector regulator
- Independent regulatory agency

Legal basis

FPSC Law of 1887

Number of employees

Budget

3.6 million euros (water)

Contacts

Florida Public Service Commission
2540 Shumard Oak Blvd.,
Tallahassee, FL 32399-0850,
USA
Tel: +1 850 413 6100
Fax: +1 850 511 0809
http://www.psc.state.fl.us

Relevant Entities

Rural Water Association

- http://www.frwa.net

Ground Water Association

- http://67.199.22.20

National Association of Regulators

- http://www.naruc.org

American Water Works Association

- http://www.awwa.org

The Sector

State

- Area – 170,300 km^2
- Population – 16 million
- Population density – 117 inhab/km^2

Service Coverage

- Water – ≈100%
- Wastewater service – ≈100%
- Wastewater treatment – ≈100%

Number of operators (private)

- Water – 197
- Wastewater – 148

Ownership of services

	Ownership		Management	
	Pub.	Priv.	Pub.	Priv.
Water	100%	0%		
Wastewater	100%	0%		

Volume of activity

- Abstracted water volume –
- Drinking water –
- Average consumption – 770 litres per person per day
- Main source – 81% groundwater
- Water losses –

Economic and financial data

- Turnover – 77.6 million (private)
- Water service average price –
- Average annual invoice per customer –
- Employment –

USA (INDIANA)

Regulatory Model

In Indiana, there is an independent multisector regulatory agency (Indiana Utility Regulatory Commission – IURC) with the duties of economic regulation, quality of service regulation and supervision of public service obligations. Atypical in the USA, the regulator has regulatory authority over both public and private operators. Among its most important tasks are the setting of tariffs and service standards, the definition of areas of jurisdiction, the rules of accounting and security related aspects, the transfer, purchase and pooling of assets and the regulatory procedures. The main task of the commission is to ensure the interests of customers. WWS prices are regulated according to the rate-of-return regulation for a period of one year. The IURC has organic, functional and financial independence. The water sector is mainly regulated by the Safe Drinking Water Act of 1974 and the Clean Water Act of 1977. Up until now, the regulatory agency has not applied any method of benchmarking as a way of evaluating the quality of service or promoting the productive efficiency.

Market Structure

The ownership and management of WWS are mainly in the public sector. The WWS are organized differently throughout the state. In addition to the WWS that are provided directly by municipalities, there are also regional operators that merge all services related to the water cycle, from water abstraction to wastewater disposal. Moreover, there are private operators dedicated exclusively to the operation of the water service (e.g. Indianapolis Water).

Tariff Setting

The tariff system normally applied by the operators is a two-part system, consisting of a fixed part (meter) and a proportional part priced per decreasing block rate (volume consumed), in cases where customers have meters. Customers who do not have a meter are charged a fixed rate according to the type of property. The operator sets distinct tariffs depending on the characteristics of the customer.

Quality of Service

In general, the WWS are characterized by an acceptable quality, which includes aspects relating to continuity, good pressure conditions and compliance with current standards. However, no detailed information exists about the quality of service provided by operators and their efficiency.

The Regulator

Characteristics
- Multisector regulator
- Independent regulatory agency

Legal basis
IURC Law of 1880

Number of employees
75 employees (10 water sector)

Budget
5.2 million euros (2006)

Contacts
Indiana Utility Regulatory Commission
101 West Washington Street, Suite 1500E
Indianapolis, Indiana 46204,
USA
Tel: +1 317 232 2701
Fax: +1 317 232 6758
http://www.in.gov/iurc

Relevant Entities

Water Quality Association
- http://www.iwqa.org

Rural Water Association
- http://www.indianaruralwater.org

Indiana Office of Utility Consumer Counsellor
- http://www.in.gov/ouc

National Association of Regulators
- http://www.naruc.org

The Sector

State
- Area – 94,300 km^2
- Population – 6.3 million
- Population density – 65.5 inhab/km^2

Service Coverage
- Water – ≈100%
- Wastewater service – ≈100%
- Wastewater treatment – ≈100%

Number of operators (private)
- Water – 835
- Wastewater – 541

Ownership and management of services

	Ownership		Management	
	Pub.	Priv.	Pub.	Priv.
Water	100%	0%		
Wastewater	100%	0%		

Volume of activity
- Abstracted water volume –
- Drinking water –
- Average consumption –
- Main source –
- Water losses – 20%

Economic and financial data
- Turnover – 194.2 million euros (private)
- Water service average price –
- Average annual invoice per customer –
- Employment –

USA (MONTANA)

Regulatory Model

In Montana, there is an independent multisector regulatory agency (Montana Public Service Commission – MPSC), which has the duties of economic and quality of service regulation, specifically in the setting of tariffs for the water service. The MPSC regulates private operators in the sectors of telecommunications, electricity, natural gas, WWS and transportation, safeguarding the quality of WWS for the customers and guaranteeing sustainability for the operators. The MPSC has organic and functional independence and the capacity to set standards and impose penalties and fines in serious cases of service violations, particularly in cases of consistent non-compliance with water quality parameters. The agency acts on its own initiative or at the request of any stakeholder. The regulator reports to congress when requested and is guided by principles of transparency and by the participation of stakeholders in decision-making, by information dissemination, and by financial responsibility. It uses the rate-of-return method of economic regulation with a hybrid mechanism of revenue-cap regulation. The regulatory period is one year. The water sector is mainly regulated by the Safe Drinking Water Act of 1974 and the Clean Water Act of 1977. Currently, the regulatory agency does not apply any method of benchmarking as a form of evaluating the quality of service or promoting the productive efficiency.

Market Structure

The WWS in Montana are mostly provided by public operators within the different institutional arrangements of their services and are publicly managed. The private sector has grown rapidly. However, as a rule, private sector participation only includes management and not ownership. The WWS are heterogeneously integrated across the state. The private sector participation is noticed at the level of small local operators, and varies from the integration of WWS, as in the case of Firelight Meadows, Inc., to the integration of the production chain, as in the case of Columbia Heights Water Co.

Tariff Setting

The tariff system adopted by operators is split between the implementation of a fixed part uniformly charged to domestic customers and a proportional part according to the volume of water consumed. The tariffs implemented vary with the customers' characteristics.

Quality of Service

No detailed information exists about the quality of service provided by operators as well as their efficiency, although the WWS are characterized, in general, for an acceptable quality, which includes continuity, good pressure conditions and compliance with current standards.

The Regulator

Characteristics

- Multisector regulator
- Independent regulatory agency

Legal basis

PSC Law of 1913

Number of employees

35 employees

Budget

Contacts

Montana Public Service Commission
State of Montana, Service Commission, 1701 Prospect Ave, PO Box 202601, MT 59620-2601,
USA
Tel: +1 406 444 6199
Fax: +1 406 444 7618
http://www.psc.state.mt.us

Relevant Entities

Rural Water Association

- http://www.mrws.or

Department of Agriculture

- http://www.usda.gov

National Association of Regulators

- http://www.naruc.org

American Water Works Association

- http://www.awwa.org

The Sector

State

- Area – 381,200 km^2
- Population – 997,000
- Population density – 2.4 inhab/km^2

Service Coverage

- Water – ≈100%
- Wastewater service – ≈100%
- Wastewater treatment – ≈100%

Number of operators (private)

- Water – 14
- Wastewater – 2

Ownership of services

	Ownership		Management	
	Pub.	Priv.	Pub.	Priv.
Water	100%	0%		
Wastewater	100%	0%		

Volume of activity

- Abstracted water volume –
- Drinking water –
- Average consumption –
- Main source – 56% superficial
- Water losses –

Economic and financial data

- Turnover –
- Water service average price –
- Average annual invoice per customer –
- Employment –

USA (TENNESSEE)

Regulatory Model

In Tennessee, there is a multisector regulatory agency (Tennessee Regulatory Authority – TRA), which has the responsibility for public services regulation. In this domain, the regulator has the task of setting tariffs and rules that regulate the privately run services of telephone, natural gas, electricity and WWS. The regulator has functional and financial independence. Despite the independence declared in its statute, the government also has some powers at the level of economic regulation in the sector, as the TRA regulates only private operators. The TRA does not issue legislation, but may adopt rules that implement it. The rate-of-return method of economic regulation is used for the setting of tariffs. The water sector is mainly regulated by the Safe Drinking Water Act of 1974 and the Clean Water Act of 1977. The regulatory agency does not apply any method of benchmarking as a form of evaluating the quality of service and promoting the productive efficiency.

Market Structure

Overall, the WWS are provided by public operators within their different institutional arrangements, and are publicly managed. As a rule, private sector participation includes only management and not ownership. The WWS are heterogeneously integrated across the state. There are local public operators which are responsible for the water supply simultaneously with the wastewater service and they also have vertical integration. There are also private regional operators, such as Aqua Utilities Company, Inc., that entirely provide the WWS, and also private operators that provide services in a single area, such as Tennessee-American Water Company, which supplies the water service and Tennessee Wastewater Systems, which provides the wastewater service.

Tariff Setting

The tariff system differs among municipalities, ranging from a fixed part charged uniformly to all customers, and a two-part system, consisting of a fixed part and a proportional part (constant/increasing/decreasing) depending on the volume of water consumed. The operator sets different tariffs depending on the characteristics of the customer.

Quality of Service

The WWS are characterized as a whole by an acceptable quality, which includes continuity, good pressure conditions and compliance with current standards. However, the sector needs detailed information regarding the level of quality of service provided by the operators and their efficiency.

The Regulator

Characteristics
- Multisector regulator
- Independent regulatory agency

Legal basis
TRA Law of 1897

Number of employees
77 employees (5 water sector)

Budget
2.9 million euros (2006)

Contacts
Tennessee Regulatory Authority
460 James Robertson Parkway, Nashville,
TN 37243-0505,
USA
Tel: +1 615 741 4648
Fax: +1 615 741 5015
http://www.tennessee.gov/tra/index.htm

Relevant Entities

Water Resources Association
- http://tnawra.er.usgs.gov/

Department of Agriculture
- http://www.usda.gov

National Association of Regulators
- http://www.naruc.org

American Water Works Association
- http://www.awwa.org

The Sector

State
- Area – 109,200 km^2
- Population – 5.7 million
- Population density – 59.3 inhab/km^2

Service Coverage
- Water – ≈100%
- Wastewater service – ≈100%
- Wastewater treatment – ≈100%

Number of operators (regulated)
- Water – 8
- Wastewater – 10

Ownership of services

	Ownership		Management	
	Pub.	Priv.	Pub.	Priv.
Water	100%	0%		
Wastewater	100%	0%		

Volume of activity
- Abstracted water volume –
- Drinking water –
- Average consumption –
- Main source –
- Water losses –

Economic and financial data
- Turnover –
- Water service average price –
- Average annual invoice per customer – 282 euros (regulated)
- Employment –

USA (WEST VIRGINIA)

Regulatory Model

In West Virginia, the multisector independent regulatory agency (Western Virginia Public Service Commission – WVPSC) acts at the level of economic and quality of service regulation. Atypical for North American regulators, the WVPSC regulates some public and private service operators of telephone, electricity, natural gas, road transportation and WWS. Its primary mission is to ensure the quality of the WWS provided to the customers, helping them to deal with failures of compliance with commitment agreements. The WVPSC has organic and functional independence and is able to set standards and impose penalties and fines in cases of serious service violations, especially in cases of consistent non-compliance with water-quality parameters. As with WWS management, the regulatory process also varies depending on the WWS ownership, whereby private operators are regulated by the rate-of-return regulation and public operators according to the cash flows presented. The water sector is mainly regulated by the Safe Drinking Water Act of 1974 and the Clean Water Act of 1977. As a way of encouraging quality of service and efficiency, the WVPSC developed a system of benchmarking based on a set of performance indicators.

Market Structure

Public operators are the main providers of WWS. In general, private sector participation includes only management and not ownership. The WWS, in particular the local operators, are mainly public with horizontal and vertical integration. There are also private operators that provide the whole WWS, such as the Alpine Lake Public Utilities Company, as well as private operators that provide services in a single area, such as the West Logan Water Company, which supplies the water service, and Sewage Systems, Inc. (Barboursville), which provides the wastewater service.

Tariff Setting

The tariff system normally applied by the operators comprises a variable charge, priced per decreasing blocks according to the volume of water consumed. Domestic, commercial, industrial and irrigation customers are charged different tariffs.

Quality of Service

The WWS are characterized in general for an acceptable quality (e.g. continuity and pressure). As a way of encouraging the performance of the sector, the regulator sets an annual ranking of operators based on the cost of water sold.

The Regulator

Characteristics
- Multisector regulator
- Independent regulatory agency

Legal basis
Code of Western Virginia of 1913

Number of employees
284 employees (39 water sector)

Budget
16.9 million euros (2006)

Contacts
Public Service Commission
4201 Brooks Street, PO Box 812, Charleston, WV 25323, USA
Tel: +1 304 340 0300
Fax: +1 304 340 0325
http://www.psc.state.wv.us

Relevant Entities

Water Environment Association
- http://www.wv-wea.org

Rural Water Association
- http://www.wvrwa.org

National Association of Regulators
- http://www.naruc.org

American Water Works Association
- http://www.awwa.org

The Sector

State
- Area – 62,800 km^2
- Population – 1.8 million
- Population density – 29 inhab/km^2

Service Coverage
- Water – ≈100%
- Wastewater service – ≈100%
- Wastewater treatment – ≈100%

Number of operators
- Water – 345
- Wastewater – 298

Ownership of services

	Ownership		Management	
	Pub.	Priv.	Pub.	Priv.
Water	100%	0%	69%	31%
Wastewater	100%	0%	98.7%	

Volume of activity
- Abstracted water volume –
- Drinking water –
- Average consumption –
- Main source –
- Water losses –

Economic and financial data
- Turnover – 351.5 million euros
- Water service average price –
- Average annual invoice per customer –
- Employment –

USA (WISCONSIN)

Regulatory Model

In Wisconsin, there is a multisector independent regulatory agency (Wisconsin Public Service Commission – WPSC) with the tasks of economic and quality of service regulation and public service obligations. It is responsible for regulation of electricity, natural gas and telephone services and WWS. The regulator carries out its activities aiming at safeguarding competition and ensuring reasonable and appropriate tariffs for public services. Operators are compelled to report their activities to the regulator. Accordingly, the regulator publishes on its website all information relating to prices, as well as some benchmarks based on a set of performance indicators. The regulator has functional and organic independence and the ability to set standards. The WPSC gives special attention to those who have water losses exceeding 30%. Furthermore, the regulator makes periodic audits that mainly serve to check data concerning water abstraction and the accuracy of the annual report. At the level of economic regulation, the WPSC has adopted the rate-of-return regulation. The Safe Drinking Water Act of 1974 and the Clean Water Act of 1977 are the key legislation of the water sector.

Market Structure

The WWS are mainly managed by public operators within the different institutional arrangements of their services and are publicly managed. The private sector participation generally includes management but not ownership. The integration of the WWS is dissimilar throughout the state. The WWS in the State of Wisconsin are mostly public with horizontal and vertical integration. However, there are also private operators that provide the whole WWS such as Bayfield Water and Sewer, and private entities that provide single services, such as Medford Water Works, which provides the water service.

Tariff Setting

The tariff system adopted by operators changes between a fixed part based on the characteristics of the property, and a proportional part priced per decreasing blocks according to the volume of water consumed. The operator sets different tariffs depending on the characteristics of the customer.

Quality of Service

In general, the WWS are characterized by an acceptable quality (e.g. continuity). As a means of stimulating progress in the sector, the regulator publishes on its website benchmarks consisting of a set of performance indicators.

The Regulator

Characteristics

- Multisector regulator
- Independent regulatory agency

Legal basis

WPSC Law of 1907

Number of employees

152 employees (9 water sector)

Budget

19.9 million euros (2006)

Contacts

Public Service Commission
610 North Whitney Way, PO Box 7854, Madison,
WI 53705-2729,
USA
Tel: +1 608 266 5481
Fax: +1 608 266 3957
http://psc.wi.gov

Relevant Entities

Water Association

- http://www.wiawwa.org

Rural Water Association

- http://www.wrwa.or

National Association of Regulators

- http://www.naruc.org

American Water Works Association

- http://www.dnr.state.wi.us/org/water/dwg

The Sector

State

- Area – 169,800 km^2
- Population – 5.6 million
- Population density – 38 inhab/km^2

Service Coverage

- Water – ≈100%
- Wastewater service – ≈100%
- Wastewater treatment – ≈100%

Number of operators (regulated)

- Water – 586
- Wastewater – 15

Ownership of services

	Ownership		Management	
	Pub.	Priv.	Pub.	Priv.
Water	100%	0%		
Wastewater	100%	0%		

Volume of activity

- Abstracted water volume – 783.9 million m^3
- Drinking water – 703.2 million m^3
- Average consumption – 513 litres per person per day
- Main source –
- Water losses – 10.3%

Economic and financial data

- Turnover – 408 million euros (water)
- Water service average price – 0.58 €/m^3
- Average annual invoice per customer –
- Employment –

ZAMBIA

Regulatory Model

In Zambia, the water sector is supervised by an independent sector-specific regulatory agency (National Water Supply and Sanitation Council – NWASCO). The responsibilities of NWASCO include the licensing, approval of tariffs, supervision of activities and accounts, and the establishment of benchmarking for the regulated operators. At the level of economic regulation, the method adopted is the price-cap regulation for a period of three years. As far as customers are concerned, the duties of NWASCO consist of the comparison and publication of information from the operators (sunshine regulation) and the government, and the resolution of disputes between customers and operators. NWASCO annually reports its activities to the Ministry of Energy and Water Development (MEWD). To improve the protection of customer rights, in 2001 NWASCO created Water Watch Groups (WWG). The WWG are groups of volunteers who are responsible for safeguarding the rights of customers in relation to the water sector, and the dissemination of available information on the services provided to them. The National Water Policy of 1994 and the Law of Water and Wastewater provide the legal framework for the water sector.

Market Structure

The WWS are almost entirely provided by the public sector. Despite the existence of some private sector involvement in the provision of water services, it is quite limited and only in isolated areas (urban and rural). The WWS are characterized in general by their horizontal and vertical integration. In this context, the provision of services is generally provided by the same operator, in the case of commercial operators, local authorities or private operators. Local authorities provide these services directly to customers through their departments of water and wastewater.

Tariff Setting

The tariff system adopted by operators consists of a proportional tariff that can be either constant or priced per increasing block rate, both according to the volume of water consumed. There are special tariffs for low-income customers, and a lifeline consumption of six cubic metres of water is provided.

Quality of Service

The WWS are generally known for providing an acceptable quality of service. The level of service coverage and the level of reliability (17.6 hours per day) require enhancement. Mainly due to the degradation of water infrastructures the water losses are high, thus large investments are required in this scope. The regulator has also played a leading role in developing the sector through the use of benchmarking.

The Regulator

Characteristics
- Sector-specific regulator
- Independent regulatory agency

Legal basis
Parliament Law of 1997

Number of employees
16 employees

Budget
663,000 euros (2006)

Contacts
National Water Supply and Sanitation Council
Mulombwa Close, Off Bwinjimfumu Road Fairview Lusaka 10101, Zambia
Tel: +260 1 226 941/2
Fax: +260 1 226 904
http://www.nwasco.org.zm

Relevant Entities

Operators' Association
- http://www.zambia-water.org.zm

Ministry of Energy and Water Development
- http://www.energy.gov.zm

Ministry of Local Government and Housing
- http://www.mlgh.gov.zm

Environmental Protection Council
- http://www.necz.org.zm

The Sector

Country
- Area – 752,600 km^2
- Population – 11.7 million
- Population density – 16 inhab/km^2

Service Coverage
- Water – 68%
- Wastewater service – 34%
- Wastewater treatment – 34%

Number of operators
- Water – 29
- Wastewater – 29

Ownership and management of services

	Ownership		Management	
	Pub.	Priv.	Pub.	Priv.
Water	100%	0%	99%	1%
Wastewater	100%	0%	99%	1%

Volume of activity
- Abstracted water volume – 316 million m^3
- Drinking water – 167.5 million m^3
- Average consumption – 110 litres per person per day
- Main source – 88% superficial
- Water losses – 47%

Economic and financial data
- Turnover – 50.7 million euros
- Water service average price – 0.25 €/m^3
- Average annual invoice per customer –
- Employment –

ANNEX III – REGULATORS' CONTACTS

Region	Country	Type of entity	Entity	Address			Telephone	Fax	Email	Website
Africa	South Africa	Other entities	South Africa Water Research Commission (WRC)	Private Bag X03	GEZINA 0031	South Africa	+27 12 330 340	+27 12-331-2565	info@wrc.org.za	http://www.wrc.org.za/index.htm
Africa	Angola	Other entities	Ministério da Energia e Águas	Avenida 4 de Fevereiro	105, 4°, Luanda	Angola	+244 33 78 36 / 33 74 48	+244 39 36 87	marmuf@gamek.com	
Africa	Cameroon	Other entities	Ministry of Mines, Water and Power	PO Box 955	Yaoundé	Cameroon	+237 23 3404	+237 23 3159	spm@spm.gov.cm, minmee@camnet.cm	http://www.spm.gov.cm
Africa	Côte d'Ivoire	Other entities	Direction de l'Hydralique	17 BP 673	Abidjan	Ivory Coast	225 20 212988	225 20 216898	sakhoma@globeaccess.net	
Africa	Egypt	Sector/ multisector Regulator	Ministry of Housing, Utilities and Urban Development	1 Ismail Abaza From El Kasr El Anei st.Cairo		Egypt	+20 794 4867	+20 336 7179	info@housing-utility.gov.eg	http://www.housing-utility.gov.eg
Africa	Gambia	Sector/ multisector Regulator	Gambia Public Utilities Regulatory Authority	1 Paradise Beach Place, Bertil Harding Highway	Kololi KSMD	The Gambia	+220 446 5180	+220 4465187	pura@pura.gm	http://www.pura.gm/
Africa	Ghana	Sector/ multisector Regulator	Public Utilities Regulatory Commission (PURC)	51 Liberation Road, African Liberation Circle	PO Box CT 3095, Accra	Ghana	+233 212 44181	+233 212 44188	info@purcghana.com, patacus@yahoo.com	http://www.purc.com.gh
Africa	Guinea-Bissau	Other entities	Ministério Recursos Naturais	Secretario de Estado do Meio Ambiente	C.P 85	Bissau	+245 20 10 38	+245 20 10 12		
Africa	Cape Verde Islands	Sector/ multisector Regulator	Agência de Regulação Económica (ARE)	Avenida Cidade de Lisboa CP 785	Praia, Santiago	Cape Verde	+238 260 0424	+238 261 1440	are.ritoevora@cvtelecom.cv, are.jrlima@cvtelecom.cv	http://www.are.cv/portal/index.php
Africa	Cape Verde Islands	Other entities	Ministério do Ambiente e Agricultura	Ponta Belem, 115	Praia, Santiago	Cape Verde	+238 261 5716			
Africa	São Tomé and Principe Islands	Other entities	Ministério dos Recursos Naturais e Meio Ambiente		C.P. 130	São Tome	+239 12 16 36	+239 12 25 00		
Africa	International	Other entities	African Development Bank (ADB)	Rue Joseph Anoma	01 BP 1387 Abidjan 01	Ivory Coast	+225 2020 4444	+225 2020 4959	afdb@afdb.org	http://www.afdb.org
Africa	International	Other entities	Water Utility Partnership (WUP)	Abidjan	05 BP 2642	Côte d'Ivoire	+225 21 240 828	+225 21 75 86 56	wup@africaonline.co.ci	http://www.wupafrica.org
Africa	Morocco	Other entities	Office National de l'Eau Potable (ONEP)	6 Bis, Rue Patrice Lumumba	Rabat	Morocco	+212 7 721 281	+212 7 73 13 55	onepigi@mtds.com	www.mincom.gov.ma
Africa	Mauritania	Sector/ multisector Regulator	Authority Regulation	Ilot Z lot N° 14	BP 4908 Nouakchott	Mauritanie	+222 529 12 70	+222 529 12 79	smiddi@are.mr	www.are.mr
Africa	Mozambique	Sector/ multisector Regulator	Water Regulatory Council (CRA)	Sede do Ministério, Av. Amilcar Cabral 757	PO Box 235, Maputo	Mozambique	+258 1 312 825	+258 1 31 28 26	cra-presidente@teledata.mz	http://www.cra.org.mz/
Africa	Mozambique	Other entities	Ministério da Coordenação da Acção Ambiental	Rua Kassuende, 167		Mozambique	+258 21 492 403			http://www.micoa.gov.mz/

Region	Country	Type of entity	Entity	Address			Telephone	Fax	Email	Website
Africa	Namibia	Other entities	Ministry of Agriculture, Water and Rural Development Department of Water Affairs	Private Bag 13193	Windhoek 9000	Namibia	+264 61 208 7111	+264 61 208 7228	heynsp@mawrd.gov.na	
Africa	Niger	Sector/ multisector Regulator	L'Autorite de regulation multisectorielle (ARM)	Siège Social, 64, Rue des Bâtisseurs	EP 13 179, Niamey	Niger	+227 20 739 008	+227 20 73 85 91	ousseinibachir@yahoo.fr	http://www.arm-niger.org/
Africa	Niger	Other entities	Ministère de l'Hydraulique, de l'Environnement et de la Lutte Contre la Désertification du Niger (MHE/LCD)	Direction de l'Environnement (DE)	Niamey	Niger	+227 72 38 89 / 73 47 22	+227 73 35 88	direnv@intnet.ne, famgarba@intnet.ne	
Africa	Nigeria	Operator	Lagos State Water Corporation	Ijora Causeway	Lagos	Nigeria	+234 01 764 6674		olawale@coker.metrong.com, salou.abdou@ties.itu.int	http://www.case.ibimet.cnr.it/den/niger.php?page=niger03_03
Africa	Kenya	Sector/ multisector Regulator	Water Services Regulatory Board	9th Floor, NHIF Building	PO Box 41621 00100-GPO, Nairobi	Kenya	+254 (0) 20 273 3559/61	+254 (0) 20 273 3558	info@wsrb.or.ke	http://www.wsrb.ewater.co.ke/
Africa	Rwanda	Sector/ multisector Regulator	Rwanda Utility Regulatory Agency (RURA)		PO Box 7289 Kigali	Rwanda	+250 58 45 62	+250 58 45 63	mukasine@rura.gov.rw, arms@rwanda1.com, jbbwanakeye@yahoo.fr	http://www.rura.gov.rw/en/index.htm
Africa	Senegal	Operator	Societe Nationale des Eaux du Senegal (SONES)	97, Avenue Andre Peytavin, BP 400	Dakar	Senegal	+221 839 7800	+221 832 2038	sones@primature.sn, aniang@sones.sn	
Africa	Tanzania	Sector/ multisector Regulator	Energy and Water Regulatory Authority (EWURA)	Samora Avenue, 6th Floor, Harbor View Towers (Former JM Mall Building)	PO Box 72175, Dar es Salaam	Tanzania	+255 (0) 22 2130759	+255 (0) 754 780 557	info@ewura.com, ssayore@cats-net.com	http://www.ewura.com/
Africa	Togo	Other entities	Centre Régional pour l'Eau Potable et l'Assainissement à faible coût (CREPA)	121 bretelle Bèklikamè	BP 3689 Lomé	Togo	+226 50 36 62 10/11	+226 50 36 62 08	crepa-togo@reseaucrepa.org, crepatg@togo-imet.com	http://www.reseaucrepa.org/
Africa	Uganda	Other entities	Directorate of Water Development	PO Box 20026	Kampala	Uganda	+256 41 220 374		director@dwd.co.ug	www.dwd.co.ug
Africa	Zambia	Sector/multisector Regulator	National Water Supply and Sanitation Council (NWASCO)	Mulombwa Close, Off Bwinjimfumu Road	Fairview Lusaka, 10101	Zambia	+260 1 226941/2	+260 1 226904	nwasco@zamnet.zm	http://www.nwasco.org.zm/
Asia	Bangladesh	Operator	Dhaka Water Supply and Sewerage Authority	98, Kazi Nazrul Islam Avenue, Kawran Bazar	Dahaka – 1215	Bangladesh	8116792 201 (E-11), 202 (E-12)		secretary@dwasa.org.bd	http://www.dwasa.org.bd/
Asia	Bhutan	Sector Association	Bhutan Water Partnership	c/o Department of Energy	Post Box 106, Thimpu	Kingdom of Bhutan	+975 2 322 505	+975 2 331 522	ulhendup@rspnbhutan.org	
Asia	Korea	Other entities	Ministry of Environment	Government Complex-Gwacheon 1, Joongang-dong, Gwacheon-si	Gyeonggi-do, 427–729	Korea	+82 2 2110 6561		minister@me.go.kr, chanchan@me.go.kr, omarmahir_1980@yahoo.com	http://eng.me.go.kr

Region	Country	Type of entity	Entity	Address			Telephone	Fax	Email	Website
Asia	United Arab Emirates	Sector/ multisector Regulator	Abu Dhabi regulation and supervison Bureau	PO Box 32800	Abu Dhabi	United Arab Emirates	+971 2 642 6777	+971 2 642 4217	water@rsb.gov.ae, bureau@rsb.gov.ae, nali@rsb.gov.ae, saadia@rsb.gov.ae	http://www.rsb.gov.ae/English/Index.aspx
Asia	Philippines	Sector/ multisector Regulator	Public Assessment of Water Services (PAWS)	3/F UP National Engineering Center, UP Diliman	Quezon City	Philippines	+63 920 5301	+63 929 5601		http://www.paws-info.org/index.php
Asia	Philippines	Operator	Metropolitan Waterworks and Sewerage System	4th Floor, Adminsitration Building, MWSS Complex	489 Katipunan Road, Balara, Quezon City 1105-	Philippines	+632 922 3757	+632 921 2887	info@mwssro.org.ph	http://www.mwss.gov.ph/
Asia	India	Sector Association	Indian Water Works Association (IWWA)	Pipeline Road, Vakola, Santacruz (East)	Mumbai – 400 076	India	+91 22 2614 0926	+91 22 2616 8613	info@iwwa.info, iwwa@rediffmail.com	http://www.iwwa.info/
Asia	India	Other entities	The Energy and Resources Institute (TERI)	Darbari Seth Block, IHC Complex, Lodhi Road	New Delhi – 110 003	India	+91 11 2468 2100	+91 11 2468 2144	mailbox@teri.res.in, pmc@teri.res.in	http://www.teriin.org/
Asia	Indonesia	Sector/ multisector Regulator	Jakarta Water Supply Regulatory Body	JL PEJOMPONGAN RAYA No. 57	JAKARTA PUSAT	Indonesia	+62 570 9732	+62 570 9723	a_lanti@cbn.net.id	http://jakartawater.org/eng/index.php
Asia	International	Other entities	Asian Development Bank (ADB)	PO Box 789	0980 Manila	Philippines	+632 632 4444	+632 636 2444	information@adb.org	http://www.adb.org/
Asia	International	Other entities	East Asia and Pacific Infrastructure Regulatory Forum (EAPIRF)	The Mira, G 1, First Floor, Plot 1, Ishwar Nagar	New Delhi 110065	India	+40 2330 0750	+40 2332 6117	harry.koller@au.pwc.com; secretariat@eapirf.org	http://www.eapirf.org/Index.asp
Asia	Japan	Sector Association	Japan Water Works Association	4-8-9 Kudan Minami	Chiyoda-ku, Tokyo 102-0074	Japan	+81 3 3264 2281	+81 3 3262 2244	shomu@jwwa.or.jp	http://www.jwwa.or.jp/home_index.html
Asia	Malaysia	Sector Association	Malaysian Water Association (MWA)	No 24, 2nd Floor, Jalan Sri Hartamas 8	Taman Sri Hartamas, 50480 Kuala Lumpur	Malaysia	+60 03 6201 2250	+60 03 6201 5801	mwmgtsb@tm.net.my, tmwa@streamyx.com	http://www.malaysianwater.org.my/
Asia	Maldives	Sector/ multisector Regulator	Maldives Water and Sanitation Authority (MWSA)			Maldives	+960 31 75 63	+960 31 75 69	mwsa@health.gov.mv, shaheedha@mwsa.gov.mv	http://www.mwsa.gov.mv/
Asia	Singapore	Operator	Public Utilities Board	40 SCOTTS ROAD #22-01	228231	Singapore	+65 6235 8888	+65 6731 3020	pub_tender@pub.gov.sg	http://www.pub.gov.sg/home/index.aspx
Asia	Sri Lanka	Sector/ multisector Regulator	Public Utilities Commission of Sri Lanka	6th Floor, BOC Merchant Tower	St. Michael's Road, Colombo 3	Sri Lanka	+94 112 392 608	+94 112 392 641	info@pucsl.gov.lk	http://www.pucsl.gov.lk/
Europe	Germany	Sector Association	Bundesverband der deutschen Gas und Wassewirtschaft – BGW	Reinhardstrasse 14	10117 Berlin	Germany	+49 30 280 410	+49 30 280 41 520	info@bgw.de	www.bgw.de
Europe	Germany	Sector Association	Deutsche Vereinigung des Gas- und Wasserfaches – DVGW	Postfach 14 03 62	D-53058 Bonn	Germany	+49 228 91 885	+49 228 91 88 990	info@dvgw.de	www.dvgw.de

Region	Country	Type of entity	Entity	Address			Telephone	Fax	Email	Website
Europe	Germany	Sector Association	Deutsche Vereinigung für Wasserwirtschaft, Abwasser und Abfall (ATV-DVWK)	Niedersedlitzer Platz 13	01259 Dresden	Germany	(03 51) 2 03 20 25	(03 51) 2 03 20 26	schnaufer@dwa-st.de	http://www.atv-dvwk-st.de/index.htm
Europe	Germany	Competition Authority	Bundeskartellamt	Kaiser-Friedrich-Str. 16	D – 53113 Bonn	Germany	0228/-9499-0	0228/-9499-400	info@bundeskartellamt.bund.de	http://www.bundes kartellamt.de/wEnglisch/index.php
Europe	Germany	Consumer Protection	Federation of German Consumer Organizations	Markgrafenstrasse 66	10969 Berlin	Germany	+49 30/25800-0	+49 30/25800-218	schaefer@vzbv.de	www.vzbv.de
Europe	Armenia	Sector/ multisector Regulator	Public Services Regulatory Commission of the Republic of Armenia (PSRC)	22 Saryan St	375002 Yerevan	Armenia	+374 1 522 522	+374 1 525 563	psrc@psrc.am	http://www.psrc.am/en/
Europe	Austria	Sector Association	Österreichische Vereinigung für das Gas – und Wasserfach – ÖVGW	Schubertring 14	A-1010 Wien	Austria	+43 1 5131 5880	+43 1 5131 58825	office@ovgw.at	http://www.ovgw.at
Europe	Austria	Sector Association	Österreichischer Wasser-und Abfallwirtschaftsverband – ÖWAV	Marc-Aurel-Strasse 5	A-1010 Wien	Austria	+43 1 535 57 20	+43 1 535 40 64	buero@oewav.at	http://www.oewav.at
Europe	Austria	Competition Authority	Bundeswettbewerbsbe-hörde (BWB)	Praterstraße 31	A-1020 Wien	Austria	+43 1 245 080		wettbewerb@bwb.gv.at	http://www.bwb.gv.at/BWB/English/default.htm
Europe	Austria	Consumer Protection	Verein für Konsumenten Information (VKI)	Mariahilfer Str.81	A-1061 Wien	Austria	+43 1 588 770	+43 1 588 7771	konsument@konsument.at	http://www.konsument.at
Europe	Belgium	Sector Association	Fédération Belge du Secteur de l'Eau – BELGAQUA	127 rue Colonel Bourg	B-1140 Brussels	Belgium	+32 2 706 40 90	+32 2 706 40 99	info@belgaqua.be	http://www.belgaqua.be
Europe	Belgium	Consumer Protection	Association des consommateurs Test-Achats (SCRL)	Rue de Hollande 13	1060 Bruxelles	Belgium			membres@test-achats.be	http://www.test-achats.be/
Europe	Belgium		Vlaamse Milieumaatschappij (VMM)	Alfons Van de Maelestraat 96	9320 Erembodegem	Belgium	+32 053 72 62 10	+32 053 77 7168	info@vmm.be	http://www.vmm.be/
Europe	Belgium	Sector/ multisector Regulator	Comité de contrôle de l'eau (CCE) Valónia	Rue du Vertbois, 13c	4000 LIEGE	Belgium	+32 04 232 9869	+32 04 232 9881	info.cceau@cesrw.be, Info@cesrw.be, mireille.begon@cesrw.be, nathalie.hounje@cesrw.be	http://www.cesrw.be
Europe	Bulgaria	Sector/ multisector Regulator	State Energy and Water Regulatory Commission (SEWRC)	Sofia 1000	Dondukov 8-10	Bulgaria	+359 2 988 87 30	+359 2 988 87 82	dker@dker.bg	http://www.dker.bg/index_en.htm
Europe	Bulgaria	Sector Association	Bulgarian Water Association (BWA)	Block 1, room 111, 1 'Christo Smirnenski' Av.	1046 Sofia	Bulgaria	+359 2 963 2669	+359 2 963 2669	bawk@mail.prosoft.bg	http://www.bawk-bg.com
Europe	Bulgaria	Consumer Protection	Bulgarian National Consumer Association	10, '11th August' Str.	1000 Sofia	Bulgaria	+359 2 9890106	+359 2 9890107	bnap@bnap.org	http://www.bnap.org/en/

Region	Country	Type of entity	Entity	Address			Telephone	Fax	Email	Website
Europe	Kazakhstan	Sector/ multisector Regulator	Agency of the Republic of Kazakhstan on Regulation of Natural Monopolies (AREM)	14, Bukeykhan str.	Astana, 010000	Kazakhstan	+7 3172 591637	+7 3172 215473	Kense@arem.kz, regulator@regulator.kz, info@arem.kz	http://www.regulator.kz/
Europe	Cyprus	Sector Association	Water Board of Nicosia	84 Athalassa Ave. Strovolos POB 21943	CY 1515 Nicosia	Cyprus	+357 2 42 22 91	+357 2 31 60 33	waterbrd@cytanet.com.cy	
Europe	Cyprus	Consumer Protection	Cyprus Consumers' Association	5 Acropolis Ave., (Flat 21) 2000 Strovolos	PO Box 24874, 1304 – Nicosia	Cyprus	+357 22 516112	+357 22 516118	cyconsas@spidernet.net, cca@spidernet.net	http://www.cyprus consumers.org.cy/
Europe	Croatia	Sector Association	Croatian Water and Wastewater Association (GVIK)	Folnegoviceva 1	10 000 Zagreb	Croatia	+385 1 618 71 52	+385 1 618 71 52	gvik@gvik.hr	http://www.gvik.hr
Europe	Denmark	Sector Association	Danish Water and Wastewater Association (DANVA)	Vandhuset, Danmarksvej 26	DK-8660 Skanderborg	Denmark	+45 7021 0055	+45 7021 0056	jba@danva.dk	http://www.danva.dk
Europe	Denmark	Competition Authority	Danish Competition Authority (KS)	Nyropsgade 30	1780 København V	Denmark	+45 7226 8000	+45 3332 6144	ks@ks.dk	http://www.ks.dk/english/
Europe	Denmark	Consumer Protection	National Consumer Agency of Denmark	Amagerfaelledvej 56	2300 Copenhagen S	Denmark	+45 3266 9000		forbrug@forbrug.dk	http://www.forbrug.dk/ english/
Europe	Scotland	Sector/ multisector Regulator	Water Industry Commission For Scotland (WICS)	Ochil House, Springkerse Business Park	Stirling FK7 7XE	Scotland	+44 01786 430200	+44 01786 462018	enquiries@watercommission. co.uk, src10-14@watercommission. co.uk	http://www.water commissioner.co.uk/ default.asp
Europe	Scotland	Consumer Protection	The Scottish Consumer Council	Royal Exchange House, 100 Queen Street	Glasgow, G1 3DN	Scotland	+44 0141 226 5261	+44 0141 221 0731	mevans@scotconsumer.org. uk	http://www.scotconsumer. org.uk/
Europe	Slovakia	Sector/ multisector Regulator	Regulatory Office for Network Industries (URSO)	Bajkalská 27, PO Box 12	820 07 Bratislava 27	Slovak Republic	+421 2 581 004 36	+421 2 581 004 79	urso@urso.gov.sk, hudcovicova@urso.gov.sk, abraham@urso.gov	http://www.urso.gov.sk/ index_en.html
Europe	Slovakia	Sector Association	Asociacia vodarenskych spolocnosti (AVS)	Jasikova 2	821 03 Bratislava	Slovak Republic	+421 2 48 291 376	+421 2 48 291 237	avs@kmba.sk	
Europe	Slovakia	Consumer Protection	Association of Slovak Consumers (ZSS)	Palisády 22	811 06 Bratislava	Slovak Republic	+421 2 54411 148	+421 2 54411 148	zss@zss.sk	http://www.isnet.sk/zss
Europe	Slovenia	Consumer Protection	Slovene Consumers' Association (SCA)	Frankopanska 5	1000 Ljubljana	Slovenia	+386 1 474 06 00	+386 1 433 33 71	zps@zps-zveza.si	http://www.zps-zveza.si
Europe	Spain	Sector Association	Asociacion Espanola de Abastecimientos de Agua y Saneamiento (AEAS)	Sor Angela de la Cruz, 2- 13a planta	ES-28020 Madrid	Spain	+34 914 490 010	+34 915 713 523	aeas@aeas.es	http://www.aeas.es
Europe	Spain	Consumer Protection	Confederación de Consumidores y Usuarios (CECU)	C/Mayor n° 45, 2°	28013, Madrid	Spain	+34 91 364 13 84	+34 91 366 90 00	cecu@cecu.es	http://www.cecu.es/
Europe	Estonia	Sector Association	Estonian Water Works Association (EWWA)	Adala 10	10614 Tallinn Eesti	Estonia	+372 62 62 458	+372 62 21 063	evel@tvesi.ee	http://www.sea.ee
Europe	Estonia	Consumer Protection	Eesti Tarbijakaitse Liit (ETL)	Kiriku 4	15071 Tallinn	Estonia	+372 6 201 700	+372 6 201 701	info@consumer.ee	http://www.tka.riik.ee

Region	Country	Type of entity	Entity	Address			Telephone	Fax	Email	Website
Europe	Finland	Sector Association	Finnish Water and Wastewater Works Association (VVY)	Asemapäällikönkatu 7	00520 Helsinki	Finland	+358 9 868 90 10	+358 9 8688 0190	vvy@vvy.fi	http://www.vvy.fi
Europe	Finland	Other entities	Finnish Environment Institute (SYKE)	Mechelininkatu 34a, PO Box 140	FI-00251 Helsinki	Finland	+358 20 490 123	+358 20 490 2190	kirjaamo.syke@ymparisto.fi	http://www.environment.fi/syke
Europe	Finland	Competition Authority	Finnish Competition Authority (FCA)	Pitkänsillanranta 3 A	00530 Helsinki, P.O.B. 332, FI-00531 Helsinki	Finland	+358 9 73141	+358 9 7314 3328	kirjaamo@kilpailuvirasto.fi	http://www.kilpailuvirasto.fi/cgi-bin/english.cgi
Europe	Finland	Consumer Protection	Kuluttajat-Konsumenterna	Kasöörinkatu 3	00520 HELSINKI	Finland	+358 9 877 50120	+358 9 877 50120	kkry@kuluttajat-konsumenterna.fi	http://www.kuluttajat-konsumenterna.fi/index.html
Europe	Finland	Sector Association	Association of Finnish Local and Regional Authorities (AFLRA)	Kuntatalo Toinen linja 14	00530 HELSINKI	Finland	+358 9 771 2528	+358 9 771 2291	erja.horttanainen@kuntaliitto.fi	http://www.kunnat.net/k_etusivu.asp?path=1
Europe	Finland	Consumer Protection	The Finnish Consumer Agency and Consumer Ombudsman	Haapaniemenkatu 4 A, Box 5	00531 Helsinki	Finland	+358 9 7726 7571	+358 9 7726 7557	tilaukset@kuluttajavirasto.fi, posti@kuluttajavirasto.fi	http://www.kuluttajavirasto.fi
Europe	France	Sector Association	Fédération Professionnelle des Entreprises de l'Eau (FP2E)	83, avenue Foch	F-75116 Paris	France	+33 1 5370 1358	+33 1 5370 1341	fp2e@fp2e.org	http://www.fp2e.org/
Europe	France	Competition Authority	Conseil de la concurrence	11, rue de l'Echelle	75 001 Paris	France	+33 1 55 04 00 00	+33 1 55 04 02 35	communication@conseil-concurrence.fr	http://www.conseil-concurrence.fr
Europe	France	Consumer Protection	Association defense de comsomateurs (CLCV)	CLCV – 17, rue Monsieur	75007 Paris	France	+33 1 56 54 32 10	+33 1 43 20 72 02	communication@clcv.org	http://www.clcv.org/
Europe	France	Other entities	Cour des comptes	13 rue Cambon	75001 Paris	France	+33 1 42 98 95 00	+33 1 42 60 01 59		http://www.ccomptes.fr/
Europe	France	Other entities	Organization for Economic Co-operation and Development (OECD)	2, rue André Pascal	F-75775 Paris Cedex 16	France	+33 1 45 24 82 00	+33 1 45 24 85 00	webmaster@oecd.org	http://www.oecd.org/
Europe	France	Sector Association	Fédération nationale des collectivités concédantes et régies (FNCCR)	20, Boulevard de Latour-Maubourg	75007 Paris	France	+33 1 40 62 16 40	+33 1 40 62 16 41	fnccr@fnccr.asso.fr	http://www.fnccr.asso.fr/
Europe	Greece	Sector Association	Hellenic Union of Municipal Enterprises for Water Supply and Sewage (EDEYA)	37–43 Papakyriazi Str.	41222 Larissa	Greece	+30 4125 8261	+30 4153 2347	info@edeya.gr	http://www.edeya.gr
Europe	Greece	Consumer Protection	Consumers' Protection Center (KEPKA)	54, Tsimiski Str.	Thessaloniki, GR 54 623	Greece	+30 2310 233 333	+30 2310 242 211	consumers@kepka.org	http://www.kepka.org
Europe	Netherlands	Sector Association	Unie van Waterschappen	Postbus 93218	2509 AE Den Haag	The Netherlands	+31 70 351 97 51	+31 70 354 46 42	info@uvw.nl	http://www.uvw.nl
Europe	Netherlands	Sector Association	The Association of Dutch Water Companies (VEWIN)	Postbus 1019	2280 CA Rijswijk	The Netherlands	+31 70 414 47 50	+31 70 414 44 20	vewin@vewin.nl, geudens@vewin.nl	http://www.vewin.nl
Europe	Netherlands	Competition Authority	Netherlands Competition Authority (NMA)	PO Box 16326	2500 BH The Hague	The Netherlands	+31 70 330 33 30	+31 70 330 33 70	info@nmanet.nl	http://www.nmanet.nl/engels/home/Index.asp

Region	Country	Type of entity	Entity	Address			Telephone	Fax	Email	Website
Europe	Netherlands	Consumer Protection	Consumentenbond (CB)	Postbus 1000	2500 BA Den Haag	The Netherlands	+31 70 445 45 45	+31 70 445 45 96	kpeters@consumentenbond.nl, klantenservice@consumentenbond.nl	http://www.consumentenbond.nl/
Europe	Netherlands	Foundation	RIONED	Stichting RIONED, Postbus 133	6710 BC Ede	The Netherlands	+31 318 631 111	+31 318 633 337	info@rioned.org	http://www.riool.net
Europe	Netherlands	Sector Association	Vereniging van Nederlandse Gemeenten	Koningin Marialaan 15–17	2595 GA The Hague	The Netherlands	+31 70 373 8393	+31 70 363 5682	informatiecentrum@vng.nl, vng@vng.nl	http://www.vng.nl/
Europe	Hungary	Sector Association	Hungarian Water Utility Association (MAVIZ)	Sas u. 25. floor 4	1051 Budapest	Hungary	+36 1 312 3066	+36 1 302 7600	info@maviz.org	http://www.maviz.org
Europe	Hungary	Consumer Protection	National Association for Consumer Protection in Hungary	1012 Budapeste	Logodi u. 22–24	Hungary	+36 1 311 7030	+36 1 331 7386	ofeiroda@t-online.hu	http://www.ofe.hu/
Europe	E&W	Sector/ multisector Regulator	Water Services Regulation Authority (OFWAT)	Centre City Tower, 7 Hill Street	Birmingham B5 4UA	United Kingdom	+44 121 625 1300 /1373	+44 121 625 1400	philip.fletcher@ofwat.gsi.gov.uk, regina.finn@ofwat.gsi.gov.uk	http://www.ofwat.gov.uk/
Europe	England	Other entities	Drinking Water I nspectorate (DWI)	Room M03, 55 Whitehall	London SW1A 2EY	United Kingdom	+44 20 7082 8024	+44 20 7082 8028	dwi.enquiries@defra.gsi.gov.uk	http://www.dwi.gov.uk/
Europe	England	Competition Authority	Competition Commission (UK)	Victoria House, Southampton Row	London WC1B 4AD	United Kingdom	+44 20 7271 0243	+44 20 7271 0367	info@cc.gsi.gov.uk	http://www.competition-commission.org.uk/
Europe	England	Sector Association	Water UK	1 Queen Anne's Gate	London SW1H 9BT	United Kingdom	+44 20 7344 1844	+44 20 7344 1866	contact@water.org.uk	http://www.water.org.uk
Europe	England	Consumer Protection	National Consumer Council	20 Grosvenor Gardens	London SW1W 0DH	United Kingdom	+44 20 7730 3469	+44 20 7730 0191	info@ncc.org.uk	http://www.ncc.org.uk
Europe	England	Consumer Protection	Consumer Council for Water	1st floor, Victoria Square House, Victoria Square	Birmingham B2 4AJ	United Kingdom	+44 121 345 1000	+44 121 345 1001	enquiries@ccwater.org.uk	http://www.ccwater.org.uk/
Europe	International	Other entities	European Water Association (EWA)	Theodor-Heuss-Allee 17	D-53773 Hennef	Germany	+49 2242/872-189	+49 2242/872-135	EWA@dwa.de	http://www.ewaonline.de
Europe	International	Other entities	International Water Association (IWA)	Alliance House, 12 Caxton Street	London SW1H 0QS	United Kingdom	+44 207 654 5500	+44 207 654 5555	water@iwahq.org.uk	http://www.iwahq.org
Europe	International	Other entities	European Union of National Associations of Water Suppliers and Wastewater Services (EUREAU)	Eureau secretariat	1140, Rue Colonel Bourg 127	Brussels	+32 2 706 4080	+32 2 706 4081	secretariat@eureau.org	http://www.eureau.org/index.htm
Europe	International	Other entities	European Centre of Enterprises (CEEP)	Rue de la charité, 15 bte 12	B-1210 Bruxelles	Belgium	+32 2 219 27 98			http://www.ceep.org/
Europe	Ireland	Consumer Protection	Consumers' Association of Ireland	43–44 Chelmsford	Road Dublin 6	Ireland	+353 1 497 8600	+353 1 497 8601	cai@consumerassociation.ie	http://www.consumerassociation.ie/
Europe	Ireland	Other entities	Department of the Environment and Local Government (DELG)	Custom House, Dublin 1	ENFO, 17 St. Andrew's St., Dublin 2	Ireland	+353 1 888 2000		minister@environ.ie, qcsofficer@environ.ie	http://www.environ.ie/en/
Europe	Northern Ireland	Sector/ multisector Regulator	Northern Ireland Authority for Utility Regulation (NIAUR)	Queens House, 14 Queen Street	Belfast BT1 6ER	Northern Ireland	+44 28 9031 1575	+44 28 9031 1740	peter.naylor@niaur.gov.uk; david.flinn@niaur.gov.uk	http://ofreg.nics.gov.uk/

Region	Country	Type of entity	Entity	Address			Telephone	Fax	Email	Website
Europe	Northern Ireland	Consumer Protection	The Consumer Council for Northern Ireland	116 Holywood Road	Belfast BT4 1NY	Northern Ireland	+44 28 9067 2488	+44 28 9065 7701	info@consumercouncil.org.uk	http://www.consumer council.org.uk/
Europe	Iceland	Sector Association	Federation of Icelandic Energy and Waterworks (SAMORKA)	Sudurlandsbraut 48	108 Reykjavik	Iceland	+354 588 44 30	+354 588 44 31	samorka@samorka.is	http://www.samorka.is
Europe	Iceland	Consumer Protection	Consumers' Association of Iceland (NEYTENDASAMTOKIN)	PO Box 8160	IS-128 Reykjavik	Iceland	+354 545 1200	+354 545 1212	ns@ns.is	http://www.ns.is/ns/ english/
Europe	Italy	Sector/ multisector Regulator	Comitato per la Vigilanza sull'uso delle Risorse Idriche (COVIRI)	Via Cristoforo Colombo, 44	00147 Roma	Italy	+39 6 572 25244	+39 6 5722 5290	coviri.segreteria@minam biente.it	http://www.minambiente.it/ index.php?id_sezione- =266
Europe	Italy	Sector Association	FEDERUTILITY	Via Cavour 179/a	00184 Roma	Italy	+39 6 4786 5620	+39 6 4786 5605	acqua@federgasacqua.it	http://www.federutility.it
Europe	Italy	Consumer Protection	Associazione Difesa Consumatori e Ambiente (ADICONSUM)	via G. M. Lancisi	25 – 00161 Roma	Italy	+39 6 441 7021	+39 6 441 7023	adiconsum@adiconsum.it	http://www.adiconsum.it/
Europe	Italy	Sector Association	Associazione Nazionale Autorità e Enti di Ambito (ANEA)	Via Verdi, 16	50122 – Firenze	Italia	+39 55 263 2930	+39 55 263 2940	info@associazioneanea.it	http://www.associazionea nea.it/index.php
Europe	Kosovo	Sector/ multisector Regulator	Water and Waste Regulatory Office	Bregu i Diellit/Sunny Hill, Rr: Ferat Dragaj nr: 68	1000 Prishtinë	Kosovo	+38 249 165 101	+38 249 165	syle.syla@wwro-ks.org	http://www.wwro-ks.org/
Europe	Latvia	Sector/ multisector Regulator	Public Utilities Commission (PUC)	Brivibas street 55	Riga LV-1010	Latvia	+371 7097200	+371 7097277	sprk@sprk.gov.lv	http://www.sprk.gov.lv/
Europe	Latvia	Consumer Protection	Latvia Consumer Association (PIAA)	55 Brivibas Street, 2nd entrance	Riga LV-1010	Latvia	+371 7 296 884		piaa@delfi.lv	http://www.consumer-guide.lv/english/index.html
Europe	Lithuania	Sector/ multisector Regulator	National Control Commission for prices and energy (NCC)	Algirdo g. 31	LT-03219, Vilnius, Lietuva	Lithuania	+370 213 5166	+370 213 5270	info@regula.lt	http://www.regula.lt/lit/ English
Europe	Lithuania	Sector Association	Lithuanian Water Suppliers Association	Laisves pr. 117A	LT-2022 VILNIUS	Lithuania	+370 5 230 13 70	+370 5 230 13 80	vanduo@lvta.lt	http://www.lvta.lt
Europe	Luxembourg	Sector Association	Association Luxembourgeoise des Services d'Eau – ALUSEAU	c/o Administration des Services Techniques de la Ville de Luxembourg	2449, Boulevard Royal 51	Luxembourg	+352 325 860 20	+352 325 863	aluseau@aluseau.lu	http://www.aluseau.lu
Europe	Luxembourg	Consumer Protection	Union Luxembourgeoise des Consommateurs	55, rue des Bruyères	L-1274 Howald	Luxembourg	+352 49 60 22-1	+352 49 49 57	ulc@pt.lu	http://www.ulc.lu/
Europe	Malta	Sector Association	Water Services Corporation	Qormi Road	LQA 05 Luqa	Malta	+356 21 249 851 7	+356 21 223 016	water@wsc.com.mt	http://www.wsc.com.mt
Europe	Malta	Consumer Protection	Ghaqda tal-Konsumaturi – CA Malta	Consumers' Association – Malta, PO Box 464,	Valletta, CMR 01	Malta	+356 21 239 091	+356 21 221 210	info@camalta.org	http://www.camalta.org/
Europe	Norway	Sector Association	Norwegian Water and Wastewater BA – NORVAR	Vangsvegen 143	2317 Hamar	Norway	+47 6255 3030	+47 6255 3031	post@norvar.no, ivar.helleberg@norvar.no	http://www.norvar.no

Region	Country	Type of entity	Entity	Address			Telephone	Fax	Email	Website
Europe	Norway	Consumer Protection	Norwegian Consumer Council Forbrukerrådet	Pb 4594 Nydalen	0404 Oslo	Norway	815 58 200 (0900-1500)			http://forbrukerportalen.no/Emner/engelsk_fransk
Europe	Norway	Other entities	Norwegian Water Resources and Energy Directorate (NVE)	PO Box 5091 Majorstua	N-0301 Oslo	Norway	+47 22 95 95 95	+47 22 95 90 00	nve@nve.no	http://webb2.nve.no/
Europe	Norway	Consumer Protection	Norwegian Food Safety Authority	Ullevålsveien 76	0454 Oslo	Norway	+47 23 21 68 00	+47 23 21 68 01	postmottak@mattilsynet.no	http://www.mattilsynet.no
Europe	Norway	Other entities	Norwegian Pollution Control Authority (SFT)	Postboks 8100 Dep	0032 Oslo	Norway	+47 22 57 34 00	+47 22 67 67 06	postmottak@sft.no	http://www.sft.no
Europe	Wales	Consumer Protection	Welsh Consumer Council	Welsh Consumer Council 5th Floor	Longcross Court, Newport Road, Cardiff CF24 0WL	Wales	+44 29 2025 5454	+44 29 2025 5464	info@wales-consumer.org.uk	http://www.wales-consumer.org.uk
Europe	Poland	Sector Association	Izba Gospodarcza 'Wodociagi Polskie'	ul. Jana Kasprowicza 2	85-073 Bydgoszcz	Poland	+48 52 376 89 10	+48 52 376 89 20	sekretariat@igwp.org.pl	http://www.igwp.org.pl
Europe	Poland	Consumer Protection	Association of Polish Consumers (APC)	6 Gizow Street	01-249 Warsaw	Poland	+48 22 634 06 68/9	+48 22 634 06 67	consumer@skp.pl	http://www.skp.pl/eng/
Europe	Portugal	Sector/ multisector Regulator	Instituto Regulador de Águas e Resíduos (IRAR)	Centro Empresarial Torres de Lisboa	Rua Tomás da Fonseca, Torre G – 8°, 1600-209 LISBOA	Portugal	+351 210 052 200	+351 210 052 259	irar.geral@irar.pt	http://www.irar.pt
Europe	Portugal	Sector Association	Associacao Portuguesa de Distribuicao e Drenagem de Aguas – APDA	Av. de Berlim 15	1800-031 Lisboa	Portugal	+351 21 855 1359	+351 21 855 1360	geral@apda.pt	http://www.apda.pt
Europe	Portugal	Consumer Protection	Associação Portuguesa para a Defesa do Consumidor (DECO)	Av. Eng. Arantes e Oliveira, n. 13 – 1°B Olaias	1900-221 Lisboa	Portugal	+351 21 841 0800	+35121 841 0802	info@deco.proteste.pt	http://www.deco.proteste.pt
Europe	Portugal	Consumer Protection	Associação Portuguesa do Centro Europeu das Empresas com Participação e/ou de Interesse Económico Geral (APOCEEP)	Rua Laura Alves N°7	1° Lisboa 1050-138	Portugal	+351 21 780 7983	+351 21 780 7780	ceep@cgd.pt	http://www.apoceep.pt/index.php
Europe	Czech Republic	Sector Association	Czech Water Supply and Sewer Association (SOVAK)	Novotneho lavka 5	116 68 Praha 1	Czech Republic	+420 221 082 207	+420 221 082 646	sovak@csvts.cz	http://www.sovak.cz
Europe	Czech Republic	Consumer Protection	Consumers Defence Association of the Czech Republic (SOS)	SDRUZENI OBRANY SPOTREBITELU CESKE REPUBLIKY	Novakovych 8, 18000 Praha 8	Czech Republic	+420 2 9621 2004	+420 2 9621 2004	sos@spotrebitele.cz	http://www.spotrebitele.cz
Europe	Romania	Sector/ multisector Regulator	National Regulatory Authority for Municipal Services (ANRSC)	Str. Romulus nr.6	Sector 2, Bucuresti	Romania	+40 21 326 17 81	+40 21 326 17 82	cutulea@anrsc.ro, dciliescu@anrsc.ro, biroubresa@anrsc.ro	http://www.anrsc.ro/
Europe	Romania	Sector Association	Romanian Water Association (ARA)	202A Splaiul Independentei, 9th floor, 6th district	060022 Bucharest	Romania	+40 21 316 27 88	+40 21 316 27 88 (id)	secretariat@ara.ro	http://www.ara.ro

Region	Country	Type of entity	Entity	Address			Telephone	Fax	Email	Website
Europe	Russia	Consumer Protection	Inter Republican Confederation of Consumer Societies (Konfop)	Varvarka 14	103 690	Moscow	7095 298 4649	7095 298 4718	info@spros.ru	http://www.spros.ru
Europe	Sweden	Sector Association	Swedish Water and Wastewater Association (Svenskt Vatten AB)	Box 47607	117 94 Stockholm	Sweden	+46 8 506 002 00	+46 8 506 002 10	svensktvatten@svenskt vatten.se	http://www.svenskt vatten.se
Europe	Sweden	Sector Association	Swedish Association of Local Authorities and Regions, SALAR	Sveriges Kommuner och Landsting	118 82 Stockholm	Sweden	+46 8 452 7000	+46 8 452 7050	info@skl.se	http://www.skl.se
Europe	Sweden	Competition Authority	Swedish Competition Authority (KKV)	Konkurrensverket	SE-103 85 Stockholm	Sweden	+46 8 700 16 00	+46 8 24 55 43	konkurrensverket@kkv.se	http://www.kkv.se
Europe	Sweden	Consumer Protection	The Swedish Consumers' Association	Box 6086	102 32 Stockholm	Sweden	+46 8 674 4300	+46 8 674 4329	info@sverigeskonsumenter.se	http://www.sverigeskon sumenter.se/
Europe	Sweden		The National Food Administration	Box 622	SE – 751 26 Uppsala	Sweden	+46 1 817 5500	+46 1 810 5848	livsmedelsverket@slv.se, karg@slv.se	http://www.slv.se
Europe	Switzerland	Sector Association	Société Suisse de l'Industrie du Gaz et des Eaux – SSIGE	Grütlistrasser 44 – Case Postale 658	CH-8027 Zürich	Switzerland	+41 1 228 33 33	+41 1 202 16 33	info@svgw.ch	http://www.svgw.ch
Europe	Switzerland	Consumer Protection	Fédération Romande des Consommateurs (FRC)	Rue de Genève 7	CP 6151, 1002 LAUSANNE	Switzerland	+41 21 331 00 90	+41 21 331 00 91	info@frc.ch	http://www.frc.ch/
Europe	Ukraine	Sector/ multisector Regulator	State Committee for Water Management of Ukraine (SCWRM)	V. Vasylkovska Str. 8	Kiev 01601	Ukraine	+380 44 235 4102	+380 50 357 3435	scwm@scwm.gov.ua	http://scwm.kmu.gov.ua/ water/control/uk/index
North America	Canada	Other entities	Canadian Water and Wastewater Association (CWWA)	Unit 11, 1010 Polytek Street	Ottawa, ON K1J 9H9	Canada	+1 506 648 4612	+1 506 658 4740	admin@cwwa.ca	http://www.cwwa.ca/
North America	Canada – Alberta	Sector/ multisector Regulator	Alberta Energy and Utilities Board	640 – 5th Avenue SW	T2P 3G4, Calgary, Alberta	Canada	+1 403 297 8311	+1 403 297 7336	EUB.Inquiries@eub.ca, Utilities.Concerns@eub.ca	http://www.eub.ca
North America	Canada – Manitoba	Sector/ multisector Regulator	The Public Utilities Board	Room 400, 330 Portage Avenue	Winnipeg, MB R3C 0C4	Canada	+1 204 945 2638	+1 204 945 2643	publicutilities@gov.mb.ca	http://www.pub.gov.mb.ca
North America	Canada – New Scotia	Sector/ multisector Regulator	Nova Scotia Utility and Review Board	Box 1692, Unit 'M'	Halifax, Nova Scotia B3J 3S3	Canada	+1 902 424 4448	+1 902 424 4448	uarb.board@gov.ns.ca	http://www.nsuarb.ca/ index.html
North America	Canada – Prince Edward Island	Sector/ multisector Regulator	Prince Edward Island Regulatory and Appeals Commission	5th floor, Suite 501	National Bank Tower, 134 Kent Street, Charlottetown	Canada	+1 902 892 3501	+1 902 566 4076	info@irac.pe.ca	http://www.irac.pe.ca/
North America	International	Other entities	World Bank (WB)	1818 H Street, NW	Washington, DC 20433 USA	USA	+1 202 473 1000	+1 202 477 6391	pic@worldbank.org	http://www.worldbank.org/
North America	USA	Other entities	American Water Works Association (AWWA)	6666 W. Quincy Ave.	Denver CO 80235	USA	+1 303 794 7711	+1 303 347 0804	gkail@awwa.org; khedding@awwa.org	http://www.awwa.org/

Region	Country	Type of entity	Entity	Address			Telephone	Fax	Email	Website
North America	USA	Other entities	Water Environment Federation (WEF)	601 Wythe Street	Alexandria, VA 22314-1994	USA	+1 800 666 0206	+1 703 684 2492	juboerner@wef.org	http://www.wef.org/Home
North America	USA	Other entities	National Association of Regulatory Utility Commissioners (NARUC)	1101 Vermont Avenue, N.W., Suite 200	Washington, DC 20005	USA	+1 202 898 2200	+1 202 898 2213	admin@naruc.org	http://www.naruc.org/
North America	USA	Other entities	Inter-American Development Bank (IADB)	1300 New York Avenue, NW	Washington, DC 20577	USA	+1 202 623 1000	+1 202 623-3096	pic@iadb.org	http://www.iadb.org/
North America	USA – Alabama	Sector/ multisector Regulator	Alabama Public Service Commission	PO Box 304260, 100 N. Union St., RSA Union, Suite 850	Montegomery, AL 36130	USA	+1 334 242 5207	+1 334 242 0921	Jim.Sullivan@psc.alabama.gov	http://www.psc.state.al.us/
North America	USA – Alaska	Sector/ multisector Regulator	Alaska Regulatory Commission	701 W. 8th Avenue, Suite 300	Anchorage, AK 99501-3469	USA	+1 907 276 6222	+1 907 276 0160	webmaster.rca@alaska.gov, rca.mail@alaska.gov, cp.mail@alaska.gov	http://www.state.ak.us/rca/
North America	USA – Arizona	Sector/ multisector Regulator	Arizona Corporation Commission	Commissioners Wing, 1200 West Washington	Phoenix, AZ 85007-2996	USA	+1 602 542 4140	+1 602 542 4111	admindiv@azcc.gov	http://www.cc.state.az.us
North America	USA – Arkansas	Sector/ multisector Regulator	Arkansas Public Service Commission	1000 Center Street, PO Box 400, 72203-0400	Little Rock, AR 72201	USA	+1 501 682 2051		gas.water@psc.state.ar.us	http://www.accessar kansas.org/psc/
North America	USA – California	Sector/ multisector Regulator	California Public Utilities Commission	Headquarters Office, 505 Van Ness Avenue	San Francisco, CA 94102	USA	+1 415 703 2782	+1 415 703 1758	public.advisor@cpuc.ca.gov	http://www.cpuc.ca.gov
North America	USA – Colorado	Sector/ multisector Regulator	Colorado Public Utilities Commission			USA	+1 303 894 2000	+1 303 894 2065	puc@dora.state.co.us	http://www.dora.state.co.us/PUC/
North America	USA – Connecticut	Sector/ multisector Regulator	Connecticut Department of Public Utility Control	Ten Franklin Square	New Britain, CT 06051	USA	+1 860 827 1553	+1 860 827 2613	louise.rickard@po.state.ct.us	http://www.state.ct.us/dpuc/
North America	USA – Delaware	Sector/ multisector Regulator	Delaware Public Service Commission	Cannon Building, Suite 100, 861 Silver Lake Boulevard	Dover, DE 19904	USA	+1 302 736 7500	+1 302 739 4849	karen.nickerson@state.de.us	http://depsc.delaware.gov/
North America	USA – Florida	Sector/ multisector Regulator	Florida Public Service Commission	2540 Shumard Oak Blvd.	Tallahassee, FL 32399-0850	USA	+1 850 413 6044		Chairman@psc.state.fl.us	http://www.psc.state.fl.us/
North America	USA – Guam	Sector/ multisector Regulator	Guam Public Utilities Commission	Suite 207 GCIC Building,414 West Soledad Avenue	Hagatña, Guam 96910	USA	+1 671 472 1907	+1 671 472 1917	info@guampuc.com	http://www.guampuc.com/main/
North America	USA – Hawaii	Sector/ multisector Regulator	Hawaii Public Utilities Commission	465 South King Street, Room 103	Honolulu, Hawaii 96813	USA	+1 808 586 2020	+1 808 586 2066	Hawaii.PUC@hawaii.gov	www.hawaii.gov/budget
North America	USA – Idaho	Sector/ multisector Regulator	Idaho Public Utilities Commission	PO Box 83720	Boise, ID 83720-0074	USA	+1 208 334 0300	+1 208 334 3762	tclark@puc.state.id.us	www.puc.state.id.us

Region	Country	Type of entity	Entity	Address			Telephone	Fax	Email	Website
North America	USA – Illinois	Sector/ multisector Regulator	Illinois Commerce Commission	527 East Capitol Ave	Springfield, IL 62701	USA	+1 217 785 8596	+1 217 785 5431	jfeeley@icc.illinois.gov	http://www.icc.illinois.gov/
North America	USA – Indiana	Sector/ multisector Regulator	Indiana Utility Regulatory Commission	National City Center, 101 West Washington Street Suite 1500E	Indianapolis, Indiana 46204	USA	+1 317 232 2701	+1 317 232 6758	info@urc.state.in.us	http://www.ai.org/iurc/ index.html
North America	USA – Iowa	Sector/ multisector Regulator	Iowa Utilities Board	350 Maple Street, Des Moines	Iowa 50319-0069	USA	+1 515 281 5979	+1 515 281 5329	iub@max.state.ia.us	http://www.state.ia.us/iub
North America	USA – Kentucky	Sector/ multisector Regulator	Kentucky Public Service Commission	PO Box 615, 211 Sower Boulevard, Frankfort	Kentucky 40602-0615	USA	+1 502 564 3940	+1 502 564 3460	psc.info@ky.gov; kathys.lewis@ky.gov	http://psc.ky.gov/
North America	USA – Louisiana	Sector/ multisector Regulator	Louisiana Public Service Commission	Post Office Box 91154	Baton Rouge, Louisiana 70821-9154	USA	+1 225 342 4404	+1 225 342 2831	blossman@lpsc.org	http://www.lpsc.org/ default.asp
North America	USA – Maine	Sector/ multisector Regulator	Maine Public Utilities Commission	242 State Street, State House Station 18	Augusta, ME 04333	USA	+1 225-342-4404	+1 225 342 2831	maine.puc@maine.gov	http://www.maine.gov/ mpuc/
North America	USA – Maryland	Sector/ multisector Regulator	Maryland Public Service Commission	William Donald Schaefer Tower, 6 St. Paul St., 16th Floor	Baltimore, MD 21202	USA	+1 400 767 8000			www.psc.state.md.us
North America	USA – Massachusetts	Sector/ multisector Regulator	Department of Public Utilities (DPU)	One South Station	Boston, MA 02110	USA	+1 617 305 3500	+1 617 345 9101	robert.keating@state.ma.us	www.mass.gov/dpu/
North America	USA – Mississippi	Sector/ multisector Regulator	Mississippi Public Service Commission	PO Box 1174	Jackson, MS 39215-1174	USA	+1 601 961 5430	+1 601 961 5824	central.district@psc.state.ms. us	http://www.psc.state.ms. us/
North America	USA – Missouri	Sector/ multisector Regulator	Missouri Public Service Commission	PO Box 360	Jefferson City, MO 65102-0360	USA	+1 573 751 3234	+1 573 751 1847	pscinfo@psc.mo.gov	http://www.psc.mo.gov/
North America	USA – Montana	Sector/ multisector Regulator	Montana Public Service Commission	PO Box 202601	Helena, MT 59620-2601	USA	+1 406 444 6199	+1 406 444-7618	kwhitney@mt.gov	http://www.psc.state. mt.us/
North America	USA – Nevada	Sector/ multisector Regulator	Nevada Public Utilities Commission	101 Convention Center Drive, Suite 250	Las Vegas, Nevada 89109	USA	+1 702 486 2600	+1 702 486-7206	alevig@puc.state.nv.us; scrano@puc.state.nv.us	http://pucweb1.state.nv.us/ PUCN/PUCHome.aspx
North America	USA – New Hampshire	Sector/ multisector Regulator	New Hampshire Public Utilities Commission	21 South Fruit Street, Suite 10	Concord, N.H. 03301-2429	USA	+1 603 271 2431	+1 603 271-3878	puc@puc.nh.gov	http://www.puc.state.nh. us/
North America	USA – New Jersey	Sector/ multisector Regulator	New Jersey Board of Public Utilities	Two Gateway Center (8th Floor)	Newark, NJ 07102	USA	+1 973 648 2026		benjamin.hunter@bpu.state. nj.us, eric.wilkinson@bpu.state.nj.us	http://www.bpu.state.nj.us/ bpu/index.shtml
North America	USA – New York State	Sector/ multisector Regulator	New York State Public Service Commission	Empire State Plaza, Agency Building 3	Albany, NY 12223-1350	USA	+1 518 474 7080	+1 518 474 0421	Web_Questions@dps.state. ny.us, roberta_worden@dps.state. ny.us	http://www.dps.state.ny.us/

Region	Country	Type of entity	Entity	Address			Telephone	Fax	Email	Website
North America	USA – North Carolina	Sector/ multisector Regulator	North Carolina Utilities Commission	4325 Mail Service Center	Raleigh, NC 27699-4325	USA	+1 919 733 4249	+1 919 733 7300	James.McLawhorn@ncmail.net, swatson@ncuc.net	http://www.ncuc.commerce.state.nc.us/index.htm
North America	USA – Ohio	Sector/ multisector Regulator	Ohio Public Utilities Commission	180 E. Broad St.	Columbus, OH 43215	USA	+1 800 686 7826	+1 614 752 8351	Leon.Winget@puc.state.oh.us, kim.wissman@puc.state.oh.us	http://www.puco.ohio.gov/
North America	USA – Oregon	Sector/ multisector Regulator	Oregon Public Utility Commission	PO Box 2148	Salem OR 97308-2148	USA	+1 503 378 6611		puc.consumer@state.or.us, puc.commission@state.or.us	http://www.puc.state.or.us/PUC/index.shtml
North America	USA – Pennsylvania	Sector/ multisector Regulator	Pennsylvania Public Utility Commission	PO Box 3265	Harrisburg, PA 17105-3265	USA	+1 717 772 8883	+1 717 783 0698	browns@puc.state.pa.us, carlson@puc.state.pa.us	http://www.puc.state.pa.us/home.aspx
North America	USA – Rhode Island	Sector/ multisector Regulator	Rhode Island Public Utilities Commission	RI Public Utilities Commission, 89 Jefferson Boulevard	Warwick, RI 02888	USA	+1 401 941 4500		mary.kent@ripuc.org	http://www.ripuc.state.ri.us/index.html
North America	USA – South Carolina	Sector/ multisector Regulator	South Carolina Office of Regulatory Staff	Office of Regulatory Staff, 1441 Main Street, Suite 300	Columbia, SC 29201	USA	+1 803 737 0805		cdscott@regstaff.sc.gov	http://www.regulatorystaff.sc.gov/
North America	USA – Utah	Sector/ multisector Regulator	Utah Public Service Commission	PO Box 45585	Salt Lake City, UT 84145-0585	USA	+1 801 530 6716	+1 801 530 6796	phicken@utah.gov, mlivingston@utah.gov	http://www.psc.state.ut.us/
North America	USA – Vermont	Sector/ multisector Regulator	Vermont Public Service Board	112 State Street, Drawer 20	Montpelier, VT 05620-2601	USA	+1 802 828 2811	+1 802 828 2342	vtdps@state.vt.us, consumer@state.vt.us	http://publicservice.vermont.gov/
North America	USA – Virgin Islands	Sector/ multisector Regulator	Virgin Islands Public Services Commission	PO Box 40	St. Thomas, USVI 00804	USA	+1 340 776 1291	+1 340 774 4971	awells@psc.gov.vi, jboschulte@psc.gov.vi, dcole@psc.gov.vi	http://www.psc.gov.vi/index.htm
North America	USA – Virginia State	Sector/ multisector Regulator	Virginia State Corporation Commission	PO Box 1197	Richmond, VA 23218	USA	+1 804 371 9967		Commissioners@scc.virginia.gov	http://www.scc.virginia.gov/index.htm
North America	USA – West Virginia	Sector/ multisector Regulator	West Virginia Public Service Commission	201 Brooks Street, PO Box 812	Charleston, WV 25323	USA	+1 304 340 0300	+1 304 340 0325	webmaster@psc.state.wv.us, NPauley@psc.state.wv.us	http://www.psc.state.wv.us/
North America	USA – Wisconsin	Sector/ multisector Regulator	Public Service Commission of Wisconsin	PO Box 7854	Madison, Wisconsin 53707-7854	USA	+1 608 266 5481	+1 (608)266-3957	Sandra.Paske@psc.state.wi.us	http://psc.wi.gov/
North America	USA – Wyoming	Sector/ multisector Regulator	Wyoming Public Service Commission	Hansen Building Suite 300	2515 Warren Avenue Cheyenne, WY 82002	USA	+1 307 777 7427	+1 307 777 5700	klewis@state.wy.us	http://psc.state.wy.us/default.htm
South and Central America	Anguilla	Sector/ multisector Regulator	Public Utilities Commission of Anguilla	Fair Play Commercial Complex	PO Box 1400, The Valley	Anguilla, B.W.I.	+264 497 7374	+1 264 497 2782	comments@pucanguilla.org	http://pucanguilla.org/
South and Central America	Argentina	Sector/ multisector Regulator	Ente Regulador de Agua y Saneamiento (ERAS)	Callao 982 – 2 piso	Buenos Aires CF 1023	Argentina	+54 11 4815 9229	+54 11 4812 6726	carlos.vilas@eras.gov.ar	

Region	Country	Type of entity	Entity	Address			Telephone	Fax	Email	Website
South and Central America	Argentina	Sector/ multisector Regulator	Superintendencia de Servicios Públicos y Otras Concesiones (SUSEPU)	Gral. Alvear N° 1.243 – C.P. (4600)	San Salvador De Jujuy – Provincia De Jujuy	Argentina	+54 388 422 1442		susepu@jujuy.gov.ar, susepu@susepu.gov.ar	http://www.susepu.gov.ar/
South and Central America	Argentina	Sector/ multisector Regulator	Ente Regulador de los Servicios Públicos de Córdoba	Sede Central	Rosario de Santa Fe 238 Córdoba	Argentina			roberto.avalle@cba.gov.ar	
South and Central America	Argentina	Sector/ multisector Regulator	Ente Unico Regulador de los Servicios Públicos de Buenos Aires	Bartolomé Mitre 760 9 piso	(C.P. C1082aap) Ciudad Autónoma de Buenos Aires	Argentina	+54 351 434 2111		ccampolongo@entedela-ciudad.gov.ar	http://www.entedelaciudad.gov.ar/
South and Central America	Argentina	Sector/ multisector Regulator	Ente Regulador De Obras Y Servicios Públicos	Av. Napoleón Uriburu N° 57 Este	(C.P. P3600bua) Provincia De Formosa	Argentina	+54 371 743 5291		Entereguladorfsa@Infovia.Com.Ar	
South and Central America	Argentina	Sector/ multisector Regulator	Ente Provincial del Agua y el Saneamiento	San Juan 825 (M5500AFQ), Ciudad de Mendoza	Provincia de Mendoza	Argentina	+54 261 423 4500		emunoz@mendoza.gov.ar	http://www.epas.mendoza.gov.ar/
South and Central America	Argentina	Sector/ multisector Regulator	Departamento Provincial de Aguas	San Martin 249 (Viedma)	Provincia de Río Negro	Argentina	+54 292 042 0432	+54 292 042 3456	mmanuel@dpa.rionegro.gov.ar	http://www.dpa.gov.ar/
South and Central America	Argentina	Sector/ multisector Regulator	Ente Regulador de los Servicios Publicos	Mitre 1231(Salta)	Provincia de Salta	Argentina	+54 387 421 4579	+54 387 421 3021	cbonduri@salta.gov.ar	http://www.salta.gov.ar/ministerios/produccion/ente/estructura.htm
South and Central America	Argentina	Sector/ multisector Regulator	Ente Regulador de Servicios Sanitarios de Santa Fe (ENRESS)	25 de Mayo 1951	Santa Fe	Argentina	+54 342 459 4072		sdiripodas@enress.gov.ar, jmkerz@enress.gov.ar	http://www.enress.gov.ar
South and Central America	Argentina	Sector/ multisector Regulator	Ente Regulador de los Servicios de Aguas y Cloacas de Santiago del Estero (ERSAC)	Rivadadavia N°602	Sgo. Del Estero	Argentina			rferreiro@ersac.com.ar, sfranichevich@ersac.com.ar	http://www.ersac.com.ar/
South and Central America	Argentina	Sector/ multisector Regulator	Ente Regulador del Servicio de Aguas y Cloacas de Tucumán (ERSACT)	San Martín 3347		Argentina	+54 381 432 0054		ersact@arnet.com.ar, ersactgg@arnet.com.ar	http://www.producciontucuman.gov.ar/ersact.asp
South and Central America	Argentina	Sector/ multisector Regulator	Ente Único De Control De Privatizacionres (EUCOP)	Edificio Federación – San Martín N° 117 2° Piso (C.P. F5300dhc)	Ciudad De Todos Los Santos De La Nueva Rioja – Provincia De La Rioja	Argentina	+54 382 245 3624		estebandiaz@arnet.com.ar, Eucop@Larioja.Gov.Ar	
North America	Bahamas	Sector/ multisector Regulator	Public Utilities Commission	Agape House, 4th Terrace East Collins Avenue	PO Box N 4860, Nassau	Bahamas	+242 322 4437	+242 323 7288	info@PUCBahamas.gov.bs, acrolle@pucbahamas.gov.bs	http://www.pucbahamas.gov.bs/

Region	Country	Type of entity	Entity	Address			Telephone	Fax	Email	Website
North America	Barbados	Sector/ multisector Regulator	Fair Trading Commission	CEO, Manor Lodge, Lodge Hill	St. Michael	Barbados	+246 424 0260	+246 424 0300	info@ftc.gov.bb	http://www.ftc.gov.bb/index.htm
North America	Belize	Sector/ multisector Regulator	Public Utilities Commission	63 Regent Street	PO Box: 300, Belize City	Belize	+501 227 1176	+501 227 1149	consumeraffairs@puc.bz	http://www.puc.bz/default.asp
South and Central America	Bolivia	Sector/ multisector Regulator	Superintendencia de Saneamiento Básico (SISAB)	Av. Mariscal Santa Cruz Nro 1392, Edificio Camara de Comercio Piso 16 y 4	Cajon Postal 4245, La Paz	Bolivia	+591 2 310 801	+591 2 310 554	contactos@sisab.gov.bo	http://www.sisab.gov.bo
South and Central America	Brazil	Sector/ multisector Regulator	Agência Reguladora de Águas e Saneamento Distrital (ADASA)	SGAS Avenida W5, Quadra 904, Edifício CEB	CEP: 70300-040 Brasília – DF	Brazil	3961-4956/3961-4957		carlos.neto@adasa.df.gov.br, antonio.padua@adasa.df.gov.br	http://www.adasa.df.gov.br/
South and Central America	Brazil	Other entities	Associação Brazileira de Agências de Regulação	Rua 08, n° 192 – sala 07 – Centro	CEP 74.013-030 – Goiânia – GO	Brazil	55(62)3224-2560	55(62)3224-6079	alvaromachado@abar.org.br	www.abar.org.br
North America	Brazil	Sector/ multisector Regulator	Agência Municipal de Regulação dos Serviços de Água e Esgotos de Joinville	Rua Paraná, 420 – Anita Garibaldi	CEP: 89202-160 – Joinville -Santa Catariana	Brazil	+47 3433 1158		amaesc@amae.sc.gov.br, anp.riva@amae.sc.gov.br	http://www.amae.sc.gov.br/pt/home/
South and Central America	Brazil	Sector/ multisector Regulator	Agência Municipal de Regulação dos Serviços de Saneamento de Cachoeiro de Itapemirim (AGERSA)	Rua Prof. Quintiliano de Azevedo, 31, Edf. Guandu – Center – 6° andar	Guandu – Cachoeiro de Itapemirim – ES	Brazil	+28 3511 7077		agersa@agersa.com.br	
South and Central America	Brazil	Sector/ multisector Regulator	Agência Reguladora de Serviços Públicos de Santa Catarina (AGESC)	Rua General Bittencourt, 327 – Ed. José Boabaid – Centro	Florianópolis – SC – CEP 88020-100	Brazil	+48 3229 5700	+48 3229 5713	agesc@agesc.sc.gov.br, marchewsky@agesc.sc.gov.br	http://www.agesc.sc.gov.br
South and Central America	Brazil	Sector/ multisector Regulator	Agência Reguladora de Serviços Públicos do estado do Acre (AGEAC)	Rua Valério Magalhães, 172 – Bosque	Cep: 69.909-710 Rio Branco – Acre – Brazil	Brazil	+68 3214 2600	+68 3223 7470	faleconosco@ac.gov.br	http://www.ac.gov.br
South and Central America	Brazil	Sector/ multisector Regulator	Agência Reguladora de Serviços Públicos do Estado de Alagoas (ARSAL)	Rua Cincinato Pinto, 226, 2° andar	Edf. Ipaseal – Centro – Maceió – AL. CEP: 57020-050	Brazil	+800 284 0429	+82 3315 2501	consultapublica@arsal.al.gov.br, alvaromachado@arsal.al.gov.br	http://www.arsal.al.gov.br
South and Central America	Brazil	Sector/ multisector Regulator	Agência Reguladora dos Serviços Públicos Concedidos do Estado do Amazonas	Rua Prof. Ernani Simão, 1314 – Cachoeirinha	CEP: 69020-050 Manaus – AM	Brazil	+92 3234 8787		arsam_projur@hotmail.com	http://www.arsam.am.gov.br
South and Central America	Brazil	Sector/ multisector Regulator	Agência Reguladora de Serviços Públicos do Estado do Ceará (ARCE)	Av. Santos Dimant 1789 14° andar	CEP: 60750-160 Fortaleza – CE Brazil	Brazil	+85 3101-1027	+85 3101 1000	arce@arce.ce.gov.br, uciocl@arce.ce.gov.br	http://www.arce.ce.gov.br

Region	Country	Type of entity	Entity	Address			Telephone	Fax	Email	Website
South and Central America	Brazil	Sector/ multisector Regulator	Agência Goiana de Regulação, Controle e Fiscalização de Serviços Públicos (AGR)	Av. Goiás, n° 305, Edf. Visconde de Maná	Centro, 74005-010, Goiânia – GO	Brazil	+62 3226 6400		agr@agr.go.gov.br	http://www.agr.go.gov.br
South and Central America	Brazil	Sector/ multisector Regulator	Agência Estadual de Regulação de Serviços Públicos do Mato Grosso do Sul (AGEPAN)	Av. Afonso Pena, 3026, Centro	CEP: 79002-73 Campo Grande/MS	Brazil	+67 3025 9500	+67 3025 9503	marceloamaral@agepan.ms.gov.br, atiago@agepan.ms.gov.br	http://www.agepan.ms.gov.br
South and Central America	Brazil	Sector/ multisector Regulator	Agência Estadual de Regulação dos Serviços Públicos Delegados do Estado do Mato Grosso (AGER)	Av. Carmindo de Campos, 329 -	CEP : 78070-100, Bairro Shangri-lá – Cuiabá – MT	Brazil	+65 3618 6100	+65 3618 6104	francisval@ager.mt.gov.br, marco@ager.mt.gov.br	http://www.ager.mt.gov.br
South and Central America	Brazil	Sector/ multisector Regulator	Agência Estadual de Regulação e Controle dos Serviços Públicos do Pará (ARCON)	Rua dos Tarnoios, n° 1578	Batista Campos, CEP: 66025-540, Belém/PA	Brazil			ouvidoria@arcon.pa.gov.br, alfredobarros@arcon.pa.gov.br	http://www.arcon.pa.gov.br
South and Central America	Brazil	Sector/ multisector Regulator	Agência de Regulação do Estado da Paraíba (ARPB)	Av. Ministro José Américo de Almeida, s/n	Edf. DER – 1° andar – Centro, CEP: 58040-300 – João Pessoa – PB	Brazil	+83 3218 6681	+83 3218 6644	arpb@arpb.pb.gov.br, xikufranca@hotmail.com	http://www.arpb.pb.gov.br
South and Central America	Brazil	Sector/ multisector Regulator	Agência de Regulação dos Serviços Públicos Delegados do Estado de Pernambuco (ARPE)	Praça Ministro João Gonçalvez Sousa s/n, Edf. Sudene, Engenho do Meio	CEP: 50670-500	Brazil	+81 3303 4712	+81 3303 4725	ouvidoria@arpe.pe.gov.br	http://www.arpe.pe.gov.br
South and Central America	Brazil	Sector/ multisector Regulator	Agência Reguladora de Energia e Saneamento Básico do Estado do Rio de Janeiro (AGENERS)	Rua São Bento, 8, 13° andar – Centro	Rio Janeiro/RJ CEP: 20090-010	Brazil	+21 2516 3760	+21 2283 1352	ouvidoria@agenersa.rj.gov.br	http://www.agenersa.rj.gov.br
South and Central America	Brazil	Sector/ multisector Regulator	Agência Estadual de Regulação dos Serviços Públicos Delegados do Rio Grande do Sul (AGERGS)	Av. Borges de Medeiros, 659, 14° andar	Porto Alegre/ RS CEP: 90020-023	Brazil	+51 3288 8800		agergs@agergs.rs.gov.br	http://www.agergs.rs.gov.br
South and Central America	Brazil	Sector/ multisector Regulator	Agência Reguladora dos Serviços de Água e Esgotos de Mauá/SP (ARSAE)	Av. Queiroz Pedroso, 235 –	Jardim Pedroso – Mauá – SP	Brazil	+11 4513 5972		soares.ari@gmail.com	http://www.arsae.maua.sp.gov.br
South and Central America	Brazil	Sector/ multisector Regulator	Agência Tocantinense De Regulação, Controle e Fiscalização Dos Serviços Públicos (ATR)	501 Sul, Av. Theotônio Segurado, Lt. 6	Ed. Amazônia CEP: 77016-002	Brazil	+63 3218 2300	+63 3218 2395	ascom.atr@uol.com.br	http://web.secom.to.gov.br/atr/index.php
South and Central America	Brazil	Sector/ multisector Regulator	ARESESP – Agência Reguladora de Saneamento e Energia do Estado de São Paulo	Rua Boa Vista, 170 – 3° andar	Centro – SãoPaulo/SP– CEP: 01014-000	Brazil	+11 3293 5100	+11 3293 5114	cspe@sp.gov.br	http://www.arsesp.sp.gov.br

Region	Country	Type of entity	Entity	Address			Telephone	Fax	Email	Website
South and Central America	Chile	Sector/ multisector Regulator	Superintendencia de Servicios Sanitarios (SISS)	Oficina Región Metropolitana	Moneda N°673, piso 1, Santiago	Chile	+56 2 382 40 00	+56 2 382 40 03	siss@siss.cl, mespinosa@siss.cl	http://www.siss.cl/
South and Central America	Colombia	Sector/ multisector Regulator	Comisión de Regulación de água potable y Saneamento básico (CRA)	Edificio Palma Real Cra.13 No. 28-01	Pisos 5 y 6, Bogotá	Colombia	+571 327 2800	+571 350 9393	correo@cra.gov.co, agualy@cra.gov.co, curibe@cra.gov.co	http://www.cra.gov.co
North America	Costa Rica	Sector/ multisector Regulator	Autoridad Reguladora de los Servicios Públicos (ARESEP)	Apartado 936-1000	Sabana Sur, San jose	Costa Rica	+506 220 0102	+506 290 2010	attencionusuario@aresep.go.cr, gchavez@aresep.go.cr	http://www.aresep.go.cr/menu.htm
North America	El Salvador	Other entities	Administración Nacional de Acueductos y Alcantarillados (ANDA)	Centro de San Salvador	Calle Rúben Dário, entre 11 y 13 Av. Sur, Centro Comercial D Chico	San Salvador	+503 2514 3284		sugerencias@anda.gob.sv	http://www.anda.gob.sv/2007/home.asp
South and Central America	Ecuador	Sector/ multisector Regulator	Organismo Regulador y de Control de la Concesión de Agua Potable y Saneamiento de Guayaquil (ECAPAG)		Guayaquil	Ecuador	+593 4 268 1315	+593 4 268 2225	secretariageneral@ecapag.gov.ec, presidenciad@ecapab.gob.ec	
South and Central America	Guyana	Sector/ multisector Regulator	Public Utilities Commission (PUC)			Guiana	+592 226 7042	+592 227-3534	pucsecretary@yahoo.com	
North America	Honduras	Sector/ multisector Regulator	Ente Regulador De Los Servicios De Agua Potable Y Saneamiento (ERSAPS)	Colonia Palmira, Edificio ROMA No. 1601,	Tegucigalpa, M.D.C.	Honduras	+504 238 8151	+504 238 2598	ersaps@cablecolor.hn, ersaps@gmail.com	http://www.ersaps.gob.hn/portal/informacion/mensaje/
North America	International	Other entities	Asociacion de entes reguladores de agua potable e saneamiento de las americas (ADERASA)				+55 51 3288 8869	+55 51 3288 8868	presidente@aderasa.org	http://www.aderasa.org/es/index.htm
North America	International	Other entities	Comisión Económica para América Latina y el Caribe (CEPAL)	Casilla 179-D, Santiago de Chile		Chile	+56 2 471 2000	+56 2 208 0252	secepal@cepal.org	http://www.eclac.cl/
North America	Jamaica	Sector/ multisector Regulator	The Office of Utilities Regulation (OUR)	3rd Floor PCJ Resource center	36 Trafalgar Road, Kingston 10	Jamaica	+876 968 6053	+876 929 3635	consumer@our.org.jm	http://www.our.org.jm
North America	Mexico	Sector Association	Asociación Nacional De Empresas De Agua Y Saneamiento (ANEAS)	Palenque # 287	Col. Narvarte	México, D.F.	+55 5543 6600	+55 5543 6605	aneas@aneas.com.mx, roberto.olivares@aneas.com.mx	http://www.aneas.com.mx/main.htm
North America	Nicaragua	Sector/ multisector Regulator	Instituto Nicaragüense de Acueductos y Alcantarillado Sanitario (INAA)	De la Mansión Teolinda 3 cuadras al sur	Managua	Nicaragua	+505 266 7882	+505 266 7917	cschutze@inaa.gob.ni	http://www.inaa.gob.ni
South and Central America	Paraguay	Sector/ multisector Regulator	Ente Regulador de Servicios Sanitarios (ERSSAN)	Avda. Artigas 1921	Edificio Automotor 1er. Piso	Paraguay	+595 21 283 556	+595 21 283-557	deconomica@erssan.gov.py, presidencia@erssan.gov.py	www.erssan.gov.py

Region	Country	Type of entity	Entity	Address			Telephone	Fax	Email	Website
South and Central America	Panama	Sector/ multisector Regulator	Ente Regulador De Los Servicios Públicos (ERSP)	Apartado Postal: 0816-01235	Zona 5	Panamá	+507 508 4500	+507 508-4600	mtroitino@asep.gob.pa, atencionalusuario@asep.gob.pa	http://www.enteregulador.gob.pa
South and Central America	Peru	Sector/ multisector Regulator	Superintendencia Nacional De Servicios De Saneamiento (SUNASS)	Sector Saneamiento : Av. Bernardo Monteagudo 210–216	Magdalena del Mar	Peru	+51 1 264 1440	+51 1 264 4020	sunass@sunass.gob.pe	http://www.sunass.gob.pe
South and Central America	Trinidad and Tobago	Sector/ multisector Regulator	Regulated Industries Commission (RIC)	1st and 3rd Floors Furness House, Corner Wrightson Road and Independence Square	Port of Spain	Republic of Trinidad and Tobago	+868 625 5384	+868 624 2027	complaints@ric.org.tt	http://www.ric.org.tt/cms/
Oceania	Australia	Competition Authority	Australian Competition and Consumer Commission (ACCC)	GPO Box 3131	Canberra, ACT 2601	Australia	+61 2 6243 1111	+61 2 6243 1199	foi@accc.gov.au	http://www.accc.gov.au
Oceania	Australia	Competition Authority	National Competition Council	GPO Box 250	Melbourne, VIC 3001	Australia	+61 3 9285 7474	+61 3 9285 7477	info@ncc.gov.au	http://www.ncc.gov.au/
Oceania	Australia	Other entities	Council of Australian Governments	Secretariat, Department of the Prime Minister and Cabinet	PO Box 6500, CANBERRA, ACT 2600	Australia	+61 2 6271 5343	+61 2 6271 5540		http://www.coag.gov.au/
Oceania	Australia	Other entities	Water Services Association of Australia (WSAA)	PO Box 13172,Law Courts Post Office	Melbourne, VIC 8010	Australia	+61 3 9606 0678	+61 3 9606 0376	patrice.belcher@wsaa.asn.au	https://www.wsaa.asn.au
Oceania	Australia	Sector Association	Australian Water Association (AWA)	Level 2, 44 Hampden Road	Artarmon, NSW 2065	Australia	+61 2 9495 9911	+61 2 9413 1047	info@awa.asn.au	http://www.awa.asn.au
Oceania	Australia – New South Wales	Sector/ multisector Regulator	Independent Competition and Regulatory Commission (IPART)	PO Box Q290	QVB POST OFFICE NSW 1230	Australia	+61 2 9290 8400	+61 2 9290 2061	ipart@ipart.nsw.gov.au, richard_warner@ipart.nsw.gov.au	http://www.ipart.nsw.gov.au/welcome.asp
Oceania	Australia – New South Wales	Other entities	Energy and Water Ombudsman NSW	Reply Paid K1343	Haymarket NSW 1239	Australia	+61 1 8002 46545	+61 1 8008 12291	omb@ewon.com.au	http://www.ewon.com.au/
Oceania	Australia – Queensland	Sector/ multisector Regulator	The Queensland Water Commission	GPO Box 2257	Brisbane QLD 4001	Australia	+61 7 3222 0555	+61 7 3222 0599	water.enquiries@qca.org.au, water@qca.org.au	http://www.qca.org.au/
Oceania	Australia – Queensland	Other entities	Queensland Ombudsman	GPO Box 3314	Brisbane QLD 4001	Australia	+61 7 3005 7000	+61 7 3005 7067	ombudsman@ombudsman.qld.gov.au	http://www.ombudsman.qld.gov.au/cms/
Oceania	Australia – South Australia	Sector/ multisector Regulator	Essential Services Commission of South Australia	GPO Box 2605	Adelaide SA 5001	Australia	+61 8 8463 4444	+61 8 8463 4449	escosa@escosa.sa.gov.au	http://www.escosa.sa.gov.au/site/page.cfm
Oceania	Australia – South Australia	Other entities	South Australia Ombudsman	Level 5 East Wing, 50 Grenfell St	Adelaide 5000	Australia	+61 8 8226 8699	+61 8 8226 8602	Ombudsman@Ombudsman.sa.gov.au	http://www.ombudsman.sa.gov.au/
Oceania	Australia – Tasmania	Sector/ multisector Regulator	Government Prices Oversight Commission	GPO Box 770	Hobart TAS 7001	Australia	+61 3 6233 5665	+61 3 6233 5666	office@gpoc.tas.gov.au	http://www.gpoc.tas.gov.au

Region	Country	Type of entity	Entity	Address			Telephone	Fax	Email	Website
Oceania	Australia – Tasmania	Other entities	Ombudsman Tasmania	GPO Box 960	Hobart Tas 7001	Australia	+61 3 6233 3364	+61 3 6233 8966	ombudsman@justice.tas.gov.au	http://www.ombudsman.tas.gov.au/
Oceania	Australia – Victoria	Sector/ multisector Regulator	Victorian Essential Services Commission (ESC)	Level 2, 35 Spring Street	Melbourne Victoria 3000	Australia	+61 3 9651 0222	+61 3 9651 3688	reception@esc.vic.gov.au	http://www.esc.vic.gov.au/public/
Oceania	Australia – Victoria	Sector Association	Victorian Water Industry Association	Suite 1, Level 6, 2 Collins Street	MELBOURNE VICTORIA 3000	Australia	+61 3 9639 8868	+61 3 9639 8860	vicwater@vicwater.org.au	http://www.vicwater.org.au
Oceania	Australia – Victoria	Other entities	Victorian Ombudsman	Level 9, 459 Collins Street (North Tower)	Melbourne Victoria 3000	Australia	+61 3 9613 6222	+61 3 9614 0246	ombudvic@ombudsman.vic.gov.au	http://www.ombudsman.vic.gov.au/
Oceania	Australia – Western Australia	Sector/ multisector Regulator	The Economic Regulation Authority	PO Box 8469	Perth Business Centre 6849	Australia	+61 8 9213 1925	+61 8 9213 1999	paul.kelly@era.wa.gov.au	http://www.era.wa.gov.au/
Oceania	Australia – Western Australia	Other entities	Ombudsman Western Australia	PO Box Z5386, St Georges Terrace	PERTH, WA 6831	Australia	+61 8 9220 7555	+61 8 9325 1107	mail@ombudsman.wa.gov.au	http://www.ombudsman.wa.gov.au/
Oceania	Australia – Australian Capital Territory	Sector/ multisector Regulator	Independent Competition and Regulatory Commission	GPO Box 296	Canberra City, ACT 2601	Australia	+61 2 6205 0779	+61 2 6207 5887	icrc@act.gov.au, Krish.goundar@act.gov.au	http://www.icrc.act.gov.au/
Oceania	Australia – Australian Capital Territory	Other entities	Australian Capital Territory Ombudsman	GPO Box 442	Canberra, ACT 2601	Australia		+61 2 6249 7829	ombudsman@ombudsman.gov.au	http://www.ombudsman.act.gov.au/
Oceania	Australia – Northern Territory	Sector/ multisector Regulator	Northern Territory Utilities Commission	GPO Box 915	Darwin, NT 0801	Australia	+61 8 8999 5480	+61 8 8999 6262	AnneMarie.Hart@nt.gov.au	http://www.utilicom.nt.gov.au/
Oceania	Australia – Northern Territory	Other entities	Ombudsman Northern Territory	PO Box 1344	DARWIN, NT 0801	Australia	+61 8 8999 1818	+61 8 8999 1828	nt.ombudsman@nt.gov.au	http://www.omb-hcscc.nt.gov.au/
Oceania	New Zealand	Competition Authority	New Zealand Commerce Commission	PO Box 105-222	Auckland	New Zealand	+64 9 920 3480	+64 9 920 3481	contact@comcom.govt.nz	http://www.comcom.govt.nz/
Oceania	Papua New Guinea	Sector/ multisector Regulator	Papua New Guinea Waterboard	PO Box 2779	Boroko, NCD Papua New Guinea	Papua New Guinea	+675 325 6866	+675 325 8733	wwalo@pngwater.com.pg, awalne@pngwater.com.pg, jla-hui@pngwater.com.pg	http://www.pngwater.com.pg

IWA Publishing's authorised EU representative for General Product Safety Regulations is Diane D'Arras, 15 rue Duret, 75116 Paris, France, e-mail: safety@iwap.co.uk.

Printed and bound by CPI Group (UK) Ltd, Croydon, CR0 4YY
17/09/2026
02218802-0019